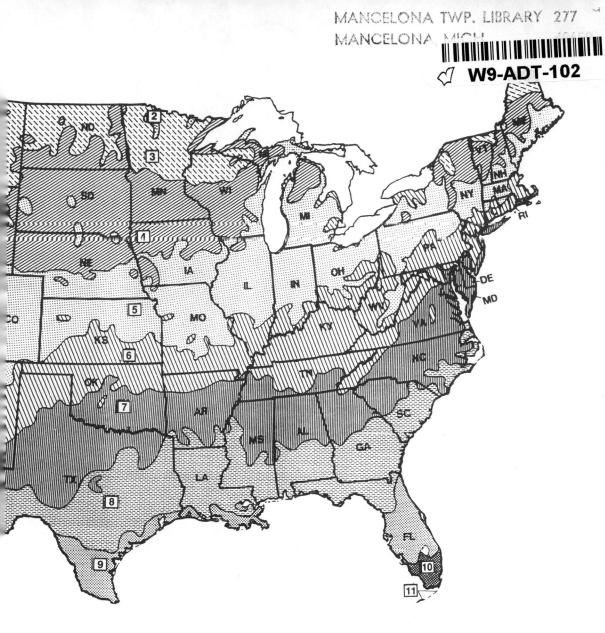

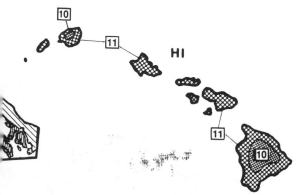

HI

**RANGE OF AVERAGE ANNUAL MINIMUM
TEMPERATURES FOR EACH ZONE**

ZONE 1	BELOW −50° F	
ZONE 2	−50° TO −40°	
ZONE 3	−40° TO −30°	
ZONE 4	−30° TO −20°	
ZONE 5	−20° TO −10°	
ZONE 6	−10° TO 0°	
ZONE 7	0° TO 10°	
ZONE 8	10° TO 20°	
ZONE 9	20° TO 30°	
ZONE 10	30° TO 40°	
ZONE 11	ABOVE 40°	

SECOND EDITION

ARBORICULTURE

Integrated Management of Landscape Trees, Shrubs, and Vines

Richard W. Harris

Professor Emeritus
Department of Environmental Horticulture
University of California at Davis

Illustrations by Vera M. Harris

REGENTS/PRENTICE HALL, Englewood Cliffs, New Jersey 07632

Library of Congress Cataloging-in-Publication Data

Harris, Richard Wilson
   Arboriculture : integrated management of landscape trees, shrubs,
and vines / Richard W. Harris : illustrations by Vera M. Harris.—
2nd ed.
    p. cm.
  Includes bibliographical references (p.   ) and index.
  ISBN 0-13-044280-1
  1. Arboriculture. 2. Ornamental woody plants. I. Title.
SB435.H317 1992
634.9′77—dc20   91–19477
                                      CIP

Cover: The tree depicted is to emphasize that roots usually extend two to three times the radius or the height of the crown of a tree. Most absorbing roots are in the surface 30 cm (12 in.) of soil and few roots grow below one meter (40 in.). Most roots are so slender that it is not possible to represent them to scale on the cover (Perry 1982).

Acquisitions editor: Robin Baliszewski
Editorial/production supervision
   and interior design: Joan L. Stone
Copy editor: Linda Pawelchak
Cover designer: Wanda Lubelska
Cover illustration: Vera Harris
Prepress buyer: Mary McCartney
Manufacturing buyer: Ed O'Dougherty
Editorial assistant: RoseMary Florio

©1992, 1983 by Prentice-Hall, Inc.
A Simon & Schuster Company
Englewood Cliffs, New Jersey 07632

Printed in the United States of America

10 9 8 7 6 5 4 3

ISBN 0-13-044280-1

PRENTICE-HALL INTERNATIONAL (UK) LIMITED, *London*
PRENTICE-HALL OF AUSTRALIA PTY. LIMITED, *Sydney*
PRENTICE-HALL CANADA INC., *Toronto*
PRENTICE-HALL OF HISPANOAMERICANA, S.A., *Mexico*
PRENTICE-HALL OF INDIA PRIVATE LIMITED, *New Delhi*
PRENTICE-HALL OF JAPAN, INC., *Tokyo*
SIMON & SCHUSTER ASIA PTE. LTD., *Singapore*
EDITORA PRENTICE-HALL DO BRASIL, LTDA., *Rio de Janeiro*

Contents _____

Contents   v

Preface

Research findings, application of research information, innovative ideas of practitioners, new and improved equipment and products, and a better informed and more concerned public bode well for the future of arboriculture. The revision of *Arboriculture* attempts to evaluate new information and maintenance practices and, where appropriate, to reassess current practices in light of new information.

Improved approaches to integrated pest management have dramatically demonstrated the interrelatedness of maintenance practices and physical, chemical, and biological environment to a plant's health and well being. The subtitle of the book has been changed to reflect the opportunity and the need to integrate the management of landscape plants. The aim of this book is to increase understanding of plants and plant processes, to improve the ability of horticulturists to analyze problems and practices, and to help them more effectively integrate the management of the environment and maintenance of landscape plants.

The common and botanical names used in this book generally conform to those listed in *Hortus Third* (Bailey and others 1976) and the *Annotated Checklist of Woody Ornamental Plants of California, Oregon, and Washington* (McClintock and Leiser 1979).

Measurements are given in metric units, followed by nonmetric equivalents in parentheses. In many situations, approximate values are accurate enough, so conversions between the two systems are rounded for simplicity.

Certain statements in the text are printed in boldface in order to emphasize their importance.

Due to space limitations, certain topics are only briefly discussed in this edition; some were covered in more detail in the first edition. These topics are printed in boldface in the index so that you can refer to the first edition should you wish to pursue the subject further.

The last five years have been extremely interesting and challenging as I reviewed literature, attended meetings, and talked with researchers and others involved in landscape planning and maintenance. Even more rewarding was the exhilarating feeling experienced from the warmth and generosity of researchers and practioners in sharing photographs, ideas, and expertise. I am greatly indebted to many, many people, some of whom are credited in the text for supplying illustrations, photographs, and information new to me.

As in the first edition, I am deeply indebted to my wife, Vera, for her patience and understanding, for her help in preliminary and proof editing, and for her skill as an illustrator.

In addition to the many who helped with the first edition, the following were particularly helpful with this revision. Colleagues generously assisted by sharing their experience and in many cases for reviewing portions of the manuscript. These include Kevin L. Blaze, Victoria College of Agriculture and Horticulture, Australia; James R. Clark, University of Washington; Henry Donselman, Rancho Soledad Nurseries, Rancho Santa Fe, CA; James R. Feucht, Colorado State University; Edward F. Gilman, University of Florida; Jitze Kopinga, Research Institute for Forestry and Landscape Planning, Wageningen, The Netherlands; Glen P. Lumis, University of Guelph, Canada; Dan Neely, University of Illinois; Derek Patch, Forestry Commission Research Station, Farnham, England; Thomas O. Perry, University of North Carolina; Michael J. Raupp, University of Maryland; Frederick Roth, California State Polytechnic University, Pomona; Alex L. Shigo, Shigo and Trees, Associates, Durham, NH; and Gary W. Watson, The Morton Arboretum, Lisle, IL.

Among colleagues of the University of California who helped, I am particularly indebted to Arthur H. McCain at Berkeley; Alison M. Berry, David W. Burger, Steve H. Dreistadt, Clyde L. Elmore, Andrew T. Leiser, James D. MacDonald, Jack L. Paul, Roy M. Sachs, and Richard F. Walters at Davis; and Laurence R. Costello, W. Douglas Hamilton, and Pavel Svihra of the Cooperative Extension Service.

Firsthand experience and examples that have been useful were provided by many arborists, including Stephen Bakken, California Department of Parks and Recreation; John C. Britton, John Britton Tree Service, St. Helena, CA; Guido H. Ciardi, City of San Francisco; Steve Clark, SC&A, Brentwood, TN; Eugene Eyerly, Eyerly & Associates, Denver, CO; Niels Hvass, SITAS, Ballerup, Denmark; Richard A. Johnstone, Delmarva Power and Light Company, Salisbury, MD; Keith R. Jones, Central Illinois Public Service, Springfield; Gordon W. Mann, City of Redwood City, CA; Nelda Matheny, HortScience, Pleasanton, CA; John McNeary, McNeary Arborists, Charlotte, NC; Kenneth C. Meyer, Mayne Tree Expert Company, San Mateo, CA; Jack Siebenthaler, Consultant, Clearwater, FL; Robert W. Skiera, City of Milwaukee, WI; Michael W. Watson, Potomac Edison Company, Hagerstown, MD; and Richard H. Wells, Philadelphia Electric Company, PA.

Although I have been helped by many, I take full responsibility for the ideas presented, their interpretation, and their accuracy. Research, practices, and observations that postdate this book or that have escaped my attention may warrant modification of the recommendations or explanations that appear here. I would appreciate receiving such information so that all may benefit from the most current advances in the field.

I am particularly indebted to Linda Pawelchak for editing the manuscript, helping ideas flow logically, and ensuring accuracy of cross references. Joan L. Stone and Robin Baliszewski of Prentice Hall were extremely patient and helpful throughout the preparation of the book.

Richard W. Harris

CHAPTER 1

Landscape Trees, Shrubs, and Vines

While plants contribute to our pleasure, comfort, and well-being, they also help conserve energy and the quality of air, water, and soil. Plants, particularly trees, are an important part of our lives—around homes, schools, shopping centers, and places of work, along streets and highways, in the central city, parks, and other landscaped areas.

Trees have been held in high esteem since earliest times. More than 4000 years ago, Egyptians wrote of trees being transplanted with a ball of soil around the roots of each (Chadwick 1971); some were moved 2400 km (1500 miles) by boat. In Greece, Theophrastus (370–285 B.C.) and Pliny (A.D. 23–79) gave rather complete directions for tree planting and care. Many books on the care of plants, including trees, have been written since these times.

In the Middle Ages, botanical gardens primarily contained plants that had medicinal potential. Later, the gardens of private estates boasted exotic plants brought in through trade and travel. Many of these gardens have since become public and are great sources of information and pleasure.

By the 1700s, trees were being planted with some frequency in the cities and estates of Europe. In the early settlements of North America, trees were cut to make room for farms and towns, but by the late 1700s, trees were being planted in town squares. After planting, however, few of these trees received much care, except on large estates. As settlers moved west onto the prairies, they planted the seeds of fruit and shelter trees around their homes.

In the early 1900s, national research agencies in Europe and North America and state agricultural experiment stations began to study fruit and forest trees. By the 1950s, state and national research stations had begun working specifically on landscape tree problems. The devastation caused by chestnut blight, Dutch elm disease, phloem necrosis, Gypsy moths (*Porthetria dispar*), and Japanese beetles (*Popillia japonica*) in the northeastern and midwestern United States was largely responsible for the increased interest in tree research, though this research consequently focused on disease and insect control. Experiment stations, botanical gardens, arboreta, and some large nurseries have long been involved in landscape plant introduction; increasing efforts have sought trees that will be better able to withstand the rigors of the urban environment.

Governmental agencies sponsor extensive parks and landscape tree plantings. Many cities require street-tree plantings in new residential and commercial developments. People are more aware of trees and their value. Many cities in the United States have ordinances controlling the removal of trees, even on private property. One of the stipulations of England's Civic Amenities Act of 1967 addresses the preservation and planting of trees on private property (Chadwick 1971).

Several professional organizations are concerned with tree planting and care. The National Shade Tree Conference, organized in 1924 in Connecticut, later became the International Society of Arboriculture, with current headquarters in Urbana, Illinois. The Society has members on every continent except Antarctica. Other United States organizations include the National Arborists Association (commercial arborists), the Society of Municipal Arborists, and the American Society of Consulting Arborists. The Arboricultural Association in Great Britain is similar to the International Society of Arboriculture. Many other organizations are concerned with landscape trees; among them are the American Association of Botanic Gardens and Arboreta, the American Entomological Society, the American Forestry Association, the American Horticultural Society, the American Society for Horticultural Science, the American Society of Landscape Architects, the National Arbor Day Foundation, the Society of American Foresters, and the American Phytopathological Society.

Many technical schools, colleges, and universities offer courses in arboriculture, landscape horticulture, and urban forestry, with supporting courses in botany, entomology, plant pathology, landscape design, and soils.

ARBORICULTURE

Arboriculture, as herein defined, is primarily concerned with the planting and care of trees and more peripherally concerned with shrubs and woody vines and ground-cover plants. Arboriculture is commonly defined as the cultivation of trees and shrubs only (Bailey and others 1976, *Webster* 1976), but woody plants that are called *vines* in the United States are called *wall shrubs* in England (Brown 1972, Halliwell, Turpin, and Wright 1979) and *climbing shrubs* in Australia (Mullins 1979). It seems reasonable to assume that the common definition of arboriculture includes woody vines. All of the plants mentioned are woody, perennial plants that have many common needs and characteristics and differ primarily in their form and training requirements. Even so, a wisteria vine can be trained into a tree, and the mature form of English ivy is shrub-like (see Fig. 2–18). *Hortus Third* (Bailey and others 1976) further defines arboriculture as the cultivation of plants as individuals rather than as elements in a forest or orchard.

Arboriculture is one of the branches of horticulture within the plant sciences. Following are the fields within plant science, their definitions, and their relationships to one another.

Plant Science

Agronomy and *Range Science* deal primarily with the cultivation of field crops and range and pasture plants

Forestry concerns the commercial production and utilization of timber

Silviculture is the practice of raising forests

Urban Forestry is the management of trees in urban areas on larger than an individual basis

Horticulture concerns plants that are intensively grown for food and aesthetics

Pomology is the cultivation of perennial fruiting plants, primarily woody trees and vines

Vegetable Crops (Olericulture) is the growing of herbaceous plants for human consumption

Environmental Horticulture is the cultivation of plants to enhance our surroundings

Floriculture is the production of cut flowers and potted plants

Nursery Production (Ornamental Horticulture) is the production of primarily woody plants for landscape plantings and fruit production

Landscape Horticulture is the care of plants in the landscape

Arboriculture concerns the cultivation of woody plants, particularly trees

Landscape Construction involves the installation of structural and plant materials according to a landscape plan

Landscape Maintenance (*Gardening* or *Grounds Maintenance*) specializes in the planting and care of a wide variety of plants used in the landscape

Turfgrass Culture concerns the growing of turf for landscape and sports use (some consider it agronomic)

Landscape Architecture concerns the planning and design of outdoor space for human use and enjoyment

Park Management concerns the total responsibility of planning, developing, and managing public and private landscaped areas, from housing developments and city parks to heavily used national parks

Even though these areas of specialization are often thought to be mutually exclusive, the lines of demarcation between them have blurred. A company or agency may engage in landscape design, nursery production, and all phases of landscape horticulture. Another may specialize in one or two. Many organizations find that by engaging in more than one specialization, they are able to maintain a more stable labor force and to spread financial risks.

PLANTS IN THE LANDSCAPE

As public and private plantings become more common and more expensive to maintain, people responsible for such plantings are becoming more concerned about the proper selection and care of landscape plants. The landscape comprises a palette of plant growth habits: trees, shrubs, vines, grass, and herbaceous flowering and foliage plants. This book focuses on the selection and care of woody perennial plants, primarily trees, but also devotes some attention to shrubs, vines, and ground-cover plants (see Fig. 2–1). Perennial plants with these habits of growth have much in common as regards selection and care in the landscape. Because of the ultimate size and long life of trees, their proper selection and care is of greater importance. Woody plants respond much as trees do to most environmental conditions and cultural practices.

Plants not only provide food and fiber but enhance our surroundings in a variety of ways: physical, aesthetic, economic, and psychological. Some influences are important to our immediate surroundings; others are significant only on a more extensive scale. Cited most often when the value of landscape plantings is extolled are *physical* attributes: influence on climate, air purification, noise reduction, and erosion control. The microclimate can be greatly enhanced by plants, particularly trees, when they are properly selected, placed, and maintained.

PHYSICAL BENEFITS

Microclimate Enhancement

Plants absorb heat as they transpire, provide shade that reduces solar radiation and reflection (Fig. 1–1), can reduce or increase wind speed, and can increase fog precipitation and snow deposition. Trees are frequently called nature's air conditioners. A hectare (2.47 acres) of vegetation transpires about 17,000 liters (4000 gal) of water on a sunny summer day. Oke (1972) estimates that a 30 percent cover of vegetation will give two-thirds as much cooling as will a complete plant cover. Thus, in a 0.08

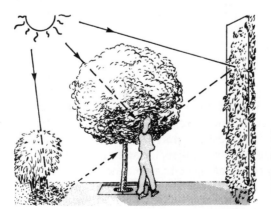

Figure 1-1 In summer, plants provide comfortable surroundings by intercepting direct solar radiation and reradiation from heated surfaces. (Adapted from Robinette 1972)

hectare (one-fifth acre) lot with 30 percent plant cover, transpiration would absorb 1.2 million kilogram-calories of energy that would otherwise warm the environment. That is equivalent to the amount of cooling necessary to air condition two moderately sized houses 12 hours a day in the summer.

This evaporative cooling, however, is quickly diffused so that even on a hot summer day, the air temperature of an area of plants is not much lower than that of an area with few or no plants. Herrington, Bertolin, and Leonard (1972) found that shaded sites in a 30-hectare (75-acre) park in Syracuse, New York, averaged only 1.3° C (2.5° F) cooler than urban sites outside the park, even though solar radiation under the trees was only one-fourth of that in the urban sites. Temperatures outside the park would certainly have been higher had the park and other vegetation not been there, since temperature differences are evened out fairly effectively by air movement and diffusion. Fortunately, thermal comfort results from the interaction of the human body with a number of factors in the environment, including air temperature, relative humidity, wind speed, solar radiation, and infrared radiation. **The most important influence of trees on the microclimate is their control of solar radiation in both winter and summer.**

Trees minimize heat reflection and reradiation by shading pavement and buildings and by reducing direct rays of the sun (Fig. 1–1). Shade from trees can reduce room temperatures in poorly insulated houses by as much as 11° C (20° F) in summer (Deering 1955). Deering and Brooks (1953) found that bare-ground surface temperatures of 56° to 67° C (130°–150° F) were cooled an average of 20° C (35° F) in five minutes after being shaded. Any barefoot youngster knows the value of shade on a summer afternoon. In Memphis, Tennessee, Dunn (1975) found that property owners, in appraising the value of a tree, were able to use the cost of constructing a structure that would provide equivalent shade, and they had not been successfully challenged by the U.S. Internal Revenue Service. At the time, a modest sturdy structure would have placed the value of a shade tree at $27.50 to $33.00 per square meter ($2.50–$3.00/ft²) of shade. In the tropics, evergreen trees provide year-round protection from the sun. In temperate regions, deciduous trees provide summer shade and permit the winter sun to warm the ground, although the sun can be considerably impeded: Light intensity under dormant trees is reduced 20 to 74 percent (Geiger 1961). Unless properly sited in temperate regions, deciduous trees can intercept enough sunlight to increase winter-heating costs more than they decrease summer-cooling costs of solar and most conventional houses (see Chapter 5) (Thayer 1986a).

Plants modify wind by obstructing, guiding, deflecting, and filtering air flow (Robinette 1972). Air movement influences real and perceived temperatures; for example, with a chill factor of 30 kmph (20 mph), a wintry blast of 0° C (32° F) air will cool as if it were −14° C (7° F). Windbreak and shelterbelt plantings shelter smaller plants, animals, and property by reducing wind speed on both the windward and leeward sides (Geiger 1961). Nearby shrubs help protect buildings from the cold by reducing radiation loss and providing an insulating zone of relatively still air. A dense planting across a slope can create a frost pocket on its uphill side, however, by slowing the downward flow of cold night air (Fig. 4–3).

Plants, by their form and placement, can guide or funnel wind through openings or over their tops and in so doing can increase wind speeds by as much as 20 percent (Robinette 1972). A row of trees with dense heads can create quite a wind tunnel between the ground and the lowest branches.

Trees in particular can reduce fog density by condensing moisture on leaf and twig surfaces. Geiger (1961) reported that woods along the coasts of Japan protect inland areas from fog; six to ten times as much moisture was deposited under trees as on nearby open grassland. Similarly, the windward edge of a woods had 20 times as much fog precipitation as did the lee side. The summer fogs common in many coastal sites greatly influence the moisture economy of these areas. Not only does summer fog reduce evapotranspiration, but fog precipitation can account for 20 percent of the annual precipitation in wooded areas (Geiger 1961).

By slowing air, a windbreak with a density of 50 percent can quite effectively cause snow to accumulate in front of, within, and behind the barrier. Conversely, plantings that channel air flow will keep the areas of increased wind relatively free of snow and other airborne particles.

Air Purification

In many urban areas, the concentrations of air pollutants are so great that plants are not able to grow at their best, much less reduce pollution to acceptable levels. Air currents and the diverse sources of contaminants make air pollution a regional concern. The major effort must be to reduce emissions from vehicular and industrial sources (Schmid 1975). When air pollutants are at reasonably low levels, plants will be healthier, more active, and more effective in further reducing air impurities. Schmid found that ozone concentrations were reduced several times faster during the day when plants were transpiring rapidly (stomates open) than at night, when transpiration was low.

Gaseous pollutants are absorbed into active plant tissue, primarily within leaves, and are adsorbed on plant surfaces. Lanphear (1971) estimated that 50 million Douglas fir trees 300 mm (1 ft) in trunk diameter would be required to mitigate the 410,000 metric tons (452,000 tons) of sulfur dioxide released each year in St. Louis. That many trees would cover about 5 percent of the city's land surface. Bernatzky (1978), however, cites two German reports of deciduous forests having no perceptible sulfur dioxide filter effect compared with coniferous forests. In addition, many of the gaseous pollutants are exhausted or rise above even the tallest trees. Little sulfur dioxide released in a city remains there to be absorbed by trees (Bernatzky 1978, see Chapter 20).

Even though plants absorb carbon dioxide from the air and release oxygen, plants in a city have little effect on carbon dioxide and oxygen levels there. In fact, Weidensaul (1973) states that "It is not accurate to say that [land] plants really play a significant role in maintaining the concentration of oxygen and carbon dioxide in the atmosphere." Photosynthesis in the oceans supplies 70 percent (Cole 1968) to 90 percent (Bonner and Galston 1952) of the world's total oxygen. Winds and convection currents help maintain these gases at fairly uniform levels. **Protecting oceans from pollution is indeed critical to preserving a viable CO_2–O_2 exchange.**

Even though the world's oxygen supply is exceedingly large and well buffered

(Broecker 1970), a slight reduction in the level of oxygen could result in a sizeable percentage increase in carbon dioxide. Carbon dioxide and other gases in the atmosphere intensify the so-called "greenhouse" effect which would tend to increase the earth's temperature. Most all-natural systems, however, are well buffered. For example, should the earth warm, more moisture vaporizes from the oceans forming clouds which in turn reduce the amount of sunlight (Cosgrove 1989). This is a simplistic example but illustrates the point. This is not to imply plants are of little value in maintaining our environment, but to caution against over reaction.

Vegetation ameliorates air pollution most effectively through its ability to reduce airborne particulates. This is primarily a windbreak effect, reducing wind speed so that heavier particles settle out. In addition, particles are adsorbed on plant surfaces, primarily the leaves. Evergreens are recommended for this purpose since they are equally effective the year around. Foliage can become so coated, however, particularly along highways, that it must be spray-washed for both health and appearance. Bernatzky (1978) reports that in Frankfurt/Main streets with trees had 3000 particles per liter of air, compared with 10,000 to 12,000 in streets without trees.

Plants also reduce the content of heavy metals polluting air. Summarizing several studies, Schmid (1975) estimates that in Connecticut, one sugar maple tree of 300 mm (1 ft) trunk diameter at 1.4 m (4.5 ft) above the ground removes from the air in one growing season 60 mg of cadmium, 140 mg of chromium, 820 mg of nickel, and 5200 mg of lead. He further states that lead concentrations in plants taper off quickly with distance from major highways. In spite of the several ways that plants reduce air pollution, their presence near polluting sources is not generally considered to be of particular importance to urban air quality (Schmid 1975).

Noise Reduction

Plants are not very effective in reducing noise. "Out of sight, out of mind" applies somewhat and may be of importance in confined spaces. However, **in order to reduce noise levels appreciably, plantings must be dense, tall, and wide** (25–35 m; 80–115 ft) (Fig. 1–2). Plantings close to the noise source are more effective than similar plantings further from the source. Fleshly leaves and numerous branches increase excess attenuation; that is, they reduce (attenuate) noise more than simple distance will. Most species, however, do not differ greatly in their ability to reduce noise

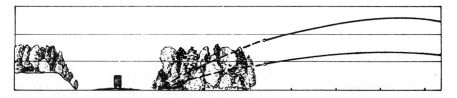

Figure 1–2   Thirty meters (100 ft) of trees and shrubs reduce truck noise about as effectively as would a similar area of bare cultivated ground. (Cook and Van Haverbeke 1971)

levels, although evergreen plants are best for year-round effectiveness (Cook and Van Haverbeke 1971).

Reports conflict in their assessment of sounds of differing frequencies. Robinette (1972) indicates that excess attenuation by plants increases with frequency, which would be desirable, since the higher frequencies are most annoying to humans. Herrington (1974), however, finds no consistent correlation between excess attenuation and frequency. No ready explanation of these contradictory findings is available.

Plants can not only hide a noise source but can also diffuse noise so that it is more steady. Leaf and branch movement and wind mask other sounds by raising the background level of noise, much as music in stores and offices masks external sound.

Erosion Control

Bare soil can be seriously eroded if exposed to rain, flowing water, and wind. Water accounts for most topsoil erosion (Fig. 1–3). Plants intercept rain and thereby soften its impact, which can otherwise loosen soil particles and get them into suspension to be carried with the water. Plant litter on the soil further reduces raindrop impact. A mulch increases the rate of water infiltration and slows movement on the surface, so that water tends to enter the soil near where it falls, either as rain or as sprinkler irrigation. Fibrous roots hold soil and further reduce erosion.

The faster the speed of wind or water, the larger the particles that can be carried away. Wind erosion can be minimized by plantings that slow the movement of air.

Plants also conserve water by reducing runoff and by accumulating snow near windbreaks for later melting and infiltration into the soil. This latter phenomenon is important in areas of winter snow when precipitation has been low.

Other Uses

Plants can be used to direct pedestrian and vehicular traffic as well as to improve the appearance of roadways. Shrubs can screen headlight glare from oncoming traffic and can serve as barriers to slow down vehicles that are out of control. Plants can reduce bothersome glare and reflection from the sun or artificial lights, just as they can reduce solar radiation, reradiation, and reflection.

VISUAL BENEFITS

Landscape Aesthetics

In our increasingly man-made world, the landscape aesthetics of plants are becoming more highly valued. Plants provide a basic contact with nature and heighten pleasure in our surroundings. Their aesthetic value is more difficult to quantify than the values already discussed and is difficult to describe, in fact, without seeming trite

Figure 1-3 Sloping bare soil is easily eroded when the application of water exceeds infiltration (left); ground-cover plants or a mulch could have prevented this erosion. **Plants, however, cannot prevent soil slumps and slides** (right); **an engineering solution is necessary.** (Photos courtesy U.S. Soil Conservation Service)

or overly sentimental. Without elaborating on the principles of landscape design, I list below some of the aesthetic advantages plants can provide (adapted from Bingham 1968):

Plants provide a variety of color, form, texture, and pattern in the landscape
Plants soften architectural lines and accentuate structural details
Plants can form vistas, frame views, provide focal points, and define spaces (Fig. 1–4)

Visual Benefits

Figure 1-4 Plants can form vistas or screen unsightly views.

Plants relieve the monotony of pavement and masonry

Plants, particularly trees, make enticing play areas

Plants offer cooling shade, pleasant fragrances, intriguing sounds, and serene settings

Plants create the impression of a well-established place in new residential areas and minimize the raw unfinished look

Plants unify, giving coherence to visually chaotic scenes

Plants can emphasize the seasons

ECONOMIC BENEFITS

Economic Value

Landscape plantings in America have not been thought of as an economic resource. Placing a realistic value on the benefits of landscape plants is difficult. Wisely designed landscapes can reduce heating and cooling costs of buildings. Erosion-control plantings conserve storm water and can prevent or reduce maintenance costs and loss of property. Outdoor landscaping and interior plants have been shown to increase worker productivity and to shorten hospital stays (see section on Psychological and Health Benefits). Thorton (1971) estimates the tree prunings and removals in Philadelphia are sufficient to keep a small sawmill busy all year.

The American Forestry Association (Ebenreck 1989) estimated the annual value of the ecological contributions of an average fifty-year-old urban tree to be $270. These 1985 values would vary depending on the region and site.

U.S. Forest Service studies estimate that the presence of trees increases the market value of homes from 7 to 20 percent (Ebenreck 1989). Trees increased the appraised value of undeveloped land in New England by 27 percent (Payne 1973). Waller (1965) reported that an offer of $24,000 for a house and lot with a 300-year-old elm in its front yard was withdrawn after a hurricane destroyed the tree. The house eventually sold for $15,000.

The Council of Tree and Landscape Appraisers (CTLA 1992), representing five landscape and horticultural industry groups, revised a guide for estimating the monetary value of woody landscape plants. Two methods are recommended for trees depending on size. The *Replacement Cost* method is for plants that can be replaced by one of the same size and species. The *Trunk Formula* method is for trees considered too large to replace. A *Basic Value* is calculated by adding: the cost of replacing the largest commonly-available tree of the same or comparable species in the region to the increase in value based on the difference in trunk areas of the appraised tree and the replacement tree, the cost per unit trunk area of the replacement tree, and the appraised tree's *Species* rating. In both, the value obtained is adjusted by *Condition* and *Location* ratings of the tree to obtain its *Appraised Value*. The costs are based on the growing and handling practices of plants in the region.

Appraisals based on the CTLA procedures are used for settling insurance claims, civil suits, and property condemnations and for establishing the value of plants for inventory purposes or for replacement should they be damaged during construction. The U.S. Internal Revenue Service has accepted tree casualty losses of replacement-size trees but has yet to fully recognize losses based on the *basic value* formula for larger trees.

The Arboricultural Association in Great Britain updated its *Amenity Valuation of Trees and Woodlands* in 1990. Six factors are identified for a tree, plus any special factors such as historical value. Each of these factors is given a score from one to four points. The monetary value of an individual tree is determined by multiplying the factors together and then multiplying that product by the monetary value (£10 in 1990) assigned to one unit of that product. Tree size, one of the factors, is determined by the product of the *mean crown diameter* and tree height. McAlister (1985) proposed a similar rating scheme for Australia.

PSYCHOLOGICAL AND HEALTH BENEFITS

A number of studies verify the psychological and health benefits of plants. Talbott and others (1976) found that hospitalized psychiatric patients spent more time eating when flowering plants were on their dining tables. They also ate more food and talked more with other patients than when flowering plants were not present.

Surgical patients who could see plants from their windows, compared to those who could not, had shorter hospital stays, required fewer potent medications, had fewer postoperative complications, and were reported to have a more positive hospital experience (Ulrich 1984). If window views of plants could shorten postoperative hospitalization by the 8.5 percent experienced in Ulrich's study, annual health cost savings in the United States would total several hundred million dollars (Ulrich 1986). Hospitals increasingly include plants and their care in rehabilitation programs to speed patient well-being.

Inmates at a large federal prison who viewed the prison interior from their cells sought health care more often than prisoners who had a view of farmland (Moore 1981). Ulrich (1986) reports the benefits of visual encounters with trees and other vegetation may be greatest for individuals experiencing stress or anxiety.

A National Aeronautics and Space Administration study finds that interior landscape plants have the potential for improving indoor air quality by removing trace organic pollutants from the air in energy-efficient buildings (Wolverton, Johnson, and Bounds 1989). Eight species of plants were each able to reduce normal levels of benzene 50 to 90 percent within 24 hours. Two species were able to reduce the level of trichloroethylene about 20 percent. Using an activated carbon-enriched rooting medium with positive air flow, golden pothos (*Scindapsus aureus*) reduced normal levels and 100 times the normal levels of both gases to almost zero within two hours. The plant root-soil zone appears to be the most effective area for removing volatile organic chemicals. The authors speculate that the activated carbon adsorbs air pollutants (those not absorbed directly by the plant) and holds them until the plant roots and microorganisms can transform them to non- or less-toxic forms, thereby, bioregenerating the carbon.

Business and industry executives have found that attractive buildings and landscapes result in above-average labor productivity, lower absenteeism, and easier recruitment of workers with hard-to-find skills (Lovelady 1965). Equally important, handsome factories and offices build good community relations.

Many articles state that people have a basic desire for contact with plants, even that plants have a strong positive influence on human behavior. More studies are needed to verify such relationships and to find out how plants can be used more effectively for human well-being.

COSTS OF LANDSCAPE PLANTINGS

Although landscape plantings in the United States perform some of the functions described in the preceding pages, Schmid (1975) states that "the plants which occupy Anglo-American front yards are present, neither because they enhance the mesoclimatic or acoustic environment, nor because they attract birds and small mammals, nor because they provide food or firewood, but rather because they are traditional ornaments necessary to maintain a public image of the appropriate setting for single family houses." To support that notion, he cites Gans (1967): "Homeowners are urged to support neighborhood norms and deviate neither toward shabbiness (which might lower neighborhood values) nor toward elaborate plantings (which could tend to push upward the general standard of maintenance and require more effort by others)." For all their values to our well-being and enjoyment, plants in the landscape have their price. If these plant problems are recognized, it is possible to eliminate or at least minimize them.

Plants not only have an initial cost, but they must be maintained throughout their life—some plants grow easily from seed and seem to need little or no care; others require a king's ransom to plant and maintain. The cost in 1990 was about $100 to install a 20-liter (5-gal) plant in an irrigated highway landscape and $10 annually to maintain it until the plant was established. A 1975 survey (Irvine 1975) of 16 southern California cities found that the cost of maintaining a hectare of landscape park ranged from $4750 to $17,500 ($1900–$7000/acre).

Because of their large size, trees can be dangerous and costly. Trees and their limbs fall in fair weather or foul, injuring people and damaging property. A tree is a formidable object when struck by a vehicle or by lightning. Tree roots clog sewers (Fig. 1–5) and break and lift sidewalks, curbs, and pavement and damage house foundations (see Fig. 18–19). Electric power companies and districts in the United States spent an estimated one billion dollars in 1990 to keep trees and shrubs from interfering with power lines and rights of way (Ng 1990).

Trees can seriously affect UHF (Ultra High Frequency) television reception (BBC 1979). Screening by trees can vary greatly with seasonal and weather conditions and is a special problem where television signals are weak. Trees can also reflect a signal so that a delayed image or ghost appears on the screen. Evergreen

Figure 1–5 These sections of sewer pipe were replaced because roots clogged them. Austin Carroll (with hat) of Sacramento, however, is touching two fruitless mulberry roots that crushed the 100-mm (4-in.) diameter orangeburg (asphalt impregnated fiber) sewer line (to which the roots are wired to show their positions). The tree was planted in the backfill of a sewer trench that had been dug into hardpan. The expanding roots crushed the sewer line.

trees block television signals more completely than do deciduous trees, and conifers more so than broadleaved evergreens. When bare, deciduous trees have little effect except during wet weather; when covered with moisture, all trees can appreciably affect signals, particularly higher-frequency signals. UHF reception can be protected by mounting aerials above the tops of trees or so that no trees intervene between transmitter antennae and aerials. Cable television eliminates the problem altogether.

Many people are allergic to the pollen or leaf pubescence of many plants. Poison oak and ivy are dermally toxic when they are touched or their smoke encountered. The leaves and fruits of many landscape plants are poisonous, although the number of fatalities is extremely low considering the number of poisonous plants.

Leaves, flowers, and fruit can be an expense and nuisance to clean up. Many plants seed profusely or root sucker and become weeds. People have requested that street trees be removed because of the litter they produce.

For whatever reasons, landscape plantings are important features of private and public properties and must be selected and cared for with skill. By understanding how plants grow and respond to cultural practices and their environment, we can increase our enjoyment of landscapes. To that end, the following chapters are designed to provide the basis for a more complete understanding of the care of woody landscape plants.

FURTHER READING

BERNATZKY, A. 1978. *Tree Ecology and Preservation*. New York: Elsevier Scientific Publishing Co.

GEIGER, R. 1961. *The Climate Near the Ground*. Cambridge: Harvard University Press.

MILLER, R. W. 1988. *Urban Forestry: Planning and Managing Urban Greenspaces*. Englewood Cliffs, NJ: Prentice-Hall.

ROBINETTE, G. O. 1972. *Plants/People/and Environmental Quality*. Washington, DC: U.S. Government Printing Office.

SCHMID, J. A. 1975. *Urban Vegetation*. Univ. of Chicago, Dept. of Geography Research Paper 161.

CHAPTER 2

Plant Structure and Function

Arboriculture concerns the planting and care of woody perennial plants, primarily trees, but also shrubs, vines, and ground-cover plants (Fig. 2–1). A tree is usually described as having one dominant vertical trunk and a height greater than 5 m (15 ft). In contrast, a shrub has a number of vertical or semiupright branches originating at or near the ground and is smaller than a tree. A vine or wall shrub has a slender, flexible stem that requires support. Prostrate, spreading shrubs and vines are often considered to be ground-cover plants. These are arbitrary definitions but afford some convenient characterizations of plant growth and form. The words *tree, shrub, vine,* and *ground cover* bring certain ready images to mind, although the differences among them are not always easy to specify. Multistemmed trees, when young, might be considered shrubs, and many plants we regard as shrubs eventually grow to be 10 m (30 ft) tall or more. The distinguishing feature is form rather than size.

Knowing the potential form and size of plants and the stages of their growth is essential to anyone who plans landscape plantings and maintenance practices. The genetic makeup of a plant not only determines many obvious characteristics but also influences bud formation and break, shoot elongation, and caliper growth in response to environmental conditions. The above-ground shoots and stems of woody perennial plants must survive the winter and have buds grow the following spring.

Even though woody landscape plants vary greatly in size and form, they are vascular plants (have conducting tissues) similar in composition and internal functioning. Like all plants, they are composed of individual cells grouped into specialized tissues, such as parenchyma (simple thin-walled cells) and xylem (water-

15

Figure 2-1 Trees, shrubs, and vines offer a wide variety of landscape forms.

conducting tissue). Stems, roots, leaves, flowers, and fruits are composites of various tissues. Meristematic tissue contains actively dividing cells which develop and differentiate into new tissues as well as perpetuate themselves.

Trees, shrubs, and vines compete more effectively with other plants and even with animals by getting their leaves up in the air for more exposure to the sun and protection from land-bound animals. Upright growth and branching of trees increase the amount of leaf surface exposed to sunlight. In order to grow upright, plants must resist the pull of gravity; this response is called *geotropism*. In addition, if plants are to grow tall without support, they need a strong structure.

THE SHOOT

At the tip of each shoot in a tissue volume no larger than a pinpoint, all of the aerial parts of a plant are determined, that is, the form of the mature plant is programmed. This region, called the shoot apical meristem, is a hemispherical mass of rapidly dividing and expanding small cells. Flower, leaf, and axillary bud primordia and the conductive tissues of the stem are initiated in this region. **Usually the entire zone of stem elongation, in which cells divide and enlarge to their mature dimensions, is in the most apical 20 to 50 mm (1–2 in.) part of the stem.** This is true for branches as well as for the main stem. In some vines, stem elongation may be distributed over the terminal 100 mm (4 in.), but this is uncommon.

The Bud

A bud is an unexpanded shoot or flower. A shoot elongates by growth of its terminal bud and forms branches from lateral buds formed in leaf axils at nodes along the shoot (Fig. 2–2). When a terminal bud is removed by pruning or breakage, a bud or buds grow from below the break. In addition, *adventitious* buds may form on a root, a leaf, or occasionally an internodal region of a shoot or branch. Unlike terminal and lateral buds, adventitious buds have no pattern to their location.

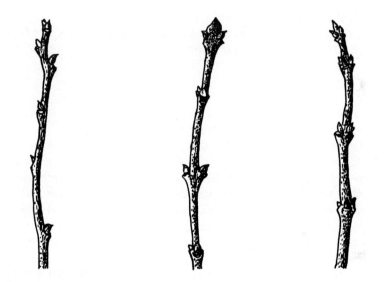

Figure 2–2 Buds are located either terminally or laterally (axillary = in a leaf axil). Lateral bud arrangement is alternate (left), opposite (center), or whorled (right).

Buds may produce leaves (and a shoot) or flowers or both; they are respectively called leaf, flower, and mixed buds. More than one or one type of bud may be present in the same leaf axil.

The arrangement of leaves and buds on a shoot is called *phyllotaxy*. Each species, and sometimes a genus or a family, has a specific pattern. When two or more buds are opposed or spaced around the same node, their arrangement is said to be *opposite* (as in the ash or maple) or *whorled* (as in a catalpa) (Fig. 2–2). An imaginary line between two opposite buds at a node will usually not be parallel to similar lines at the nodes immediately above and below but rather will be nearly at right angles to such lines. Buds may be spirally arranged around the shoot when only one bud is at a node (as in an apple tree), or all the buds may be on one side at a node (as in a peach tree). The phyllotaxy of this *alternate* bud arrangement is usually expressed as a fraction, such as $\frac{1}{2}$, $\frac{1}{3}$, $\frac{2}{5}$, or $\frac{3}{8}$. The numerator is the number of revolutions around the shoot it will take to reach a bud directly above or below another; the denominator is the number of buds passed. The phyllotaxy of a plant is helpful in identifying it, particularly when the plant is not in leaf or flower.

Buds vary in their state of activity. The shoot apical meristem is an actively growing bud from which a shoot with leaves, buds, and possibly flowers develops. A *dormant* bud is an inactive bud on a shoot from late autumn or after leaf drop and the start of new growth in the next season. Pruning, spraying, and transplanting may be scheduled for this dormant period, when the plant is most tolerant to most stresses. A *latent* or *suppressed* bud is one that is more than a year old but has grown enough each year so that its growing point is at or near the surface of the bark. When a terminal bud is removed by pruning or breakage, one or more buds near the tip of the remaining branch usually grow to take over the terminal role. **Most shoots that grow from old branches, from the trunk, or near large pruning cuts are likely to be from latent buds, not from** adventitious **buds,** as many people think. These are called *epicormic* shoots.

Shoot Growth: Length

Annual shoot length is a good measure of plant vigor. Recent annual growth can be determined in plants that have one period of growth a year and set dormant terminal buds (Fig. 2–3). On deciduous plants, the bud scale scars of the terminal buds and on conifers each whorl of branches indicate a year's growth when species have only one flush of growth a year or grow only from preformed initials. Current growth of deciduous plants can be determined from the leaves or buds that are present on the current season's shoot. On the basis of external appearances alone, however, current growth and the growth of previous years may be almost impossible to distinguish on broad-leaved evergreens and plants that make multiple flushes of growth.

Many woody species form short as well as normal shoots (Fig. 2–3). Short shoots are usually less than 100 mm (4 in.) long and are called *spurs* by fruit grow-

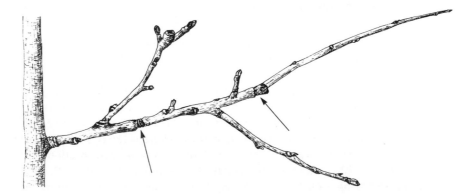

Figure 2–3 Three years of shoot and spur growth from the vertical branch; terminal bud scale scars (arrows) separate each year's growth. Two short shoots and two short spurs have grown from the second-year spur.

Chap. 2   Plant Structure and Function

ers. Nodes are much closer together on spurs than on normal shoots and usually produce a higher proportion of flower buds. Most spurs have wide angles of attachment to the branch on which they arise. Annual spur growth may be terminated by a thorn (as on pyracantha and black locust), a normal vegetative bud, or a mixed bud that will produce both leaf and flower (as on apple). Spur-forming plants of a given size will usually have more flowers and fruit than species of a comparable size that lack spurs.

Watersprouts and *suckers* are vigorous, upright, epicormic shoots that grow from latent buds in older wood. They can be a problem in a number of fruit and landscape plants (Fig. 2-4) and can grow 3 to 5 m (10–15 ft) in one season. Often they outgrow the main leader of young trees. In some species, they can be an annual problem for almost the entire life of a plant. If previous growths of this kind have not been pruned off close to the trunk or roots from which they arise, they can become particularly numerous.

Watersprouts and suckers differ primarily in their location or origin: Watersprouts arise above the ground or graft union, primarily from latent buds on the trunk or older branches; suckers arise below the graft union or the ground from the trunk or roots. Almost without exception, both should be removed as early as possible. In a few situations, one might want to develop a watersprout into a main branch to fill a void in the plant; but on grafted plants, suckers can be quite different from the grafted top.

Watersprouts are often forced into growth just below large pruning cuts, particularly when branches have been cut to stubs. Watersprouts are seldom firmly attached to the trunk or branch from which they arise. Only the new layers of bark (phloem) and of wood (xylem) formed since the sprout started growth unite it with the tree. Consequently, they can break out relatively easily for a number of years.

Watersprouts and suckers indicate that a plant is in vigorous condition, at least that part of the plant below the vigorous growth. They may be forced into growth by injury or death of the top of a plant. When a tree is stressed by drought, nutrient deficiency, or disease, many shoots grow along the main branches and trunk, but they are of low vigor, usually less than 600 mm (2 ft) long (see Fig. 19-4). It almost seems as if they were the tree's last effort to stay alive, and something more than pruning is needed. Similar shoots may grow if a branch or trunk is exposed to sunlight because of defoliation, opening up of the tree crown, severe pruning, or death of branches due to fire or a similar catastrophe.

Control of Shoot Elongation

The rate of stem elongation depends on the rate and duration of cell division and expansion. Daily and annual elongation rates differ widely among different species. Faster-growing plants usually have longer zones of cell division and expansion at their shoot tips than do slower-growing plants.

Pines constitute an excellent example of genetic diversity in growth rates, often within a single species. Dwarf members may appear and grow at only a fraction of the rate for more vigorous plants in the species. A major task for a nursery person working with seed-propagated plants is to *rogue out* or set aside the seedlings with

Figure 2–4 Vigorous shoots that grow from above ground or above the graft union are called watersprouts (top left); those from below the graft union (top right) or the ground (lower left) are called suckers. Note the black walnut leaves and bark of the rootstock differ from those of the English walnut top (top right).

extraordinarily high or low growth rates. When selecting young plants that have been seed-propagated, one should choose those of uniform size and shape to better ensure mature plants with uniform growth characteristics.

Some plants elongate for a short period (three to four weeks) in the spring, others in flushes during the growing season, while still others may elongate nearly continuously as long as conditions are favorable. The duration of shoot elongation may be controlled in several ways.

Initiation of Terminal Flowers. An inherent control of elongation occurs in species, such as crape myrtle and southern magnolia, that flower terminally on the current season's shoots (see Fig. 2–16). Such control of elongation is similar to that of many herbaceous annuals whose growth habit is said to be *determinate.* This contrasts with plants of *indeterminate* habit (like peach and mulberry trees) that have lateral flower buds and a terminal that remains vegetative. The terms *determinate* and *indeterminate* are, however, usually confined to herbaceous growth habits.

Some woody species, such as apple and pear trees, form terminal mixed buds (with both leaf and flower primordia), usually on short shoots, which do not grow until the next spring. Although the terminal mixed bud does not control elongation in the way terminal flowering does, the effect on the next year's growth is similar. In both cases, growth begins from lateral buds, giving a somewhat zigzag (*sympodial*) pattern to the branches.

Preformation of Nodes in the Terminal Bud. In all deciduous and some evergreen plants, bud and node initials are formed in the terminal bud, although some cell division and internode elongation occur the following spring or when stimulated to do so. These are called *preformed shoot initials.* Some plants continue shoot growth after their shoot initials have expanded, while species like the fir, the spruce, most pine, and many oak stop shoot elongation after the preformed initials have expanded. No matter how favorable the growing conditions, these plants seldom have more than one flush of growth. Although no additional buds or nodes will be formed in the same season, favorable growing conditions will result in longer internode length and in more buds and possibly more vigorous growth the following season.

Species such as the juniper, the sequoia, and most angiosperms (plants with seeds within an ovary) continue shoot growth as long as conditions are favorable. Within a single season, these plants respond to environmental conditions and to cultural practices which influence growth. Still other species, like some of the pines, are intermediate in their shoot growth. Under normal or stress conditions, these plants have a single period of growth from preformed initials. Under favorable conditions, shoot terminals remain active and continue growing.

Response to Environmental Signals. Environmental conditions, such as temperature and daylength, must be satisfactory for growth to occur (see Chapter 4). Many plants slow or stop growth when day-length shortens. Species native to higher latitudes, where late spring and early summer days are especially long, may be dwarfed when grown closer to the equator because there are fewer long days.

Response to Internal Controls. Even though buds form on currently growing shoots, many do not grow until the following spring. *Apical dominance* (see the latter part of this chapter) prevents newly formed buds from growing. Later in the growing season and into the winter, a condition called *rest* (see Chapter 4) keeps the buds of many temperate-climate plants from growing even if favorable growing conditions prevail. Bud growth will not occur until these internal controls are removed or satisfied. Finally, the plant may be ready to grow but may encounter late winter temperatures that are too cool.

Shoot Growth: Caliper

In conifers (gymnosperms, naked seeds) and dicotyledonous plants (two-seed leaves, angiosperms, covered seeds), trunk, branch, and root are strengthened by secondary growth of the vascular cambium. Monocots (angiosperms, one-seed leaf) such as palms may produce large stems by a special thickening meristem that is most active immediately below the top of the plant and achieve root stability by increasing the number of roots but not their individual size.

Radial growth of shoots in woody plants depends upon cell division and expansion in a thin cylinder of cells, the vascular *cambium,* which extends from the base to the tip of most plants. New cells are formed on both sides of the cambium (Fig. 2–5): The cells to the inside give rise to the woody water- and nutrient-conducting tissues of the stem, the *xylem;* the outside cells become part of another transport system, the *phloem,* which conducts organic substances (such as sugars, amino acids, vitamins, hormones, and stored food) from the mature leaves to other portions of the plant. **The oldest xylem is nearest the center, the oldest phloem on the outside.**

In some plants, the phloem tissue alone forms the bark; but in many species, another layer of dividing and expanding cells, the cork cambium or *phellogen,* forms outside the phloem, adding bulk to the bark and protection to the trunk. Species with smooth bark either have cortical tissue and epidermis that keep pace

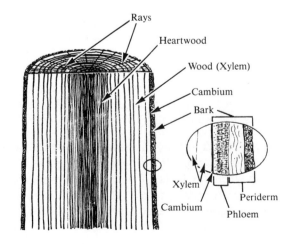

Figure 2–5 Cross and longitudinal sections of a tree trunk. The cambium is a thin layer of meristematic cells which gives rise to phloem to the outside and to concentric rings of xylem to the inside. Exterior to the phloem, the periderm is the outer bark and may be continuous (smooth) or discontinuous (rough). The vascular rays provide pathways and storage for food and other materials in the xylem and pathways between the xylem and phloem.

with the expanding stem by cell division and enlargement for the life of the plant or the first formed phelloderm remains active for years (Zimmermann and Brown 1971). In most trees, however, as stems increase in girth, phelloderm does not keep pace and ruptures; successively deeper phelloderm tissues are formed over the years giving rise to rough and fissured bark.

Xylem provides mechanical support to hold plants upright and is the pathway for water and nutrients from roots to the top. In addition, beginning after midsummer the living xylem stores organic materials (primarily starch) for later use (Wargo 1979). When injured, xylem forms protective chemicals that may retard or restrict infection and wood decay.

Annual rings of xylem growth are evident because earlywood, produced in spring, usually has larger, lighter-colored cells with thinner walls than does latewood, formed later in the growing season. *Sapwood* is xylem immediately interior to the cambium. The number of annual rings included which remain active in water conduction varies depending on species; those studied by Sargent (1926) ranged from one to 20 annual rings. In contrast, the entire xylem in roots is able to transport water and nutrients. **The conducting xylem elements are nonliving.**

In gymnosperms (primarily conifers), the xylem is composed mainly of uniform-size *tracheids* which are the water-conducting elements. The xylem is called *nonporous* because the openings (*pits*) between the tracheids slow water movement. The thick-walled small-diameter tracheids of the latewood of one year are in marked contrast to the earlywood of the following spring (Fig. 2–6). Water moves in the earlywood tracheids of several of the most recent annual rings of conifer xylem.

In hardwoods (angiosperms), the xylem contains both tracheids and *vessels*. Wood is described as *diffuse porous* if the vessels, larger in diameter than tracheids, are scattered among the tracheids and other cells of each annual growth ring (Fig. 2–6). Water moves primarily in the vessels in several of the youngest annual rings (Chaney 1986). Maple, walnut, and poplar have diffuse-porous xylem.

Oak, ash, and black locust have *ring-porous* xylem in which large diameter vessels are concentrated in the earlywood of each annual ring (Fig. 2–6). Water moves primarily in the large vessels of only the current growth increment. The vessels of older growth rings are blocked by gases, tyloses, or gums (Chaney and Kozlowski 1977).

Xylem rays (living) in the *sapwood* are the pathways for food movement to and from the phloem. The axial (longitudinal) and ray parenchyma cells in the sapwood are the primary storage sites of food reserves in the trunk and branches. In most species, the center portion of the xylem becomes *heartwood* as ray parenchyma cells die therein. It is "age-altered" and may or may not be different in color from sapwood (Shigo 1986b). Heartwood is essentially physiologically inactive, but it retains structural strength and infection-restricting ability.

Cambial activity is markedly seasonal. In temperate climates, it begins shortly before bud break in the spring and proceeds down the stem in an advancing wave that has been correlated with the movement of auxin and other hormones. As with shoot elongation, the duration of cambial activity differs considerably among different species and different climatic conditions, and especially with variations in soil moisture. Wet and dry years are clearly recorded in the xylem of many woody plants

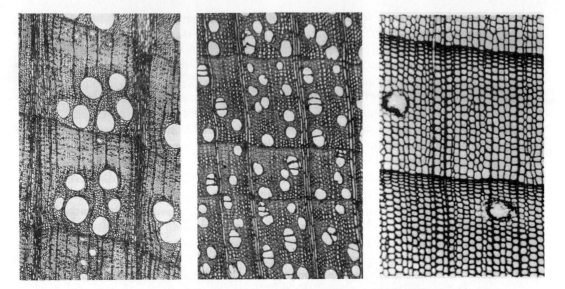

Figure 2-6 Variation in vessel and tracheid sizes and distribution within cross-sections of annual growth rings. Large vessels of ring-porous xylem of oak are concentrated in earlywood (left). Large vessels of diffuse-porous xylem of maple are smaller and more evenly distributed throughout an annual ring (center). Tracheids of nonporous xylem of gymnosperms, such as pine, spruce, and fir, occur in longitudinal stacks of spindle-shaped cells interconnected by pits in the sidewalls (right). Earlywood in each photo is toward the bottom. (Two left photos courtesy Alex Shigo and Kenneth Dudzik, U.S. Forest Service; right photo courtesy William Chaney, Purdue University)

(Glock and Agerter 1962). **The width of annual rings can be used as a measure of vigor similar to shoot length.**

Lumber with a relatively large proportion of latewood is considered the most desirable. Not only are there relatively more small cells, but the cell walls are thicker, creating a harder, stronger wood. For a given thickness, large-celled wood in vigorous trees is not as strong, but overall strength will usually be greater in a more vigorous plant because its trunk makes more caliper growth. The relative strength of these two types may be difficult to assess in living plants.

Compartmentalization

Perennial woody plants are highly compartmented: Cells (compartments themselves) are grouped between adjacent xylem rays and the latewood of two annual rings; these in turn are grouped within a growth ring that circles the stem (Shigo 1986b). These are existing boundaries.

When wounded, perennial plants are able to react to form further physical and chemical restraints to wall off (contain) the injury and its effects (Hepting 1935, Shigo 1979b). (Chapter 17 has more about wounds.) When wounded, xylem cells adjacent to the wound begin to form a six-sided compartment with four walls (Shigo and Marks 1977): Wall(s) 1, the plugging of xylem vessels (angiosperm) and tracheids (gymnosperms) above and below the wound (two sides of the compartment); Wall 2, the thick-walled cells of latewood of an interior annual ring; Wall(s) 3, radial xylem rays (two more walls); and Wall 4, new xylem formed after wounding that is anatomically and chemically different from that formed at other times (Fig. 2–7). Walls 2 and 3 undergo chemical changes (primarily oxidation and polymerization of phenolic compounds in angiosperms and terpenes in gymnosperms) that further increase their resistance to discoloration and decay. Shigo termed this process CODIT, an acronym for *Compartmentalization Of Decay In Trees*.

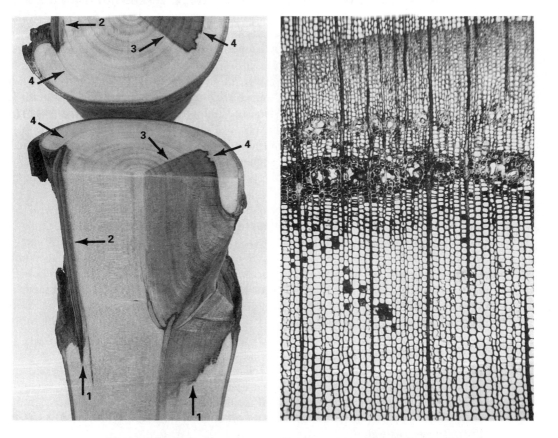

Figure 2–7 Left: Compartmentalization of two wounds in a red maple trunk, one a two-year-old pruning cut into the branch collar (right) and the other a larger but shallow wound of the same age (left). The numbers identify the four walls of compartmentalization which are formed following wounding (Shigo 1989). Walls 1, 2, and 3 form a reaction zone in wood present at the time of wounding; wall 4 creates a barrier zone that separates tissue formed after wounding from that formed earlier.

Right: Note in the photomicrograph the darker barrier zone (wall 4) formed late in the springwood within an annual ring of sugar maple. (Photos courtesy Alex Shigo and Kenneth Dudzik, U.S. Forest Service)

Boody and Rayner (1983) and Cooke and Rayner (1984) question the concept of CODIT that plants can actively respond to injury to prevent decay. They maintain that the existing "moisture content may be the principal determinant of decay location," and that various (normal) tree functions which "preserve the moisture content of living sapwood can limit accessibility of decay organisms" (Boody and Rayner 1983). They agree as to the anatomical development of decay and that wounding should be avoided or minimized. Most tree-care–practice recommendations based on these interpretations would be the same.

Wall 1 is least resistant to the spread of wood discoloration and decay. Wall 4, also called the *barrier zone* (Sharon 1973), is pathologically the strongest of the walls and can usually confine injured and decay-infected wood to that present at the time of wounding. **Although Wall 4 is quite resistant to decay, it is structurally weak and can lead to ring shakes** (circumferential separation of xylem along an annual ring), which are undesirable in timber trees and can weaken landscape trees.

Injured plant tissue is not replaced or repaired to its previous healthy state (Shigo 1979b) but instead is usually confined so that the rest of the plant does not become infected. The closest a tree comes to "healing" occurs when cambium is able to survive loss of bark; new xylem and phloem continue to be formed (Fred Roth, California State Polytechic University, Pomona, 1990 pers. comm.).

Genetic makeup and vigor determine a plant's ability to compartmentalize. Clonal selections within certain tree species are being made on their ability to compartmentalize wounds. Plant survival largely depends on the ability to compartmentalize and to form new tissue.

When a branch dies, breaks, or is properly pruned off, discoloration and decay (should they occur) are usually confined to the branch stub zone within the trunk by physical and chemical barriers. On the other hand, if a pruning cut is within or inside of the branch bark ridge (a flush cut) (Figs. 2–7 and 15–7), trunk tissue is exposed to the possibility of infection (Shigo and others 1979). There is some question, however, about whether all of the collar at a branch attachment is trunk tissue (Neely 1991). Resolution of this question may influence where the lower edge of a pruning cut should occur.

FACTORS THAT INFLUENCE SHOOT GROWTH

Climatic influences on shoot growth will be discussed in a later section. Direct internal factors and responses to conditions other than climate, however, can affect shoot growth as well.

Dwarf Plants

Mature plants of small size have been long prized by plant growers (Fig. 2–8). Dwarf plants are more compact, more symmetrical, and usually more floriferous and fruitful than plants of normal size. Trees may be dwarfed in several ways.

A small genetic change may dwarf a plant in relation to its normal counterpart. Dwarfs of perennial plants almost always have to be propagated vegetatively

Figure 2–8 Dwarf plants, such as the conifers in the left foreground, appeal to many people. Normal-sized mature conifers are in the background at Strybing Arboretum, San Francisco.

to maintain this characteristic. A few fruit species (such as peach) have dwarf varieties that will propagate from seed. In almost all of these plants, the internodes are much shorter than normal, accounting for their compact, symmetrical appearance.

A number of species can be dwarfed by grafting normal-sized varieties onto rootstocks that reduce the growth of the scion. Many citrus species and apples can be successfully dwarfed by proper rootstock selection; cherries and pears have less success. A number of landscape plants, such as lilac, can be similarly dwarfed.

Bonsai is based on top and root restriction achieved by pruning and the use of small containers. Some plants from the extreme northern or southern latitudes can be dwarfed under the short-day growing conditions of lower latitudes. With some species, gibberellins (growth-regulating hormones) will overcome dwarfism and, along with other substances, probably regulate stem elongation and account for normal or dwarf habits of growth. Horticulturists continue to seek more genetic dwarfs and more techniques to manipulate plant size.

Intraplant Competition

Interior leaves and those on lower branches, particularly those on branches that have been severely pruned, can be weakened and die due to dense shade of foliage above. Conversely, lower and interior leaves can be nurtured by wise thinning of the top.

Vegetative growth is reduced when plants flower and produce fruit. This is particularly noticeable if most of the blossoms are removed from a large limb of a fruit tree. Shoots will be much more vigorous on that branch than on branches that continue to nourish flowers and fruit.

Several branches arising near the same level on a trunk will often grow less than if one of them were dominant. This is probably due to competition for water and nutrients. More than one branch arising from the same level on the main leader of a broadleaved tree is undesirable, particularly if the tree is young. The main leader may be subdued and weak branch attachments produced.

Trunk Development

A strong, upright trunk is desirable in many trees for street, patio, and park use. By understanding the factors that influence trunk development, an arborist can train trees with such a structure, or with multiple stems or a more sculptured form.

Lateral branches along its length encourage caliper growth of the trunk, but height growth will be inhibited if the laterals are not kept shortened (Chandler and Cornell 1952, Leiser and others 1972). Horizontally growing laterals usually grow slowly and are easily kept in bounds. As indicated earlier, vigorous shoots are more upright and usually form more laterals than do shoots of low vigor.

Reaction Wood

When an upright leader of a tree is forced from the vertical, the new xylem formed tends to return the leader to the upright. This is true not only for new growth at the shoot apex, but even back on the leader where secondary wood has already formed (Fig. 2–9).

When a main stem is grown at an angle away from the vertical, it produces so-called *reaction wood* to counteract the lean. In broadleaved trees, reaction wood is termed *tension wood* and forms on the upper surfaces of branches and leaning trunks. It exerts an internal contraction that tends to pull the trunk to the vertical or a branch to its original angle of growth. The corresponding type of wood in conifers, *compression wood,* develops on the underside of branches and leaning trunks. Compression wood exerts a compressive force due to the greater elongation of xylem cells on the lower side of the trunk that tends to force the trunk upright or a branch to its original angle of growth. This contraction (pull or tension) or expansion (push or compression) of the new xylem can be great, bending trunks 300 mm (1 ft) or more in diameter toward the vertical. For those who like memory aids, some distinguish between the tension wood formed in broadleaved plants and

Figure 2-9 The young growth of the trunks of many trees, such as this deodar cedar at Arundel castle, England, will become as vertically straight as the lower trunk due to reaction wood forcing it more upright.

the compression wood of conifers by the words that start with the letter **c** and remembering that tension wood forms on the top side of branches or leaning trunks.

Christmas-tree growers may capitalize on this phenomenon by leaving one or more whorls of branches on the stump when they harvest trees. The strongest growing branch of the remaining whorl will usually bend upright. A tree that grows from such an upright branch or a bud will develop more quickly than will a young seedling. Its trunk will not be as straight or its branches as symmetrical as growth that would come from a latent bud, but there are few if any such buds in most conifers.

When dry, reaction wood warps and twists and is weaker than straight-grained wood. Reaction wood is therefore undesirable for lumber, and forests are managed to minimize its formation. In the live state, however, reaction wood appears to be as strong or stronger than normal wood.

Each branch's particular angle of growth appears to be determined by its upward response to geotropism and the weight of the branch. Branches attempt to maintain their inherent angle by the formation of reaction wood. In most cases, more wood is formed on the lower than the upper side of a branch. If a branch is forced from its normal angle, whether up or down, reaction wood will form in an attempt to restore the original angle of growth. Arborists will do well to keep this response in mind when they cable or brace tree limbs so that they do not prevent normal reaction wood from forming to help support the branches. In fact, pulling a branch out of its inherent angle of growth may actually increase the load on the cables and inhibit the branch's ability to maintain its natural position.

In most cases, the formation of reaction wood cannot completely overcome diversions of the trunk from the vertical or of branches from their normal angle of growth. If a trunk or a branch is held for a year or two so that it cannot correct the angle at which it is growing, the new wood laid down makes it more difficult for it to return to its former position when the external force is removed.

The differences in compression and tension wood can be observed under the microscope. Cells of compression wood are more elongated, while cells of tension wood are shorter than normal. Auxin concentrates on the undersides of branches or leaning trunks in some species and, along with other substances, may account for these differences in xylem elements. Differences can also be seen when a branch dies. In many conifers, a branch that was horizontal when alive will curve down as the wood dries because the lower side of the branch shrinks more than the upper side. In contrast, the branches of many angiosperms curve upward when they die; this is particularly noticeable in the branches that were more upright originally.

Trunk Movement

When a trunk has been kept from moving by guying, staking, or dense planting, the tree grows taller, exhibits less caliper growth near the ground, and develops with less taper (a decrease in trunk diameter with height) and a smaller root system. Each of these growth responses lessens the ability of a staked or guyed tree to stand upright when its support is removed. The small trunk base is less able to support the top of the tree. The higher branches encounter more wind resistance and exert greater leverage on the trunk. The lack of taper tends to concentrate stress close to the ground instead of distributing it more uniformly along the trunk. These points are discussed further in Chapter 9 in the staking section.

The primary effect of staking and dense plantings seems to be the reduction of trunk movement. When the upper stems of greenhouse sweetgum and corn are moved as little as 30 seconds a day, height growth is reduced by 25 percent or more (Fig. 2–10) (Neel and Harris 1971). Bud dormancy is also hastened. Trunk movement seems to function somewhat similarly to diversion from the vertical, except that movement causes reaction wood to form more uniformly around the trunk.

Lack of movement may be one of the main causes of the tall, slender trunks on trees and shrubs grown close together in nurseries (see Fig. 4–10) (Harris and others 1972). While such trunks are undesirable for nursery trees, they are ideal in forests where trunks should be straight, have little taper along their main trunks, and have few or no side branches in the lower portion. This produces lumber of high quality, straight grain, and with relatively few knots. When a forested area is partially cleared for development, as for a road, subdivision, or recreation site, trees adjacent to the clearings may not be able to withstand the storms from which they had previously been protected. Such trees can be extremely hazardous.

Unidirectional Shading of the Trunk

Many times the trunk of a young tree is tied closely to a single stake. **When the ties are removed, the tree almost invariably bends away from the stake.** In extreme cases, the treetop may actually lie on the ground. Originally, it was thought that this was due to a weak trunk, but when some of these trees, still in containers, were

Figure 2–10 The two liquidambar trees growing in a greenhouse were the same height (arrows) before the upper portion of the left tree was gently shaken for 30 seconds daily during the 21 days before the photograph was taken. (Neely and Harris 1971)

turned upside down, the trunk still bent away from what had been the vertical (Fig. 2–11) (Neel 1971). Examination of the wood showed that the cells of the xylem next to the stake were more elongated than the cells away from the stake. This asymmetrical development did not occur when transparent stakes were used. Studies have shown that the auxin content on the shaded side of a shoot tip is higher than that on the lighted side, causing the shoot tip to grow toward the light (Janick 1986).

Figure 2–11 Young Japanese zelkova tied closely to opaque stakes (left tree in left photo), when untied, bend away from the vertical (center), even when turned upside down (right). The control trees (on the right in each photo) were tied to transparent stakes.

Factors That Influence Shoot Growth

The tendency of the trunk to bend away from an opaque stake is probably due to a redistribution of growth substances between the lighted and the shaded sides of the trunk.

BRANCH ATTACHMENT

A branch has little or no direct structural or conductive connection above its attachment to the trunk or branch to which it is attached (Shigo 1985). In 20 species studied, Shigo elegantly shows that branch tissues turn abruptly at the branch base and extend down the trunk (Fig. 2-12, upper left). Dye injections by Neely (1991) support Shigo's observation that xylem vessels on the upper side of a branch turn downward at the branch base (Fig. 2-12, lower right). These findings are contrary to information published before 1985 (Eames and McDaniels 1947, and others).

Neely (1991), however, questions Shigo's hypothesis that successive annual layers of branch xylem and trunk xylem occur around and under branch attachments in trees as interpreted from Fig. 2-12 (top two photos). Neely reasons that if such lamination of branch and trunk xylem occurs, the flow of xylem sap into and around a branch attachment should change if the branch xylem around a branch base is overlain by trunk xylem. He did not find that to happen (Neely 1991).

Neely (1991 pers. comm.) interprets the xylem pattern of the red-oak in the upper right of Fig. 2-12 as typical of a trunk collar surrounding a declining branch. Both Shigo (1989) and Neely (1991) agree that if a branch base is completely surrounded by a collar, trunk tissue is overgrowing a declining or dead branch.

Neely calls the enlarged area at the base of a branch the *branch shoulder;* Davey (1967) called it *shoulder rings.* Shigo (1986a) calls the area the *branch collar,* considering it trunk tissue that flows around a branch and over branch tissue below. Whether this area is all trunk tissue or both trunk and branch is not clear.

No matter whether branch and trunk xylem tissues are laminated at a branch or are intermingled, **as successive annual rings of xylem are laid down by the cambium, the tissues of both the trunk and its branch are buried more and more deeply in the trunk** (Fig. 2-12, lower left). Such a branch is held by the trunk like a dowel in a chair leg.

Whichever interpretation is chosen, it should be emphasized that **the only way trunk xylem can grow around a branch is for the trunk to be larger in diameter than the branch at its attachment.** It is instructive that, in the photographs that Shigo (1985, 1989) and Neely (1991) use to show branch attachments, almost all the branch diameters are 40 to 50 percent smaller than the trunk diameters (Fig. 2-12).*

*More information about Shigo's and Neely's interpretations of branch and trunk xylem formation at branch attachments may be of interest. Shigo (1985) proposed that cambium near a branch attachment becomes active before the trunk cambium. In spring, cambial activity begins at shoot tips and progresses toward the trunk base (Zimmermann and Brown 1971). From this and many observations Shigo reasons that collars of overlapping xylems are ensured because the cambium at a branch base should become active in the spring before that of the trunk. As a result, a shield of branch xylem is laid down on the trunk below the branch. Later, trunk xylem appears to flow around the branch base and

It has been shown that **to have a strong attachment, a branch must be smaller than the trunk or limb from which it arises** (MacDaniels 1932; Ruth and Kelley 1932; Miller 1959). If the branch and trunk are near the same size, their attachment will be weak because their xylem tissues are essentially parallel—neither able to grow around the other. These would be considered double leaders or codominant stems (Fig. 2–13 and 17–3).

Relative branch size is more critical in determining the strength of an attachment than is the angle of attachment. When Miller (1959) subjected 60 apple branches of the same variety, age, and size to stress until they split, there was no correlation between the angle of attachment and the force at which the branch split. Their angle has little direct effect on the strength of branch attachments, but a branch will grow more upright and vigorously when the angle of attachment is more acute. Not all upright-growing branches have acute angles of attachment, but almost all sharply-attached branches are upright. Upright branches are usually larger in relation to the trunk than are more horizontal limbs. With time, the height, spread, and weight of vertical branches increase the stress at the crotch, which is weaker than if the trunk were larger in relation to branch diameter.

MacDaniels (1932) states that the angle of branch attachment does not affect strength except when there is included (embedded) bark. In his studies, included bark weakened the branch attachments of apple trees. Miller (1959), however, found that included bark does not influence crotch strength consistently. Bark-included crotches that have split out certainly appear to be structurally weak compared with those that have connective wood in the crotch (Fig. 2–13). Bark-included crotches are easy to detect, and it seems wise to eliminate them as soon as possible to avoid the growth of large branches from them.

Vigorous leaders tend to grow more upright and produce laterals with wider angles of attachment than do leaders of lower vigor. Angle-of-branch attachment, however, is primarily influenced by the genetic makeup of the plant. Certain species or cultivars, like Modesto ash, are noted for their narrow angles of branch attachment with included bark.

On young apple whips, the angles of attachment increase from the terminal toward the base (Verncr 1955). The branch angle formed by a new shoot will be

---

over branch xylem on the trunk below the branch (Fig. 2–12, upper right). As successive annual growth increments occur, a strong, laminated branch attachment is thought to result.

However, the two upper photographs in Fig. 2–12, which were used to show alternating branch and trunk growth, are of red oak, which is ring-porous. Zimmermann and Brown (1971) report that in ring-porous trees, the cambium becomes active in the spring "almost simultaneously throughout the whole stem" and, depending on species, is active for 8 to 19 weeks. If this is so, the thesis of overlapping branch and trunk xylem is difficult to support on the basis of the difference in times of the initiation and duration of cambial activity.

Neely injected a water soluble dye into the most recently formed xylem in stems beneath branches of eight deciduous tree species. Dye was injected eight times during a growing season from before leaf emergence until the first frost. The stems and branches were later excised, the bark removed, and the pattern of dye movement observed (Fig. 2–12, lower right). In none of the eight species, however, did the dye distribution patterns vary during the growing season as might be expected if there were alternating layers of branch and trunk xylem below branch attachments. Neeley concluded that these results do not support the notion of lamination of branch and trunk xylem at branch attachments.

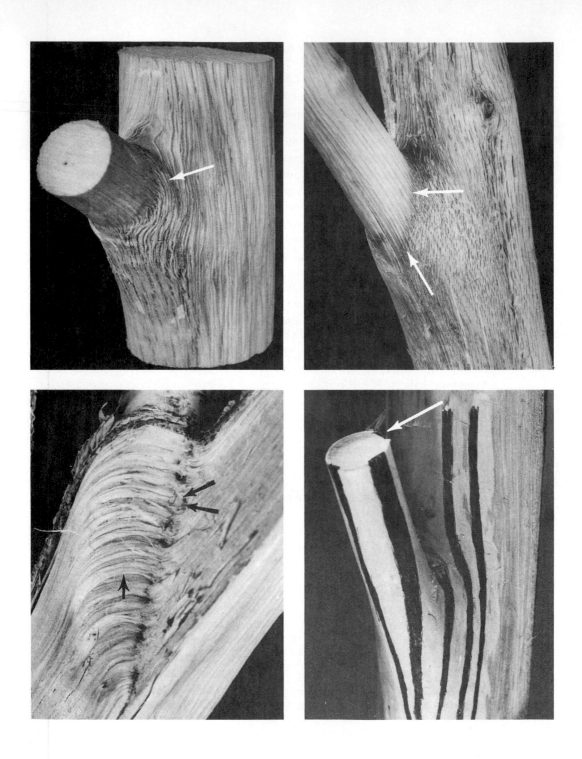

Figure 2-12 The bark was peeled from this young red oak branch attachment in late spring (upper left). Xylem vessels on the top of the branch turn downward at the branch base (arrow) and appear to flow down the trunk below the branch.

A red oak trunk section debarked in midsummer shows trunk xylem flowing around and over branch xylem laid down earlier (upper right). Shigo (1986a) observes that alternating branch and trunk xylem at branch attachments is normal for healthy branches. Neely (1991 pers. comm.) questions this interpretation of the photograph; he agrees that trunk tissue is encircling the branch but thinks the reason for doing so is that the branch is declining in vigor.

The trunk of a paper birch (lower left) has been split, revealing what appears to be alternating layers of dark-colored branch xylem and light-colored trunk xylem (arrow) (Shigo 1989). This interpretation supports Shigo's hypothesis. It would seem, however, that if branch xylem were one of the two alternating layers, some branch xylem should be observed turning toward the branch terminal as it encircles the branch base. Without seeing the specimen, it is difficult to be sure what the xylem patterns are. (Photos courtesy Alex Shigo and Kenneth Dudzik, U.S. Forest Service)

Dye was injected into the stem of *Tilia americana* at several sites beneath the branch (lower right). (Dye patterns were highlighted with ink for photographic purposes.) Dye injected directly below the branch (left injection) moved into the lower side of the branch. Dye injected somewhat to the right of directly below the branch (right of left injection) moved along the side of the branch. Dye injected slightly farther to the right (center injection) moved around to the top of the branch (see arrow). Dye injected to the right of the branch shoulder (the two injections to the right of center) moved only in stem xylem. If branch and then trunk xylem were laid down at the branch attachment, Neely (1991) expected to see a change in the dye flow patterns during the growing season. Similar dye flow patterns, however, were seen from before leaf initiation to after the first frost in eight tree species. (Photo courtesy Dan Neely, Illinois Natural History Survey)

more acute if the main stem is broken or cut back, if buds are removed, or if the bark is scored (cut to the cambium) above the bud that gave rise to the new shoot. Fruit growers learned this when they disbudded newly planted apple and pear whips (with no side branches) and left well-spaced buds to develop into what they hoped would be well-distributed main branches. These practices apparently decrease the level of hormones (primarily auxin) or at least the balance that favors a wide angle of branch attachment. Conversely, if the bark is scored below a bud, the branch angle will be wider.

Removing a lateral with a sharp angle of attachment from a young leader may force one or more accessory buds to grow from the same node (Harris and Balics 1963). The new shoots will be much smaller than the trunk, and their angles of attachment will usually be greater than that of the original shoot (see Fig. 15-18). By applying this information, one can train young trees to develop stronger branch attachments.

Figure 2-13 Wide angles of branch attachment (top left) and narrower angles with bark ridged up in the crotch (top right) are stronger than sharp angles of attachment with included bark (bottom left). As limbs increase in spread and weight, they can split at this point of weakness (bottom right); note the lack of attachment in the upper part of the tight crotch.

PLANT FORM

The form or shape of woody plants, particularly of trees and shrubs, is determined by (1) the location of leaf and flower buds (terminal or lateral); (2) the pattern of bud break along the trunk and branches; (3) the angle at which branches grow; and (4) the differential elongation of buds and branches. For example, the absence of lateral buds in most of the arborescent monocots leads to a columnar growth habit in which an unbranched trunk ends in a tuft of leaves, as in palms. In most of the conifers and a few angiosperms, the main stem or leader outgrows and subdues the lateral branches beneath, giving rise to cone-shaped crowns with a central trunk. This branching habit is called *excurrent* (Fig. 2-14). In contrast, most angiosperm trees and shrubs are more round-headed and spreading, with no main leader to the

Figure 2-14 A typical excurrent tree, such as liquidambar, after one (left), three (center), and many years (right). The terminal usually outgrows and subdues the branches below. Not to scale.

Plant Form   **37**

top of the plant. This habit is called *decurrent,* deliquescent, or diffuse. The lateral branches of such plants grow almost as fast as the terminal shoot or may even outgrow it so that the central leader becomes lost among other branches (Fig. 2–15).

Apical Dominance and Control

Lateral buds may be inhibited by the active growth of a terminal bud cluster on a currently elongating shoot; this is known as *apical dominance.* Trees with decurrent (spreading) branching habit have been thought to have weak apical dominance, while those of excurrent habit were thought to have strong apical dominance. Just the opposite is found to be true (Brown, McAlpine, and Kormanik 1967). Decurrent species such as oak, elm, and maple exhibit strong apical dominance while the shoots are growing; in the year this growth occurs, few or no lateral buds grow on the developing shoots. The next year, however, one or more lateral buds are released

Figure 2–15   A typical decurrent tree, such as ash, after one (left), three (center), and many years (right); the terminal bud usually prevents lower buds from growing during the first season, but the terminal shoot usually is outgrown in succeeding years. Not to scale.

and sometimes develop into branches that outgrow the original leader, thus producing a round-headed plant. The leader is not able to subdue and outgrow the laterals that develop after the first year. Such a tree is said to have weak *apical control* (Fig. 2–15).

On the other hand, some excurrent species, such as sweetgum, tulip tree, and random-branching conifers (e.g., cedar and cypress) have weak apical dominance. In such trees, varying numbers of lateral buds grow during the same season as the shoot on which they are formed. In most cases, however, the leader is able to keep ahead of the lateral branches, resulting in a typical "central leader" or conical form. In these cases, the leader has weak apical dominance but strong apical control (Fig. 2–14).

Conifers, such as fir, pine, and spruce, have their branches in whorls on the trunk because those species have both strong apical dominance and strong apical control (James Clark, University of Washington, 1990 pers. comm.). Seldom do laterals grow on current growth, but the next year, laterals grow from buds that formed just below the terminal the year before. These conifers have an excurrent form, at least while young. Occasionally, some hardwoods with strong apical dominance also exhibit strong apical control, especially while young. Initially, some sweet cherry varieties are excurrent but become decurrent as they mature.

To understand the relationship between apical dominance and apical control, it may help to remember that strong dominance is confined primarily to shoot growth of the current season. During the following season, lateral buds and the terminal bud formed the year before will start growth. The apex of each new shoot will have strong dominance over buds that form on current growth but, except for whorl-branching conifers, little or no control over shoots that are growing below. If the new lateral shoots outgrow the original terminal shoot year after year, a round-headed (decurrent) tree will result. Except for whorl-branching conifers, **strong apical dominance of current growth usually leads to weak apical control of subsequent growth, and vice versa.** Plants do not always fit neatly into one growth-habit category but range between the easily distinguished extremes.

Plant vigor influences the expression of apical dominance and control. Within a single species, **vigorous shoots exhibit less apical dominance than those of low vigor.** Therefore, some of the lateral buds on more vigorous shoots may be released from shoot tip dominance and produce a more excurrent growth habit. In contrast, when plants mature or grow under conditions of stress or low fertility, they become less vigorous. Shoots on such trees increase in apical dominance, which then leads to a loss of apical control and gives rise to a more round-headed plant. A tree that has an excurrent growth habit while young may become round-headed as it reaches maturity.

Young liquidambar seedlings, which normally have an upright central leader, completely change their habit when grown in dense shade (Brown, McAlpine, and Kormanik 1967). They become almost prostrate with a decurrent growth habit because their vigor is so low that few or no lateral buds develop on the current year's growth.

Removal of the terminal bud cluster will not immediately change the excurrent or decurrent habit of an individual tree. Only by repeated pruning can excurrent forms be made to approximate decurrent forms, and vice versa. It is true, however, that removal of the terminal bud cluster on a shoot with strong apical dominance will release one or more buds immediately below the point of removal and keep buds further below the cut from growing. Particularly if a dormant shoot has been cut back before growth begins, the new shoots are clustered just below the cut. On the other hand, if a shoot with no lateral branches is not cut back, more buds will develop the following season and will usually be distributed more evenly along the shoot.

Inhibition of lateral buds is apparently determined by a balance of growth factors. These relations are complex and vary with species. No single explanation is yet available, but dominance seems to reflect auxin levels produced by the growing point and by young leaves.

Other Factors

Species with terminal flowers or flower inflorescences are normally round-headed, since growth after blossoming comes from lateral buds below the flowering terminal (Fig. 2–16). More than one bud will usually break, thus speeding the process that leads to round-headedness. Competition between two or more shoots results in co-dominant stems and reduces the growth of each, leading to a more compact plant.

Plant form can also be influenced by other plants growing nearby. Plants that are close together grow taller, with less spread and fewer or no branches along their lower trunks. In contrast, a similar plant in the open will be shorter and develop

Figure 2–16 Species that flower terminally, such as *Koelreuteria bipinnata,* often fork below the previous year's terminal flowers (note terminal flower stalks). This hastens a decurrent form and can lead to weak structure.

more branches with greater spread. The branches of a plant in the open will be more uniformly distributed from the ground upward.

In a mixed planting of trees, certain species dominate and are less affected by their neighbors. The more sensitive trees may be quite misshapen or devoid of lower branches (Fig. 2–17). Some large trees may form a canopy over slower-growing species without greatly affecting the symmetry or normal growth of the lower plants (Fig. 2–17). In fact, some species when young require such a protective canopy. Even in close plantings of a single species, certain trees dominate and outcompete others. Under severe stress, the smaller trees may not survive.

Figure 2–17   A young Chinese pistache misshapen by growing in the shade on the west side of a large incense cedar (left); on the other hand, *Pinus patula* grows upright even though it receives most of its light from the west (right).

Juvenility and Maturity.   A number of morphological and physiological changes occur in the development from seedling stage to maturity. These changes have been of particular interest to propagators, breeders, and those who grow plants for their leaves or flowers and fruit; some plants may not flower until they are more than 10 years old. During its early life, a plant is considered *juvenile* if it will not flower even though all environmental conditions are favorable. A plant is considered sexually *mature* when it will respond to flower-inducing conditions.

The entire plant does not change from juvenile to mature; only the apical meristem and its tissues. Even though they are actually older, the lower and inner portions of woody plants may retain juvenile characteristics while the upper and outer growing points produce mature shoots and flowers. These upper meristems tend to produce mature shoots as they age or when levels of moisture, nutrients, and other substances that come from other parts of the plant are reduced (Zimmermann and Brown 1971).

Morphological traits associated with juvenility may include vigorous upright growth; vine form, leaf shape and phyllotaxy that differ from mature forms; excurrent growth habit; smooth bark, spines, or thorns; and retention of leaves in deciduous plants (Fig. 2–18) (Zimmermann and Brown 1971). The most common juvenile physiological traits are easy rooting of cuttings and lack of flowering response.

Practices that will increase the size of a juvenile plant most rapidly will hasten

Figure 2–18 The juvenile form of English ivy is a vine with palmately lobed leaves (right, inset); the mature form is shrub-like and upright with cordate leaves (left, inset) and fruit.

flowering and fruiting; on the other hand, young plants that have been propagated from tissue of a mature plant, such as most fruit trees and many woody ornamentals, will be delayed in flowering if they are extremely vigorous (Hackett 1976). Young juvenile plants are much slower to mature and flower than are young plants propagated from mature tissue. To retain the juvenile form and foliage of a plant, low to moderate plant vigor should be maintained and any mature growth pruned out as soon as it appears. If one realizes that some plants have distinctively different juvenile and mature characteristics, one may avoid some incorrect assumptions about their well-being. In vegetatively propagating plants, the proper selection of cutting or grafting wood can be quite important to rooting success and structural development. Not only can juvenile and mature tissues differ, but propagating wood taken from flattened horizontal branches may impart the flattened characteristic to the plant produced; this is true for some of the conifers, particularly araucaria.

Allelopathy is the term given to the phenomenon in which one plant directly or indirectly harms another, particularly when young, through the production of chemical compounds released into the environment. Roots and leaves are the most common source of such inhibitors. Roots of one plant can be inhibited or repulsed by either genetically identical or different plants (Del Moral and Muller 1970). Allelopathy can keep root systems from intermixing or seedlings from growing near or under established plants. Competition from other plants is thereby reduced, enhancing chances for survival, particularly in arid conditions.

Fescue sod has been shown to inhibit the growth and survival of young liquidambar (Walters and Gilmore 1976) and black walnut (Todhunter and Beineke 1979) (Fig. 2–19). Leachates from live fescue roots, from dead roots, and from dead leaves have been filtered through quartz sand in which month-old liquidambar trees were growing; they reduced the trees' dry weights by 19, 48, and 60 percent, respectively. This could raise some question about using lawn clippings for mulching young plants (see Chapter 14). On the other hand, growth of young, woody, perennial lupine growing in sand with tall fescue was not affected by recyling the leachate (Wu University of California, Davis, 1989, pers. comm.).

Even though an increasing number of herbaceous and woody plants are reported to have allelopathic effects, not all species are equally inhibited by the same plant. Caution should also be taken concerning reports of allelopathic effects, because it is difficult in the field to separate such influences from those due to nitrogen, moisture, aeration, microorganisms, or other soil influences. Some allelopathic toxins are reported to remain in soil or dead roots for many years after a large plant has been removed, thus arguing against the use of certain replant species. Some toxins, however, are rendered harmless by being adsorbed or altered by the soil or soil organisms.

Controlling or at least reducing unwanted tree species, particularly in utility rights-of-way (ROW), appears promising. Several studies identify plant-cover types that greatly reduce the number of tree seedlings that are able to invade an area. In Pennsylvania, Bramble, Byrnes, and Hutnik (1990) found three types of grass-shrub, herb-grass, and shrub-grass plant covers that had no tree species invade during a three-year period. Allelopathy along with competition may have been responsible.

Figure 2-19 The allelopathic effect of fescue sod on black walnut seedling growth (right) is compared to that of a ground cover composed of mixed herbaceous species (left). Photos are same scale and trees the same age. (Photos courtesy M. N. Todhunter and W. F. Beineke, Purdue University)

THE ROOT

The above-ground parts of a plant depend upon roots for anchorage, absorption of water and mineral nutrients, storage of food reserves, and synthesis of certain organic materials, including those that may regulate activities in the top of the plant. The development, size, form, and function of root systems are controlled by the genetic makeup of a plant and by environmental and management conditions that modify the expression of these characteristics. Roots differ from shoots in a number of ways. Roots have no regular branching pattern like that of stems whose branches develop at nodes from buds in the axils of leaves. Lateral roots develop from tissue several layers inside the epidermis but outside the phloem (Fig. 2-20); lateral branches grow from buds whose tips are usually exposed beyond the epidermis

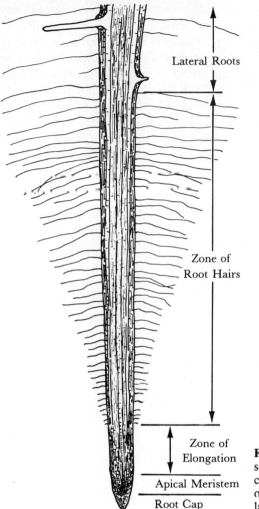

Lateral Roots

Zone of
Root Hairs

Zone of
Elongation

Apical Meristem

Root Cap

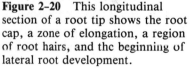

Figure 2-20 This longitudinal
section of a root tip shows the root
cap, a zone of elongation, a region
of root hairs, and the beginning of
lateral root development.

though they have vascular connections within the main stem. A root cap protects
each root tip as it is forced through the soil by the elongating tissue behind; a growing shoot tip is composed of tender, expanding shoot, leaves, and buds with little
or no protection from the elements. At its tip, the oldest cells of the root cap slough
off and lubricate the movement through the soil.

Behind the root cap, the root apical meristem replenishes the root cap through
cell division and forms undifferentiated cells away from the tip for later development into specialized root tissue. Behind the apical meristem of active cell division
is the region of most intense cell elongation, which forces the terminal 1 to 5 mm
(0.05–0.2 in.) through the soil (Fig. 2–20). In and behind the zone of cell elongation,
cells differentiate into xylem, cambium, phloem, pericycle, endodermis, and cortex.
The epidermal cells in this region of primary tissue formation form root hairs up

to 8 mm (0.3 in.) long. Water is absorbed from the soil primarily through root hairs and the region of cell elongation, although the absorbing surface of the latter is much smaller than that of the root hairs. The root hairs of a single mature rye plant have been estimated to number more than 14 billion, with a surface area greater than 370 sq m (4000 sq ft), and a rate of increase that exceeds 100 million per day (Robbins, Weier, and Stocking 1950). Root hairs do not become roots; in fact, those of most species live only a few weeks. The root hairs of plants like the honey locust and redbud, however, may persist for more than a year, while some of the conifers have few or no root hairs (Weier, Stocking, and Barbour 1974).

Lateral roots, which arise in the pericycle, and secondary growth of xylem and phloem from the root cambium occur in the region behind the root hairs. Adventitious roots, on the other hand, can arise from stems, rhizomes (underground stems), stolons, and other organs.

Root Form

Seedlings tend to have either one main *tap root* or a number of primary roots that form a *fibrous root system* (Fig. 2–21). In some species and soil conditions, primary roots may be almost devoid of secondary roots until they have grown a considerable distance or encountered an obstacle. In other plants, primary roots may readily form laterals as they grow. The tap roots of young seedlings can penetrate the soil to a considerable depth, while growth of the top is considerably less. Six months after a blue-oak acorn was planted in central California, the tap root of one seedling was excavated to a depth of one meter (40 in.) before it broke; the top of the plant was only 75 mm (3 in.) above the ground. Under arid conditions, such root growth significantly aids the establishment of young plants. Tap-rooted plants, however, are considered difficult to transplant, either bare root or balled and burlapped, because they lack a fibrous root system to hold the soil ball together and have few small roots to absorb moisture after transplanting.

Figure 2–21 Roots of young plants vary from a single tap root with few or no lateral roots (left) to a fibrous root system with most roots originating near the crown (right).

The primary seedling roots, particularly tap roots, are strongly geotropic and grow vertically if soil conditions are favorable. Lateral roots usually grow more horizontally. If the primary root tip is pinched or damaged, lateral roots grow more vertically. If the tip of an unbranched tap root is pinched, laterals develop above the pinch and grow in almost the same direction as the parent root (Fig. 2-22). As root systems enlarge, the primary roots have less influence over the growth of the other roots.

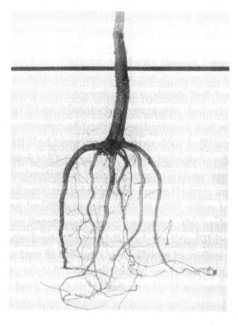

Figure 2-22 Branching of an oak tap root. The root tip was pinched when the seedling was transplanted from a seed flat into a small container.

Roots other than the tap root may enlarge as they grow downward, giving stability to the top of the tree. These are called *heart roots*. Some trees have primarily horizontal roots near the surface and *sinker* or *striker roots* that grow vertically from the horizontal ones. Sinker roots usually occur near the tree trunk, adding stability to the tree and increasing the volume of deeper soil exploited by the roots (Fig. 2-23). In most mature trees, the tap root is outgrown by heart and lateral roots and is difficult to find. Root systems are much closer to the surface than is commonly thought.

Although root growth is greatly influenced by soil conditions, individual roots seem to have an inherent guidance mechanism. Large roots with vigorous tips usually grow horizontally (Wilson 1984). Smaller roots lateral to the large roots grow at many angles to the vertical, and some grow up into the surface soil. On the other hand, few roots in a root system actually grow down (Wilson 1984). The growing tips of large, horizontal roots are often *plagiotropic;* that is, if the tip is displaced up, it grows down; and if it is displaced down, it grows up. In addition, these roots tend to grow at about the same soil depth regardless of the slope of the soil surface.

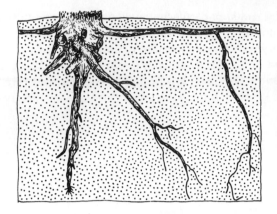

Figure 2–23 Types of roots that may occur on a mature tree. The vertical tap root is usually choked out by roots above. The heart root grows at an angle from the buttress of the trunk (root collar). Horizontal lateral roots are usually near the surface. Sinker (striker) roots grow downward from lateral roots. (Adapted from Fayle 1968)

More study of this phenomenon may help to minimize root-heaving of pavement and sidewalks.

The roots of some plants are able to maintain compass direction of growth. If a root grows into an obstacle and is deflected, it will tend to resume its original direction when the obstacle has been passed (Wilson 1984). This is called *exotropy*.

Size of Root Systems

Roots of most plants, including large trees, grow primarily in the top meter (3 ft) of soil. **Most plants concentrate the majority of their small absorbing roots in the upper 150 mm (6 in.) of soil if the surface is protected by a mulch or forest litter** (Zimmermann and Brown 1971, Perry 1982). Exposed bare soil can become so hot near the surface that roots do not grow in the upper 200 to 250 mm (8–10 in.).

Under forest and many landscape situations, however, the soil surface is not directly exposed to the sun, and soil near the surface is most favorable for root growth. Management practices, such as mulching, that favor exploration by feeder roots near the soil surface foster good plant growth, particularly in shallow or fine-textured soils. The surface rooting of trees in lawn areas or near buildings and pavement, however, can be a serious problem, more especially with vigorous young trees than with mature trees that have almost reached their ultimate size (Morling 1963). Surface rooting will be increased if poor irrigation practices keep the soil too wet on the surface and too dry at lower depths or if inadequate drainage keeps the soil too wet throughout.

Roots may extend laterally for considerable distances, depending on the plant and the soil conditions (see book cover). Roots growing in moist, fertile, well-aerated soil branch more, and each root grows less than if there were fewer roots. Roots will be more numerous, more fibrous, and closer to the trunk than those that grow in less desirable soil. In infertile soil inadequately supplied with moisture, roots are fewer in number but larger and able to grow greater distances from the plant. This is certainly a good survival tactic.

Roots of trees grown in the open often extend two to three times the radius of the crown (Zimmermann and Brown 1971, Perry 1982), while those growing in a close planting will be more confined. If there are no internal inhibitions, roots grow wherever aeration, moisture, temperature, nutrition, and soil tilth (structure and openness) are favorable. These factors are discussed in more detail in Chapters 6 and 14.

Root-Top Relationships

The factors that improve growing conditions, such as more favorable weather, fertilization, irrigation, or aeration, result in proportionately less root growth than top growth; that is, the *root-shoot* ratio is reduced. Many arborists believe that a reduction in the root-shoot ratio of a plant is undesirable.

The relative size of the top and root system can vary considerably depending on the size of the plant and the conditions under which it grows. The root-shoot ratio is usually given as the ratio of the weight of the root to that of the top. For most trees under normal conditions, the root-shoot ratio is $\frac{1}{5}$ to $\frac{1}{6}$ (Kramer 1969, Perry 1982); the top is five to six times heavier than the roots. If it were not for the trunk, however, the top and roots would weigh about the same (Perry 1982).

An effective root system is extremely important in order to adequately supply water; a water deficit can quickly kill or severely damage a plant. The water-absorbing ability of roots compared to water loss from leaves can be critical. Relative root and top weights are not a valid measure of the adequacy of the roots to supply water and nutrients; a more useful comparison is the root surface or length to the transpiring leaf surface of a plant. The roots within the dripline of a tree are estimated to have 2.5 to 4.5 times more surface area than do the leaves (one side) of the same tree (Perry 1982). Deciduous-plant leaves have stomata primarily on the lower surfaces.

Length is probably a better measure than surface area of the absorbing ability of roots. Water moves slowly in soil so that a small root is almost as effective as a larger one in absorbing water. Roots have to grow in length to explore soil for water and nutrients. In addition, root hairs and the mycorrhizal roots (root-fungus associations; see Chapter 6) of most tree species greatly increase moisture absorption. According to Perry (1982), mycorrhizae "functionally amplify the effective surface of the finer roots a hundred times or more." Under normal conditions, the roots of a plant are well able to supply the top with water, nutrients, stored carbohydrates, and certain growth regulators.

A root-shoot–ratio reduction might be of concern if the reduction were marked and water in the root zone became limited. Except for injury to the roots, **a reduction in the root-shoot ratio is almost always in response to more favorable growing conditions.** An increase in the root-shoot ratio, on the other hand, would indicate that a plant probably was growing under less favorable conditions.

Most plants are able to adapt to changing conditions if the changes are not too drastic or rapid. If the top is pruned, more carbohydrates are utilized to restore the top and less are available for the roots. Conversely, if roots are damaged or nutrients and water become limited, more carbohydrates go to the roots and less to the top. If nitrogen is limited, roots grow proportionately more than the top; and more nitrogen is shunted to leaves which may keep chlorophyll higher than otherwise (see Chapter 12).

The root-shoot ratio or changes in it would appear to be good management information with which to evaluate the health of plants. Actually, it is more a concept than a tool. A plant would have to be sacrificed in order to determine such a ratio and just obtaining the necessary samples is tedious. It is mainly a research tool to help understand plant responses to different conditions and practices.

The root-shoot concept can be useful for an arborist. When roots are damaged or reduced as during construction and transplanting, the functioning roots may not be able to keep the tree upright or supplied with adequate water **(to a root-injured tree water is more critical than are nutrients)**. If soil is compacted around a tree, root effectiveness will be reduced. Aeration and wise water management will be critical to help restore root activity. A tree left when clearing a forested area requires support because its limited roots seldom can maintain the tree in a wind.

The size of the top in relation to the volume of the root system is an important factor in the failure of many container-grown plants when they are set out in the landscape. In many nurseries, plants are irrigated and fertilized almost daily during the growing season and develop proportionately large tops, particularly when their roots are confined in containers. Even though no roots may be removed during transplanting, the moisture-holding capacity of the root balls is relatively small in relation to the transpiring surface of the plant tops. In addition, a root ball in the soil retains less water following irrigation than it did in the container. Plants of moderate size in relation to their containers may better survive in an unirrigated landscape. After any transplant, frequent watering is usually necessary.

Shigo (1989) states, "Tree design came about in the forest where trees grew close to other trees. [But we] have . . . planted it as an individual. This is a new condition for the tree." Trees are much more adaptable than these statements imply. Trees in a forest support one another, shade lower foliage, compete with one another, and protect each other from storms. A tree in the open, however, retains branches lower on the trunk; the trunk develops taper; the top does not grow as tall as it would in a forest; and it develops an extensive root system. All of these responses to its surroundings result in a tree well able to withstand most storms (see Chapter 9). It is indeed remarkable that **most trees can adapt to the situation in which they grow, whether it be forest or savannah!**

What is a "new condition for the tree" is that all too often we grow a tree as an individual, cut the top back in the nursery to force branches near that height, remove all the branches below the forced branches so it "looks like a tree," plant it in a hostile environment, and leave it to fend for itself. It is amazing that as many trees survive as do. Hopefully, information in this book will provide understanding to grow and maintain better plants in our landscapes. I like and underscore part of Shigo's (1989) quote from *The Little Prince:* **"You become responsible, forever, for what you have tamed."**

THE LEAF

Leaves convert light energy into chemical energy that is utilized within the plant and by animals and certain other plants for life processes. Although variable in many ways, leaves of hardwoods (broadleaved) have a thin, broad blade and a slender *petiole* (stalk). Except for ginkgo, leaves of softwoods (conifers) are needle- or scale-like.

Seen in cross-section (Fig. 2–24), a leaf blade has an upper and lower *epidermis* covered with a waxy *cuticle* that protects and reduces water loss. *Stomata* (singular *stoma*), usually only on the lower surface, are minute openings in the epidermis that control water loss (transpiration). Other gases, mainly carbon dioxide and oxygen, enter or exit a leaf through stomata. The *mesophyll* (middle of the leaf) is composed mostly of palisade and spongy parenchyma cells. Parenchyma cells, particularly the palisade just below the upper epidermis, contain *chlorophyll* in granules called *chloroplasts*. Xylem and phloem in *veins* in the mesophyll conduct water, minerals, and organic compounds.

Conifer needles and scales (Fig. 2–24) contain essentially the same tissues and perform similar functions as leaves of broadleaved trees do.

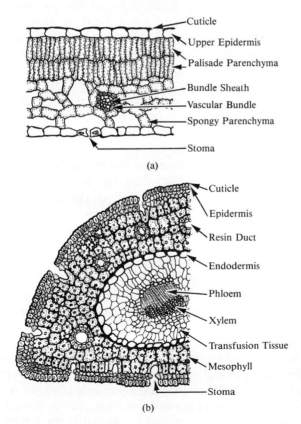

(a)

Cuticle
Upper Epidermis
Palisade Parenchyma
Bundle Sheath
Vascular Bundle
Spongy Parenchyma
Stoma

(b)

Cuticle
Epidermis
Resin Duct
Endodermis
Phloem
Xylem
Transfusion Tissue
Mesophyll
Stoma

Figure 2-24 Cross section of a leaf of a broadleaved plant (top). Cross section of half of a needle of a two needle black pine (bottom). (Adapted from C. J. Chamberlain, *Gymnosperms: Structure and Evolution,* 1935. University of Chicago Press)

Leaves of a deciduous plant last but one season, while those of evergreen plants persist longer, usually three to four years. An evergreen plant drops its older leaves each year, which may occur over a long period and at almost any season. Leaf drop of an evergreen may be more of a nuisance than that of a deciduous tree.

Photosynthesis

Photosynthesis is the transformation, in the presence of chlorophyll and light, of carbon dioxide (from air) and water (primarily from soil) into a simple carbohydrate and oxygen. Simple carbohydrates are quickly converted into more complex carbohydrates and other organic substances. Leaves are the primary site for photosynthesis although some occurs in immature fruit and in young stems (a significant factor for deciduous plants during the winter). Chlorophyll gives plants their green color.

Growth and the well-being of plants ultimately depend on an adequate rate of photosynthesis; which, in turn, is governed by an interacting complex of internal and environmental factors. Internal factors that favor photosynthesis include young leaves with large spongy mesophyll space; numerous, large, and responsive stomata; high chlorophyll content; and adequate water (turgor) and carbohydrate availability. Resistance to the entrance of carbon dioxide due to stoma size is a major limitation to photosynthesis. The two guard cells that form a stoma opening increase in sugar in the presence of light, resulting in an increase in turgor and opening of the stoma aperture. A localized water deficit (which often happens on summer afternoons) or insufficient light results in closure of stomata.

Many external factors influence both chlorophyll level and rate of photosynthesis. Almost anything that upsets a plant's metabolism will decrease the amount of chlorophyll and its rate of photosynthesis.

Sun leaves are smaller, thicker, and have more spongy mesophyll surface than do shade leaves. Sun leaves of most hardwoods become light saturated (light intensity at which photosynthesis is no longer increased by higher intensities) at 20 to 40 percent of full summer sunlight. Photosynthesis of conifers increases with increasing light intensities up to 40 to 85 percent of full summer sun (Kramer and Kozlowski 1979). Light intensity and photosynthetic rate of sun leaves are about 15 and 3.5 times respectively as much as those of shade (interior) leaves of a mature apple tree (Heinicke 1966). Plants that have been in full sun commonly defoliate when subjected to low light intensity for 8 to 10 weeks (see Fig. 11–9).

Temperature effects on photosynthesis are as marked and complex as those of light. Photosynthesis of woody plants can occur from near freezing to over 40°C (104°F) (Kramer and Kozlowski 1979). At low temperatures, photosynthesis increases markedly with rising temperatures; at higher temperatures, not only does photosynthesis slow but respiration increases rapidly.

Carbon dioxide is usually the limiting factor in photosynthesis when there is sufficient light. Not only is the carbon dioxide concentration low in air (0.036 percent in 1991), but its diffusion into leaves is markedly reduced by stoma closure due to water deficits. Even when stomata are open, carbon dioxide diffusion into mesophyll cells within a leaf is limited (Kramer and Kozlowski 1979). Carbon dioxide concentration is increased by auto and industrial activity in urban areas. In wooded areas, respiration at night increases carbon dioxide concentration and photosynthesis during the day decreases it. The soil gives off considerable carbon dioxide.

Water in adequate amounts is essential for photosynthesis. Not only is water a raw material for photosynthesis, but, equally important, it is needed to maintain leaf turgor. Either a soil-moisture deficiency or excess can result in leaf-water deficits. Water deficits cause stomata to close, leaves to wilt (exposing less surface to the sun), and protoplasm to dehydrate; these effects slow photosynthesis. Also, without adequate water, leaves will be smaller and shed earlier.

Nutrients are needed to ensure a large leaf area having extensive mesophyll tissue for gaseous exchange and adequate exposure to light. Nutrients, particularly nitrogen and iron, are essential for synthesis of chlorophyll; but even at low nitrogen levels, chlorophyll may remain high (Ingested and Lund, 1979) (see Chapter 12).

Air pollutants, pesticides, herbicides, antitranspirants, and insects and diseases directly and indirectly reduce the rate of photosynthesis.

Respiration

Respiration is the reverse of photosynthesis. It is the oxidation of organic compounds in living cells, thereby releasing energy for the many life processes of the plant. Oxygen is usually required and the products, in addition to the energy released, are mainly water and carbon dioxide. Of the total carbohydrates produced by a beech forest, it is estimated that 45 percent are consumed in respiration by the forest trees; for a tropical rain forest, 78 percent of the carbohydrates produced are respired by the trees (Kramer and Kozlowski 1979).

Respiration is most rapid in the most metabolically active tissues, such as leaves, shoot- and root-growing points, cambium, and developing fruit. Young leaves and shoots have higher rates of respiration than older leaves and branches. Respiration rate increases with temperature faster than does photosynthesis. Soil moisture much above or below the optimum is reported to increase respiration. Low levels of oxygen (below 10 percent) in the soil greatly reduce root respiration, thereby limiting root growth and activity. Inadequate soil aeration is one of the most seriously limiting factors in growing plants in the landscape.

A plant's compartmentalization responses to an injury require energy (respiration) and at the same time wall off stored energy that would otherwise be available to the plant in the future.

Transpiration

Transpiration is the evaporation of water from plants. Almost all transpiration occurs from leaves, primarily through stomata, with some occurring from lenticels (openings) in the bark. About 95 percent of the water absorbed by a plant is transpired. One isolated tree with a canopy spread of 12 m (36 ft) may transpire 2000 liters (525 gal) per summer day.

Although not essential for plant survival, transpiration cools leaves and increases the rates of water and mineral absorption and ascent. Transpiration and the water used in photosynthesis and other processes are responsible for the rise of water to the tops of trees. Water is literally pulled to the tops of trees by transpiration.

Transpiration will be greatest when internal and environmental factors create the lowest *vapor pressure* around leaves and the highest within them. Stomata are only open when leaves are turgid and in light. The size of stomata openings, through which about 90 percent of transpiration takes place, and their density determine to a large extent the rate of transpiration of a leaf. Plants with large leaf areas transpire more than similar plants with smaller leaf areas; **removing leaves, however, will not decrease transpiration in direct proportion to the leaves removed.** Certain deciduous trees use more water during a year because they retain leaves longer. Plants of the same leaf area with small, thin leaves will transpire more than those with large, thick leaves. Dark-green, nonpubescent leaves with thin cuticle usually transpire more than gray or light-colored leaves with dense pubescence and thick cuticle (Kramer and Kozlowski 1979).

The rate of transpiration will be high when there is light, high temperature, low humidity, a slight breeze, and adequate water. Transpiration is usually greater during the long days of late spring than the hotter, shorter days of midsummer.

FLOWER AND FRUIT

Woody plants are often featured in landscapes for their interesting, colorful, and beautiful flowers and fruit. Some species, however, are not appreciated because people may be allergic to their pollen or consider their fruit messy, foul smelling, or dangerous—a cone near the top of a 20 m (65 ft) tall Coulter pine can weigh up to 4.5 kg (10 lbs).

Angiosperm Flowers

The flower of an angiosperm is a shoot bearing floral leaves. Flowers that bloom in the spring develop from buds differentiated the previous spring and early summer, such as peach and elm; the flowers are borne on one-year-old wood. Flowers that bloom in late spring or summer usually do so from buds developed on the current season's shoots, such as crape myrtle and golden rain tree (see Fig. 15–2b).

A complete flower consists of a lower ring of *sepals* called the *calyx* which are usually green (Fig. 2–25). Above or interior to the calyx is a ring of usually showy *petals* known as the *corolla*. Interior to the corolla are the pollen-bearing *stamens* (male) which surround the *ovule*-containing *pistil* (female). Flowers which contain both stamens and pistil are said to be *perfect*. Some species, however, have *staminate* and *pistillate* flowers in separate flowers in separate structures (*imperfect*) on the same tree (walnut, birch) or in different trees (poplar, willow). In the latter case, it is possible to have fruitless trees by vegetatively propagating from plants which

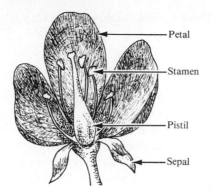

Petal

Stamen

Pistil

Sepal

Figure 2-25 A typical angiosperm flower. Most pistils require fertilization in order to develop fruit and seed.

bear only staminate flowers. Many of these plants produce abundant pollen and are wind pollinated.

Most plants with perfect flowers or with both sexes on the same tree will set fruit by the pistil being pollinated and the ovule fertilized with pollen from a flower on the same or genetically identical (same variety) plant (*self-fruitful*). Other plants may have different fertilization patterns that require their flowers to be fertilized by pollen from a genetically different plant though of the same or closely related species (*self-unfruitful*). A plant may set few or no fruit for many reasons:

> The pollen is not compatible with the pistil
> Only male or female trees are in the vicinity
> The pistil is not receptive when the pollen is shed
> There are few pollinating insects, especially bees
> It is too cold, windy, or rainy for insects to pollinate
> Freezing temperatures kill the flowers or fruit
> Few flowers present due to heavy crop the previous year

The result of fertilization is the development of a fruit. Depending on the species and vigor of a plant, usually more fruit are set than the plant can mature. Many developing fruit drop within a month or two after bloom. Even then, so many fruit usually remain that they are small and break branches. Although it may take more than a year for some species to mature their fruit, most do so in three to six months.

Most fruit contain one or more seeds. In some species, some individual trees in some years produce few viable seeds. Viable seeds of many species will not germinate when taken from mature fruit and planted due to *seed dormancy*. This is an important survival mechanism for a species. Seed dormancy may be due to the seed coat being impermeable to water or gases, particularly oxygen, and/or the *embryo* (the fertilized ovule) being dormant. Though specific for a particular species, various methods for softening seed coats can be used to enhance germination. Embryo dormancy, common in seeds of woody perennials, usually can be overcome by subjecting the seeds to 3 to 12 weeks of chilling temperatures (1°-7°C, 35°-45°F)

Flower and Fruit

(Hartmann, Kester, and Davies 1990). Embryo dormancy is naturally overcome by winter cold.

Gymnosperm Flowers

Flowers of gymnosperms are either *pollen* (*staminate*) *cones* or *seed* (*ovulate*) *cones* and are usually produced on the same tree (Fig. 2–26). Pollen cones average 10 to 12 mm (0.5 in.) in length and 5 to 6 mm (0.25 in.) in diameter. They are usually borne in the lower part of trees. Seed cones are always larger than the pollen cones of the same species. At maturity, seed cones range from less than 25 to 500 mm (1 to 20 in.) in length. Seed cones are borne singly or in doubles on branches usually in the tops of trees; such cone placement ensures cross pollination since conifers are wind pollinated.

Depending on the species and climate, cone primordia of pine, for example, begin to form on currently growing shoots in spring but do not become visible until the next season. Pollination takes place the second spring. The fertilized seed cones develop for two growing seasons, maturing in late summer, three growing seasons after the cone primordia began to develop.

Figure 2–26 Flowering in pine: Pollen cones are clustered below an expanding terminal and lateral shoots (left); three seed cones are below the large terminal shoot bud (right). (Photos courtesy Elliot Weier also appear in *Morphology and Evolution of Vascular Plants,* 3rd ed. 1989. E. M. Gifford and A. S. Foster. New York: W. H. Freeman.

FURTHER READING

KOZLOWSKI, T. T., P. J. KRAMER, and S. G. PALLARDY. 1991. *The Physiological Ecology of Woody Plants*. San Diego, CA: Academic Press.

KRAMER, P. J., and T. T. KOZLOWSKI. 1979. *Physiology of Woody Plants*. San Diego, CA: Academic Press.

PERRY, T. O. 1982. The ecology of tree roots and the practical significance thereof. *J. Arboriculture* 8:197–211.

RICE, E. L. 1974. *Allelopathy*. New York: Academic Press.

WEIER, T. E., C. R. STOCKING, M. G. BARBOUR, and T. L. ROST. 1982. *Botany: An Introduction to Plant Biology*. 6th ed. New York: John Wiley.

WILSON, B. F. 1984. *The Growing Tree,* rev. ed. Amherst: University of Massachusetts Press.

ZIMMERMANN, M. H., and C. L. BROWN. 1974. *Trees: Structure and Function*. New York: Springer-Verlag.

CHAPTER 3

Plant Selection _____

Selecting plants that will adapt well to their intended site and will fulfill their intended landscape function is extremely important to the success of a planting and the ease with which it can be maintained. The quality of young plants is also a crucial factor.

SELECTING THE RIGHT PLANT

From earliest times, people have selected, propagated, and grown plants they thought superior to others. These included food plants, plants that produced fiber and other useful materials, and plants that enhanced their surroundings. Migrating people took seeds and cuttings of desirable plants to new areas. Early explorers and travelers brought back plants for ornamental purposes as well as for food and fiber. Estates of the nobility and early botanical gardens became repositories of a vast array of interesting plants, many of which have landscape value. Ornamental plants were selected for particular qualities of size, form, texture, leaf color, flowers, fruit, and many other characteristics.

Flowering shrubs and small trees like the rose, camellia, and lilac were bred by amateur and professional horticulturists. Almost all new trees and shrubs (without showy flowers), however, have come from native and locally developed plantings. The new cultivars were selected primarily because of aesthetic characteristics rather than for adaptability to unfavorable environments and resistance to pests. Even so, many of them are as vigorous as the species from which they were selected, or more so.

Plant explorers from arboreta and governmental agencies have looked primarily for food and fiber plants but have also collected seed from many plants with landscape potential. The U.S. Department of Agriculture (USDA), through its plant introduction stations and contacts in foreign embassies, has acquired many promising ornamental plants. Research workers, plant pathologists in particular, began screening landscape plants for their resistance to specific serious diseases, pests, and disorders. Nursery persons, botanical garden horticulturists, arborists, landscape architects, and amateur gardeners found plants that grew better than others of their kind. These they introduced or brought to the attention of research workers so the plants could be tested or observed.

The first efforts to breed superior landscape trees began in the late 1920s in an attempt to develop trees that would be resistant to Dutch elm disease (in Holland) and chestnut blight (at the Brooklyn Botanical Garden). Tree-breeding is a slow process; it took the Dutch 48 years to develop elms through the F_3 generation (Shurtleff 1980).

At the National Arboretum, Santamour (1972) began the first large-scale effort in the United States to develop genetically superior shade trees for use in urban areas. The goal is to produce superior trees in a wide range of genera with different growth characteristics to fit urban environments. Resistance to major diseases and insect pests, tolerance to environmental stress factors, and ability to compartmentalize are basic requirements for these trees. Townsend (1977), Santamour (1979b), and others have used provenance and progeny testing, long used by foresters, to test variation within species that are native to different regions and to select superior individuals for further evaluation and possible breeding. For example, Bey (1980) found that black walnut seedlings grown in the midwestern United States from seeds collected 320 km (200 mi) closer to the equator grew better than seedlings from local sources. Seedlings from seed collected yet closer to the equator were more likely to be injured by spring frosts and cold winters.

As these efforts continue and more people become involved in research and observation, we will be in a much better position to select trees and other plants that can perform well. Even then, **the best way to estimate how a plant will perform and look when mature is to observe mature specimens in nearby parks, botanical gardens, arboreta, street plantings, and private gardens.** People who care for the plants and live near them should be consulted.

Inherent Characteristics

Besides the many visual and functional characteristics of woody plants, other attributes must be considered if they are to satisfy their owner and/or caretaker. **Plant species should be selected on the basis of its functional uses, its adaptation to the site, and the amount of care it will require.** The relative importance of a particular characteristic will vary. Table 3–1 rates plant characteristics in relation to function, adaptation, and care requirements. For example, the size of a tree and its leaves are important to all of the tree's landscape functions, its adaptation to a specific site, and the care it may need; the bark, on the other hand, is of primarily aesthetic

TABLE 3-1 CHARACTERISTICS OF WOODY PLANTS AND SUGGESTED RELATIVE IMPORTANCE OF THEIR INFLUENCE ON LANDSCAPE FUNCTION, SITE ADAPTATION, AND PLANT CARE

Plant characteristics	Function			Adaptation to site	Plant care
	Architectural and engineering	Climate and human comfort	Aesthetic		
Growth Habit					
Tree, shrub, vine	***	***	***	**	**
Size	***	***	**	**	***
Form	**	**	***	*	
Growth rate	*			*	**
Branching	**	*	*	*	**
Wood strength	**			*	*
Rooting	**			**	***
Plant Features					
Leaves	**	***	***	**	***
Thorns	**	*			**
Flowers		*	***		*
Fruit			**		**
Bark			**		
Environmental Tolerances					
Temperature				****	**
Drought				**	**
Wind				**	**
Light				**	*
Soil				***	**
Air				**	*
Pests				***	***
Fire	*				**

*** = major influence; no * = little or no influence.

interest. Environmental tolerances mostly influence plant adaptation and the care needed.

Plants have several landscape functions. Their *architectural features* can afford privacy, progressively reveal a view, or articulate space. In an *engineering* sense, plants can be used to reduce glare, direct traffic, filter air, reduce soil erosion, and attenuate noise. Plants influence the *meso-* and *microclimates* by transpirational cooling; interception of solar radiation, reflection, and reradiation; and modification of rain, fog, and snow deposition. Plants can be placed to increase, decrease, or direct wind. On the other hand, certain plants irritate some people with their allergenic pollen and leaf pubescence, toxic sap or exudations, and odorous flowers and fruit. In our surroundings, however, the primary use of landscape plants has been aesthetic. Their diversity of form, color, and texture lend charm, grace, and interest to our landscapes.

Many growth habits and features of functional value may influence a plant's *adaptation* to a particular site, its care requirements, and the ease of giving this care. Species vary greatly in the amount of care required; for example, elm and oak on a California campus required 5.5 and 2.8 hours per year respectively for maintenance compared to 0.1 hour or less for liquidambar and crape myrtle (Van Dam, Mamer, and Wood 1981). Certain stress *tolerances* of plants directly influence their ability to adapt to a site and the amount of care they will require.

Plants will affect and be affected by a specific landscape site. Seldom will a plant have all the qualities desired; as with most decisions, the selection of a plant will usually be a compromise. Certain site-adaptation or maintenance problems may be accepted for the sake of a prized aesthetic effect or a special functional need. Some people in warm temperate regions put up with the pollen and messy fruit of the olive in order to have a gray-foliaged small tree that will grow with little maintenance under difficult conditions.

The important plant characteristics will be discussed with regard to their influence on landscape function, site adaptation, and plant care.

Growth Habit and Size

The *growth habit* of a plant, whether tree, shrub, vine, or ground cover, and its ultimate *size* should be considered in terms of intended landscape use and the nature of the site. Is the plant to shade, screen, enclose, soften, accentuate, direct, or protect?

Mature size is extremely important, because a plant should not outgrow the allotted space. Overgrown shrubs are a universal problem: Designs are spoiled, windows and views blocked, walks and drives crowded, and lower branches bereft of foliage (Fig. 3–1). Trees that become too large for their niche can break curbs, gutters, sidewalks, and paving; can unduly shade plants beneath them; are difficult to care for when mature; can be hazardous in storms; and are costly to remove. A large tree may be a poor selection for a shallow soil or a windy site. Vines can require special support and can damage structures if they grow beyond a certain point.

Considerable interest in smaller trees began in the 1940s as Dutch elm disease began to take its toll in the eastern United States, as home landscapes became more modest, as trees planted along downtown streets blocked store signs, and as utility companies experienced increasing costs in keeping trees out of their overhead lines. Some arborists, however, think that small trees have been too heavily planted in public and private landscapes.

More attention is now being given to landscape function. Size, shade intensity, leaf persistence, and plant placement greatly affect a tree's ability to shade paved areas and reduce air-conditioning costs in the summer, to allow proper functioning of solar heating and power units, and to allow sunlight to warm buildings directly in winter. Species or even clonal selection are the most satisfactory approaches; using pruning or chemicals to control plant size is a bother and a recurring expense.

Selecting the Right Plant

Figure 3-1 All too often trees and shrubs become too large for their intended use.

Plant Form. Tree and shrub species and their clones may have distinctive forms that provide the landscape architect and arborist with an almost unlimited range of architectural and functional elements (see Fig. 2–1). The bark of many plants develops interesting patterns as it stretches and cracks with increases in trunk caliper. Bark color, though subtle, can accent an intimate landscape scene. Branching structure not only provides aesthetic potential but can be particularly important in determining the structural strength of trees and their daily and seasonal shade patterns. A tree will be inherently stronger if its branches are well-spaced along the trunk, if the trunk is larger in girth than its branches at the point where each branch arises, and if the branches have laterals. These characteristics are inherent in certain species, such as most conifers, pin oak, and, to a lesser extent, plane trees, but not in others, such as Modesto ash and black locust. Undesirable traits can be overcome by proper training when the tree is young, but such early training is often overlooked or done improperly.

Rate of Growth. People usually want a fast-growing tree (growing more than 1.5 m or 5 ft per year) that will quickly provide shade, screen unpleasant views, protect from the wind, or just do its job in the landscape. **Fast-growing trees are usually tolerant of difficult sites and neglect** and will also grow rapidly in the nursery. As a result, most nursery workers, landscape architects, developers, homeowners, and landscape maintenance people like them. How could one go wrong with such attributes?

In a number of ways! Most fast-growing trees become large and incur the related disadvantages; **fast-growing trees are usually weak-wooded and subject to limb breakage,** especially in storms; fast-growing trees usually live a shorter time than trees with more modest growth rates.

Species with moderate growth rates (300 to 600 mm or 1 to 2 ft per yr) can usually be encouraged to grow more rapidly than normal if plants are selected for good quality, are planted properly, and are given adequate water and adequate nitrogen fertilization. Even so, in especially difficult landscape situations, fast-growing trees may be the best solution for at least part of the overall plan. A mix of trees for rapid effect and long-term results may be a reasonable compromise. Care in evaluating the landscape site and developing a design solution is essential in deciding the plant characteristics that will be best.

Wood Strength. Branch structure and wood strength are closely allied in determining the inherent ability of a tree to withstand wind, ice storms, and limited vandalism. Again, the faster-growing species tend to have weaker wood. The wood of some conifers is not as strong as that of hardwoods. In their ability to withstand the elements, however, the form and the branch structure of conifers often more than compensate for any lack of wood strength. Branches of trees from a nonsnow area are more likely to break if grown where it snows.

The inside limbs of some trees die and may be hazards unless removed. Dead branches up to 50 mm (2 in.) in diameter are common in London plane, Japanese zelkova, and some elms and can cause injury when they fall, although they are not large. The ground under these trees is usually littered with twigs and small branches after a strong wind.

Rooting. Roots provide anchorage, nutrients, and water for plants. Deep and **particularly spreading root systems are essential to hold large plants against strong winds,** particularly where soil is wet. Wet sites require species that are tolerant of low oxygen or high carbon dioxide levels in the soil and are resistant to root rots.

Shallow soils, soils with differing textural strata, and rainfall or irrigation practices can contribute to shallow rooting, but certain species (some ash, mulberry, and elm, for example) exhibit surface rooting more frequently than do other species (such as oak, pistache, and pear). Large, shallow-rooted trees are often subject to windfall. The surface roots of some species (elm and poplar) sucker heavily, particularly when injured, and can be an unsightly maintenance problem. Species with shallow roots that frequently raise the soil around the tree trunk (elm, Japanese pagoda, and mulberry) are usually poor choices for narrow planting strips and tree wells in pavement (Fig. 3–2).

Willow, poplar, and silver maple are renowned for invasive roots. Rapidly expanding roots can crack intact sewer lines and provide entry for finer roots, which can quickly form a fibrous mass inside the lines and plug them (see Fig. 1–5). These roots can be a problem near septic leach lines. In Australia, the regulations of the Adelaide Waterworks Act of 1986 list 100 species of trees and shrubs that can be planted no closer than 2 m (6.6 ft) to a sewer main in a street, 95 species that can be located no closer than 3.5 m (11.5 ft) to a sewer main, and 30 species that cannot

Figure 3–2 The soil is often raised near the trunks of large trees that have shallow roots.

be planted in a street in any drainage area. On the other hand, such plants may be of value for transpiring water that is being discharged from water treatment facilities. The structure and growth rate of tree roots can pose special problems where utility pipes and conduits are close together in easements in or near the tree planting strip along streets.

Plant Features

Leaves. Leaves are our primary source of food and fiber; in addition, they add interest and beauty to plants. Leaves also influence a plant's functional effectiveness and, to a certain extent, the maintenance it requires.

The many sizes, shapes, and colors of leaves provide *aesthetic* enrichment for landscapes. *Leaf size* offers a range of textures, from light and airy to bold and dark, and can influence the scale of a landscape. *Leaf shape* can add interest and character to a landscape. *Leaf color* can supplement the architect's palette of flower and fruit colors by providing ranges over a growing season, over the entire year, during the first burst of growth, or during the final days of autumn only. Leaf color influences the amount of radiation reflected by the leaf canopy. Gray and light-colored leaves reflect more than dark green and red leaves. Color, however, may have more effect on the perception of temperature than on the temperature itself. Red foliage produces warm feelings and gray, cooler ones.

Thick, pubescent, and waxy leaves enhance the drought tolerance of plants. *Large, numerous, or dense leaves* enhance the depth of shading under a tree. Trees with large, overlapping leaves, such as mulberry, magnolia, and catalpa, cast a shade so dense that it is difficult to grow lawn or flowering plants under them. Other species, such as honey locust, silk tree, and acacia, have more open branching and small leaves or leaflets that produce light, filtered shade. *Fine leaves* may not need raking when they grow in open areas or over large shrubs or coarse grass but can make cleanup difficult on rough surfaces, around patio furniture, and in pools. Large leaves may be more voluminous, noticeable, and easily blown but are usually easy to rake and pick up.

Leaf persistence determines whether a plant will be deciduous or evergreen. Evergreen trees and shrubs provide year-round foliage and windbreak protection;

their leaves normally stay attached for three years or more, depending on the species and the vigor of the plant. In contrast to deciduous plants, broadleaved evergreens may be more subject to winter wind burning and snow breakage. Leafless trees let only 26 to 80 percent of winter sunlight through (Geiger 1961), and species vary greatly as to when they produce and lose their leaves. Honey locust leafs out late but defoliates early, while Bradford pear does the opposite. Near Portland, Oregon, the Bradford pear is in leaf about four months longer than are some of the honey-locust cultivars (Ticknor 1981). Such information can be helpful in selecting plants where summer shade and winter sun are wanted to moderate landscape and building temperatures (see Chapter 5).

The *season and duration of leaf fall,* in both deciduous and evergreen trees, affect landscape appearance and maintenance. Many evergreens drop their leaves over a longer period than deciduous trees, and not necessarily in the fall. Cork oak, an evergreen, sheds leaves in early spring before new growth begins. London plane begins dropping leaves in midsummer and may continue until after the first hard frost. The leaf fall of southern magnolia varies greatly within the species: Some trees may shed leaves for only two weeks while trees close-by lose leaves for six weeks or longer. Under dry summer conditions, trees such as honey locust and Japanese pagoda drop most of their leaves before the early fall when shade is still desired. The fronds of some palms do not fall when they die, but hang appressed to the trunk. Some think these palms are attractive, but dead fronds on the trunk can be a fire hazard. The base of the live fronds can be a haven for rats and birds. On occasion, palm trees thus infested have been removed as hazardous to public health.

Thorns and Prickly Foliage. Most people consider thorns and prickly foliage to be undesirable on landscape plants. In recent years, however, such plants are being recommended to deter burglaries and trespassing (Pratt 1982). Lists of local *Plants for Security* are distributed by law-enforcement agencies in a number of areas. Species need to be wisely selected to perform properly without being a maintenance problem.

Flowers and Fruit. A landscape is brought to life with the bright hues and varied forms of flowers and fruit. The flowers of many plants add fragrance to the garden and some cause allergies. In some locations, it is possible to have flowers or fruit year round. Witch hazel and forsythia flower in winter and early spring. Prunus species bloom late winter to late spring. Buddleia and crape myrtle flower in the summer. In general, flowering may be slim during the late fall and early winter.

Wildlife may be attracted by shrub berries. This is often a source of enjoyment but may have its negative aspects as well. Mistletoe can afflict susceptible trees more seriously near extensive plantings of fruiting shrubs. If other mistletoe is in the vicinity, birds attracted to the berries carry mistletoe seeds and leave some on nearby trees when they perch between feedings. The areas below these trees can also become quite messy with bird droppings.

The fruit of some plants, such as ginkgo and tree of heaven, is foul-smelling when crushed. The fruit of large shrubs and trees can be a maintenance problem if it falls on paved or turf areas. Such plants should be underplanted with shrubs or thick ground cover, so that the fruit is concealed when it drops. To eliminate the fruit problem, fruitless clones have been selected and propagated for a number of species. Fruitless trees may still produce heavy bloom and abundant pollen and therefore present problems for hayfever sufferers.

Environmental Tolerances

Climatic Adaptation. Semitemperate regions sustain a wide range of plants and plant characteristics. Most broadleaved evergreens and palms are limited to the semitemperate and warmer regions, while the colder areas are populated primarily by conifers and deciduous broadleaved plants.

Before planting exotic (nonindigenous) trees, check local weather records to learn the lowest temperatures on record to be sure the species could withstand such low temperatures. It is wise to observe plants in nearby areas similar to your planting site and talk to knowledgeable growers about which plants in the area may reach or exceed their hardiness range. It would be a serious loss to have a fifteen-year-old southern magnolia frozen to the ground during an unusually severe winter just when the tree was beginning to fulfill its landscape function. Rehder (1940), the USDA (1988), and Dunmire (1988) present plant hardiness zones that identify the climatic adaptation of many plants. Plants indigenous to areas having cold winters are usually more hardy than plants of the same species indigenous to warmer areas. Yet a few fine plants survive in areas much colder than their native locations.

Occasionally, low temperatures above the lowest recorded can be devastating if they occur in the fall or after a winter warm spell when plants are not at their maximum hardiness. In 1972, thousands of eucalyptus in the San Francisco Bay Area were killed or lost branches back to the trunk because of temperatures of $-5°C$ (23°F) during December. More than $10 million of public funds were spent to clear dead branches and trees to reduce the serious fire hazard. Spring or early fall frosts also can damage plants that begin growth early or do not harden soon enough in the fall.

Within an area or even within a single landscape site, microclimates may exert particularly favorable or unfavorable influences on certain plants. If tender vegetation in riskier portions of generally protected sites is small and inexpensive, injury or death will not be serious. It may, however, be worth growing certain plants for their unique beauty and taking a calculated risk that they might be killed or damaged once in 10 or 20 years. Without this kind of risk-taking, many landscapes would be poor indeed.

Just as the temperature can become too cold for certain plants, areas with mild winters may not provide enough chilling to allow some plants to resume proper growth the following spring (see Chapter 4). Some varieties of apple and peach have been bred for mild-winter areas and require less chilling than many of the older varieties. The source (provenance) of seed and propagating stock can be quite important in the performance of species that naturally have a wide geographical range.

Moisture is another crucial factor. Landscape sites must be analyzed as to water reservoir potential, evapotranspiration, and the amount of supplemental water available. Drought tolerance, although not of great concern in many areas, can be crucial in unirrigated landscapes that have longer than normal periods with little or no rain. Woody plants can survive drought periods only through adaptations that enable them to obtain or conserve water. These plants can be classed into three groups.

>*Water spenders* use water freely but many are considered drought tolerant because they have extensive root systems that absorb water from a large volume of soil. As long as some of their roots are in moist soil, they can survive. Many common landscape plants are of this type: black walnut, plane tree, and mulberry.
>
>*Drought evaders* avoid water stress by drying up or dropping their leaves, in some cases by shedding twigs and branches, or by becoming virtually dormant during the dry period. Examples are California buckeye and palo verde. While they have leaves, however, these plants usually transpire as rapidly as the water spenders. Most of them have limited value in our landscapes.
>
>*Water conservers* have ways of reducing water loss. Their leaves may be small, gray-colored, leathery, or arranged to reduce the amount of sunlight that strikes them. Their stomates may be structured to conserve moisture. Many plants from desert and Mediterranean climates are of this type.

Conversely, some planting sites may be waterlogged or flooded for varying periods during the year. Many reservoir shorelines and river levees that used to be kept free of vegetation are being planted for erosion control, fish habitat, and aesthetic reasons. Observation and experiments have identified many woody plants with inundation tolerance (Whitlow and Harris 1979).

Prevailing *wind* may sculpture or deform some trees and shrubs more than others. If the planting site is in an exposed location, plants that can withstand such conditions should be sought. For information, survey trees in the area and talk with local experts. Planting techniques may be able to maximize protection so that a wider range of plants can be used (see Chapter 10).

Light (radiation) is essential for green plants, although full sunlight may cause many of the so-called shade plants to have scorched leaves or grow poorly. Full sunlight warms both the exposed leaves and the air. If transpiration becomes so great that the plant loses water faster than it is absorbed, the tips, margins, or entire leaves may be killed. Leaves that have grown in low-light conditions may become bleached under full sun through a loss of chlorophyll. It is high transpiration, not high irradiation or high temperature, that causes many shade plants to do poorly in full sun.

Certain plants perform well under low-light situations and are well-suited for plantings indoors, on the shady side of structures, or under large trees (see Fig. 11-8). Many plants lose their leaves when moved from high to low levels of radiation. Others become deformed when grown near a building or a larger plant. Their branches grow toward the light and mold the plants into asymmetrical forms. Plants should be observed in similar sites before selected for such use.

Unfavorable Soil. Poor soil can spell disaster for many plants. Soil can often be modified or amended to improve plant growth, but that may not be feasible in large landscape developments, so it is well in such situations to use plants, like fast-growing trees, that are fairly tolerant of poor soil conditions.

Shallow, poorly drained soil dooms many trees and shrubs to poor growth or to an uncertain future unless they are tolerant to poor aeration. Small to medium trees will be less subject to uprooting by wind in wet, shallow soil.

Alkaline soils can reduce the availability of iron and manganese so much that some landscape plants develop small, pale, chlorotic foliage that browns along the margins; growth may also be stunted. Particularly susceptible are liquidambar, camphor, and pin oak. Certain plants, on the other hand, are more tolerant of soils that are saline or high in certain minerals and some laboratories in the southwestern United States are studying this phenomenon.

Air Pollution.   Many species weaken, reduce their growth, and suffer leaf injury because of air pollution; other species are more tolerant. Air pollution may involve many chemicals from a number of sources. Fluorine or sulfur dioxide, especially toxic to plants, may come from metal, ceramic, and chemical manufacturers or from petroleum refineries. Levels of these compounds should be checked at locations downwind from such sources. Pollution from autos and other industries is widespread in some areas. Many pines are quite sensitive to air pollutants, whereas ginkgo, London plane, and pin oak are fairly tolerant.

Pest Resistance.   Genetic resistance to pests is the ultimate weapon. A few woody plants, like the ginkgo and the goldenrain tree, are practically pest-free. Most plants, however, can fall victim to one or more maladies. Even limited tolerance to serious pests helps plants to look and perform their best and reduces maintenance. Some insects and diseases cause little damage, whereas others can kill within a season. Mildew may be unsightly only on some plants, but Dutch elm disease can be fatal. A pest can be serious in one region but not in another: Mimosa webworm infests honey locust in Pennsylvania and further south but is not a problem in northern regions where winters are cold.

A few plant breeders are beginning to breed for pest and disease resistance. An apple-breeding program in the Midwest, for example, has produced a number of fine ornamental crabs that are resistant to most common apple diseases. The National Arboretum has made selections of crape myrtle that are more resistant to mildew than those more commonly planted.

Inquiry among local cooperative extension agents, nursery workers, or arborists, along with careful observation, can afford valuable information about the local pest problems of plants you would like to grow.

Longevity.   Long life is more important in trees than in shrubs and vines, which can be replaced more easily and regrown to landscape effectiveness more quickly than trees. Generally, fast-growing trees and shrubs are not as long-lived as slower-growing ones. As a consequence, trees with different growth rates are sometimes planted in a mixture: A rapid landscape effect is obtained with the fast-growing trees, which can be successively pruned back and finally removed as the slower-growing trees become large. All too often, though, an owner or supervisor waits too long to prune or does not have the heart to cut down good-sized trees in their prime.

Certain trees and shrubs are known to be fairly short-lived, and others are noted for their longevity. Even though site and care are important to a plant's longevity, its survival potential ought to influence selection for landscape use. Most trees, however, will outlive their owners.

Fire. Fire is a threat in chaparral areas, particularly during hot, dry, windy weather. To reduce the jeopardy to homes in such areas, people have recommended landscaping around homes with so-called fire-resistant plants. Of course, there is no such thing as a plant that will not burn if there is enough heat.

The main objective is to minimize the accumulation of dry fuel (Maire 1976) and to concentrate on low, slow-growing plants with high water content. Tall-growing conifers that retain dry needles and high volumes of resins should be avoided close to buildings and to one another. See Chapter 11 for more information.

Native Plants. In the effort to select plants that do well with minimum care, native plants are usually recommended. **Often, however, native or indigenous plants do not perform as well as exotic or non-native species.** Most urban landscape sites are no longer "native"—usually the soil has been greatly modified, the microclimate has been affected by pavement and buildings, the water regime has changed, pollution has increased, and on and on. "Adapted" species are what is needed; some native species may perform well, but many more exotics will do equally well or better (see Fig. 4–6 and accompanying discussion).

SOURCES OF PLANTING STOCK

A few individuals and agencies propagate and grow plants for their own use, but almost all plants for public and private landscapes come from wholesale and retail nurseries. Wholesale nurseries usually grow plants in the field or in containers. Some nurseries specialize in a few kinds of plants, whereas others grow a wide variety. Some propagate and grow plants for retail sale, while others buy small plants and grow them for resale. Most retail nurseries and garden centers sell a complete selection of landscape plants as well as landscape supplies and equipment. Many retail nurseries grow some of the plants they sell.

It is wise to get acquainted with one or more nursery people in your area: Not only are they a source of plants, but the better ones can offer a wealth of information about plant adaptation and maintenance. They also can benefit from your experience and knowledge. Because many trees take several years to produce, you should contact a nursery several months or even a year before you plan to plant in order to be assured of the species, sizes, and quality you want. For recently introduced plants, in the quantity and size wanted, you may have to contract with a nursery to grow them. If the number of plants is large enough, most wholesale nurseries will grow by contract for future delivery.

Woody plants are usually sold or transplanted as *bare root, balled and burlapped* (B & B), or *container* plants (Fig. 3–3). Most deciduous plants are undercut, dug, and handled with little or no soil on their roots (bare root) during late fall and winter. The plants are stored so the roots are kept moist and the tops dormant. They should be planted in the landscape before growth begins.

Many trees and shrubs are field grown. Field-grown evergreens and some de-

Figure 3-3 Young trees are available for planting either as bare root (left), balled and burlapped (B & B, center), or container (right). (B & B photo courtesy William Flemer, Princeton Nurseries, NJ)

ciduous plants are dug with a ball of soil around the roots, which is then wrapped in burlap or other material. Plants with balls that are larger than 500 mm (20 in.) in diameter are usually packed into boxes that are 600 mm (24 in.) or larger on a side; such plants are handled somewhat like container-grown plants until sold. B & B plants have a longer planting season than bare root plants but should be planted before spring.

An increasing number of plants are being grown in wire baskets or in fabric bags in nursery fields, especially where plants in containers are subject to winter freezing. After planting, the plants are treated the same as field-grown plants. The basket or bag holds the root ball together, making it easier to dig in the nursery and handle until planted in the landscape.

Ease of handling is the main advantage of a plant grown in a wire basket. Otherwise it is cared for as a B & B plant: The root ball must be wrapped, kept moist, and protected from breakage.

Whitcomb (1987) developed a porous fabric bag that could be placed in the soil and a small tree planted in it to be grown as a field-grown tree. It is marketed as a Root Control® bag. The roots penetrate the fabric and become constricted by the fabric as they enlarge. The roots within a bag increase more in diameter and carbohydrates than the roots of field-grown trees in the same volume of soil (Fig. 3-4). Even though constricted, the roots outside a bag actively absorb and transport nutrients and water. Bag-grown plants are reported to establish more quickly than conventionally grown plants (Whitcomb 1987), but no studies compare the establishment of trees that had been container-grown, field-grown, or bag-grown.

A first-year report that compares these three production methods indicates,

Figure 3-4 The roots of a conventionally field-grown white ash (left) can be compared with those grown in the field in a Root Control® bag (center); the bag-grown tree has more small absorbing roots. Root Control® bag removed, revealing enlarged root ends on the root-ball surface (white circles) due to girdling of roots which penetrated the bag (right). The fabric caused few if any roots to circle inside the bag. A fairly fine-textured soil needs to be used in Root Control® bags in order for the root ball to remain intact when the bag is removed when the tree is planted. (Photos courtesy Root Control, Inc., Oklahoma City)

after two years in the nursery, that more than 90 percent of the total root weight was contained within the root ball of field-grown and bag-grown trees (Harris and Gilman 1991). However, less than 15 percent of the weight of roots smaller than 2 mm (0.1 in.) in diameter (the main absorbing roots) was within their root balls, while all of the roots were in the container root ball. When planted out, the field- and bag-grown trees became water stressed before those that had been grown in containers. These results, however, are not unexpected; the results underscore the importance of wise irrigation following planting in the landscape. The future growth of the trees in this experiment will be of great interest. The observations of many horticulturists that bag-grown trees establish quickly probably will be borne out, primarily because of the carbohydrate reserves that are planted with the trees.

A number of deciduous plants cannot be handled bare root but transplant well from containers. Container-plant production makes year-round planting possible in many areas and makes greater uniformity and mechanization possible in nursery operations. Since the 1950s, an increasing number of plants have been grown in containers. A 20-liter (5-gal) plant will start as a seed in a seedflat, where it germinates and grows from 40 to 100 mm (1.5–4 in.) tall. It is transferred (hopefully, with root pruning and care) to a liner pot 55 to 75 mm (2.25–3 in.) in diameter. After several weeks, the plant is transferred to a 4-liter (1-gal) can and moved from the greenhouse to the nursery yard. Later it is moved into a 20-liter can.

Because of increasing interest in planting large specimens, particularly trees, more plants are now grown to large size in nursery fields, are dug mechanically, and are placed directly in the landscape or into large containers for transport and later sale (see Chapter 10).

SELECTING QUALITY PLANTS

The quality of the trees and shrubs selected for planting can be as important in determining their success in the landscape as proper selection of species, planting, and maintenance. Certain characteristics have considerable influence on plant survival and growth, while other characteristics, although desirable, may not affect performance greatly. Branching pattern and size determine the immediate landscape effect of a plant but may have little influence on subsequent growth. Variations in vigor, laterals on the trunk, and height-to-caliper ratio will not jeopardize survival but can materially affect the effort necessary to obtain satisfactory growth. **Root and shoot quality can determine not only performance but even survival.**

Root Characteristics

A well-developed, healthy root system is essential to a vigorous plant, particularly to a tree that lives many years and becomes large. In a well-formed root system, branching is symmetrical and main roots grow down and out to provide trunk support. Container-grown and balled-and-burlapped plants should have fibrous roots sufficiently developed that the root mass will retain its shape and hold together when removed from the container or when handled during planting. The main roots should be free of kinks and circles. The main roots of bare root plants should be sound and free from breaks, torn or bruised bark, crown gall, and nematodes.

 Root quality at planting cannot be overstressed, because certain defects can doom a plant to poor growth, breaking of the trunk at the ground, or death (Barrows 1970, Long 1961). The top of a young plant seldom is a good indicator of the quality of the root system (Harris and others 1971). Although root defects are thought to be primarily associated with container-grown plants, field-grown plants propagated in seedflats and liners or seeded directly in the field can also develop root problems.

 Types of Root Defects. Two types of defects may occur: (1) *kinked roots,* in which the taproot, major branched roots, or both are sharply bent, and (2) *circling* or *girdling roots,* which form circles, generally horizontal, around the trunk or other roots (Fig. 3–5) (Harris, Long, and Davis 1967, Harris and others 1971). Both types of defects often occur within the same root system.

 Location of Root Defects. Roots must be inspected to evaluate their condition. Root deformities that are serious enough to affect growth and survival may occur in any or all of three zones within the root ball of a container-grown plant (Fig. 3–6).

> *The trunk-surface root zone* is within 75 mm (3 in.) of the soil surface and 50 mm (2 in.) from the trunk of the plant; this zone is generally that of the liner pot
> *The center root zone* is within the root ball outside the trunk-surface root zone, but not at the periphery of the root ball; the location(s) depends on the size of the intermediate container(s), if one or more have been used
> *The peripheral root zone* occurs at the edge and bottom of the root ball

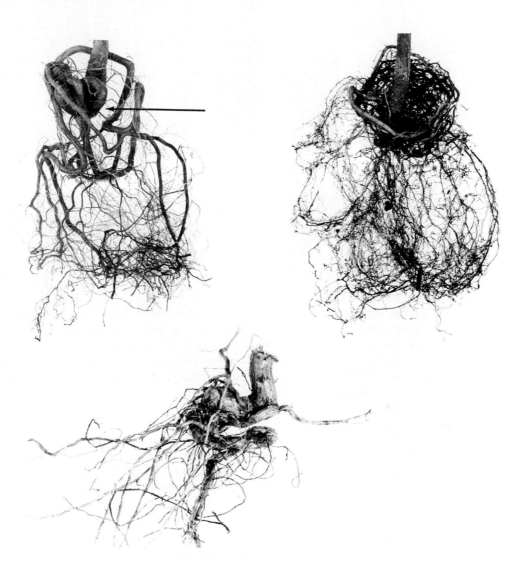

Figure 3–5 Kinked and circling roots are exposed when soil mix is washed from the roots of trees in 4-l (1-gal) cans. The oak root (top left) was kinked (arrow) when transplanted from a seed flat to a 50-mm (2-in.) peat pot; the shape of the peat pot is revealed by the roots that were not pinched back when the peat pot was put in a 4-l can. Pine roots (top right) still outline the shape of a 100-mm (4-in.) square pot because they were not pruned and straightened when moved up to the 4-l can. The root of a field-grown Monterey pine (bottom) is severely kinked. Kinking occurred at the first transplanting in the greenhouse. None of these root systems could now be pruned to make planting in the landscape worthwhile. (Harris and others 1971)

Selecting Quality Plants

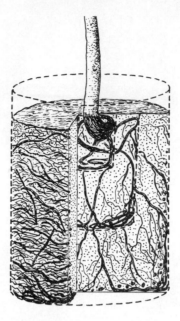

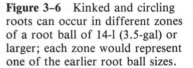

Figure 3–6 Kinked and circling roots can occur in different zones of a root ball of 14-l (3.5-gal) or larger; each zone would represent one of the earlier root ball sizes.

Trunk-surface roots are often visible on or above the soil surface; if not, you may examine them by washing the soil from the crown to a depth of 50 to 75 mm (2–3 in.) below the crown and 25 to 50 mm (1–2 in.) from the trunk. If no branch roots from the main root are found, at least one plant should be sacrificed by washing deeper until branch roots are found. If plants are planted too deep originally, crown rot or other problems can result when they are planted out (Pirone and others 1988). Plants with taproots that have not been pruned during transplantings might also have main roots without laterals near the surface. Unpruned taproots may become twisted during transplantings, so that they later become girdling.

You may see center roots that are extremely circled when you examine trunk-surface roots; otherwise, you may inspect them by washing out more soil from around the crown to a depth of 100 to 125 mm (4–5 in.) and 50 to 75 mm (2–3 in.) from the trunk. To examine the center root zone completely, however, you must remove the container and wash the soil from the roots. If such destructive inspection is necessary, it is done on a sample of the plants.

Peripheral roots can be seen by removing the container from the root ball. This is particularly easy when containers have tapered sides. Since this inspection is primarily designed to compare general root condition to the vigor of the tops, only a few plants usually need be inspected.

Except for those inspected destructively, plants whose roots have been partially exposed will usually not be affected adversely if the roots do not dry out and a soil mix equal to that removed is replaced, firmed, and moistened after inspection. When plants are accepted after such a root inspection, however, their performance in the landscape should be the responsibility of the purchaser, not of the supplier. Handling, planting (including judicious pruning), and irrigation should proceed with care.

Evaluating Root Defects. Observations of landscape plants indicate that many are restricted in growth by kinked and circling roots (Barrows 1970, Long 1961). **The following root conditions may not only reduce growth but may kill plants,** particularly trees:

> *Kinked roots* with a sharp bend in the main root(s) of 90 degrees or more and less than 20 percent of the root system originating above the kink; a plant with such roots would probably bend at or below the soil line if untied from its supporting stake
>
> *Circling roots* at the *trunk-surface* or the *center* that circle 80 percent or more of the root system by 360 degrees or more; a plant with such roots would have less than 20 percent of its roots free for support should girdling occur

Even though there is considerable evidence that kinked and circling roots cause girdling or constricting of the main root(s) and trunk, no conclusive studies have indicated how severe the kinking or circling has to be before a plant is jeopardized. A small portion (20 percent or more) of free roots may be adequate to maintain the plant. Agencies, landscape architects, and arborists are more frequently specifying that trees be inspected for root defects as a condition of acceptance. See Appendix 4 for tree specifications.

Circling peripheral roots are usually not a problem, since they can be corrected at planting; container-grown plants with adequate root development will inevitably exhibit some circling of this sort. Peripheral roots, however, should be checked for indications that the plant has been in the container too long. Masses of large, entwined roots at the bottom and around the root ball may indicate an abnormally potbound condition (Bancroft-Whitney 1967). If the top of the plant is vigorous, however, the root system is not yet abnormally potbound.

The most difficult root defects to correct occur in the trunk-surface and center root zones. Since main roots are usually involved, correction will probably seriously weaken the plant. Even if the top is still vigorous, correcting entwined and matted roots at transplanting may be too time-consuming to be worthwhile. On the other hand, circling and matted peripheral roots can and should be thinned and straightened when the plant is transferred to a larger container or into the landscape. If the peripheral roots are large, entwined, and matted, cutting and straightening them may further retard growth that has already been slowed by prolonged confinement in the container. In some cases, however, if the top is severely cut back, the plant may be invigorated. If the circling is not corrected, the plant will probably continue its poor growth until the circling roots finally girdle it.

Certain trees, such as eucalyptus and redwood, are known to form new roots from portions of the trunks that become covered with soil or rock. Hamilton (1983) found that if following planting *Eucalyptus sideroxylon* had or developed more than five roots above a severely circling root system, the trees grew normally. The fact that many plants, including eucalyptus, are lost from trunk-root breakage and root girdling would indicate, however, that not enough new roots grow fast enough on many trees to be of much importance. More information is needed.

Preliminary studies and field observations indicate that plants will grow rapidly when planted out if their growth has not been slowed during production. Even some plants that appear healthy have been left too long in containers that have begun to check growth. At any transplanting, the ideal root system is developed just enough to hold the root ball together. Any less development would jeopardize tree survival by loss of soil from around roots and root breakage; much further development at that particular stage might reduce future vigor.

Top and Trunk Characteristics

Most trees grown with adequate space and without staking or severe pruning are capable of supporting themselves even in high winds (Harris and Hamilton 1969, Leiser and others 1972). Most trees grow with certain height-to-caliper relationships, fairly uniform trunk taper, and crown configurations that distribute wind loading along the trunk (Leiser and Kemper 1973). Growing practices can modify these characteristics greatly.

Height-to-Caliper Ratio. Field-grown trees, whether they are sold bare root or balled and burlapped, have long been specified according to height and trunk caliper. The American National Standards Institute (ANSI Z60.1) (1986) specifies height-to-caliper values for bare root trees (Fig. 3–7). Heights that are greater than those listed for a given caliper indicate a spindly trunk that may not be capable of standing upright when in full leaf. Trees with heights considerably less than those listed for a particular caliper should be checked for adequate plant vigor.

Height-to-caliper values have not been established for container-grown trees. Experiments with trees growing in 20-l (5-gal) containers in three nursery areas in California, however, showed that certain staking and pruning practices will produce trees that meet or exceed the standards for field-grown trees (Leiser and others 1972). The calipers of these 2-m (6-ft) trees averaged 5 mm (0.2 in.) or 30 percent greater than those of trees that were staked and had their low trunk laterals removed. Most of the staked trees were not able to stand upright when untied from their stakes (see Fig. 9–9).

The ANSI Z60.1 standards for height-to-caliper values (Fig. 3–7) have an upper caliper limit for a given tree height as well, though most of the emphasis is on lower caliper limits. Certain species normally develop large caliper for their height. If a tree is too short for given caliper, however, its growth may be so poor that the tree will stagnate. Such a tree probably would not do well when planted in the landscape.

Taper. Taper is the decrease in trunk caliper with increasing height. Under wind stress, the tapered trunk of an unstaked tree bends uniformly along the stem, but a trunk with little or no taper bends essentially along one section near the ground (see Fig. 9–9) (Leiser and Kemper 1973). In the first case, stress is evenly distributed along the lower two-thirds of the trunk and decreases toward the tip, thus reducing the possibility of trunk deformation or breakage. In the latter case, most of the stress is concentrated, causing the trunk to break or remain bent even after the wind has stopped. Leiser and others (1972) discovered that staking and severe pruning produce a tree trunk with little or no taper or even with reverse taper; that is, trunk caliper may be greater at 2 m (6 ft) than at a lower level.

In the same study, the trunk below the main branches of trees grown in 20-l

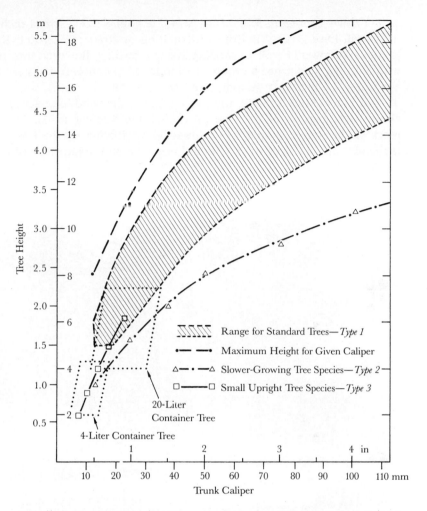

Figure 3–7 Height and caliper standards for different types of field-grown trees (ANSI Z60.1, 1986). The shaded area indicates range in acceptable heights and calipers for standard trees (Type 1). Note that the height-to-caliper ratio for small trees (Type 3) is almost a continuation of the lower portion of the curve for standard trees. The height of a slower-growing tree is two-thirds the height of a standard tree for the same caliper. The two parallelograms represent the height-to-caliper relationships for unstaked container-grown trees that are given adequate space. (Harris, Leiser, and Davis 1976)

(5-gal) cans without stakes exhibited from 5 to 10 mm reduction in caliper for each meter ($\frac{1}{16}$ to $\frac{1}{8}$ in./ft) of trunk height. A trunk with this degree of taper would have a caliper at half the tree height equal to 50 to 80 percent of the trunk caliper at 50 mm (2 in.) from the ground. Trunks with branches along their length normally increase more in caliper at their base than at higher levels, so that as a young tree grows its taper increases (Harris and Hamilton 1969).

Selecting Quality Plants

Crown Configuration.   In a wind, a young tree with branching along its trunk will have a fairly uniform distribution of stress (Leiser and Kemper 1973). Removal of lower laterals or shading due to crowding, however, may produce a tree with a crown (and hence a center of wind loading) concentrated near the top. Such a tree may be unable to support itself without staking.

An ideal branch distribution would center the wind load acting on the tree at about two-thirds of the total height (Leiser and Kemper 1973). Thus, **one-half or more of the foliage should be on branches originating on the lower two-thirds of the trunk,** and one-half or less should originate on the upper one-third (Fig. 3–8).

Figure 3–8   The wind load on the main trunk of the plane tree (left) will be more evenly distributed than on the magnolia because more than one-half of the foliage is on branches that originate along the lower two-thirds of the plane-tree trunk (the stake was never needed). Most of the stress on the magnolia will be concentrated just below the lowest branch.

In addition to equalizing stress distribution, lower limbs nourish the trunk and shade it, resulting in a tree with greater caliper and taper. Low foliage may help obscure the trunk and protect the tree somewhat from vandalism. The lower laterals should be pinched (headed) to keep them small in relation to the higher permanent branches and the trunk. These laterals can be removed when the top more completely shades the trunk, when the trunk is larger and the bark thicker. Pruning wounds on young trees usually callus quickly.

If a tree can stand without support and can return to an upright position after being deflected by wind or hand, it will usually meet the aforementioned guidelines as to height-to-caliper ratio, taper, and top configuration. Such trees usually result when they are grown without staking, with lightly headed laterals left along their lower trunks, and with adequate space in the nursery.

Since such trees may not always be available, however, landscape staking and pruning can be used when trees are not able to stand upright by themselves. These practices are discussed in Chapter 9. The guidelines set forth here should be followed whenever possible, and suppliers should be encouraged to produce plants that meet these criteria. When such trees are available, support staking can be virtually eliminated, though protective or anchor staking may be needed.

Branching Pattern. Plants should have a branching pattern or shape that is in keeping with their intended use. Low main branches may be used as a windbreak, visual screen, or specimen in a shrub bed but may be unacceptable in situations where traffic or visual clearances are desired. Early removal of lower laterals will be at the expense of trunk development and wind stress distribution (Chandler and Cornell 1952, Leiser and Kemper 1973). When trees have been headed (their main leader cut back) (see Fig. 15–4) branches with close vertical spacing will usually develop immediately below the cut; if these are not corrected, they can lead to structural weakness as they become large (Harris and others 1969). Also, many trees that have been headed become misshapen when lower branches outgrow the branches above. **For most landscape purposes, the main leader should not be headed,** but if it is, a new leader should be selected to fill its place (see Appendix 5).

If a tree is tall enough to have branches that could be selected for permanent scaffolds, such branches should be smaller in diameter than the trunk and free of included bark at their attachments. If a potential scaffold branch similar in size to the trunk and/or with an included-bark attachment is removed from a young tree, another bud at the same node will grow into a new shoot smaller in diameter and with a wide angle free of included bark (Harris and Balics 1963). Such a branch will be strongly attached.

A young tree may not be tall enough to develop branches that will be the permanent ones. Once formed, a branch does not increase in height above the ground. As the tree develops, laterals below the lowest permanent branch should be handled as described in the section on Crown Configuration and main branches selected.

Top-to-Root Relationships

More public and private landscaped areas, including city streets, are being planted with 60-l (15-gal) and boxed trees in an effort to maximize desirable landscape effects and reduce vandalism and theft. These larger trees are not only more expensive and difficult to plant, but they usually require more intensive initial maintenance.

The size of a tree should not necessarily affect its performance in the landscape but, practically, it usually does. The larger the tree, particularly if it has been grown in a series of containers, the more critical is the vigor of its top. Skill and careful scheduling must foster tree growth during production. At the other extreme, the frequent watering and high fertility that are common in most nurseries may result in a larger top than the root ball can support without continued high maintenance in the more severe landscape environment. Such a condition is not necessarily undesirable, but it should be recognized and handled accordingly: High initial maintenance must be provided even if the top is thinned in an effort to reduce transpiration.

A tree whose top is of moderate size in proportion to the root system will usually grow more vigorously than a tree with a large top (Hendrickson 1918). The size of the root system or the volume of soil available influences the interval between irrigations. Reducing the volume of the top by thinning will improve moisture and nutritional relationships somewhat but not in proportion to the thinning. **It should be remembered that too much water can be as detrimental as too little.**

The American National Standards Institute (ANSI Z60.1, 1986) recommends that the root spread of bare root nursery-grown trees should be about one-sixth the height of trees up to 3 m (10 ft) tall. The ANSI Z60.1 also specifies minimum ball sizes for balled-and-burlapped trees, depending on tree caliper or height (Table 3–2, Fig. 3–9). Until other information is available, these can also be used as guidelines for trees that are moved mechanically. The specifications in Table 3–2 may be modified in that plants whose roots are coarse or widespreading because of natural habits of growth, soil condition, or infrequent transplanting and plants that are moved out of season would require balls in excess of the recommended sizes. Also, when stock has been grown in pots or other containers, when field plants have recently been planted out from containers or with smaller balls, and when material

Figure 3–9 The standards of the American National Standards Institute (ANSI Z60.1, 1986) for balled-and-burlapped trees base the size of the root ball on trunk caliper for single-stemmed standard and slower-growing trees (left) and on height for small trees (right).

TABLE 3-2 RECOMMENDED MINIMUM ROOT BALL SIZES FOR DIFFERENT TYPES OF FIELD-GROWN TREES (ANSI Z60.1 1986)

Shade trees types 1 and 2[a]			
Caliper	Minimum[b] diameter ball	Caliper	Minimum diameter ball
mm	cm	in.	in.
15[c]	30	0.5	12
20	35	0.75	14
25	40	1.0	16
30	45	1.25	18
35	50	1.5	20
40	55	1.75	22
50	60	2.0	24
65	70	2.5	28
80	80	3.0	32
95	95	3.5	38
105	105	4.0	42
115	120	4.5	48
125	135	5.0	54

Smaller trees types 3 and 4			
Height -------- caliper	Minimum diameter ball	Height -------- caliper	Minimum diameter ball
m/mm	cm	ft/in.	in.
0.5	20	2	10
1.0	30	3	12
1.25	35	4	14
1.5	40	5	16
------[d]		------	
20	40	0.75	16
25	45	1.0	18
35	50	1.5	20
40	55	1.75	22
50	60	2.0	24
65	70	2.5	28
80	80	3.0	32
95	95	3.5	38
105	105	4.0	42
115	120	4.5	48
125	135	5.0	54

[a] Types 1 and 2 are standard and slower-growing shade trees. Type 3 is a small upright tree. Type 4 is a small spreading tree.

[b] Trees having a coarse or wide-spreading root system or that are moved out of season would require a root ball in excess of the above recommended sizes.

[c] The metric measurements have been rounded for convenience.

[d] Figures above the broken lines are height (m or ft); those below the line are caliper (mm or in.).

has been frequently transplanted or root pruned, the sizes recommended may be excessive (ANSI Z60.1, 1986).

Plants grown in containers are usually sold by the size of the container: One might buy a 20-l (5-gal) liquidambar. **Container size does not necessarily indicate the size of the plant.** In regions where year-round production is common, plants transplanted to larger containers in the fall may by spring have relatively small tops but extensive root systems, which have been able to develop during cool weather. On the other hand, plants transplanted in the spring may by fall have relatively large tops but few roots. Seasonal variation should be taken into account when one evaluates plant size.

Health

The health of a tree is characterized by its vigor and freedom from injury and pests.

Vigor. Vigor is a good measure of a plant's ability to perform when planted out. A knowledge of the species is of value in making this rather subjective evaluation. Leaf color should be green to dark green, depending on the species and time of year. Large leaves and dense foliage denote vigor. Shoot growth is a more quantitative indicator, but also varies greatly with species. Almost all trees to be used in the landscape should grow at least 300 mm (1 ft) a year; many kinds will grow 1 to 2 m (4–6 ft). This is not undesirable if the trunk is large and tapered enough to hold the top upright. Certain dwarf plants and species that grow only from preformed shoot initials may grow less vigorously. Shoot growth that has been pruned will be more difficult to evaluate, but the age and size of the pruned shoots can be used in estimating vigor.

Smooth, bright bark on the trunk and main branches of young plants usually signifies good vigor. Rough, cracked, dull, and dark bark, although characteristic of some species, could be indicative of poor vigor, so other indicators should be checked.

Young roots of most plants will be light in color. Their diameters vary greatly with species and are not good indicators of vigor. Large, dark-colored roots, however, may be an indication of low vigor; a plant with such roots may be old for its size.

Injury. Select plants free from injured bark, but be sure to distinguish proper pruning wounds from actual injury. Bark injury can be a problem when trees are staked (Harris and others 1969, Leiser and others 1972); tree ties and labels sometimes girdle the trunk or main branches so that growth is reduced and slow to recover even when the restriction is removed.

Sunburn is common when plants, particularly trees of low vigor, are exposed to the afternoon sun or have trunks with few or no leafy shoots. Sunburned trunks are extremely slow to recover and are subject to borer infestation.

Pests. Pests should seldom be a problem because most nursery stock is inspected or comes from nurseries that are periodically inspected. Delivery of plants should be in compliance with nursery inspection and government nursery stock grades, standards, laws, and regulations. If there is reason to suspect a problem, the purchaser should call the local county agricultural commissioner's office or other responsible nursery regulatory official.

It should be remembered that some organisms or indications of their presence are not necessarily harmful. Many organisms have little or no effect on plants; some are beneficial.

Galls or swellings on roots and trunks are normal on a number of plants (Fig. 3–10). Sometimes these swellings are confused with those caused by nematodes or crown gall bacteria. Mycorrhizal fungi can cause enlarged feeder roots with areas that are lobed, beaded, or brain-like in shape. The roots of legumes may have nodules with nitrogen-fixing bacteria. Swellings may occur at the trunk base of redwood, manzanita, eucalyptus (Kelly 1969), and other species; these are referred to as *burls* or *lignotubers*. These swollen masses of woody tissue form reservoirs of latent buds. A plant with well-developed lignotubers is able to sprout quickly if its top is injured. In contrast, crown galls and knots caused by nematodes do not, to our knowledge, produce sprouts.

Figure 3–10 Lignotuber on eucalyptus (arrow). Lignotubers on some eucalyptus and swellings on other species are normal and should not cause concern. Some trees may have lignotubers while others of the same species do not.

Inspecting Plants

All plants, or a representative sample, should be carefully inspected before they are accepted to be sure the plants meet criteria for size and quality. When a large number of plants, particularly trees, are involved, arborists should inspect them at the nursery before delivery. It is much easier to hold to the specifications or make substitutions at the nursery than at a planting site. Many times it is possible to select the plants you want at the growing grounds, just as you would in a retail nursery.

The value of clearly stated specifications is most evident at the time of acceptance. Realistic specifications should ensure good growth, ease of maintenance, and landscape fulfillment but should minimize unnecessary restrictions. Some characteristics of young plants are essential for future vigorous growth or even for survival. Others may encourage better performance or easier maintenance, but their absence will not necessarily jeopardize the life of the plants. Depending on the use of the plants, other characteristics may be desirable but not critical for survival or well-being. These characteristics are categorized by their probable importance.

Characteristics that may jeopardize plant life and growth	Characteristics that may improve growth or ease maintenance	Characteristics desirable but not critical or characteristics that can be corrected
Severely kinked roots	Moderate to good vigor	Correct species
Severely circling roots	Moderate top-to-root ratio	Branching pattern
Noxious pests	Caliper and taper (ability to stand without stake)	
Desiccation of shoots or roots	Laterals along trunk	
	Slight or no bark indentation below branch	

The portions of trees and shrubs that are easiest and most obvious to inspect are the tops: Is the plant the size you want? Is it vigorous? Are shoots and leaves turgid and healthy? Will the plant stand without support? Is it free of pests and injuries? Does it have the desired shape and structure? These characteristics are fairly easy to determine.

If the top is acceptable, then check the roots. For bare root plants this is easy; for container and B & B plants it is more difficult. First check the trunk-surface zone for kinking and circling roots, no matter how the plants have been grown. That is about as far as you can go with B & B plants: If the trunk-surface roots are acceptable, the rest of the root structure is probably free of problems because the plants have been dug from field plantings. Poor root quality of a container plant may be obvious (Figs. 3–11 and and 3–12): (1) Circling or kinked roots may be visible on the root ball surface, or when the plant is untied from its support; (2) the trunk bends sharply at the soil line; (3) when the trunk is lifted slowly, it moves up 25 to 50 mm (1–2 in.) before the container and soil do. Each of these indicates circling, kinked, and/or underdeveloped roots. Such plants should be rejected. Tree specifications are summarized in Appendix 5.

Figure 3–11 Simple tests to check the root system of a container-grown tree. Circling or kinked roots on the surface indicate serious root problems (left). If the tree is staked, untie it from the stake; if the trunk bends sharply from the top of the container (right), suspect kinked or circling roots.

Figure 3–12 If the trunk arches above the container (left), the roots may be all right. Alternately, lift gently on the trunk (right); the container should rise with it. If the trunk can be raised 25 to 50 mm (1–2 in.) before the container moves, the roots may be circling or poorly developed.

Additional Thoughts

Species.   At least one plant of each clone, cultivar, or species should be labeled with the correct botanical name when delivered to the purchaser. If more than one clone of a species is being ordered, every plant should be labeled to ensure against mix-ups in planting.

If the particular kind of plant is important for a specific landscape purpose and is to be used in quantity, its correct identification should be verified by the purchaser or a qualified representative. A number of species have strains with differing flowering and growth characteristics. Replacement guarantees usually cover only the size originally supplied. Depending on the expected losses, a certain percentage of extra plants should be purchased and moved to larger containers or planted out to be used as replacements should plants suffer vandalism, do poorly, die, or turn out to be of the wrong kind.

Acceptance Delivery. Plants are usually ordered for delivery on a certain date. If the purchaser wants to delay delivery, the plants should be inspected on the original delivery date or soon thereafter to determine their acceptability. If acceptable at that time, they should not be turned down later for being potbound, overgrown, or of low vigor. The retailer or wholesaler can reasonably charge additional fees for the extended care of the plants.

Plant Protection. During handling, plants must be protected from injury and exposure to desiccation and temperature extremes. The roots should be kept moist and cool, particularly if plants are bare root or in dark containers.

FURTHER READING

BAILEY, L. H. 1949. *Manual of Cultivated Plants.* New York: Macmillan.

——, E. Z. BAILEY, and STAFF OF THE L. H. BAILEY HORTORIUM. 1976. *Hortus Third.* New York: Macmillan.

BEAN, W. J. 1950. *Trees and Shrubs Hardy in the British Isles.* 3 vols. London: Butler and Tanner.

BENSON, L., and R. A. DARROW. 1954. *Trees and Shrubs of the Southwestern Deserts.* Tucson: University of Arizona Press.

CHITTENDON, F. J., ed. 1951. *Dictionary of Gardening* (Royal Horticultural Society). Oxford: Clarendon Press.

DUNMIRE, J. R., ed. 1988. *Sunset New Western Garden Book.* Menlo Park, CA: Lane.

CLOUSTON, B., ed. 1977. *Landscape Design with Plants.* London: Heinemann.

CREASY, R. 1982. *Complete Book of Edible Landscaping.* San Francisco: Sierra Club.

FERGUSSON, B. 1982. *All About Trees.* San Francisco: Ortho.

FISHER, M. E., E. SATCHELL, and J. M. WATKINS. 1975. *Gardening with New Zealand Plants, Shrubs and Trees.* Auckland and London: William Collins N.Z.

GRANT, J. A., and C. L. GRANT. 1955. *Trees and Shrubs for Pacific Northwest Gardens.* San Carlos, CA: Brown and Nourse.

HILLIER, H. C. 1973. *Hilliers' Manual of Trees and Shrubs.* London: David and Charles.

HOYT, R. S. 1978. *Ornamental Plants for Subtropical Regions.* Anaheim, CA: Livingston Press.

KELLY, S. 1969. *Eucalypts.* Melbourne: Thomas Nelson.

LORD, E. E. 1979. *Shrubs and Trees for Australian Gardens.* Melbourne: Lothian.

McMINN, H. E. 1939. *An Illustrated Manual of California Shrubs.* San Francisco: J. W. Stacy.

McMINN, H. E., and E. MAINO. 1963. *An Illustrated Manual of Pacific Coast Trees.* Berkeley: University of California Press.

MENNEINGER, E. A. 1964. *Seaside Plants of the World.* New York: Hearthside Press.

METCALF, L. J. 1975. *The Cultivation of New Zealand Trees and Shrubs.* Wellington, N.Z.: A. H. and A. W. Reed.

MULLINS, M. G., ed. 1979. *Reader's Digest Illustrated Guide to Gardening.* Sydney: Reader's Digest Services.

OSBORNE, R. 1975. *Garden Trees.* Menlo Park, CA: Lane.

REHDER, A. 1940. *Manual of Cultivated Trees and Shrubs Hardy in North America,* 2nd ed. New York: Macmillan.

RUSHFORTH, K. 1987. *The Hillier Book of Tree Planting and Management.* London: David and Charles.

WATKINS, J. V. 1961. *Your Guide to Florida Landscape Plants.* Gainesville: University of Florida Press.

WHITCOMB, C. E. 1987. *Establishment and Maintenance of Landscape Plants.* Stillwater, OK: Lacebark Publications.

WIGGINTON, B. E. 1963. *Trees and Shrubs for the Southeast.* Athens: University of Georgia Press.

WYMAN, D. 1965. *Trees for American Gardens.* New York: Macmillan.

———. 1969. *Shrubs and Vines for American Gardens.* New York: Macmillan.

———. 1971. *Wyman's Gardening Encyclopedia.* New York: Macmillan.

CHAPTER 4

Planting Site: Climate _____

Less than 8 percent of the earth's land surface is cultivated. Most of the rest is either too cold, too hot, too dry, too salty, too rocky, or too steep to support cultivation. Climate is probably more important than soil in determining the growth and well-being of plants. In fact, climate is an important factor in the development of soil. To the old cliche "Everybody talks about the weather, but nobody does anything about it" could be added "Very few people understand the weather and its effects on plants." A horticulturist should be one of those few who understand and use it to advantage.

Climatic factors include temperature, radiation, light, precipitation, fog, humidity, wind, and air pressure. Given a particular site, climate is a key factor in deciding what plants are best adapted to it and how they can best be nurtured by spacing, irrigation, timing of fertilizer application, transplanting, pruning, and pest control. In fact, some cultural practices can modify the microclimate to benefit the plants to be grown or the people who will frequent the area. Woody perennials must be able to withstand the many vicissitudes of the weather not only for an entire year but for a succession of years.

An individual plant or group of plants undergoes climatic influences on three different scales. The *macroclimate* extends over a relatively large area defined by fairly uniform climatic conditions. These conditions are largely determined by the air masses moving over the earth's surface and modified by latitude, elevation, large bodies of water, mountain ranges, and the season of the year.

On a much smaller scale, the *mesoclimate* is the local weather of a neighborhood or a city, a large park, one or more farms in a similar setting, or a woods.

The mesoclimate is the macroclimate as modified by local influences of terrain, bodies of water, cloud cover, wind, and land cover.

On an even smaller scale, the *microclimate* includes the conditions around or within an individual plant or planting. It is a further refinement of the mesoclimate by influences in the immediate area. The term microclimate is often misapplied to the mesoclimate, but you can avoid that error if you know the scale with which you are dealing.

Precipitation and temperature are most commonly the limiting climatic factors in plant distribution and performance. In many landscape situations, the hazards of drought or too much water can be minimized by irrigation, drainage, or other cultural practices. Even when favorable sites are chosen, adaptable plant species selected, and the hazards of cold injury reduced by proper cultural practices, these efforts toward providing a favorable microclimate remain at the mercy of the macroclimate. **Temperature remains the least controllable of environmental factors in the landscape.**

TEMPERATURE

The effects of temperature vary with plant species, stage of growth, age, condition, and even with particular plant parts or tissues. Although plants may grow poorly or be killed by high temperatures, low temperatures are much more commonly a problem in all areas but the tropics.

Cold Temperatures

Plants can be injured or killed by low temperatures at almost any season of the year. The most critical periods, however, are (1) spring or autumn, (2) the coldest portion of winter, and (3) occasions when minimum temperatures occur after a warm winter period, even though plants had been at maximum hardiness before the warm period.

The more common types of cold injury include (Weiser 1970b)

Black heart in stems of trees and shrubs
Winter kill of dormant flower buds
Sun scald and frost splitting of tree trunks
Soil-heaving damage of young plants
Winter burn of conifer foliage
Dieback of overwintering broadleaved plants
Spring and fall frost damage of tender shoots, flowers, and fruit

These types of injury are especially prominent in the higher latitudes and altitudes. Several of them, however, limit plant distribution or impair performance in all but the warmest regions. Even in regions with mild winters, dieback and frost injury constitute real dangers.

Some species are more hardy than others or develop sufficient hardiness early in the autumn and retain it long enough to withstand critical temperatures. When healthy plants of moderate vigor produce a moderate fruit crop, they are usually hardier than excessively vigorous or heavily fruiting plants. Not only do species vary considerably in their inherent capacity to resist freezing injury, but the resistance of individual plants changes during the year. A plant that would be killed at temperatures slightly below freezing in summer may be able to survive the coldest of winter temperatures. This change from a susceptible (tender) to a resistant (hardy) condition is referred to as *acclimation* or *hardening*.

A further complication is the difference in hardiness between adjacent tissues or parts of a plant. In stems, for example, the living xylem is hardier than phloem in early winter but is often several degrees less resistant than are adjacent cambium or phloem tissues in midwinter. The xylem of the smaller branches on trees and shrubs may be damaged by low temperatures, creating a condition called *black heart* because of the darkening of the xylem. The problem is not evident until shoots are cut for examination, but it is signalled by poor growth, reduced flowering, and possibly death of some shoots the following spring.

Plant organs also differ in hardiness. Leaves of deciduous plants acclimate little and abscise when subjected to cold. Flower buds are usually less hardy than leaf buds and thus may suffer winter-kill before vegetative buds do. **Roots are less resistant to freezing than are overwintering stems but are usually adequately insulated by the soil.**

On sunny days in winter, a tree trunk may be warmed as much as 10°C (18°F) above air temperature. If a trunk or branch so warmed becomes shaded by a dense cloud or opaque object, the bark temperature may drop quickly to a critical level, causing injury or death to the bark and cambium; this is generally called *sun scald* even though it is actually a freezing injury. It has also been called a *frost canker.*

Wide fluctuations above and below the freezing temperatures of wood may also cause *cup shakes,* or separations of the wood along one or more annual rings. These occur when a frozen trunk warms quickly upon exposure to the sun; the warmed outer wood expands and separates from the inner wood, which has expanded less rapidly. Cup shakes are not evident until the trunk is cut or breaks, but they cause serious defects in lumber and may weaken a living tree.

Longitudinal *frost cracks* usually occur in the bark and wood parallel to the grain and extend to the center of the trunk (Fig. 4-1). Frost cracks (radial shakes), most common in oaks, are often associated with a variety of wounds and stubs of branches and basal sprouts (Butin and Shigo 1981). Himelick (1970) found that frost cracks opened on London plane trees when the air temperature dropped to −13°C (8°F). The colder the temperature, the wider these cracks opened. Once these frost cracks occurred, at least on London plane, they continued to open each winter and close in the spring. It did not seem to matter whether the temperature went down suddenly or gradually, but when it dropped to −13°C, all the old cracks reopened within half an hour of one another. After repeated splitting, considerable callus forms along the edges of the cracks, and wood decay may also begin. Trees 150 to 450 mm (6–18 in.) in diameter are more likely to be affected than trees that are either smaller or larger (Pirone and others 1988).

Figure 4–1 Frost cracks on London plane tree may occur when the temperature drops to $-13°$ C ($8°$ F) or below (left). (Photo courtesy of Eugene Himelick, Illinois Natural History Survey) Radial shake or frost crack is associated with an old wound (right). (Butin and Shigo 1981)

Even in cold-winter areas, frost cracking can be minimized by careful species selection, planting, and maintenance. Shigo (1986a) states that "We can prevent frost cracks by preventing wounds, by making proper pruning cuts, and by preventing root injuries." Kubler (1988) concedes that wounds may be a factor in frost cracks but are not the primary cause.

Evergreens are less subject to cracking than are deciduous trees. The most commonly affected species are London plane, elm, horse chestnut, linden, oak, willow (Himelick 1970), apple, beech, crabapple, goldenrain tree, and walnut (Pirone and others 1988). Trees protected by thick planting or other structures are less subject to cracking than are lone trees.

Soil moisture levels in the fall and winter do not seem to influence the occurrence of frost cracking (Himelick 1970), but shading a trunk or painting it with white latex exterior paint will tend to moderate daily temperature extremes and may reduce frost cracks as well as sun scald and cup shakes.

Himelick (1970) found that 12 mm ($\frac{1}{2}$ in.) rods screwed into 10 mm ($\frac{7}{16}$ in.) holes drilled through the crack every 300 mm (12 in.) stopped most cracks from reopening the following winter. He found it necessary to use an oval washer recessed against the wood on each end of the rod.

Surface soil that has little snow or mulch cover may alternately freeze and thaw in winter, particularly at the beginning and near the end of the usual cold period. Recently planted small trees, shrubs, and other plants can be lifted by freezing soil, returned part way to their original position during a thaw, and then lifted further by the next freeze. This movement can break roots and partially unearth the plant so that it can no longer stand by itself. Damage from *soil heaving* can be avoided by spring planting or by use of a mulch cover to moderate temperature extremes in the soil. Plant parts immediately above the mulch, however, may be injured by greater temperature extremes.

Cold weather that freezes the soil may reduce water absorption to such a low level that it cannot keep up with the transpiration of conifers, whose more exposed foliage may die. Wind can aggravate the situation by increasing heat loss from the soil and water loss from the leaves. Because the dead foliage appears as if it had been burned, such injury is called *winter burn* although it is caused by desiccation. The needles begin to turn brown at the tip and brown further backward depending on severity.

Even though roots are not as hardy as the tops of trees and shrubs, they are less often injured by low temperatures. Dry soil, organic mulch, or snow can insulate roots from winter cold. Trees are more likely to be injured by frost in poorly drained soils than in well-drained soils. The damage to roots will not be apparent until the first warm days of spring, when new shoots and leaves wilt and die.

During cold winters, trees in raised individual planters or planting containers in elevated malls may be subject to temperatures that injure roots directly or reduce water absorption to injurious levels. Roots themselves do not harden more than a few degrees Celsius (Kramer and Kozlowski 1979). Injury is more likely in small and poorly insulated containers. Where winter cold is severe and prolonged, heating cables are sometimes put in raised planters to keep roots from freezing. The soil should be well-drained (see Chapter 11) and should be moist but not wet. The soil surface should be mulched to minimize temperature fluctuations (see Chapter 14). Antitranspirant spray on foliage and bark can reduce desiccation (see Chapter 13), and sensitive plants or those in small planters can be moved to protected storage.

Attempts to reduce freezing injury to marginally hardy plants have generally involved breeding for adapted varieties, modifying the weather, providing mechanical protection, attempting to lengthen the dormant period, and hoping for the best. Although these attempts have been only partially successful, it is tempting to think that we may eventually find ways to manipulate plants so they can resist cold injury.

Lowest Temperature Expectable

In selecting and planting woody plants for a given area, one should know that they can withstand the coldest temperature that might be expected. Local horticulturists or weather bureau records will be able to supply information on expectable temperatures. General information for the United States can be obtained from the Hardiness Map prepared by the U.S. Department of Agriculture (see inside front cover) (USDA 1988). State-by-state statistics are compiled by the National Climatic Center in Asheville, North Carolina, and are available in most large libraries. Information for most foreign countries is available from a number of sources (World Meteorological Organization 1971). Although a region may have relatively consistent temperature regimes overall, meso- and microenvironments can vary considerably. Weather stations may be in the center of a city, where temperatures are often higher than in the surrounding countryside (Peterson 1969). Hardiness maps must be used with discretion, however, particularly near boundaries.

Temperatures several degrees above the winter minimums can be devastating if they occur in late fall before plants have reached maximum hardiness—even dormant deciduous plants. Every few years, certain plants may be killed or seriously injured by early cold weather. A warmer than normal period in late fall or winter will cause a plant to lose some of its ability to withstand low temperatures. Somewhat similarly, plants are more susceptible to injury if the temperature drops quickly rather than slowly for several days. Plant tissues adjust to decreasing temperatures by water moving out of the cells into the intercellular spaces and thus concentrating cell protoplast contents.

Conditions that favor early cessation of growth, such as low levels of nitrogen and water in late summer, favor early development of hardiness. A plant that is vigorous early in the season will usually have a higher food supply at the onset of the cold weather and will be better able to withstand cold. Older plants are usually more hardy than young plants of the same cultivar.

Hardiness cannot always be determined by noting which species are growing in an area. Temperatures may not have reached their coldest possible levels since the plants were introduced, or plants may have regrown after being seriously damaged. Research horticulturists are testing the relative hardiness of the more important landscape plants so that more intelligent selections can be made. This information will be particularly important for large, long-lived trees.

Spring and Fall Cold

Temperatures at or below freezing can cause wilting or flagging, darkening, and the death of new shoots and leaves. Opening buds, blossoms, and young fruit can be killed by spring frost, and late-growing shoots and maturing fruit can be damaged in autumn. Needles of conifers may redden and drop if injured by cold. If the cold occurs when there is little air movement, the portions first and most seriously injured will be the lower parts of plants and parts exposed to the sky. In the spring and fall, even the older tissues are less hardy than they are in midwinter. Intelligent selection and care of plants depends on understanding the factors that contribute to low temperatures.

First, it is well to understand how heat is transferred. *Conduction* is the transfer of heat through contact of one molecule with another, as when heat moves up the handle of a spoon in a hot cup of coffee. *Convection* is the transfer of heat by currents in gases and liquids; this may be caused by differences in density, as when heated air rises from a hot surface, or by an outside force, such as wind. *Radiation* is the transfer of radiant energy, such as heat from the sun or an open fire. Radiant energy passes through the atmosphere with a negligible direct effect on the air temperature. Conduction and convection influence air temperature more directly. Radiant energy heats (or its loss cools) objects, which in turn influence air temperature by conduction and convection. Although heat is usually transferred by all three pathways at the same time, one may predominate.

Local weather conditions (largely determined by the macroclimate) are modified by interaction with the meso- and microclimates. This interaction is particularly important to plant survival and well-being when the macroclimate approaches critical temperatures at either end of the growing season when shoots, flowers, and fruit are sensitive to cold. If a particularly intense cold front moves into an area, local conditions may not provide an adequate buffer to protect plants. This type of cold is called an *advective freeze.* Midcontinental areas are more susceptible to polar air masses than are western coastal areas. Freezes are usually associated with wind, low humidity, and no visible frost deposition. If temperatures drop rapidly, they can quickly cool exposed plants. In such situations, sites exposed to the afternoon sun and protected from the wind will be the warmest. In coastal areas where freezes are uncommon, any freezes that do occur can be particularly devastating: In December 1972, a cold front killed the branches and small trunks of hundreds of thousands of eucalyptus trees in the Oakland and Berkeley hills along the central California coast. The damage created a potential fire hazard of catastrophic proportions, and the removal of dead branches and trees cost millions of dollars.

In the spring, when plants begin to grow, flower, and set fruit, the radiant loss of heat accompanied by little or no air movement can depress near-critical temperatures to injurious levels close to the ground. Depending on the temperature and humidity, moisture condenses on exposed surfaces and may freeze, resulting in a visible *frost.* Since the exposed surfaces radiate more energy than they receive, they become colder than the air, and frost forms on them before air temperature reaches the freezing point of water. Plant parts exposed to the sky can, through radiation loss, become as much as 10°C (18°F) colder than the surrounding air. On the other hand, **if the humidity is low, freezing temperatures due to radiation cooling can occur without a visible frost;** this is called a *black frost.*

Some plants are susceptible to temperatures just above freezing; all but the youngest shoots of most temperate and subtropical plants can withstand temperatures of −1° to −3°C (26°–30°F). The amount of injury a plant sustains will depend on the extremity, duration, and speed of a temperature drop. The longer the exposure to low temperatures, the greater the damage and, usually, the more slowly the temperature drops, the lower the temperature a plant can withstand before injury occurs. A calm liquid, including cellular contents, can be *undercooled* (supercooled) 1° to 4°C (2°–8°F) below the temperature at which freezing would be expected. If the temperature rises above the freezing point before freezing occurs, the tissue will not be injured. If undercooled tissue is jarred, however, or an ice crystal is introduced, the tissue freezes immediately.

Radiation Frost Conditions. A number of factors determine whether a radiation frost will occur. The temperature at the start of the evening must be in the critical range (near 5°C or 40°F) so that the air can be cooled to or below the freezing point. **Almost without exception, the sky must be clear and the air calm during the night.** Under clear skies, surfaces that were warmed by the sun during the day radiate more energy to the cold surroundings at night than they receive. Air in contact with the cold surfaces is cooled primarily by conduction from air molecule to air molecule. Since air is a poor conductor of heat, the process is slow and air near the ground can become several degrees colder than the air a short distance above.

Clouds at night, however, absorb and reradiate heat back to the earth and thus reduce net heat loss (Fig. 4-2).

Above level ground during the night, the air will be cooled to ever higher elevations and, as it cools, become heavier and tend to settle. The air will be coldest near the ground and progressively warmer with height until, at a certain elevation, the air has not been cooled by the air below. The air cools in place and does not involve warm air rising from near the ground. Air temperatures can be affected to heights of 60 to 180 m (200-600 ft) (Chang 1968). This increase in temperature with increasing elevation contrasts to the normal daytime situation, in which the air temperature decreases 0.6°C with every 100 m (1°F/300 ft) of altitude, and is therefore called a *temperature inversion* (Fig. 4-2, left).

Under radiation conditions, the air will be coldest and the height to which it is cooled the greatest just before sunrise, since radiation loss and conductive cooling continue all night. A temperature inversion is considered "large" or "strong" if there is a marked increase in temperature with height, if the air, for example, is 6°C (10°F) warmer at 12 to 15 m (40-50 ft) than at ground level. If the change is less than 3°C (5°F), the inversion is considered "weak" (Gardner, Bradford, and Hooker 1939). A strong inversion is most likely to occur when the air is warm, still, clear, and humid and surfaces are cold with little heat reserve at the beginning of the night. Wind machines to mix the warmer air above with the cold air near the

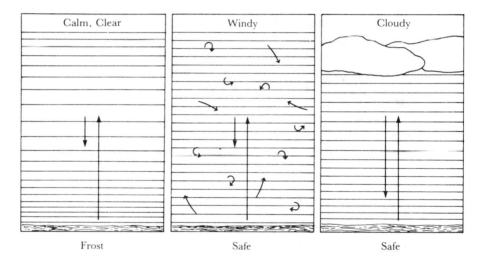

Figure 4-2 Radiation conditions before sunrise can result in frost if temperatures are in the critical range and conditions are clear and calm. Conductive cooling from the soil or other surfaces may create a temperature inversion, with colder air (horizontal lines close together) near the ground and the warmer air (lines farther apart) higher (left). A breeze of 10 km/hr (6 mph) or more mixes warmer air above with the colder air below to produce an average temperature that is warmer than if there were no wind (center). Clouds reradiate most of the heat received and thereby slow the cooling process (right). The straight vertical arrows represent the radiation received and lost by the soil and other surfaces.

ground and heaters are more effective under strong than weak inversion conditions. Calm conditions allow the coldest air to remain near the surface and reach the critical temperature more quickly (Fig. 4–2, left).

Just as clouds or wind at night reduce the likelihood of a radiation frost, cloud cover or wind during the day reduces radiation input that would warm the air, plants, the soil, and other heat reservoirs. After such a day, the night begins at a lower temperature, with less heat available to slow cooling should the skies clear and the wind die during the night. A frost is more likely the next morning than if it had been clear and calm during the day.

Other factors also influence air temperature under radiation cooling conditions. Dry air favors more rapid temperature changes than does moist air, since water vapor is the largest absorber of long-wave radiation. Even with no visible fog or clouds, high moisture content in the air will somewhat reduce radiation loss from the earth's surface. More important, as the moisture is cooled, it must change from a vapor to a liquid and then to a solid. Relatively large amounts of heat are liberated when vapor turns to liquid, and even more when water freezes. The amount of moisture in the air is commonly expressed in terms of a *dew point,* **the temperature at which the air is saturated with water vapor** (100 percent relative humidity). Vapor pressure, or humidity, can be calculated from the dew point and the temperature. If the air temperature drops below the dew point, moisture in the air condenses as water droplets (fog or rain). The higher the dew point, the more moisture there is in the air and the more slowly the temperature will drop, other things being equal.

Dew will condense on many surfaces because they cool to the dew point sooner than does the air. If the dew point is above freezing, the moisture that has condensed on surfaces will change to ice crystals as these surfaces reach the freezing point and will form typical frost. The higher the dew point, the more moisture will condense and the thicker the frost will be at a given temperature.

The heat capacity and conductivity of the material exposed to radiation cooling greatly influence surface temperature and, in turn, the degree to which the air is cooled. Soil is commonly exposed to sky, and its surface condition can have an important influence on air temperatures. A soil that is firm and moist will absorb more energy from the sun during the day and reradiate more during the night than will a soil that is drier or less firm. While there is cooling due to evaporation from moist soil surfaces (similar to the action of an evaporative cooler), it is more than offset by the heat capacity and conductivity of moist soil. Compact soil is a more efficient conductor of heat than loose soil. Therefore, the surface of a moist, firm soil will cool less rapidly than will one that is drier and looser: As heat is radiated from the surface of a moist soil, it is replaced with heat that has been absorbed in lower depths during the day.

A soil cover such as an organic mulch, a wooden deck, turf, weeds, or ground cover has poor heat conductivity and low heat capacity. A mulch surface absorbs energy but because of poor conduction reemits much of it and stores less heat during the day than does bare soil.

Weeds are often plowed under in the late winter so that the soil will be exposed and can firm during later rains. Sparse grassy weeds of 25 mm (1 in.) or less in height, however, may minimize air movement and yet allow the soil to absorb and radiate heat. This can more effectively reduce the likelihood of a frost than will a more complete weed cover or recent cultivation, both of which have an insulating influence on the soil. On the other hand, lower temperatures will occur over freshly cultivated soil and may kill new shoots while warmer temperatures over uncultivated soil result in no injury. This combination of circumstances has been recorded in a number of instances.

Asphalt and concrete have higher heat capacities than do organic mulches or turf and will more readily radiate accumulated heat and conduct heat from the soil. Frost damage to plants is therefore less likely over soil with a concrete or asphalt covering than over soil covered with turf or organic mulch or over a concrete or asphalt structure without soil in contact with it, such as a bridge.

Frost Pocket. Cold air is denser than warm air; it flows down a slope and accumulates in the lowest areas, displacing the warmer air upward. When some of the warmer air replaces the cold air that has moved down (Fig. 4–3), it creates a warm zone up the slope. In many areas where spring frosts and winter killing are problems, susceptible fruits and vegetables are grown on hillsides and tops, even though the soil there may be less desirable than in the valleys.

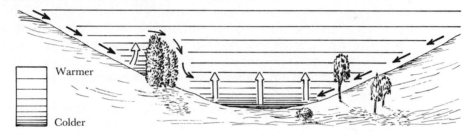

Figure 4–3 A frost pocket can result when cold (heavier) air flows down slopes and forces warmer air to rise. An obstruction on the slope can form a localized cold pocket.

A row of closely planted trees or shrubs, a solid fence, or a long building at right angles to the slope can act as a barrier to the flow of cool air and can create localized cold pockets (Fig. 4–3). If these will constitute hazards at a particular site, try not to design them into the landscape. If they are already present, plants should be chosen with this phenomenon in mind.

Protecting against Frosts

A number of measures can reduce the likelihood of frost injury to plants. Under radiation frost conditions, only 1° to 2°C (2°–4°F) can make a substantial difference to the health and survival of plants.

Select a Site to Avoid the Cold. The modifying influences of large bodies of water are well-known and can be used to advantage in reducing cold hazard. The closer the body of water, the greater will be the protection it affords. It is best to plant on the lee side, so that prevailing winds come from over the water (Fig. 4–4).

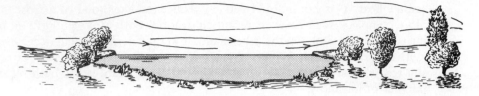

Figure 4–4 A large body of water moderates air temperatures, particularly on the leeward (downwind) side of the water.

In winter, air that is colder than the water or ice will be warmed as it passes over. In spring, warm air passing over the water is cooled, thereby delaying the time when the plants will lose their hardiness and begin to grow. Should a late cold spell develop, the water will again warm the passing air and reduce the likelihood of critical temperatures. Seldom do horticulturists have the opportunity to choose such major landscape features, but these conditions should be kept in mind. They also explain why most of the important northern fruit-growing areas of the United States are adjacent to the Great Lakes.

As has been discussed, planting sites can be chosen to avoid frost pockets. Within a given landscape, it may even be possible to select specific sites that provide more protection and warmth than the area usually affords. For example, a concrete or rock wall that faces the afternoon sun will reduce radiation loss at night and will reradiate heat stored during the day. A plant can be *espaliered* (trained to grow in one plane, see Chapter 15) against a wall to take advantage of the stored heat and reduced radiation.

Injury from low temperatures can be further minimized if one selects varieties or species that are hardy in winter cold, that begin growth or bloom late in spring, and that cease growth and mature their fruit before fall cold. When fruit matures during the time that frost may be a danger, the top and outside fruit, which is exposed to the sky and is usually riper, should be picked first. The inside fruit, protected by leaves, can be harvested later with greater safety. Plants sensitive to fall cold should be watered and fertilized sparingly so that growth will end early, enhancing the maturation of shoots and fruit. When the cold period begins, however, the soil should be moist to ensure adequate water to the plant and to improve heat absorption, transmission, and release.

Reduce Heat Loss. Plants can be protected if they are partially or completely covered to reduce the loss of long-wave radiation (heat). The larger the soil area that can be encompassed, the larger will be the heat reservoir for the plant. Canvas, plastic, wood, or evergreen boughs can provide effective covers (Fig. 4–5), as can the sunny side of a roof overhang, a wall, a solid fence, or a large evergreen tree. In extreme situations, low-growing plants and the bases of others can be protected by covering them with mulch or soil.

By reflecting and reradiating heat back toward the earth, clouds almost invariably prevent a frost. Some people think that a smoke or haze cover will work the same way. It is true that smoke can reduce incoming (short-wave) radiation from the sun, but it is fairly ineffective in preventing (long-wave) reradiation. Ten meters

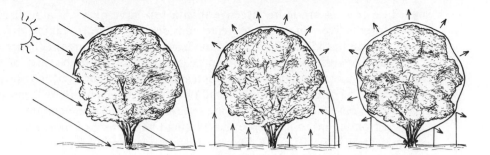

Figure 4-5 Covering a plant for frost protection. Extending the polar side of the cover to the ground and leaving the equator side partially open to the sun allows for the warming of soil and lower tree during the day (left) and conservation of heat at night (center). If the cover is tied to the trunk, night temperatures within the cover will be colder than without the cover (right).

(30 ft) of smoke will reduce transmission of infrared radiation less than 20 percent, but the same depth of fog will reduce transmission more than 90 percent (Mee and Bartholic 1979). Orchard heaters (smudge pots) are effective primarily because they radiate heat to the area not because their smoke significantly reduces radiation loss. In fact, to increase efficiency and reduce pollution, present-day heaters must emit only minimal amounts of smoke.

Utilize Heat from the Immediate Surroundings. A cover can reduce radiation loss from plants as well as from soil and other surfaces (Fig. 4-5). In order to build up as much heat as possible during the day, plants, soil, and their surroundings must be exposed to the sun. Covers are most effective if they can easily be removed from plants or at least from the soil. If a cover must be left on during the day, an opening near the top will minimize warming of the plant and consequent loss of cold hardiness. A plant cover that is gathered at the trunk often results in more cold damage than no cover at all because it excludes heat from the soil (Fig. 4-5, right).

As pointed out earlier, a firm, bare, moist soil absorbs more heat and conducts it more rapidly than soil that is loose, dry, or covered with mulch or vegetation. Walls and benches exposed to the sun during the day are good sources of heat at night.

Air near the ground can be warmed under radiation-frost conditions by mixing the warmer air from above with the cold surface air. The mixing is done naturally by a breeze—5 km (3 mi) per hour or more—or mechanically by wind machines. In some commercial fruit and vegetable areas, helicopters are used to mix the air. The effectiveness of wind machines depends upon the strength of the inversion. The stronger an inversion, the greater the protection, if the mixed air is above the freezing temperature of the plants.

Add Heat. Shrubs and small trees close to an electric source can be warmed by a 100-watt bulb hung inside the plant canopy. In some landscape situations, special orchard heaters may reduce or prevent frost injury. These heaters heat primarily by radiation from their hot stacks, not by warming the air, and a number of small heat sources are more effective than a few large ones. A large fire may be counter-productive, since it can create considerable updraft, taking the heat up above the plants and drawing in cold air from surrounding areas (Kepner 1950).

Flood or sprinkler irrigation can help to protect plants because of the heat capacity of water. **The real warming potential of water, however, is realized when ice is formed:** For example, only 1 calorie of heat is given off by 1 gm of water cooled 1° C, but 80 calories of heat are released when the same amount of water is changed from liquid to ice. Since the freezing temperature of most plant tissues is several degrees below 0° C (32° F), plants can be encased in ice without being frozen (Schultz and Lider 1968). If cold persists for several days, additional water must be added, which may cause a problem in all but the best-drained soils. For plants that can withstand the weight of ice, sprinkler irrigation may provide greater protection because of the more rapid cooling and freezing of the water so distributed. Sprinkling must be continued, however, until the air temperature is above the critical range.

These frost-protecting techniques are more effective when there is a strong inversion, that is, when temperature increases markedly with elevation above the ground. Under this condition, also referred to as a low ceiling, both heating and wind machines are most effective. In a weak inversion (high ceiling), temperature increases little with elevation and heated air will rise higher before reaching equilibrium, so more heat is dissipated above the plants (Fig. 4–2).

Inhibit Ice Formation. When certain species of bacteria (*Pseudomonas syringae, P. fluoroscens,* and *Erwinia herbicola*) commonly on plant surfaces are reduced in number or inactivated, the plant can supercool, that is, cell contents remain in a liquid state at temperatures as cold as −9° C (15° F), and avoids injury by temperatures that would normally freeze the tissue (Lindow 1980). The bacteria, up to 100 million per gram of leaf, are nuclei for ice crystal formation. Ice nucleation-active bacteria (called INA bacteria) can be greatly reduced by applying copper-containing compounds, Streptomycin, or certain other bactericides. Significant frost control has been achieved in experimental applications of bactericides on several vegetable and fruit crops. Antagonistic bacteria are another way to reduce the INA bacteria population but their action may not be rapid enough to provide protection. A third possible control is to inactivate the bacteria with chemical inhibitors that cause various physical and chemical stresses. These chemicals almost immediately inactivate the nucleus produced by the INA bacteria without necessarily killing the bacteria. Although inhibiting ice nucleation is still experimental, it holds considerable promise in protecting plants from frost. Unfortunately, registration of its use is slow in coming.

Winter Chilling and Rest

Vegetative and flower buds may not grow for a number of reasons, depending on the season of the year. While a shoot is actively growing, lateral buds will be inhibited by the strong apical dominance of the terminal. As the days begin to shorten, the buds of many temperate-zone plants begin to enter a condition of *rest* and will not grow even though all other conditions are favorable. Even after rest is no longer a factor, buds may still not begin growth if temperatures are below 10° C (50° F). To distinguish between these last two dormant conditions, we use the terms *rest* and *quiescence*. Dormant-quiescent plants will begin to grow under favorable environmental conditions, whereas dormant-resting plants will not.

A condition of rest develops within each bud or seed of certain temperate-zone plants. In most woody plants, dormant buds not only inhibit shoot elongation but slow the activity of other growth centers, roots, and cambium. Rest develops gradually during the summer as the days become shorter, intensifies in the fall, and reaches a maximum in early winter. By late winter, rest is overcome in most temperate regions.

To overcome rest, buds must be subjected for four to eight weeks to low temperatures, −4° to 10° C (24°–50° F). As might be expected, species and subspecies differ not only in the temperature but also in the length of time the temperature is necessary to overcome rest. Extremely low temperatures are little or no more effective than the threshold temperature in overcoming rest. The amount of necessary chilling is expressed as the number of hours at or below the critical temperature for a given species. The air-temperature threshold for most fruit tree species is 7° C (45° F). It is the bud temperature, however, that is important; a bud temperature of 10° C (50° F) would probably be effective in overcoming rest. Within fruit species, apples tend to require a longer chilling than do peaches, and peaches a longer chilling than almonds. Even so, peach varieties may require anywhere from 350 to 1200 hours at or below 7° C (45° F). Varieties with shorter chilling requirements have been bred to grow satisfactorily in the southern United States and other mild-winter areas.

After a certain amount of chilling, the overcoming of rest can be hastened by pruning close to a bud, massaging individual buds, pricking buds with a pin, or treating the plant with ethylene. These treatments, however, cause injury to the plant. The only practical treatment for overcoming rest is used by some pear growers, who add a dinitrophenol compound to their dormant oil sprays.

In years when the chilling period is brief, fog, cloudy weather, and wind are important in keeping buds at or near air temperature. On sunny days, buds may be warmed several degrees above air temperature. The central valley of California would not be the important fruit-growing area it is if it were not for the winter fog.

Following winters with inadequate chilling, trees are slow to leaf out and bloom is prolonged. Leaf buds are affected more seriously than flower buds. In extreme cases, so few buds grow that branches are sunburned and the tree weakened from lack of food production. Since flower buds are less affected, a weak tree may bloom and set fruit, which further weakens the tree. Trees so affected are said to be suffering from *delayed foliation*.

The setting of fruit may be enhanced by a longer bloom period. Prolonged bloom, however, increases the danger of disease, particularly on plants like the pear, pyracantha, and hawthorn. If a certain amount of chilling does not occur in the fall, the flower buds of certain species, such as apple, apricot, and pear, may die before opening the following spring. The chilling requirements of deciduous fruit trees have been extensively studied (Chandler and Brown 1951). A number of fruit-growing areas occasionally experience winters that are not cold enough for the profitable production of certain species.

The dormant-resting condition is the key to winter survival for many plants in the temperate region. Bud activity is prevented if periods of warm weather begin to induce activity that will be susceptible to later cold weather. It has been thought that, since the rest period prevents buds from starting to grow during a warm period, rest would also prevent the loss of hardiness during unseasonably warm weather in midwinter. Unfortunately, that may not be the case. Raspberry canes in a state of rest have been found to lose their hardiness under high temperatures more rapidly than equally hardy canes that were quiescent (Weiser 1970a). Whether this applies to other plants is not known.

Root dormancy or rest presents a confusing picture with regard to different species of woody plants. The roots of some pines in Northern California have been observed to grow throughout the year. The roots of other species seem to grow whenever food reserves and soil conditions are favorable. The roots of still other species exhibit considerable periodicity even during the growing season. Japanese yew roots were observed to grow four times as much after nine weeks of root chilling as roots not exposed to such cold (Lathrop and Mecklenburg 1971).

Root activity is difficult to study since it depends on the health and vigor of the top as well as the temperature, aeration, moisture, compaction, and toxicity of soil. Evidence to date suggests that the roots of most woody plants do not have a dormant-resting condition independent of the tops. A number of species must have developing buds before roots will begin growth and must have active leaves if root elongation is to continue (Richardson 1958). In addition to these hormonal controls, root growth is dependent on carbohydrates, either stored or manufactured in the top.

Other conditions being favorable, the roots of most plants will grow at lower temperatures than will the shoots. The roots of silver maple will start growth at 4° C (40° F), but leaf buds do not expand until the temperature is at least 10° C (50° F) for 20 days. This can be an important factor in determining the time to transplant trees and shrubs.

Plant Climates

It is not one particular temperature at one particular time that makes an ideal climate for a species. Many plants have different optimal temperature requirements at different stages of growth. Almost all plants grow, flower, and fruit more abundantly when day temperatures are higher than night temperatures, yet night temperature is more commonly the controlling influence during the growing season, if water is not a limiting factor (Kimball and Brooks 1959). This is particularly evident during bloom, when fruit set may be adversely affected by either low or high night temperatures.

Citrus and some deciduous fruits develop richer color and better flavor in areas where night temperatures are considerably cooler than day temperatures. The intensity of leaf color on many deciduous trees is similarly affected. Some perennials require periods of different night temperatures for different purposes. For example, camellia flower bud formation requires a period of 15° to 18°C (60–65°F) at night followed by 10°C (50°F) for best bloom. Peach tree growth and fruit bud formation occur during the summer at night temperatures of 15°C (60°F) and above, but rest is broken in the winter by periods of 7°C (45°F) or lower. In this latter case, only the night and early morning periods may be cold enough to break the rest.

Experiments under controlled conditions and observations in the field support the following statements: **Plants will make satisfactory vegetative growth over a wider range of temperatures than is satisfactory for flower bud formation and fruit setting** (Fig. 4–6); the temperature range for fruit set is usually wider than that for producing viable seed. This would imply that many exotic plants can be used for landscape purposes where vegetative growth is the prime consideration in areas where they may not be able to reproduce themselves.

For example, a number of so-called "fruitless" olive trees have been discovered over the years and have been propagated vegetatively. The lack of fruit actually

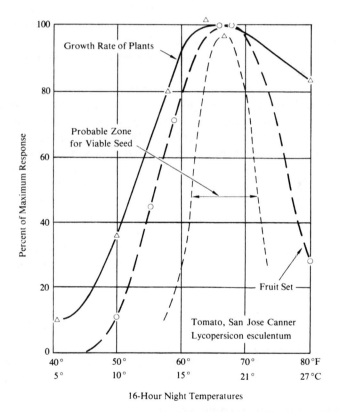

Figure 4–6 Response to tomato growing under controlled conditions demonstrates that for viable seed to be produced, the night temperature range is more limited than for satisfactory plant growth. (Went 1944, as adapted by Kimball and Brooks 1959) Most other plants probably respond similarly to temperature and other environmental constraints on growth and flowering.

enhanced their value for landscape purposes. Until recently, when these trees were planted in areas where olives were grown commercially, the trees set fruit, although less abundantly than commercial varieties. In localities where the original fruitless trees were found, conditions during flower bud formation or bloom were such that no fruit formed, even though the tree was able to make satisfactory vegetative growth. In 1961, Hartmann (1967) introduced the *Swan Hill* olive tree. This is apparently a true fruitless variety, though it does flower profusely, much to the discomfort of hayfever sufferers.

Increasing knowledge about diurnal and seasonal temperature influences on plants should be an invaluable tool in determining favorable environments for landscape plants. Even now, much worthwhile information is summarized on a county or regional basis and is available from local weather bureau offices or agricultural agencies.

High Temperatures

All plants will be killed or seriously injured above a certain maximum temperature, which varies with species, stage of development, and previous environmental history. What may be the maximum temperature for one plant may be the optimum for another.

High-temperature injury is primarily a matter of desiccation, when transpiration exceeds moisture absorption. Water content in the leaves is lowered, leaves wilt, transpiration is reduced, leaf temperatures are increased by reduced evaporative cooling, and the cycle can intensify to kill a portion or all of the leaf. Under cloudless skies, the leaves of well-watered plants in low vapor pressure (low humidity) and high net radiation (high temperatures) will be cooler than the air, while leaves in low net radiation and higher vapor pressure will be warmer than air. Leaf temperature will usually reach a temperature equilibrium below the lethal level even if the leaf comes under water stress.

Small, thick leaves with few stomates are usually more tolerant of high temperatures than are large, thin leaves. Within the same species or clone, a plant grown under cool conditions with low light intensity (irradiance) will develop larger, thinner leaves that can be quite susceptible to heat. A cool, cloudy spring may cause newly planted plants to develop large, thin leaves. Should the weather change abruptly to hot, dry, and slightly breezy conditions, transpiration may exceed the ability of a plant to absorb and transport water from the soil to the leaves. Wilting and possible injury to the exposed leaves may result. Continued high temperatures, particularly at night, can increase the rate of respiration to nearly that of photosynthesis. High night temperatures have been found to slow the translocation of carbohydrates from the leaves of many plants. If such conditions persist for several weeks, a plant may do quite poorly.

Occasionally, tree limbs up to 0.6 m (24 in.) and trunks up to 1.2 m (48 in.) in diameter break and fall during hot, calm summer and fall afternoons and subsequent evenings (Kellogg 1882, Harris 1972, Rushforth 1979). This phenomenon is known as summer branch drop (see Chapter 17).

Roots are more sensitive to temperature extremes than are the tops of plants. Wong, Harris, and Fissell (1971) found that young roots in a number of plants were killed by a single 4-hour exposure to temperatures between 40° and 45°C (104°–113°F). Temperatures in the surface 100 to 150 mm (4–6 in.) of soil exposed to the sun can become lethal to roots. The soil 25 mm (1 in.) in from the side of a 4-liter (1-gal) container exposed to the afternoon sun will reach 43° to 46°C (110°–115°F). This can occur any sunny day in spring, summer, or fall. Plants so exposed may have no roots in the west third of the container (Harris 1967) (Fig. 4-7).

Figure 4-7 Roots of European white birch were killed or prevented from growing on the side of the root ball whose container was in the outside row of a nursery block exposed to the afternoon sun.

LIGHT (RADIATION)

Plants are affected by quality (wavelength or color), intensity (irradiance), and duration of light, whether the light comes from the sun or from the artificial sources that are commonly used for interior landscaping (see Chapter 11) and night lighting.

Visible light, that portion of the electromagnetic spectrum to which the human eye is sensitive, extends from roughly 380 to 760 nanometers (nm); the wavelengths that stimulate visual sensation most effectively are between about 500 and 600 nm (green to orange). The amount of illumination incident on a surface (illuminance) is expressed as lux (lumens per m²) or foot-candles (ft-c), its English counterpart; 10.8 lux equals 1 ft-c. Plants, by contrast, respond to radiant energy instead of luminous energy. Effective wavelengths vary for the different photobiological processes in plants.

For photosynthesis—blue (400–450 nm) and red (625–700 nm)
For photoperiodic responses—red (625–700 nm) and far red (700–850 nm)
For phototropic movements—blue (400–450 nm)

Illuminance and irradiance measure the amount of energy incident on a given surface and not the intensity of the energy source. Irradiance that is photosynthetically active, called PAR (radiation between 400 and 700 nm), is most accurately measured and expressed in microeinsteins per m^2 per second ($\mu E/m^2/sec$). (Irradiance, when expressed as erg/sec/cm^2, watt/m^2, and calorie/min/cm^2, typically includes more of the electromagnetic spectrum than PAR.) Irradiance, however, is commonly expressed in lux or ft-c, which are really measures of illuminance. Measurements of these two energies could be converted from one to the other easily and accurately if we were concerned with only one or two very similar energy sources, such as sunlight or incandescent lamplight. Depending on the light source, however, the relative amounts of illuminance and irradiance can differ. Different lights have different emission spectra: One lamp gives off more light of certain wavelengths (color) than another lamp does; fluorescent light is high in blue and some red, and incandescent light is high in red and far red. Also, visual sensation is at a maximum at 550 nm of illuminance energy, whereas PAR per unit of radiant energy varies inversely with wavelength. For example, for a single wave cycle (1 photon), irradiance at 400 nm (blue) has 75 percent greater radiant energy than at 700 nm (red). Therefore, light meters are not particularly suitable for comparing irradiance from different light sources. Radiometers can measure irradiance, but they are expensive.

Even though illuminance does not accurately characterize irradiance, it has been and will continue to be used for this purpose. Therefore, measurements of illuminance or irradiance and recommendations for minimum light intensity (given in lux or ft-c) for growing plants under artificial light should include full descriptions of lamps and their irradiance characteristics.

Photosynthesis

Irradiance (intensity) during the sunny days of the growing season may be five to 10 times that needed for maximum photosynthesis of an individual leaf (Table 4-1). Within the canopy of the outer foliage of a plant, however, irradiance falls off rather quickly. Depending on branching habit, leaf density, and leaf size, irradiance levels 250 to 500 mm (10–20 in.) within the canopy surface may be so low that leaves are not able to persist; this is particularly true with sheared hedges. Many plants have a dense shell of foliage and a center devoid of leaves. Such plants often present just dense masses of green with little character or interest. It may be difficult to grow plants under trees that have a dense canopy of foliage. Training and pruning practices can either intensify or help to overcome this situation. Irradiances below the optimum will cause many plants to produce branches with long, spindly internodes, few laterals, and large, thin, pale-colored leaves.

TABLE 4-1 A. ILLUMINANCE (IRRADIANCE) FOR VARIOUS TYPES OF PLANTS, INDIVIDUAL LEAVES, AND PLANT RESPONSES

	Irradiance	Illuminance	
	Microeinsteins[a] per meter^2 per second (μE/m^2/sec)	Lux	Foot-candles (ft-c)
Adequate for most plants	1000	54,000	5000
Individual leaf	240	13,000	1200
Shade plants	200	10,800	1000
Indoor plants	10–100	540–5400	50–500
Photomorphic induction	0.06–3	3.2–160	0.3–15
Indoors (sufficient for reading)	–	200–300	20–30

B. ILLUMINANCE (IRRADIANCE) ESTIMATES FOR VARIOUS SITUATIONS

		Irradiance	Illuminance	
		μE/m^2/sec	Lux	ft-c
Full sun[b]	Summer	2400	130,000	12,000
	Winter	800	43,000	4000
Cloudy[c]	Summer	1000	54,000	5000
	Winter	200	10,800	1000
Shady side of two-story	Summer	400	21,600	2000
building (sunny)	Winter	100	5400	500
Moonlight		–	0.22	0.02
Incandescent light[d] (100 watt at 1.5 m or 5 ft)		5	240	22
Fluorescent-Deluxe cool white (150 watt at 1.5 m or 5 ft)		17	1050	100

[a] For sunlight 1 μE/m^2/sec = 54 lux = 5 ft-c
1 lux = 0.093 ft-c
1 ft-c = 10.8 lux

[b] Sunlight for noon on a typical summer and winter day in Davis, California (38.5° N).

[c] Due to cloud density, irradiance could vary 50 to 75 percent.

[d] Bulb with 250-mm (10-in.) diameter reflector.

Photoperiod

Although an individual leaf may require at least 20,000 lux (1850 ft-c) for maximum photosynthesis, photoperiod responses may be induced with less than 160 lux (15 ft-c) (Table 4–1). In plants sensitive to photoperiod, day length usually controls vegetative and reproductive activity. In some plants, photoperiod can also influence leaf shape and abscission, pigment formation, pubescence, root development, and onset of dormancy. As mentioned previously, **the irradiance required for photoperiod response is only a fraction of that required for photosynthesis.** In almost all

Light (Radiation)

sensitive woody species, long days and short nights promote vegetative growth. The shoot elongation of many woody plants native to temperate regions will stop when days are short. Shoot elongation of plants in the higher latitudes may be inhibited when the photoperiod is less than 18 hours, while others are not affected until there are less than 10 hours of light per day. As mentioned in Chapter 2, when plants are grown at lower latitudes than those to which they are native, they may experience a shortened growing season and will therefore be smaller than normal. In general, day length can be an important factor influencing plant size and form.

Experiments have shown that the length of the dark period is more important in inducing photoperiod responses than the length of the light period. When plants that are vegetative during long days are grown during short days, many can be kept growing if they are exposed to half an hour of light in the middle of the night. Nursery workers in warm areas have found that they can encourage vigorous growth in many woody plants by extending day length with incandescent lights. **If plants are to respond to increased day length, however, all growing conditions must be favorable, including temperatures above 10°C (50°F).** For many plants, growth will be most vigorous when the temperatures prevailing during the extra light period are 21° to 27°C (70°–80°F).

Night Lighting

Many people have wondered about the influence of street lights and other night lighting on the growth of nearby plants. As should be apparent from the earlier discussion, such low-intensity lighting will have little or no influence on photosynthesis but might affect plants that are sensitive to day length. Light quality, wavelength, is an important factor in plant growth. (A nanometer is a measure of the wavelength of light, as are angstroms and microns: 1 nanometer = 10 angstroms = 0.001 micron = 0.0000001 millimeter.) Phytochrome, a photo-reversible blue pigment, by means of its red (625–700 nm) and far red (700–850 nm) absorbing forms, is involved in the regulation of plant responses. A delay in dormancy induced by long days may result in continued elongation of vegetative shoots. Far red light (700–850 nm), which promotes shoot elongation, may delay dormancy if present in supplementary light. Red light (625–700 nm) promotes flowering.

Cathey and Campbell (1975a, 1975b) found that five light sources used for 16 hours at a level of 11 lux (1 ft-c) at a night temperature of 20°C (68°F) promoted vegetative growth (delayed dormancy) of woody plants. These researchers used two species each of elm and maple and one each of *Koelreuteria, Rhododendron, Rhus,* and Japanese zelkova. From most to least effective, the light sources used were: incandescent (INC), > high pressure sodium (HPS), >> metal halide, = cool white fluorescent, and >> clear mercury. The degree of delay in dormancy was dependent on the quality of light emitted by the light source. Incandescent sources, emitting the greatest amount of far red light, promoted elongation the most. At intensities higher than 11 lux (1 ft-c) or with different species, the relative effectiveness of light sources can change.

Poinsettia, Betula, Catalpa, Platanus, and *Tilia* were the most sensitive to light and continued to grow vegetatively in response to all sources. Holly and two species of pine did not respond to any of the light sources at the 11-lux level, but they did respond if the intensity of the light was increased to 200 lux; these species are said to have low sensitivity to extended photoperiod. HPS lighting had to be increased at least 4- to 8-fold (on a lux basis) to be as effective as incandescent lamps in regulating vegetative growth. **In most landscape and street situations, HPS, incandescent, and (to a lesser extent) metal halide lamps may have red and far red intensities high enough to prolong the vegetative growth of sensitive plants.**

Plants sensitive to day length may thus have their growth prolonged in the fall of the year and their first stage of acclimation for the cold winter ahead delayed. Such plants, particularly their young growth, will be more sensitive than usual to early freezing temperatures. In Maryland, European white birch whose tops are exposed to incandescent floodlights have been observed to be killed back in the top almost every year. Young plants, because of greater vigor, will be more subject than older plants to cold injury as a result of prolonged growth from illumination.

Continuous lighting depresses the formation and maintenance of chlorophyll in leaves and promotes lengthening of shoot internodes and leaf expansion. The foliage of plants grown in continuous lighting may be more susceptible to air pollution and water stress during the growing season. Cathey and Campbell (1975a, 1975b) suggest using HPS where high visibility is desired and only light-tolerant plants in the vicinity. Although less efficient than HPS lamps, metal halide lamps may be preferred in malls, parks, and residential areas with fairly dense plantings when color rendering of plants, people, and buildings is desired.

In contrast, Andresen (1977) surveyed 200 trees along streets in Chicago, Cleveland, and Milwaukee and their response to high-pressure sodium vapor (HPS) outdoor lighting. He also contacted officials in 16 other cities where HPS lamps have illuminated trees for several years and states, **"All indicators, to date, suggest that HPS has no detrimental influence on trees grown in cities of the eastern half of the United States in general and our mid-west in particular."** Andresen concludes that it is safe to install HPS lamps up to 1000 watts in the presence of transplanted or mature trees. In view of the temperature requirement (at least 10° C) for photoperiod response, it would be interesting to examine the effects of HPS lighting at lower latitudes.

If necessary, outdoor lights can be shielded to direct the light away from plants. To avoid cold injury resulting from prolonged growth, light-sensitive container plants that are still growing in late summer should not be planted until growth stops.

Transpiration

Day length can also influence the amount of water transpired by plants, particularly evergreens. At Davis, California, the average evapotranspiration for the month of June is 212 mm (8.3 in.), which is almost 30 percent more than the August rate (164 mm). Even though May in the northern hemisphere can be relatively cool, day length may cause evapotranspiration to be almost as high then as it is in August. **Where needed, irrigation should be more frequent than is generally thought for turf and plants in leaf in late spring.**

MOISTURE

Rain

Rainfall amount and seasonal distribution are important even in irrigated landscapes. The more the weather is dominated by continental air masses, the more likely precipitation will be at any season of the year. In contrast, the Mediterranean-type climates of many western continental areas in the temperate zone have rainfall limited almost exclusively to late fall, winter, and early spring. Under such conditions the moisture reservoirs of the soil become quite important to plants during the growing season.

Weather records can be central to determining the monthly rainfall and *evapotranspiration* (moisture lost by evaporation and transpiration) for an area. By estimating the amount of moisture in the soil, seasonal rainfall, the average date of last rainfall, and evapotranspiration, you can estimate the minimum irrigation needed to grow plants in a given area. Provisions for irrigation of valuable plantings are becoming more common in areas that normally receive summer rain, since periods of drought may have devastating effects in such areas.

Fog

Fog can be an important source of moisture for plant growth, particularly if it occurs during the growing season. Not only does it add to the actual soil moisture supply, but fog also adds moisture to leaf surfaces and delays increases in temperature, which decreases evapotranspiration. During the growing season, fog is most common along the coast, where warm, moist incoming air is chilled when it passes over cold ocean surfaces. Large trees in foliage can intercept considerable moisture as fog condenses on their leaf and branch surfaces. During a four-day period of heavy fog, Geiger (1961) observed about 20 mm (0.75 in.) of water to fall in open country, while over 150 mm (6 in.) fell under fir trees on the edge of a forest. At Berkeley, California, 0.7 to 1 m (30–40 in.) of fog precipitation was measured under individual eucalyptus trees during a single growing season.

Therefore, **a grove of trees can effectively protect land areas from incoming fog.** Geiger reports 20 times as much fog precipitation on the windward as on the leeward side of a wooded area. The Japanese have studied the species and grove dimensions most effective in reducing the moisture of fog from the sea.

In interior areas, ground fog may occur in winter under conditions that favor radiation loss of heat and high humidity. Winter fog can keep the buds of trees close to air temperature and make winter chilling more effective. Warm weather fog, however, may increase the incidence of certain fungal diseases.

Dew

Dew is another source of moisture that can either reduce evapotranspiration or, in some cases, add moisture to the soil reservoir. Water condenses on a surface when its temperature drops below the dew point of the surrounding air. This happens under conditions favorable for radiation loss of heat.

The maximum amount of dew that might form in a single night has been calculated to be equivalent to water about 1 mm (0.04 in.) deep (Geiger 1961). Grass will produce only 0.1 mm (0.004 in.). The maximum amount of moisture is possible only in arid and semiarid tropical climates, where rainfall is infrequent and dew formation is favored by extreme radiation cooling at night. This moisture would not be significant to the soil moisture supply, but dew on the leaf surfaces can be absorbed by a number of species and can delay leaf warming on the following day. More dew will form on an isolated plant than on an identical plant that is close to others.

Dew can partly replenish the soil reservoir in certain cases. In the deserts of North Africa, gravel mounds about 1.5 m (5 ft) in diameter and 0.6 m (2 ft) high that were built centuries ago are fairly common. They are thought to represent a method by which dew is made to condense on the radiation-cooled surfaces, allowing excess moisture to flow down to the soil. A variation of this is the desert survival technique that involves digging a hole in the soil, covering it with a plastic sheet, and placing a stone in the center to provide a sloping surface so that water may flow to the lowest point and drip into a container placed in the bottom of the hole. This has been adapted experimentally in Montana to provide moisture for new plants (see Fig. 11–13) (Jensen and Hodder 1979). Both the rock mulch and the plastic-covered hole can be effective only if good radiation cooling occurs at night. When plants grow big enough, they may reduce radiation losses from the surface of the gravel or plastic and consequently the amount of moisture condensed.

Drought

In areas that have summer rains, a period of three weeks without rain will begin to create problems for sensitive plants and those with small reservoirs of soil moisture. By contrast, regions that have Mediterranean-type climates normally receive little or no rain for four to five months of the growing season. Unless such regions have supplemental sources of water or deep, well-drained soil, their vegetation will be quite different from that in areas that receive rain in summer. Plants will be smaller but will have relatively large root systems, and they will be farther apart. Some may have special or modified structures that reduce transpiration, and others may drop most of their leaves when water is limited.

Many landscaped areas have restricted supplies of water. Paving and buildings can reduce the amount of moisture that reaches root zones and can also increase transpiration through increased radiation. This is discussed in more detail in Chapter 20.

Excess Moisture

Intense rainfall can cause serious erosion and flooding. Soil and plants can be washed away, slopes can erode or slump as plants are lost, and plants can be buried by slides and sediment. If soils become waterlogged because of long periods of rain or runoff from melted snow, all but the most hardy plants will fail or grow poorly. Landscape plantings should be designed to handle the heaviest rains that might be expected (see Chapter 8).

Hail

Hail can defoliate plants, tatter leaves, injure bark, damage or knock off fruit, and break twigs and small branches, depending on hailstone size and firmness and wind speed. Damage will be greatest at the tops and on the windward side of plants. Hail damage can be quite unsightly, but plant appearance usually returns to normal when new growth begins.

Snow and Glaze Ice

Wet snow, freezing rain, and glaze ice can build up on branches of trees and shrubs until they break under the increased weight. Wet snow sticks to branches, particularly to those of conifers, and is much heavier than drier powder snow. Young trees can be severely deformed by the weight of the first wet snows of winter, which often adhere to them until spring. Even deciduous trees and shrubs can become encased in ice by rain that freezes when it contacts cold branches (Fig. 4–8). Entire trees have been lost because of the weight of ice. Semonin (1978) reported an estimated 45 metric tons (50 tons) of ice on a 15-m (50-ft) evergreen tree and a load of ice on a 6-mm (0.25-in.) deciduous twig that was 12 to 17 times the weight of the twig. Pirone (1978a) reported an ice covered twig that weighed about 40 times as much as the twig alone.

Figure 4–8 Rain that freezes to branches can cause serious limb breakage. (Photos courtesy Kim Morse, Dallas Park Department)

Tornados, hurricanes, and strong winds can devastate landscape plantings by dismembering and uprooting trees and shrubs. Winds of lesser intensity can damage leaves and branches and deform plants. **Wind affects plant growth by increasing transpiration and carbon dioxide uptake and decreasing shoot elongation.**

Plant Damage

High prevailing winds can not only deform plants but seriously damage them, particularly if the plants are in foliage and if soil is moist (Figs 4–9 and 17–2). When winds from one direction strongly prevail, a moderately strong wind from the opposite direction can blow down trees that had withstood much stronger prevailing winds. Winds can cause more damage when trees have a full canopy of leaves than when they are without leaves. Winds in Wellington, New Zealand, are so strong in the spring that few deciduous trees are grown there, except in protected areas, because the tender new leaves are seriously damaged. In areas where the ground may freeze in winter, wind can increase the transpiration of conifers and cause winter burn of the needles.

Figure 4–9 Wind during or following heavy rains can be devastating. About half of the 100- to 125-mm (4–5 in.) diameter plane trees were blown down during an early fall California rain (left); it is interesting that the only trees that were blown down were those that were still staked. (Harris, Leiser, and Davis 1976) Many mature elms were uprooted in a severe Midwest storm (right). (Right photo courtesy Eugene Himelick, Illinois Natural History Survey)

The gale forces of hurricanes and the fierce twisting winds of tornados lay waste to thousands of trees and other plants each year. Millions of trees were uprooted and damaged in southern England by the disastrous hurricane of October 1987. Matthews (1988) observed that thinned (pruned) trees along streets and in the open stood while unthinned trees around them fell. A healthy, well-structured tree with a sound root system is best able to withstand strong winds. Trees in a grove protect one another from strong winds, but should some trees be lost or removed the survivors are more seriously endangered when their protection is reduced.

Transpiration

The direct influence of wind on transpiration is difficult to quantify, not only because transpiration varies with plant species and depth and density of leaf canopy but because changes in wind speed are usually accompanied by changes in air temperature and humidity. Reviewing several controlled experiments, Chang (1968) reported that **transpiration increases with wind speed up to a certain point,** beyond which it stabilizes or decreases slightly. Isolated tall plants will be most seriously affected, since wind increases with height above the ground and the foliage of isolated plants is more exposed to air movement. According to Chang, Copeland (1906) found that an 8-km/hr (5-mph) breeze approximately doubled the transpiration rate of coconut palms in full sunlight. A number of papers conclude that if the foliage canopy is complete and even, the effect of wind on transpiration is usually small. As wind speeds increase, however, drier and warmer air is usually brought deeper into the leaf canopy, thereby increasing transpiration. Wind influence will be greatest when humidity is low, temperatures high, and soil moisture adequate. Pruitt (1971) calculated that a 16-km/hr (10-mph) wind would increase evapotranspiration about 25 percent at noon on a clear, fairly calm midsummer day in Davis, California (a typical temperature of 30° C [87° F] and a relative humidity of 30 percent were assumed). At typical temperatures and humidities, a 16-km/hr wind would increase evapotranspiration about 50 percent in early morning and about 100 percent in late afternoon. A 32-km/hr wind would approximately double the amount of evapotranspiration.

When relative humidities are near saturation, wind has less effect on transpiration. Wind will also affect transpiration only slightly if soil moisture is limiting or if stomatal openings restrict vapor movement.

Photosynthesis

Under high light, the rate of photosynthesis increases as levels of carbon dioxide increase up to about four times the normal concentration of 0.03 percent in the atmosphere. Under calm daylight conditions, the level of CO_2 in the plant canopy drops below that in the atmosphere, but air movement will keep the CO_2 levels more nearly equal. If the CO_2 level in the leaf canopy is to be maintained, wind speed may need to be at least 20 km/hr (12 mph). This may increase photosynthesis by 10 to 20 percent (Chang 1968).

Shoot Elongation

Movement of the growing point of a plant can reduce height growth as much as 25 percent, although it will increase the growth of basal stem diameter (Leiser and others 1972) (see Chapter 9). Harris and others (1972) discovered that shoot growth of liquidambar was increased 50 percent when container trees were grown close together so that wind movement of the tops was reduced (Fig. 4–10); basal trunk growth and taper were decreased. Forest trees grow taller with less basal trunk growth and less taper than their isolated counterparts. Studies of herbaceous plants show that those grown in high-wind areas have a greater proportion of roots to top than do more protected plants (Whitehead 1963).

Figure 4–10 Effects of spacing liquidambar (in 14-l or 3.5-gal containers) during the first growing season in a nursery: 600 mm (24 in.) apart (left), 430 mm (17 in.) apart (center), and 250 mm (10 in.) apart (right). (Harris and others 1972) Close spacing restricted movement of the trees and shaded lower foliage.

Wind-Chill Factor

Wind is one of several factors that affect human comfort through the individual microclimate. In heat, air movement cools a person by causing moisture to evaporate from the skin and by removing heat directly. In cold weather, removal of body heat by wind is increased. For example, a temperature of $-15°C$ (5°F) accompanied by wind at 25 km/h (15 mph) has the same cooling effect as calm air at $-31°C$ ($-24°F$) (Foster 1978). This is true only if the object is warmer than the air. Once the object is at air temperature, wind will tend to maintain this equilibrium by removing heat if it is added by radiation or respiration and by adding heat if it is lost by evaporation or radiation.

In cold weather, the wind-chill factor is detrimental to exposed animals and plants whose comfort or well-being depends on being warmer than the air. Windbreaks can reduce wind-chill to increase animal comfort and to reduce the cost of heating buildings (see Chapter 5). Wind will not, however, increase cold injury to plants if they can withstand the air temperature. In fact, wind will keep exposed plant parts at air temperature when they might have been further cooled by radiation loss at night or warmed during the day under clear-sky conditions. On the other hand, wind can increase plant injury under freezing conditions by increasing moisture loss and causing desiccation as already discussed. Wind may also reduce the amount of undercooling before freezing occurs.

URBAN CLIMATE

Urban development certainly modifies the mesoclimate: Vegetation is reduced and pavement and buildings take up much of the denuded space. As a result, temperature (termed *urban heat island*), precipitation, and cloud cover increase relative to the surrounding rural areas; wind, humidity, and radiation decrease (Table 4–2). The extent of these influences depends on the size of the city and the amount of vegetation. Most cities have considerable vegetative cover. The surface covered by trees and grass in three Kansas cities in 1973 was 53 percent (in Kansas City, pop. 168,000), 63 percent (in Lawrence, pop. 46,000), and 85 percent (in Baldwin, pop. 2500) respectively (Marotz and Coiner 1973). Areas within a city, however, can vary greatly—Outcalt (1972) has identified what he calls an ''urban cold doughnut.'' In fact, there may be concentric areas with differing climates. For example, going from the countryside into the city center, the outer area will normally consist of farmland in crops. New residential areas follow, with little vegetation and expanses of exposed pavement, buildings, and bare ground, which will be warmer. Next, the ''cold doughnut,'' older residential areas with considerable tree and grass cover will be similar in temperature to farmland. The city center, composed almost entirely of pavement and buildings, will be the warmest area. Wind will modify the location and configuration of these mesoclimatic areas. Within the city center, areas with tall buildings usually will be cooler in summer, warmer in winter, and more protected from wind than are the areas with open pavement and low buildings of fairly uniform height (Federer 1971). On the other hand, isolated tall buildings can deflect strong winds downward and concentrate their force at the base and corners of buildings. Tall buildings also funnel wind into the streets between them, greatly increasing wind velocity. Many cities plant trees more resistant to wind deformation on streets most subject to increased wind velocity. Buildings also create permanent shade, reflect light, and radiate heat to form their own microclimates. The different sides of a single building have quite different microclimates that affect human comfort, plant growth, and plant care.

Near large areas of pavement and building surfaces, plants must endure a harsh environment. Summer temperatures will be higher, the relative humidity lower, soil moisture and aeration limited, and air quality poorer. Trees and shrubs more tolerant to such conditions should be planted in such areas, and special care should be given to get them established (see Appendix 4).

TABLE 4-2 AVERAGE CHANGES IN MESOCLIMATE ELEMENTS CAUSED BY URBANIZATION (Adapted from Landsberg, H. E., 1970. Climates and Urban Planning. In *Urban Climates.* World Meteorological Organization Tech. Note 108: 364–74.)

Element	Comparison with rural environment
Wind Speed	
Annual Mean	20 to 30% less
Extreme Gusts	10 to 20% less
Calms	5 to 20% more
Temperature	
Annual Mean	0.5° to 1.0° C (0.9°–1.8° F) higher
Winter Minima (average)	1.0° to 2.0° C (1.8°–3.6° F) higher
Heating Degree Days	10% lower
Precipitation	
Total	5 to 10% more
Days with less than 5 mm	10% more
Snowfall	5% less
Relative Humidity	
Winter	2% lower
Summer	8% lower
Cloudiness	
Cover	5 to 10% more
Fog, Winter	100% more
Fog, Summer	30% more
Radiation	
Total	15 to 20% less
Ultraviolet, Winter	30% less
Ultraviolet, Summer	5% less
Sunshine Duration	5 to 15% less
Contaminants	
Condensation Nuclei and Particulates	10 times more
Gaseous Admixtures	5 to 25 times more

FURTHER READING

BARFIELD, B. J. and J. F. GERBER, eds. 1979. *Modification of the Aerial Environment of Plants.* St. Joseph, MI: Amer. Soc. of Agr. Engineers.

CHANG, J. 1968. *Climate and Agriculture.* Chicago: Aldine.

GEIGER, R. 1961. *The Climate Near the Ground.* Cambridge: Harvard University Press.

WEISER, C. J. 1970. Cold Resistance and Injury in Woody Plants. *Science* 169:1269–78.

CHAPTER 5

Modifying Climatic Influences

Climate or matters related to climate have seldom been adequately considered when new landscapes are being planned and developed. Considerations that do not fit neatly under the traditional discussions of climate in Chapter 4 are presented here.

PLANTING FOR CLIMATE CONTROL

Plants can enhance human comfort and reduce building energy consumption by

Intercepting summer sunlight
Absorbing and reflecting summer energy
Reducing ambient summer temperature through transpiration
Shielding from cold winds
Channeling cool summer breezes in gardens and through warm buildings

Increasing energy costs and improvements in equipment and building design have generated interest in conservation through improved subdivision plans; building siting; and landscape planning, installation, and maintenance. These approaches should involve little or no inconvenience or sacrifice of comfort. The following discussion focuses on the influence of plants on energy conservation of buildings; the psychological and health-related effects of plants on people were discussed in Chapter 1.

Intercepting Sunlight

During summer, buildings become hot when early morning and late afternoon sun strikes walls that face east or west, particularly when those walls are poorly insulated or made of glass. Although a house may become warmer if its west wall is exposed to the afternoon sun, it will be heated earlier in the day and will stay hot if the east wall is equally exposed to the morning sun. Deciduous trees, hedges, and vines on trellises can shade these walls from summer sun while allowing the winter sun to penetrate. At high latitudes (>30°), evergreen plants are also suitable since the sun angles in winter are such that east and west walls will not effectively collect heat.

Walls and roofs that face the equator should be protected from the high-angle summer sun. Tall, spreading deciduous trees are most effective in screening roofs. Trees can be placed close to foundations on the east and west sides and even on the polar sides of buildings, but they should shade neither equator-facing glass in winter nor rooftop solar collectors. Bare branches of deciduous trees block substantial amounts of sunlight (Zanetto 1978, Heisler 1982); McPherson (1984) indicates that the average value of irradiance reduction by defoliated crowns is about 35 percent with 25 to 50 percent a typical range.

Screening Reflecting and Radiating Surfaces

Reflecting and radiating surfaces, such as swimming pools, concrete patios, and pavement, can heat east, west, and equator-facing glass at undesirable times, even though this glass is shielded from direct sunlight. To prevent excessive heat gain in a building, these reflecting surfaces should be screened with a hedge or a trellised vine, shaded by trees or a trellis, or replaced by a shrub or ground-cover planting.

Paved surfaces (particularly asphalt) act as heat sinks if exposed to the direct sun for long; they absorb heat and reradiate it to heat the surrounding environment. This is fine in the winter but unacceptable during hot summers. Pavement location, size, and shape and strategic placement of shade trees can be used to minimize summer heat and take advantage of winter sun. Tall, spreading deciduous trees on the equator-facing, east, and west sides of the pavement will provide the most effective summer shade but will still allow some penetration of winter sun (Fig. 5-1).

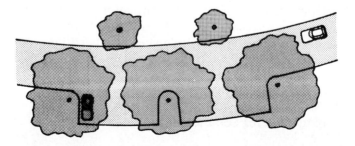

Figure 5–1 Parking bays along the equator side of narrow streets can be effectively shaded by large trees. Smaller trees on the polar side provide solar access to houses there. (Thayer 1981)

Reducing Ambient Summer Temperatures

Plants cool the air by transpiration, though breezes disperse the effect so that the temperature in a given landscape is usually lowered by only a few degrees.

SOLAR ACCESS

The use of solar energy for heating and power requires that solar collectors and areas to be passively heated have access to the sun. At the same time, as much of a building as possible should be protected from summer sun. In urban areas, not only must a property be wisely planted and maintained but governmental control of trees on abutting property may be needed to ensure solar access.

Solar Access Control Strategies

Four regulatory strategies to legally control vegetation for solar access are being used in the western United States (Thayer 1986a) (Fig. 5–2). *Nuisance law* is the basis of the California Solar Shade Control Act enacted in 1978 which declares that under certain conditions trees blocking solar collectors are a public nuisance. A public nuisance is generally regarded as "any form of annoyance or inconvenience interfering with common public rights" (Hayes 1979). Some believe this legislation is unconstitutional, reasoning that its infractions should be considered private, not public, nuisances. A state court of appeals declined to rule in favor of this nuisance law issue (Thayer 1986b).

The *police power* to regulate development through zoning and subdivision ordinances is used by cities in several states to establish solar access controls. Zoning-based solar access controls generally involve the distances that trees, buildings, and other objects of certain heights must be from property lines so as to allow a certain level of solar access to adjoining property.

A *private easement* is possible in California if a solar owner enters into a contract with a neighbor that would prohibit the neighbor from shading the solar owner's property or collector for specified times of the day, month, and/or year (Thayer 1986a). The neighbor probably would be compensated for yielding the right to develop his or her property fully. The government would be minimally involved.

Restrictive covenants and subdivision performance standards are being used to ensure solar access in new developments. Subdivision regulations can require developers to follow solar access performance standards which would apply in perpetuity to the land being developed. Similar requirements can also be attached to the deed to a property.

Developing Sites for Solar Access

Much can be done in site development to help ensure solar access and to conserve heating and cooling energy. Proper planning of a subdivision is particularly import-

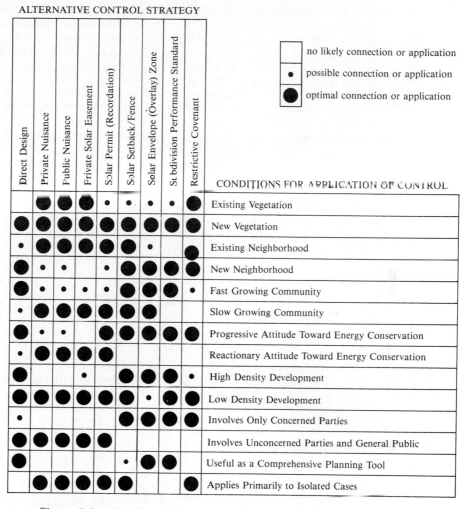
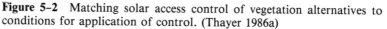

Figure 5-2 Matching solar access control of vegetation alternatives to conditions for application of control. (Thayer 1986a)

ant, because few changes are possible once the subdivision is established. Passive or natural heating and cooling can be facilitated in the following ways (Thayer 1981):

Residential streets should be oriented so that the maximum number of lots face streets that run east and west. This will maximize the number of houses with solar access to the equator end of the lots. Equator-facing wall glass is crucial for maximum passive solar heating.

Setback requirements should allow houses to be placed to the extreme polar end of a lot to allow for maximum solar access.

Usually houses should be sited so that equator-facing glass is maximized, east- and west-facing glass is minimized, and interference with the skyspace of equator-facing glass and rooftop collectors is negligible. Solar access can be provided on less-than-desirable lots if there is flexibility in siting of buildings. Attached and multistory housing units require special consideration to ensure adequate solar access.

Streets should be narrow, with little or no parallel parking; these can be shaded more easily and quickly than wide streets. Older residential areas with mature trees are several degrees cooler in summer than new subdivisions with much exposed asphalt and concrete. Perpendicular parking bays along the equator side of narrow streets can be planned so that both the parking areas and the street are shaded by trees of moderate size (Fig. 5–1).

Solar setback regulations should allow fences to be placed so that they provide privacy but do not interfere with equator-facing glass collectors.

Height restrictions on buildings and collectors may be necessary to ensure solar access to one-story houses. Taller buildings must be situated far enough away from solar collectors so that they do not shade them. A reasonable restriction is the so-called *bulk plane height,* which provides that no building or plants shall protrude above a plane in space that extends toward the equator at an angle equal to that of the sun at 12:00 noon on the first day of winter (Fig. 5–3). The angle of the plane depends on latitude.

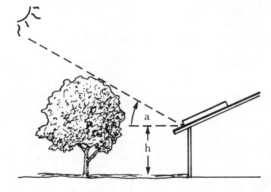

Figure 5–3 The *bulk plane height* (h) provides that no structure or plant shall protrude beyond a plane that extends toward the equator at an angle (a) equal to that of the sun at noon on the first day of winter. (Thayer 1981)

Because the slope orientation of building sites affects solar access, solar access requirements should allow for taller trees and closer spacing between buildings on equator-facing slopes. Just the opposite is true on polar-facing slopes.

No matter how good solar orientation and access are, solar houses should not be located in frost pockets, in cold-air drainages, or on ridges exposed to cold winds.

Planting to Accommodate Solar Energy Use

In all but the warmest climates, solar collectors for heating water should have unobstructed access to the sun between 9 A.M. and 3 P.M., particularly in winter. Climates differ in the amount of energy needed to maintain a building at a comfortable temperature. Certain coastal, high-elevation, and high-latitude areas require considerable heating in the winter but little or no cooling in summer. Low-elevation and low-latitude areas may need cooling in the summer but little or no heating in the winter. Much of the temperate zone requires considerable energy for both heating in winter and cooling in summer. In cold-winter areas, a windbreak can screen buildings and solar collectors from cold polar winds. In hot climates, walls and roofs

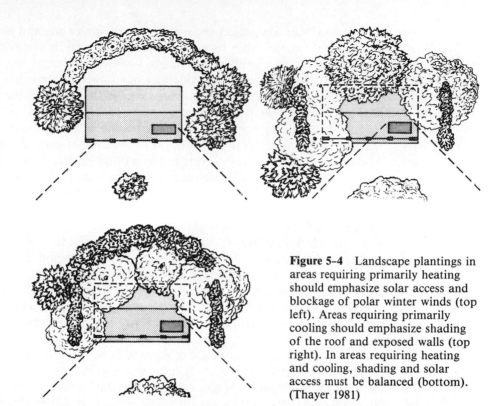

Figure 5-4 Landscape plantings in areas requiring primarily heating should emphasize solar access and blockage of polar winter winds (top left). Areas requiring primarily cooling should emphasize shading of the roof and exposed walls (top right). In areas requiring heating and cooling, shading and solar access must be balanced (bottom). (Thayer 1981)

can be screened from the summer sun. Where both heating and cooling are needed, plantings for shade must be balanced with provisions for solar access (Fig. 5-4).

The amount of heating or cooling necessary can be expressed by the number of *heating degree days* or *cooling degree days*. If the mean daily temperature is below 18°C (65°F), each degree of difference between the two is called a heating degree day unit (Berdahl and others 1978). A monthly or annual value is the sum of the degree day units for the days in the month or year. Heating degree days are a fairly good measure of the heating requirements for a conventionally constructed building. Similarly, a cooling degree day unit indicates that the mean daily temperature has been one degree above 18°C for one day. Cooling degree days are not a particularly good measure of a building's cooling requirements, because they do not account for humidity, which greatly influences human comfort, particularly at high temperatures.

Whether more effort should go into reducing the cost of heating or the cost of cooling depends on the relative amounts of energy needed for each, the form of energy used (gas, oil, coal, or electricity), and the relative cost of energy per unit of heating or cooling (Thayer and Maeda 1985). Cooling primarily requires electricity, which is usually the most expensive energy source. The amount of daytime cloud, fog, and smog cover influences the effectiveness of solar collectors and shade trees. Even though the heating requirement may be considerably greater than the cooling requirement, the wise use of shade trees for summer protection may save considerable energy in areas that are subject to long periods of winter fog and clouds and bright summer sun. On the other hand, for greatest economy in a coastal valley with relatively clear winter weather and foggy summer mornings, the property owner should maximize solar access for collectors and (in winter) the equator-facing walls. Minimizing east glass and planting a sun-screening hedge along the east wall of a building would not conserve energy under such conditions.

Even though it may be more economical for an individual property owner to strive for maximum solar access in the winter at the expense of summer shade protection, such a plan may not be in the best interests of an electric utility. The latter must accommodate peak demand, which occurs in many areas on hot summer afternoons because of the use of air conditioners. In such situations, deciduous trees should be used to block sunlight from wall and roof surfaces; even if the trees shade part of the equator-facing glass and solar collectors in the winter.

To provide maximum summer shade with minimum shading of solar collectors, structures and plants must be placed according to vertical (altitude) and horizontal (azimuth) angles of the sun at different times of the day and days of the year. Although the *bulk plane height* method is simple to use in determining locations for plants of expected mature heights (Fig. 5-3), it may not provide maximum summer shade and solar access or may unduly limit the size and placement of trees and trellised vines.

The solar access zone for a collector or equator-facing glass is most accurately described by a complex surface (Fig. 5-5) formed primarily by the lowest (winter) sunrays striking the lower and east and west edges of the collector between 9 A.M. and 3 P.M. On the east and west edges of the zone surface, however, this surface is modified so that tall structures or trees do not shade the collector at other seasons (Fig. 5-5). As long as no object projects above this imaginary surface, the collectors should have full solar access during the critical midday hours throughout the year. Solar access on the east and west edges of the zone is more important for protecting solar water heaters than for space heating. Tall obstructions along the zone edges will shade the collectors at the beginning and end of the critical period during summer but not during the cold part of the year, because of the winter azimuth of the sun.

The solar access zone can be estimated by using a *sky chart* of the sun for a particular latitude. Charts should be available from the local weather bureau or energy agency. The angles (both vertical and horizontal) of the sun are obtained from the chart for 9 A.M., noon, and 3 P.M on June 21 and December 21; by using

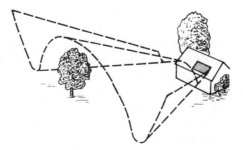

Figure 5-5 A complex surface rising toward the equator as outlined by the broken lines should not be penetrated by buildings or plants if solar access is to be provided. (Thayer 1981)

the collector dimensions and these points, a surface can be depicted that describes the solar access zone. Computer graphic analyses of solar skyspace are feasible for large housing projects and land developments and will help in analyzing spatial solar access planting and configurations. Analyses for a desired latitude should be available soon at solar equipment distributors, weather bureau offices, or planning agencies, if not currently available. Also the U.S. Forest Service has designed a computer program, called *SOLPLOT,* to maximize control of sunlight on windows, solar panels, and other features (Wagar 1985a). Such analyses may become the only accurate way to establish solar easements over adjacent properties in order to protect solar access.

Tall, upright trees with long, bare trunks can shade equator-facing glass in the summer and allow winter sun to reach the glass (Fig. 5-6). However, high crowns will shade the roof and any solar collectors there in winter. Trellises and overhangs may be more satisfactory than trees for protecting an equator-facing glass wall, particularly if a rooftop solar collector is nearby.

East-west streets planted with tall trees along the equator side of the street and shorter trees along the polar side will shade the street in the summertime but will

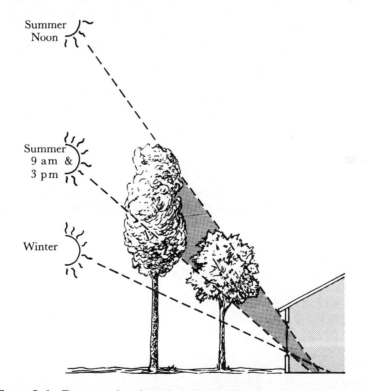

Figure 5-6 Trees can be planted and pruned to provide solar access in the winter and shade in the summer. (Adapted from Foster, R. S. 1978. *Homeowner's Guide to Landscaping That Saves Energy Dollars.* New York: David McKay. Ruth Foster is a landscape consultant in Belmont, MA).

leave solar collectors exposed on the polar side of the street (Fig. 5–1). Existing and newly planted trees can be pruned severely in the early fall to ensure that winter sun reaches collectors or areas to be heated. Appropriate species can be pollarded to minimize winter shading, provide firewood, and encourage vigorous growth during the summer (see Fig. 15–37).

CONTROLLING WIND

Air movement can be pleasant, uncomfortable, or destructive, depending on its velocity, the temperature, and its involvement with rain, snow, hail, or dust. Plants can be used alone or in combination with landforms and structures to influence airflow in the landscape and around or through buildings. This phenomenon has been studied most extensively for use in windbreaks or shelterbelts that will control strong winds, primarily in the open countryside. Much less is known about wind effects in urban areas, where there are canyon-like streets in the central city, tree-covered streets in the suburbs, and plants and structures of various sizes, shapes, and densities. More concern and study has been directed to reducing the destructive effects of wind than to enhancing the pleasant aspects of air movement, but both aspects are receiving more attention as interest in energy conservation grows. White (1945) has studied the use of landscape plantings to enhance the natural ventilation of buildings in warm regions.

Windbreaks and Shelterbelts

Windbreaks and shelterbelts can influence a wide range of activities (Caborn 1965). They can

> Reduce the destructiveness of wind
> Improve site conditions for plants, people, and other animals
> Increase growth and yields of farm and range plants
> Reduce heat loss from houses and other structures
> Mitigate salty ocean winds
> Minimize wind erosion
> Control snow accumulation to keep roads open or to enhance soil moisture supply
> Provide wildlife habitat
> Increase the ecological diversity of an area

Although the terms *windbreak* and *shelterbelt* are used somewhat interchangeably, the former is more appropriate for a structure (a fence or a screen) or a single row or narrow width of plants, while *shelterbelt* refers to several rows of trees or shrubs.

The *distance* over which wind speed is reduced is determined primarily by the height of the windbreak and secondarily by its density. The *amount* of wind reduction is controlled by the density of the windbreak. **A windbreak porosity of 40 to 50 percent will give the most acceptable combination of wind reduction and distance**

over which that reduction occurs. Protection will extend downwind for a distance up to 30 times the windbreak height, and upwind up to five times the windbreak height (Caborn 1965).

Protecting Landscape Sites

Most landscape sites are restricted in size, and many are surrounded by plantings and structures that affect air flow greatly. Buildings, which are solid windbreaks, create the greatest differentials in wind speed. Isolated tall buildings can create wind speeds on the ground that are twice the normal speed. Wind speed normally increases with elevation; when strong winds encounter the tops of tall buildings, they are deflected parallel to the building face, and some travel down the building to create strong winds. The taller the isolated building, the stronger the winds become at the bottom. On a calm, hot summer day, this phenomenon can create refreshing breezes in open areas at the base of tall buildings, but under strong wind conditions people and property can be damaged. People are frequently knocked off their feet, and once in Cleveland at the base of a 39-story building, a trailer truck was blown over (MacGregor 1971). Plants can be damaged or at least misshapen by these strong side- and down-drafts, particularly near building corners (Fig. 5–7). In a group of tall buildings, only the windward buildings are affected. The bases of isolated tall buildings and their landscaping must be designed to contend with this problem. The solution to date has been to use wind-resistant plants and handrails.

Figure 5–7 Trees near corners on the windward side of tall buildings can be misshapen. The tulip trees near the corner of a five-story building have been deformed by polar winds (left); a tulip tree planted the same time near the center of the building on the polar side has developed normally (right).

Buildings and large trees can divert wind into areas that were intended to be calm. Streets and side yards between buildings can become wind tunnels. In residential areas with moderate-to-large trees, the wind coming under the tree crowns may be the most troublesome. Plants should be placed to block winds around the side of a house and to form screens on the windward side of the house and garden to reduce wind velocities (Fig. 5–4). Depending on orientation, it may be necessary to divide protection unequally between the house and garden.

On the polar (north or south) side of the house, Caborn (1965) recommends that windbreak distance from the house be at least $2\frac{1}{2}$ and no more than 5 times the ultimate height of the barrier. If wind protection is wanted in the other quadrants but not at the sacrifice of light, Caborn suggests windbreak distances from the house of 4 to 5 times the height of the plants. Because the strongest winds usually occur in winter, evergreens will be needed for maximum protection. A deciduous windbreak is only about 60 percent as effective in winter as when it is in leaf. Summer shade, however, may be more important than full winter sun, so another compromise must be evaluated. A residential area may be so heavily populated with large trees that only low-growing windbreaks are needed. Additional protection can be provided if one wraps the ends of the main windbreak around the windward corners of the house.

As a windbreak grows in height, the area of maximum protection increases in size but moves further downwind. As a windbreak thins out, the degree of protection decreases, while the protected area increases in size and moves further downwind. Pruning, revitalization, and replacement are needed to maintain a windbreak in proper condition.

COASTAL PLANTINGS

Plantings on or near the coast, particularly along western continental shores, endure prevailing onshore winds that carry salty mists and, on occasion, salt spray for considerable distances inland. Although coastal winds may not be any stronger than continental winds, they are more continuous. Plant establishment, formation, and survival are problems because salty wind desiccates and deforms exposed plants, coats them with salt, and keeps sand dunes constantly shifting.

Wind and salt markedly curtail the number of species, particularly tree species, that can withstand direct exposure close to the shore. Small plants adapt to a shore site more quickly than large ones because they are protected by being close to the surface. If a front line defense is to deflect the wind in an upward direction, plants known to tolerate local seacoast conditions (see Appendix 4) should be chosen for planting. Tenacity is more important than appearance. Once the wind deflection has begun, less rugged plants can be introduced farther back to increase the height of the windbreak. Plants that spread by means of suckers do well in coastal sites (Caborn 1965).

Sand must be stabilized and, for most plants, enriched with nutrients and organic matter. Small areas of sand are usually protected to some degree, but large expanses often have shifting windblown dunes that can quickly bury obstacles in their path. Beach grass, ice plant, *Rosa rugosa,* and broom will usually grow well in almost sterile sand, begin to slow blowing sand, and start the development of a more permanent dune. As sand accumulates, these plants can grow up through it and stay on top. Organic matter added to sand behind the protective dune, where some protection from the wind is found, will improve conditions for other plants. John McLaren created Golden Gate Park on sand dunes by developing coastal windbreaks and enriching the sand with manure from San Francisco stables.

Sand can be slowed and most exposed plants established if temporary windbreaks are formed of staked-down brush or a low snow fence. Such protection should have a porosity of at least 40 to 50 percent. Because some upwind protection is provided, plants should be set out in front of the windbreak as well as behind it. Upwind plants will not only protect the windbreak but will replace it when it is removed or deteriorates.

Individual young trees or shrubs are often protected by almost solid screen structures set close to windward (Fig. 5-8). Plants protected in this fashion are not conditioned to the accelerated winds that overtop the screens or to the winds that will reach them when the screen is removed or becomes ineffective. Screens for single plants or groups of plants should provide only enough protection to keep shoot terminals from being killed or severely deformed, because the plants will eventually have to endure unabated winds. Screens can be placed at a distance two or three times the screen height upwind from the plants unless the wind varies considerably from a single prevailing direction. If the screen forms a shallow V with its point into the wind, it will tend to deflect more of the wind to the sides and less over the top. Establishing individual or isolated groups of plants in the brunt of coastal winds may be an exercise in frustration and futility unless the plants are hardy and the grower determined.

Figure 5-8 An almost impermeable, closely placed windbreak does not protect a plant once it grows above the windbreak. In fact, wind speed is increased at this height. The plant would be better protected if the windbreak was 50 percent permeable.

Strong coastal onshore winds, through desiccation and salt toxicity, kill exposed shoot terminals. Several laterals will appear from behind each killed twig tip, only to be killed back as they themselves grow. Soon a thick mat of leafless twigs, with leaves below, will form a barrier that forces the winds upward. Each successive plant grows taller than the one in front, following the line of wind deflection. Few hedges are sheared more evenly than plants on a wind-swept coast. The compact windward surface of exposed plants provides little of the filtering action expected of a good windbreak. The sheltered area is short and close to the plants, with little wind protection extending to leeward beyond 12 times the height of the tallest plants (Caborn 1965). The effectiveness of a wind-swept planting can be increased if clumps of vigorous trees can be grown taller than the general plant surface. Temporary protection may be necessary to allow for the taller growth.

Onshore winds of 25 km (15 mi) per hour or greater can carry considerable ocean salt spray (Boyce 1954). When rain does not accompany strong onshore winds, the winds can be particularly toxic to plants (see Chapter 20).

LIGHTNING

Lightning is an extremely powerful electrical discharge from one cloud to another or from the ground to a cloud. Thunderstorms account for the most serious lightning strikes. Raindrops and ice crystals are broken up and churned about by the turbulence in a thunderhead; electrical charges develop such that the upper portion of a thunderhead has a net positive charge and the cloud base has a net negative charge. The electrical charge at the base of a thunderhead increases until it can overcome the resistance of the air between the cloud and the earth. A typical lightning discharge has 10 million to 100 million volts at 1000 to 300,000 amperes and may travel 1.6 km (1 mi) or more (Frydenlund 1977).

Thunderstorms are counted by the number of days per year when thunder is heard. In North America, the annual number ranges from five thunderstorms on the Pacific coast and in northern Canada to more than 100 in southcentral Florida. Even though the New England and midwestern states and the Cascade, Rocky, and Sierra Nevada mountain ranges have only 20 to 40 thunderstorm days, the storms are unusually long and severe.

The trend toward increased liability litigation may mean that more deaths and injuries from lightning in areas open to the public may be blamed on management for failing to provide adequate protection. In the United States, more than 40 percent of the human deaths caused by lightning occur at recreational sites; about 20 percent occur in homes and 20 percent on farms (Frydenlund 1977). **About one-third of human deaths by lightning occur under or near unprotected trees.**

Lightning can cause property damage and death either by direct strike or by a major side flash, in which the lightning charge jumps from a stricken object (pole, tree, or structure) before being partly grounded. Step voltage (a lightning current through the soil that goes up one leg and down the other) can injure victims but is usually not fatal.

The nature and extent of lightning damage to a tree or structure can vary enormously. Branches and trunk may be blown completely asunder (Brown 1972). The crowns of trees may be killed or large limbs broken out (Fig. 5-9). Trunks can be split open. In hardwood trees, such as oak, a continuous groove of bark or bark and wood may be stripped out along the entire length of a main branch and trunk (Brown 1972). In other cases, the xylem may be burned or injured without any external evidence, or part or all of the root system may be killed (Pirone and others 1988). These variations appear to be related to the lightning intensity, the species involved, the amount of water on and in the bark, and the character of the branch and trunk tissues.

Figure 5-9 The trunk and branches of this maple tree were split asunder by lightning (left). Lightning stripped a groove of bark from the top of this oak down the trunk to the ground (right). (Photos courtesy Independent Protection Co., Goshen, Indiana)

Lightning is most likely to strike a lone tree, the tallest tree in a group, a tall tree at the end of a row or on the edge of a grove, a tree growing in moist soil or adjacent to bodies of water, and a tree closest to a building (Cripe 1978). Pirone and others (1988) note that deep-rooted or decaying trees appear more likely to be struck than shallow-rooted or healthy trees. Plants under or near lightning-struck trees are often injured or killed.

Ten species of trees have been reported to be hit by lightning more often than others (Cripe 1978).

Maple	Sycamore
Ash	Poplar
Tulip tree	Oak
Pine	Hemlock
Spruce	Elm

On the other hand, horse chestnut, birch, and beech are struck less often than average. Pirone and others (1988) suggest that the differences among species may be due to chemical constituents in the trees. Trees high in starch (ash, maple, and oak) may be better conductors than trees high in oils (beech and birch). Bernatzky (1978), however, reasons that trees with smooth bark (beech) lead rainwater from the branches to the trunk, which becomes moist; the bark then increases in conductivity so that lightning is led off down the trunk without damaging the tree. On the other hand, trees with deeply fissured bark have a poorly moistened trunk surface; lightning can penetrate into the water-saturated, conductive cambium and cleave the trunk. This would account for less injury to a smooth-bark tree, but a moist bark surface should increase the likelihood of a tree being struck.

Protection Against Lightning

People, structures, and trees can be protected against lightning if means are provided by which a lightning discharge can enter or leave the ground without causing damage or loss. Because of the danger of side flash, trees that have trunks within 3 m (10 ft) of a building and branches that extend above the structure should be equipped with a lightning protection system. Specimen trees, historic trees, and those under which people might seek shelter from a storm should also be so equipped.

The National Fire Protection Association in Boston (NFPA 1980) and the National Arborist Association (N.A.A. 1987) have published standards for protecting trees from lightning. The salient feature of these standards is a single 32-strand copper (17-gauge) conductor attached to an air terminal (tree point) installed in the highest part of a tree and then fastened along the trunk down to the ground connections (Fig. 5–10). **Aluminum should not be used in lightning-protection systems** because it corrodes in contact with moisture or decaying material (N.A.A. 1987). If the tree is round-headed, several smaller air terminals and branch conductors should be placed in the highest parts of the main limbs. If the trunk is 0.9 m (3 ft) in diameter or larger, two down conductors should be installed on opposite sides of the trunk and interconnected. The conductors should be attached securely to the tree to allow for wind sway and growth without danger of breakage. Each conductor should extend from the trunk 300 mm (1 ft) below the ground out to the dripline of the tree or 7.5 m (25 ft) from the trunk, whichever is farther. Each conductor should be attached to a copperweld ground rod 12.5 mm (0.5 in.) in diameter and 3 m (10 ft) long driven into the soil at least 3 m. In sandy soils, three conductors should be attached to each down conductor a short distance from the trunk; each of the three lateral conductors should extend at least 3 m (10 ft) from the down conductor. Special forked ground connectors should be used to ensure that the connection is secure. The end of each lateral conductor should be attached to a 3-m ground rod. In rocky areas where ground rods cannot be driven 3 m deep, two or more ground rods should be driven as deep as possible; these rods should be at least 2 m (6 ft) apart. If a grounding system is within 7.5 m (25 ft) of a water pipe or a similar grounding system for a building or another tree, the systems should be connected to provide common grounding and increased grounding power (N.A.A. 1987).

In a group of trees, only tall trees need be protected; they will protect smaller adjacent trees. In a row of tall trees, down conductors can be connected as long as a 3-m ground rod is placed at least every 25 m (80 ft).

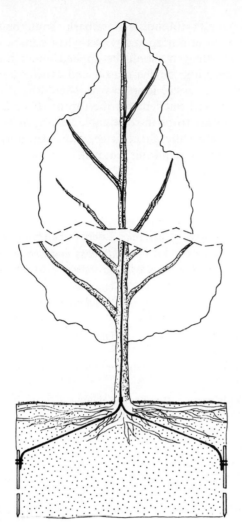

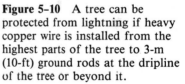

Figure 5-10 A tree can be protected from lightning if heavy copper wire is installed from the highest parts of the tree to 3-m (10-ft) ground rods at the dripline of the tree or beyond it.

Lightning protection systems should be inspected each dormant season to make sure that they are intact and that the air terminals are high enough. The air terminals may need to be raised every two or three years. None of the conductors should be bent more sharply than 90° or have a radius of less than 200 mm (8 in.).

A tree protected with a lightning rod system will actually be more likely to be struck by lightning, but a properly installed system should protect the tree and people underneath from harm.

Caring for Lightning-Struck Trees

Many trees that are struck by lightning must be removed; others may be worth attempts to save them. Since the extent of internal damage cannot be assessed immediately, repairs should be limited to safety pruning and cleanup until tree survival seems assured. If the cambium is still moist, loose bark should be tacked back into place, covered with moist cloth or burlap and polyethylene film to slow drying, and shaded to keep them cool. Nitrogen should be applied under the tree as recommended in Chapter 12. Depending on the season and expected rainfall, fertilizer should be watered into the root zone. Root growth should be stimulated so as to improve the supply of water and nutrients to the top of the tree. If, after a full growing season, the tree appears to be in reasonable health, more careful pruning and wound repair may be worthwhile. Struck trees, however, may continue to fail over a period of years, eventually requiring removal.

FURTHER READING

CABORN, J. M. 1965. *Shelterbelts and Windbreaks.* London: Faber and Faber.

DEWALLE, D. R., and E. P. FARRAND. 1978. *Windbreaks and Shade Trees: Their Use in Home Energy Conservation.* Penn. State Univ. Agr. Special Circular 245.

GEIGER, R. 1961. *The Climate Near the Ground.* Cambridge: Harvard University Press.

GEORGE, E. J. 1956. *Cultural Practices for Growing Shelterbelt Trees on the Northern Great Plains.* U.S. Dept. of Agr. Tech. Bull. 1138.

———. 1966. *Shelterbelts for the Northern Great Plains.* U.S. Dept. of Agr. Farmers Bull. 2109.

NATIONAL FIRE PROTECTION ASSOCIATION. 1980. *Lightning Protection Code 1980.* Boston: Natl. Fire Protection Assn.

THAYER, R. L., JR. 1981. *Solar Access: It's the Law.* Univ. of Calif., Davis, Inst. of Govt. Affairs and Ecology Environmental Quality Series 34.

CHAPTER 6

Planting Site: Soil

Soil provides plants with water, nutrients, and root anchorage. Soils are responsible for the poor performance of landscape plants more often than any other single factor but are often given little consideration when a site is being selected for planting. Further, the development of a site and the construction of facilities compact the soil so that the surface is practically impervious to air and water. Even when proper soils are selected and imported for special planting situations, poor plant growth often reveals that landscape developers lack the necessary understanding of soil properties and their importance to plants.

Soil is a complex physical, chemical, and biological system. About half of its volume is composed of solid matter, primarily mineral particles with some organic matter; the other half consists of pore spaces filled in varying proportions with air and water. Soils are also alive with a wide range of bacteria, fungi, and other organisms.

SOIL FORMATION

Soils develop from the partial weathering of rocks and the interaction of rock with water, climate, and organisms. Soils are dynamic and evolving: A number of changes occur over time, giving rise to more or less distinctly visible layers below and parallel to the soil surface. These layers are called *horizons,* and their sum makes up the *profile* of a soil (Fig. 6–1). Soil horizons are designated by the letters *A, B,* and *C,* moving downward from the soil surface. The *A* horizon is the surface soil from which material is being leached; even so, it is most conducive to plant

Figure 6-1 Soil profiles vary greatly: a road cut exposes a shallow layer of soil on fractured sedimentary rock (left); a loam soil with a friable surface "A" horizon, a "B" horizon of accumulation between 300 to 600 mm (1–2 ft), and a "C" horizon below (center); a clay loam profile with a dark surface (high in organic matter) layer and a light-colored "A" horizon about 300 mm (1 ft) deep, a dark "B" horizon above the "C" horizon that begins about 600 mm (2 ft) below the surface (the profile is about 1.5 m [5 ft] deep) (right). (Photos: U.S. Dept. of Agr. Soil Conservation Service)

growth because it is usually well aerated, high in nutrients from decomposing litter, and friable. The *B* horizon, or subsoil, is one of deposition, especially of fine clays and soluble material from the *A* horizon. The *C* horizon, or soil material, is less weathered and is usually similar to that from which the *A* and *B* horizons were derived.

The first distinct layer to emerge is the dark-colored layer of decomposing organic matter on the surface. Its influence on the developing soil below will depend in large measure on the climate. In colder temperate regions, where rainfall exceeds evapotranspiration and the organic matter is usually acid, the surface soil is leached of soluble material and fine clays. A light-colored *A* horizon develops under the organic layer. In this type of soil formation, called *podsolization,* calcium carbonates are leached out of well-drained soils, leaving them somewhat acidic. Poorly drained soils are alkaline and usually high in calcium.

In areas with high rainfall and high temperatures, soils form by *laterization*. Decomposing organic matter in warm climates is usually alkaline. The alkalinity combines with high rainfall to leach silica from the surface soil. Iron and aluminum compounds remain, giving the soil a reddish-brown color.

In areas where evapotranspiration exceeds rainfall for part of the year, materials from the surface soil may be leached only a short distance, depending on the amount of rainfall. Calcium carbonate is predominant among the deposited salts, and the process is called *calcification*. If sodium is the primary cation, *alkali* soils are formed. The zone of deposition can become almost impervious to water, air, and roots. This *hardpan,* which can be a few centimeters or more than 10 meters (30 ft) thick, is the extreme example of a *B* horizon, a zone of accumulation. Soil depth above a hardpan varies from less than 100 mm (4 in.) to several meters (10–15 ft), depending on rainfall and the age of the soil. Hardpan should not be confused with *plowpan* which is caused by compaction from plowing at the same depth (see the section on Structure later in this chapter). Organic muck soils and peat form in marshy areas from decomposed plants.

It is useful to know whether a soil has developed in place or has been transported some distance from the original site. *Residual* soil usually occurs on mountain and foothill slopes; depositional or *alluvial* soil occurs in valley and outwash fans. Most alluvial soils are deposited by streams that flow out of mountains, carrying sand, silt, and clay from erosion of residual soils and the breakup of rocks in stream beds. These soils are usually deeper, more level, and more permeable to water than are residual soils. In some places, soil—called *aeolian* soil—is deposited by wind.

Soil color sometimes suggests past history or soil properties that may influence plants (Wildman 1976).

> Dark soil near the surface usually indicates a higher organic matter content there. This soil often has better tilth (ease of cultivation) and higher nutrient content than the subsoil.
>
> A uniformly colored profile is typical of unweathered young soils and usually uniformity of other soil properties and thus an excellent root environment.
>
> Increasing yellow and red colors in the subsoil are typical of older, weathered soils that contain free iron oxides. These subsoils often have a high clay content, a blocky structure, and, hence, decreased permeability for roots, water, and air.
>
> Whitish or light-gray zones or white networks following soil cracks usually indicate a high content of calcium carbonate (lime), which may induce iron and manganese deficiencies, particularly in certain woody plants. If fizzing or foaming results when a drop or two of 1 percent hydrochloric acid is applied to the soil, lime content is confirmed. Small white crystals that glitter in the sun but do not foam with acid are usually gypsum (calcium sulfate).
>
> Blue or gray zones or gray-and-rust-colored mottling of the soil indicate poor drainage and a lack of good aeration. Methane gas from natural gas or a landfill may cause a similar coloration of the soil (see Chapter 20).

Soils are grouped into *series* to aid in their classification and description. A series includes soils that have developed from similar materials by similar processes and have similar horizons. The soils within a series vary primarily in the texture (particle size) of the *A* (surface) horizon. Soil series are usually designated by the geographical names of the locations in which they were first found and described. The series name is modified by the texture of the surface soil, as in *Lansing silt, Hanford fine sand,* and *Dublin clay loam.*

Soil surveys have been conducted in the agricultural and forested areas of many countries. In the United States they are usually available on a county basis.

PHYSICAL SOIL CHARACTERISTICS

A number of soil characteristics affect plant growth and well-being. These include soil texture (size distribution of particles), structure (arrangement of soil particles), and soil depth. These factors, in turn, influence water infiltration and movement in the soil, water and nutrient retention capacities, and aeration. Topography can influence soil and plants through its effect on water accumulation or runoff, erosion, and temperature.

Texture

The larger mineral particles, rock and gravel, add little to a soil chemically or physically but do take up space and affect its permeability. Sand, silt, and clay particles compose most of the solid material of a soil. The smallest sand grain is 25 times the diameter of the largest clay particle (Table 6–1). A silt particle is between the two in size. Sand and silt are spherical or cubical in shape, while clay particles are more wafer-like. This difference in shape and the differences in particle size account for the much greater surface area of clay soil than of silt or sand. Particle size and shape and the high proportion of large pores in sandy soil result in its greater permeability, aeration, workability, and its lower water- and nutrient-holding capacities.

A given soil will usually comprise particles of varying sizes, and its characteristics are determined largely by the proportion of the different sizes of particles. A soil is described by its dominant particle sizes (Fig. 6–2). Note that a soil is not considered "sandy" until it contains more than 45 percent of sand. Similarly, a "silt" soil must be 40 percent silt, but only 20 percent clay makes a "clay" soil because it takes less clay to dominate the characteristics of a soil. Thus, a soil that is 60 percent sand would nevertheless be a "clay" soil if it were 35 percent clay.

Certain soils exhibit intermediate characteristics. These *loam* soils are ideal for

TABLE 6-1 CLASSIFICATION OF SOIL TEXTURE ACCORDING TO PARTICLE SIZE (Hartmann, H. T., A. M. Kofranek, V. E. Rubatzky, and W. J. Flocker, *Plant Science: Growth, Development, and Utilization of Cultivated Plants,* © 1988, p. 173. Reprinted by permission of Prentice-Hall, Inc., Englewood Cliffs, NJ)

	Diameter size (mm)[a]			
Particle	USDA system		International system	
Gravel	>2.0	(>0.08)	– –	
Very coarse sand	2.0–1.0	(0.08–0.04)	– –	
Coarse sand	1.0–0.5	(0.04–0.02)	2.0–0.2	(0.08–0.008)
Medium sand	0.5–0.25	(0.02–0.01)	– –	
Fine sand	0.25–0.10	(0.01–0.004)	0.2–0.02	(0.008–0.0008)
Very fine sand	0.10–0.05	(0.004–0.002)	– –	
Silt	0.05–0.002	(0.002–0.00008)	0.02–0.002	(0.0008–0.00008)
Clay	<0.002	(<0.00008)	<0.002	(<0.00008)

[a] Numbers in parentheses are in inches.

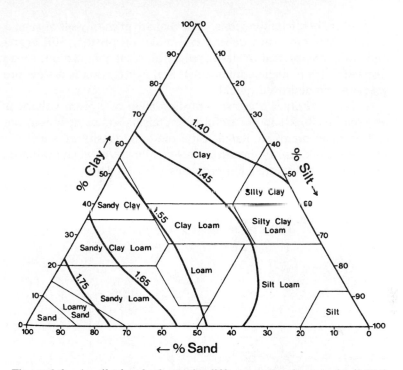

Figure 6–2 A soil triangle shows the different textural types of soil (U.S. Dept. of Agr.). Superimposed on the soil types are bulk densities (weight per volume of soil = g/cm³) that could restrict root growth and function (Morris and Lowery 1988). Combining these two measures of the physical properties of soil should better predict the suitability of a soil for plants.

growing a wide variety of plants because they combine the desirable attributes of each of the particle sizes. **Note again that it takes less than half as much clay as either silt or sand to impart its characteristics to a loam soil.**

Structure

Water and air movement and ease of root growth through soil are determined by pore size, not necessarily by the size of soil particles. Most particles in the soil are not discrete units but are grouped into aggregates or granules, each of which acts as a large particle as far as water and air movement are concerned. The water- and nutrient-holding capacities of the smaller particles, however, are still retained within each aggregate. The particles are held together by a combination of electrical and chemical bonds, gelatinous materials produced by plant roots and microorganisms, and iron and aluminum hydroxides.

Soil structure is a subjective term used to indicate the degree to which soil particles are aggregated. The more aggregated a soil, the more desirable is its structure for plants. Decomposing organic matter aids in the development and stability of soil aggregates. Soils of hot, arid regions usually have less organic matter than those of milder, humid regions. Soils that are primarily silt or sand aggregate poorly, so that their properties depend essentially on texture. Soil aggregates are fragile and easily compressed or destroyed by physical compaction, shearing, or chemical dispersion (as in high-sodium soils). Soil aggregation is a slow process, but aggregates can be destroyed quickly.

Bulk density is a measure of the weight of a given volume of soil (expressed as g/cm^3 or lb/ft^3) and is useful in appraising soil compaction. Since most mineral soil particles are of similar density, the bulk density of a soil is a measure of its porosity. Bulk density measurements can be useful when evaluating soil conditions or diagnosing suspected soil problems.

The bulk density of a soil and the extent it can be compacted depend on its structure and particle size distribution, and the compactive force exerted. However, as Figs. 6–2 and 6–3 show, two soils with similar bulk densities can differ greatly in

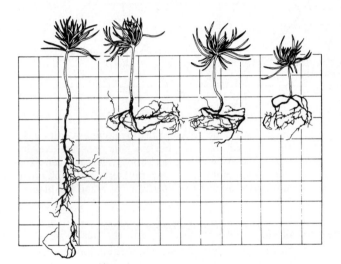

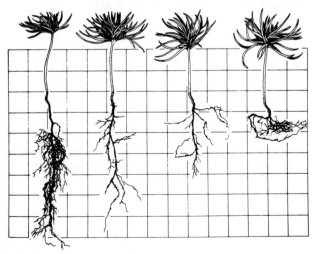

Figure 6–3 Similar bulk densities affect rooting differently in soils of different textures. The effect of bulk densities of 1.2, 1.4, 1.6, and 1.8 g/cm^3 on Austrian pine seedling growth in a sandy loam (bottom) and a silt loam (top) (20 mm [0.8 in.] grid). (Zisa, Halverson, and Stout 1980)

suitability for plant growth. The two soils (Fig. 6–3) had about the same total pore space and bulk densities, but the silt loam had fewer large pores. For a given soil, the greater the bulk density, the more that root growth and function will be restricted. Knowing the texture of a soil can help interpret bulk density measurements. The influence of texture probably accounts for much of the range in bulk densities given for each of the measurements listed:

1.0 to 1.6 g/cm^3	Normal soils (Brady 1990)
1.4 to 1.6	Root growth restricted (Alberty, Pellett, and Taylor 1984)
1.7 to 2.2	Commonly found at construction sites (Patterson 1977)
1.4 to 2.3	Brick (Gilman, Leone, and Flower 1987)

Caution is advised, however; some coarse-textured soils with seemingly acceptable bulk densities can be quite impervious because not enough of their large pores that affect bulk density measurements are connected by large enough pores.

Compaction decreases total pore space; more importantly, the large pores are severely compressed, and the resistance to root penetration is increased. The results are slow water infiltration, poor aeration, reduced drainage, impaired root growth and activity, increased erosion, and greater susceptibility to root rots but less mycorrhizal activity; in effect, plant well-being is jeopardized (Fig. 6–3).

Soil aggregates are delicate and easily damaged by cultivation, traffic, and construction activities. A loose well-aggregated soil will compact more than a soil that is less well-aggregated. Soils that have a range of particle sizes can be more severely compacted than soils of uniform texture (Fig. 6–2) regardless of particle size, because small particles fill the pores between larger particles (see Fig. 6–11). Such soils are particularly vulnerable when wet. Water causes the aggregate-cementing agents to become more fluid, so that soil particles more easily slip into positions that take up less space. Maximum compaction is most likely at field capacity, when soil particles move easily under pressure, and in soils with air-filled pores that can be compressed or filled with smaller particles.

A saturated soil cannot be further compacted, because the water in the pores does not compress. Because most pressures on soil are localized, as are those caused by feet and tires, muddy soils move out from under the area of compression and in so doing shear aggregates apart and rearrange particles. Soils that undergo this abuse are said to be *puddled* and, upon drying, shrink and become compact and practically impermeable.

A vehicle tire exerts a pressure on the surface on which it rolls approximately equal to its inflation pressure. Turning, starting, and stopping increase the force. Heavily treaded tires exert more pressure near the soil surface than do smooth tires. Compaction depends not only on the pressure exerted but also on the duration and frequency of the exertion, although the most damage occurs on the first occasion.

The pressure of a compacting force spreads with depth. The damaging effects of pedestrian and vehicular traffic are usually confined to the surface 100 mm

(4 in.) of soil (Chancellor 1976). The greatest compaction occurs about 20 mm (0.75 in.) below the surface. In a similar manner, cultivating to the same depth year after year, as is common in many farming operations, compacts the soil just below the depth of plowing. This can create a very dense layer 25 to 50 mm (1–2 in.) thick, known as a *plowpan*. Because plowpans interfere with water and air movement, they can be a problem when farmland is converted to landscape plantings, particularly if landscapers add fill soil without first breaking up the plowpan. Heavy equipment (particularly that used in road building), trucks, and tractors can affect the soil at greater depths, but the most serious compaction will usually be confined to the surface 150 to 200 mm (6–8 in.), unless the soil is so wet that it flows from under the tires.

Aggregates on the surface of bare soil can also be broken down by the impact of rain or sprinkler irrigation. Finer particles often seal the surface, thus reducing water infiltration and air exchange.

Soil Depth

The depth of a given soil determines its moisture- and nutrient-holding capacities and influences the depth of rooting. Usually, the deeper the soil, the greater will be the water and nutrient supplies for plant use. Depending on texture and structure, depth can affect the moisture content of shallow soils even when drainage is provided.

The term *effective root depth* has been used to describe the portion of the soil that is favorable for roots. In an alluvial soil with no noticeable stratification, effective root depth may be more than 1.5 m (5 ft); in a clay pan soil it may be as little as 300 mm (12 in.) or the depth of soil above the clay layer. To determine soil depth, it is necessary to determine which layers in the soil restrict root and water penetration. Table 6–2 identifies the terms used to describe soil depth in relation to average root zones.

Soil depth in landscape situations can be increased by the use of raised planters or mounds. Mounds provide surface drainage during periods of high rainfall as well as more complete internal drainage for better aeration. If the soil volume available

TABLE 6-2 TERMS USED TO DESCRIBE SOIL DEPTHS (Wildman and Gowans 1975)

Terms	Depth range according to:			
	Storie (1932)		USDA	
	mm	in.	mm	in.
Very shallow	<300	<12	<250	<10
Shallow	300–600	12–24	250–500	10–20
Moderately deep	600–900	24–36	500–900	20–36
Deep	900–1200	36–48	>900	>36
Very deep	>1200	>48	>1500	>60

in low-maintenance landscapes will not provide sufficient physical, nutrient, and moisture support, plants can be spaced farther apart to increase soil volume per plant.

Root and water penetration through a soil are altered by a layer that has a distinctly different texture from the soil above or below. If a subsoil layer has a noticeable increase in clay, water may accumulate for a time above this layer, forming a "perched" water table, and roots may be injured because of poor aeration. This condition is often called *waterlogging*. Sandy or gravelly layers can also interrupt the normal downward penetration of roots and percolation of water (Fig. 9–2). For example, water does not drain freely from a silt layer into sand until the silt layer becomes saturated for some depth above the coarser layer. This saturated layer can damage roots. The saturated zone will remain even after drainage has occurred because particle-to-particle flow of water is extremely slow and capillarity holds the water between the silt particles above the sand.

Topography

Hills and swales add variety and interest to landscapes and are often created on sites that would otherwise be fairly level. Slopes, however, offer special problems to both the landscape architect and the maintenance person and should be artificially introduced only for good reasons. Earth mounds are more effective sound barriers than plants alone on a level site. Mounds provide surface drainage during periods of high rainfall and can thus protect roots from waterlogging. Adequate drainage must be provided, however, to carry away the surface runoff so that it does not accumulate at the base of the slope. Mounds are often placed in the landscape primarily for aesthetic reasons. If these are to be covered with plants rather than turf alone, an easier solution would be to establish taller plants on level ground.

The most obvious problem with slopes is soil erosion from rain or irrigation. During irrigation, water is usually applied faster than it can enter the soil; this results not only in erosion but also in uneven water distribution, with underwatering at the top and overwatering at the base (Fig. 6–4).

Mounds should be of fairly uniform soil texture so that their full depth can be utilized by the roots of trees and large shrubs. If the surface soil is a shallow layer on top of compacted soil or soil of different texture, water will be drawn down through the top layer of soil toward the bottom of the slope, leaving little or no moisture to enter the mound for the benefit of growing plants (Fig. 6–4).

Both the soil and the air close to the surface will vary in temperature with the direction of the slope. At a highway interchange in central California, soil temperatures in mid-autumn at 25 mm (1 in.) below the surface were 34° C (93° F) on the slope facing the sun and 17° C (62° F) on the slope facing away from the sun. This range of temperatures can greatly affect the distribution of plants and their transpiration rates during the growing season. Near the top of a mound, these temperature extremes can exist within 3 m (10 ft) of each other. Two different exposures should be irrigated by different lines if they are sprinkler- or drip-irrigated.

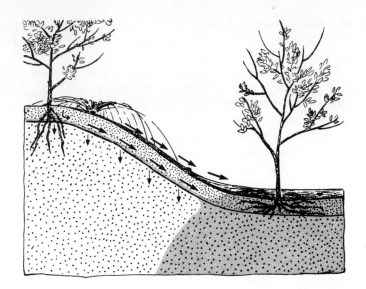

Figure 6-4 When water is applied to a slope more quickly than it can infiltrate the surface or penetrate deeper soil layers, it will flow to the bottom of the slope resulting in underwatering at the top, overwatering at the bottom, or both.

CHEMICAL PROPERTIES OF SOILS

Soil Fertility

Inorganic minerals make up about half the soil volume, organic matter usually makes up only 1 to 6 percent, and pore space makes up the balance. Mineral particles, particularly the clays, and organic matter are the principal reservoirs of 13 of the 16 elements now known to be essential for plant growth (Table 6-3).

Nutrients occur in soil in various states.

Dissolved ions in the liquid phase of the soil (soil solution)
Crystalline precipitates (calcium phosphates, lime, gypsum) and rock minerals (feldspars and micas)
Soil organic matter
Exchangeable ions adsorbed on the surface of soil particles (especially on clays and organic matter)

Nutrients in solution are available to plant roots and are also easily leached. As nutrients are absorbed by plants or are leached, more nutrients usually come into solution. Nutrients become available as soil particles weather (a slow process) and as organic matter decomposes. The capacity of a soil to adsorb nutrients on its particle surfaces, however, determines in large measure its fertility and productivity.

Cation Exchange Capacity. Nutrients most readily available to roots are those in solution and those adsorbed on the surface of clays. Soil particles are nega-

TABLE 6-3 ELEMENTS ESSENTIAL FOR PLANTS

Source and element	Symbol	Form available to plants	
Air and Water			
Carbon	C		CO_2
Oxygen	O		O_2, H_2O
Hydrogen	H		H_2O
Soil			
Macronutrients			
Nitrogen	N	Nitrate	NO_3^-
		Ammonium	NH_4^+
		Urea	$CO(NH_2)_2$
Phosphorus	P	Phosphate	H_2PO_4
Potassium	K	Potassium	K^+
Calcium	Ca	Calcium	Ca^{++}
Sulfur	S	Sulfate	$SO_4^=$
Magnesium	Mg	Magnesium	Mg^{++}
Micronutrients			
Manganese	Mn	Manganese	Mn^{++}
Zinc	Zn	Zinc	Zn^{++}
Boron	B	Borate	$H_2BO_3^-$
Copper	Cu	Copper	Cu^{++}
Iron	Fe	Iron	Fe^{+++}
Molybdenum	Mo	Molybdate	$MoO_4^=$
Chlorine	Cl	Chloride	Cl^-

tively charged and therefore attract positively charged ions, or cations. The capacity of a soil to adsorb cations is called its *cation exchange capacity*. The finer the texture of a soil, the more surface area per unit volume it has and the greater its cation exchange capacity. Fine-textured soils are usually more fertile than coarse-textured ones.

Cations held on soil are exchangeable: They go into solution and are replaced by others in solution. Even though adsorbed ions are exchangeable, they are held fairly securely against leaching unless they are replaced by ions in solution. In contrast with ions in solution, exchangeable cations are not readily leached.

Anions. The soil solution contains equivalent anions and cations. Nitrate, sulfate, and phosphate are the principal nutrient anions in a soil solution. Nitrate is almost completely soluble; it is readily available to plant roots and is also leachable. Sulfate is moderately soluble and is less subject to leaching than the nitrate ion is. Phosphates are held in most soils as precipitates, largely as insoluble iron and aluminum phosphates in acid soils and as calcium phosphates in neutral and alkaline soils (Pritchett 1979). In many soils, phosphates are not readily available to roots, nor are they readily leached.

Soil Reaction

Soil reaction, expressed as pH, refers to the acidity or alkalinity of a soil (Fig. 6-5), or the relative proportion of hydrogen (acid) and hydroxide (alkaline) ions. Equal

Chemical Properties of Soils

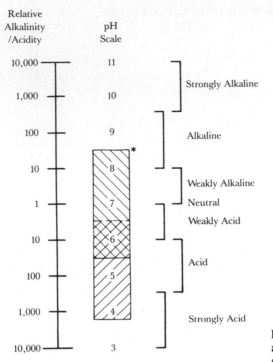

Relative
Alkalinity
/Acidity

pH
Scale

Strongly Alkaline

Alkaline

Weakly Alkaline

Neutral

Weakly Acid

Acid

Strongly Acid

*pH range of about 5.5-8.3 satisfactory for most plants.

Acid-loving plants usually grow well at pH values of 4.0-6.5.

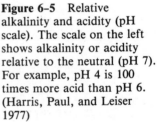

Figure 6-5 Relative alkalinity and acidity (pH scale). The scale on the left shows alkalinity or acidity relative to the neutral (pH 7). For example, pH 4 is 100 times more acid than pH 6. (Harris, Paul, and Leiser 1977)

concentrations of the two produce a neutral reaction—a pH of 7.0. As a soil becomes more acid, its pH decreases; as it becomes more alkaline, its pH increases. A one-unit change in pH indicates a tenfold change (a logarithmic relationship) in hydrogen and hydroxide ion concentrations.

Soil reaction primarily influences plant growth indirectly through its effects on the solubility of ions and the activity of microorganisms. The availability of a number of nutrients, particularly phosphorus and the micronutrients, is influenced greatly by soil pH (Fig. 6-5). Except for molybdenum and chloride, micronutrients become less available as soil alkalinity increases. In soils with high pH, usually only iron and manganese are seriously deficient. On the other hand, manganese and aluminum can reach toxic levels under moderately to highly acidic conditions. Most microbial activity in the soil, especially that of nitrogen-fixing bacteria, is reduced as soil acidity increases (below pH 5.5). Most leguminous plants grow better in neutral or alkaline soils, where nitrogen-fixing bacteria can thrive. Conversely, soils of coniferous nurseries are often kept below pH 5.5 to reduce "damping-off" (a root-killing disease) and other fungal diseases (Pritchett 1979).

Most plants that grow well in soil between pH 4.0 and 6.5 also grow best in well-aerated (well-drained) soil. Most "acid-loving" plants have developed in well-drained soil under high humidity or heavy rainfall (above 500 mm or 20 in. annually), which leach out the basic ions, primarily calcium and magnesium, and replace them with hydrogen ions. Poorly drained soils retain most basic ions and are alkaline. Even if drainage is good, most soils in arid regions (those that receive less than 500 mm of rain annually) are neutral or alkaline because rainfall is not heavy enough to leach the basic ions. In fact, ions accumulate under most arid conditions.

Soil reaction is influenced by vegetation as well as by rainfall and soil drainage. Grasses contain higher concentrations of calcium and magnesium than does conifer litter; hence, more hydrogen ions are displaced and leached from the surface soil under grass by the recycling "base" elements. Grasses tend to keep soils more alkaline than forest plants, particularly coniferous ones (Halfacre and Barden 1979). Using conifer needles as mulch is one way to help maintain acid soil.

Most plants grow well in soils with a wide range of pH—from about 5.5 to 8.3—particularly if the soil is well drained. Except for some acid-loving plants that grow well in soils as low as pH 4.0, a soil pH between 6.0 and 6.5 is thought to be best.

Adjusting Soil Reaction. If a soil is too acid or alkaline, its pH can be increased by adding lime or decreased by adding sulfur. The amount of lime or sulfur needed depends on soil texture, organic matter, material used, and how much the pH is to be changed. The finer the soil and the higher the organic matter content, the more amendment must be added to achieve the same pH change.

Ground limestone, consisting of calcium carbonate or calcium magnesium carbonate (dolomite), is commonly used to increase soil pH (Table 6-4). Quicklime (calcium and magnesium oxides) and slaked lime (calcium and magnesium hydroxides) act more quickly than limestone but are more expensive and disagreeable to

TABLE 6-4 APPROXIMATE AMOUNTS OF FINELY GROUND LIMESTONE[a] NEEDED TO INCREASE pH OF THE SURFACE 200 MM (8 IN.) OF AN ACID SOIL TO pH 6.5 (Adapted from USDA Handbooks 18 and 19 and from Dickey 1977)

	Kilograms of limestone per 100 square meters[b, c]					
Desired change in pH	Sand	Sandy loam	Loam	Silt loam	Clay loam	Muck
From 4.0 to 6.5	130	250	350	420	500	950
From 4.5 to 6.5	110	210	290	350	420	810
From 5.0 to 6.5	90	170	230	280	330	630
From 5.5 to 6.5	60	130	170	200	230	430
From 6.0 to 6.5	30	70	90	110	120	220

[a] Or dolomite lime

[b] Double the kg/100 m² to approximate the lbs/1000 ft².

[c] For soils with considerable organic matter, increase the amount of limestone applied about 15 percent for clay soils and up to 50 percent for sandy soils.

Chemical Properties of Soils

handle. Small particles of limestone will increase pH more quickly than large ones. Lime should be spread evenly on the soil and worked into the surface for 50 to 100 mm (2–4 in.) or deeper if there are no plants. When it is needed, lime should be incorporated into the soil during site preparation before planting. The increased pH will improve soil structure and will increase phosphorus and molybdenum, reduce harmful levels of soluble manganese and aluminum, and enhance microbial activity.

Wood ashes are 30 to 50 percent as effective as calcium carbonate in neutralizing acid soil, since they are about 45 percent calcium carbonate (American Horticultural Society 1980). Hardwood ash contains about one-third more calcium than does softwood ash. It also contains up to 30 percent potassium and can be used as a source of K in acid soils (see Chapter 12). Wood ash should not be applied directly to alkaline soils but can be composted along with leaves and other organic matter where organic acids will partially neutralize soil alkalinity.

Alkaline soils are usually acidified by the application of agricultural sulfur (Table 6–5) or sulfur-containing materials such as iron or aluminum sulfate (Table 6–6). The amount of sulfur required to lower pH is influenced greatly by the carbonate content of a soil. If a soil is high in carbonate, it should be analyzed to determine the precise amount of sulfur needed to produce the desired pH; how to take a soil sample is described in Chapter 12. This is a fairly simple test for a chemical laboratory. Finely ground dusting and wettable sulfurs act faster than coarser agricultural sulfur but cost more. Like lime, sulphur should be applied evenly on the soil and worked into the surface 50 to 100 mm (2–4 in.) or deeper if possible.

Alkali Soils. In arid and semiarid regions where rainfall is not adequate for leaching, soil is usually alkaline and high in salts. Corrective measures involve leaching salts from the soil (see Chapter 13). If sodium represents more than 15 percent of the exchange capacity, a medium- or fine-textured soil will lose its granular structure due to deflocculation (dispersal into separate particles) of clay. Such soils are called *alkali soils* and are exceedingly impervious to air and water. Gypsum or sulfur should be added to alkali soils before they are leached.

TABLE 6-5 THE APPROXIMATE AMOUNTS OF SOIL SULFUR REQUIRED TO INCREASE THE ACIDITY OF THE SURFACE 200 MM (8 IN.) OF A CARBONATE-FREE SOIL (Adapted from Calif. Fertilizer Assoc. 1985. *Western Fertilizer Handbook.* 7th ed.)

Desired change in pH	Kilograms of sulfur per 100 square meters[a]		
	Sand	Loam	Clay
From 8.5 to 6.5	100	125	150
From 8.0 to 6.5	60	75	100
From 7.5 to 6.5	25	40	50
From 7.0 to 6.5	5	8	15

[a] Double the kg/100 m^2 to approximate the lbs/1000 ft^2.

TABLE 6-6 THE RELATIVE ACIDIFYING EFFECTS OF CERTAIN AMENDMENTS AND FERTILIZERS (Adapted from Calif. Fertilizer Assoc. 1985. *Western Fertilizer Handbook.* 7th ed.)

Material	Grams equivalent to 100 grams of sulfur	Grams of calcium carbonate neutralized by 100 grams of the material	Grams of calcium carbonate required to neutralize material per 100 grams of nitrogen
Sulfur	100	312	–
Sulfuric acid	306	104	–
Gypsum	538	58	–
Aluminum sulfate	694	45	–
Iron sulfate	869	36	–
Anhydrous ammonia	212	148	180
Ammonium sulfate	283	110	530
Urea	439	71	160
Ammonium nitrate	503	62	180
Aqua ammonia	867	36	180
Single and treble superphosphate	–	0	–
Most potassium salts	–	0	–
Calcium nitrate	(1560)[a]	(20)[b]	(130)
Potassium nitrate	(1357)	(23)	(180)

[a] Alkaline reaction; grams of material necessary to neutralize 100 grams of sulfur.

[b] Alkaline reaction; equivalent to the same number of grams of calcium carbonate.

BIOLOGICAL PROPERTIES OF SOILS

A teaspoonful (5 ml) of soil may contain billions of living organisms (Clark 1957). Among the soil inhabitants are those that decompose organic matter, fix nitrogen from air, transform nitrogen, improve soil tilth, produce antibiotics, and otherwise affect plant welfare either positively or negatively.

Soil Animals

The usefulness of animals to the soil is essentially inversely proportional to their size (Pritchett 1979). The more important animals are insects, worms, protozoa, and nematodes. A wide range of insects and spiders chew and move organic matter into the soil.

Earthworms are probably the best known of the soil fauna and occupy the greatest volume. They contribute to moist, friable, and fertile soil by aerating, mixing, and enriching it in tremendous quantities. This does not mean, however, that earthworms can be added to a dry, compacted, or infertile soil to transform it into a decent growing medium. The worms most likely will crawl away in the night to seek a more favorable site.

Protozoa—one-celled organisms—are the most numerous of soil fauna. They decompose organic matter and bacteria, primarily in the surface soil. Nematodes, microscopic roundworms, include both saprophytic and parasitic species. The saprophytes are free-living and are generally beneficial in that they decompose organic matter (Pritchett 1979). Some of the parasitic nematodes attack other microorganisms in the soil and are considered desirable. Others, however, primarily invade the roots of plants and can inflict considerable damage (see Chapter 22).

Soil Microflora

Microflora in soil include algae, bacteria, actinomycetes, and fungi. Algae and a few species of bacteria contain chlorophyll and are able to form carbohydrates. Algae are most abundant in fertile soil that is neutral or slightly alkaline. They assist in dissolving soil minerals and in forming soil.

Bacteria, even in anaerobic soils, are involved in many biological and chemical processes that contribute to decomposition of organic matter. Certain bacteria, alone and in symbiosis with leguminous plants, fix atmospheric nitrogen; others denitrify nitrates. Still other soil bacteria may cause serious plant diseases, such as crown gall (see Chapter 21).

Actinomycetes are similar to bacteria in their environmental requirements, and they are similar to fungi in that they form mycelia, masses of hyphae (hair-like filaments), and in the method by which they decompose cellulose. Some species, in symbiosis with the roots of certain plants, such as alder and ceanothus, can fix nitrogen.

Fungi are probably the most important decomposing microflora, particularly in well-aerated, acid soils. A number of soil fungi cause serious plant diseases: damping off of seedlings, phytophthora root rot, armillaria root rot, and verticillium wilt (see Chapter 21). Fungi also form symbiotic relationships with roots, which can be vital to the well-being of a plant. These associations are called *mycorrhizae*.

Mycorrhizae

Many trees, including those on previously unforested land, fail because certain fungi are not present to form mycorrhizae with the roots of the trees. Mycorrhizae (*myco* means fungus and *rhiza* means root) are root structures created when young lateral roots are invaded by specific fungi that form symbiotic associations to the advantage of each. **Mycorrhizal plants are reported to grow more vigorously and remain healthier than do noninfected plants under stressful conditions,** as in infertile soils, at arid sites, in the presence of root-disease organisms, and in other harsh environments (Merrill and Solomonson 1977).

Mycorrhizae are numerous and common in most higher plants. Pritchett (1979) has estimated that more than 2000 species of mycorrhizal fungi exist on trees in North America. Merrill and Solomonson (1977) report that mycorrhizae occur in more than 80 percent of vascular plant taxa investigated. Although almost all researchers report that each tree species tends to produce mycorrhizae with only certain groups of fungi (Pritchett 1979), several species of fungi can infect the same species of plant; and many species of fungi can infect more than one species of plant (Bowen 1987). Iyer, Cavey, and Wilde (1980) report that all currently or previously forested soil contains fungi that can form mycorrhizae with all tree species and that prairie and other grassland soils that have never been forested do not have such fungi.

Mycorrhizae generally divide into two major types, depending on the relationship of the fungal hyphae and the root cells. The fungal hyphae of *ectomycorrhizae* grow between the cortical cells of short lateral roots and form a sheath or mantle around the root invaded. The fungal hyphae of *endomycorrhizae* (also called vesicular-arbuscular or VA mycorrhizae or VAM), on the other hand, primarily invade individual cells within the cortex and do not form a mantle around the roots. Even though fungi invade the cortex vigorously, they do not invade the endodermis or the more interior tissues of the root. Mycorrhizal fungal mycelial strands and their hyphae extend from the roots they have invaded and function much like the root hairs of normal roots, except that they are better able to extract nutrients that are otherwise not readily available (Merrill and Solomonson 1977). Ectomycorrhizae are found almost exclusively on trees, whereas endomycorrhizae are more numerous and widespread, occurring on most families of angiosperms and gymnosperms (Pritchett 1979). Ectomycorrhizae infect lateral roots, which are greatly shortened by their presence, swollen, frequently dichotomously branched, and usually devoid of root hairs (Fig. 6-6). In contrast, endomycorrhizal infection does not grossly affect root appearance. Even though it involves the invasion of roots and even of individual cells, it does not kill cortical cells but often prolongs their effectiveness.

Plants benefit from mycorrhizae in several ways. Nutrient uptake, particularly

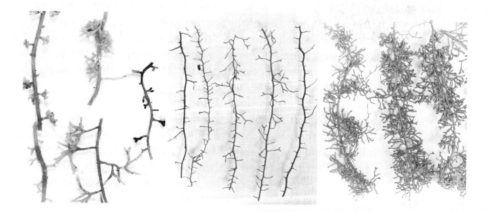

Figure 6-6 Mycorrhizae increase nutrient absorption. Examples of five ectomycorrhizal types on loblolly pine roots (each root segment is about 25 mm [1 in.] long) from a conventional nursery (left). Nonmycorrhizal roots of loblolly pine (center); loblolly pine ectomycorrhizae formed by *Pisolithus tinetorius* (right); center and right root segments about 50 mm (2 in.) long. (Photos courtesy Donald H. Marx, U.S. Forest Service)

that of phosphorus, is enhanced in infertile soils because mycelial strands and their projecting hyphae explore the soil more extensively than nonmycorrhizal roots (Stribley 1987). It has been thought that mycorrhizae improve water absorption and increase drought resistance. However, "most of these phenomena are probably a result of improved P nutrition" (Stribley 1987, quoting Safir and Nelsen 1980). It is argued that mycorrhizae are better able to absorb phosphorus in drier soil than can nonmycorrhizal roots and thereby the roots are better able to remain active. The fungal hyphae may also exude toxins, and the fungus mantle may serve as a physical barrier that reduces infection by soil-borne pathogens (Marx 1973).

Fortunately, tree mycorrhizal fungi are legion; their spores spread extensively by wind but do not form mycorrhizae unless they find host trees (Iyer, Cavey, and Wilde 1980). Attempts to improve plant performance by enhancing the variety and concentration of fungi have rarely been successful (Mikola 1973). Appropriate fungi are usually already present in the plant, the soil, or both if the site accommodates similar plants. On the other hand, mycorrhizal enrichment would be beneficial in some situations, as when exotic tree species are introduced, particularly by seed, into grasslands; when noninfected plants are planted in subsoil or sterilized soil in landscape planters or mine spoils isolated from sources of inoculum; or when noninfected plants are grown in fumigated nursery soil.

Seemingly noninfected plants will grow satisfactorily if they are wisely fertilized and watered and if diseases are controlled. Iyer, Cavey, and Wilde (1980), however, report that seed of nearly all tree species planted in grassland soils, including those of high fertility, usually produce trees that die within two years. Mycorrhizae do not develop as extensively on plant roots that are growing in fertile, moist soil as on those in infertile and inadequately watered soil. Mycorrhizae require well-aerated soil. Attempts to grow noninfected plants in an infertile soil devoid of appropriate fungi would be devastating. In some infertile soils, even abundant mycorrhizal development may not supply enough nutrients for vigorous growth, particularly if plants are young. Generous fertilization will decrease mycorrhizal development. If soils low in phosphorus are fertilized heavily with nitrogen, plant growth may actually be suppressed. This is thought to occur because fewer mycorrhizae develop under high-nitrogen conditions and thus absorb less phosphorus for the plant.

Mycorrhizal fungi can be introduced by means of soil or surface litter from woody plantings, with care taken so that pathogens are not introduced. The soil or litter should be raked into the surface under the plants to be inoculated. If plants ready for planting are not already infected with the appropriate fungi, they may have their roots dipped in a slurry of inoculated soil. Seeds can be inoculated by dipping them in a slurry of inoculum. Alternately, pure cultures of mycorrhizal fungi may be sprinkled on the soil and raked in around the plants. The most advantageous method will depend on the number of trees or seeds involved and how far the inoculum must be transported.

Even though mycorrhizae are necessary to the survival of certain plants under infertile conditions, appropriate mycorrhizal fungi are present at almost all sites. Mycorrhizal inoculation may be extremely valuable in low-maintenance landscapes known to be devoid of appropriate fungi.

Landscape plants probably suffer more from moisture-related problems than from any other cause. For them it is either feast or famine, flood or drought, air or suffocation, acceptable water or saline water.

No organic process occurs in the absence of water. Water is a primary constituent in the photosynthetic production of organic matter. Water is the solvent for nutrient and food transport within plants. Transpiration cools plants. Roots can extend into soil and shoot tips grow only when water is absorbed and the consequent turgor produced. Nevertheless, though water is vital to the well-being of a plant, excess water is often responsible for decline and death

Soil is the reservoir that supplies water to terrestrial plants. Air and water occupy the pore space between soil particles. Pore space occupies about half the total soil volume. Porosity is determined by soil texture and structure. Fine-textured soils, particularly if they are well aggregated, have more total pore space than do coarse-textured soils. Total porosity, however, is less important than the size of individual pores. Large pores allow for faster water movement and better aeration. In a well-aggregated and fine-textured soil, the soil will not only be well drained but will also have a high moisture-holding capacity. Sandy soil has a high percentage of large particles and pores and is well drained and well aerated, but it has a limited water-holding capacity.

The size and number of pores determine the amount of water that a soil can hold. Ideally, 30 to 50 percent of the pores will be large enough to allow water to drain from them after rain or irrigation. Such pores are called *noncapillary pores,* and the water that drains from them is *gravitational water.* The amount of water remaining in the soil after drainage has taken place is called the *field capacity.* The moisture remaining in the soil is held against the pull of gravity in the smaller pores by *capillarity,* on the particle surfaces by *adsorption,* and even between the layers of single clay particles. About half of this *capillary water* is available to plant roots. As water is absorbed by plants, smaller and smaller pores are drained and the films of soil water become thinner until the *tension* or force with which water is held becomes so great that the roots are no longer able to absorb water fast enough to keep the plant from wilting. The soil is said to be at the *permanent wilting point* if the plant will not recover overnight (or in 100 percent relative humidity) unless water is added to the soil (Fig. 6–7).

The water between field capacity (FC) and the permanent wilting point or percentage (PWP) is *available water.* The water that remains in a soil at PWP is essentially unavailable to plants and is termed *hygroscopic water.* Hygroscopic water is adsorbed on the surface of soil particles and is present in the tiniest capillary pores. It can be removed by oven-drying the soil or exposing thin layers to the sun, although some water will still remain chemically combined with clay particles. At the other extreme, when the soil pores are completely filled with water, the soil is said to be *saturated.*

The amounts of water held by a soil at saturation, at field capacity, and at permanent wilting, and the water available to plants are characteristics of the soil, not of the plants growing in it (Fig. 6–8). These amounts of water can be expressed

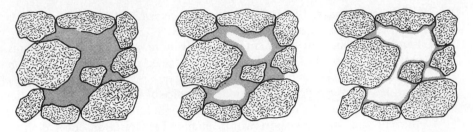

Figure 6–7 Soil moisture conditions. At saturation, pores in the soil are filled with water (left); at field capacity, water adheres to soil particles after drainage and fills smaller pores (center); at the permanent wilting point, a thin film of water remains on soil particles and roots cannot absorb water fast enough to prevent wilting (right).

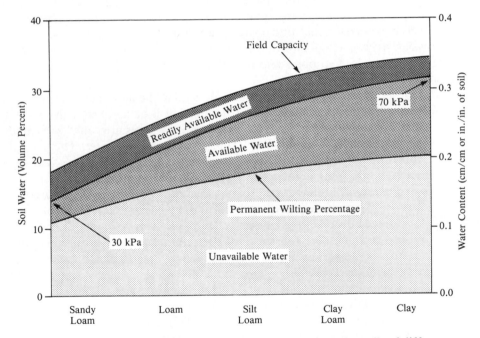

Figure 6–8 Representative soil moisture characteristics for soils of different textures are shown. Even though the permanent wilting percentage, the field capacity, and the amount of available water increase with changes in texture from a sandy loam to a clay soil, the amount of water readily available to plants remains about the same. Soil texture has little or no effect on the amount of water readily available to plants if water is applied every three to four days. More water is not readily available in a sandy loam at 30 kPa because the water films are so thin water movement is restricted (see Chapter 13). (Graph prepared from information of Grant, Goldhamer, and Gagnon 1986; Goldhamer and Snyder 1989)

in different ways. Most often, they are determined by weighing a representative sample of soil, including its water, and oven-drying it until the weight is constant for an hour or more. The water content or weight is then expressed as a percentage of the dry weight of the soil. The amount of water can also be expressed on a percent-by-volume basis, but this is more difficult to determine, since the volume of the soil sample must be measured along with the volume of water, which is calculated from the weight of the soil before and after drying. Water content is also expressed as the depth of water per unit depth of soil—for example, cm of water per meter of soil or inches of water per foot of soil. Water use or loss (evapotranspiration) is usually expressed in depth of water lost per day, week, or year. Rainfall and water application, particularly from sprinklers, are also expressed as depth of water. The water available to a plant is directly related to the depth of soil and plant roots.

Another term, *container capacity,* is useful for considering water relationships in shallow container soils. As discussed in the following section, water does not drain freely from a fine-textured soil to one that is coarser or to air. Therefore, the amount of water left in a container soil after it has stopped draining will be somewhere between field capacity and saturation, depending on soil texture and depth (Fig. 6–9).

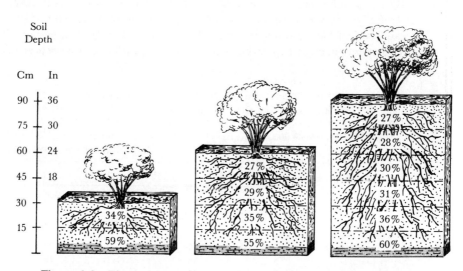

Figure 6–9 The water content of soil in a planter varies with depth. In soils of similar texture, the water content at a given distance from the bottom will be about the same.

The force with which a soil holds moisture is called the *soil moisture tension* and is most commonly expressed in kiloPascals (kPa). One hundred kPa are equal to a standard atmosphere (1030 g/cm^2 or 14.7 lb/in.^{2}). At some of the soil moisture levels, moisture tensions are (see next page)

Soil Water

	KiloPascals (kPa)
Saturation	0
Field capacity	10–35[a]
Reduced water availability	30–70[a]
Permanent wilting point	1520

[a] Tension depends on soil texture; the lower tension would be for coarse-textured soil (sand) and the higher tension for fine-textured soil (clay).

Moisture was originally thought to be equally available to plants at all tensions between field capacity and the permanent wilting point (Veihmeyer and Hendrickson 1955). For practical purposes, that is the case; even though moisture tension increases from field capacity to 1520 kPa at permanent wilting, 75 to 95 percent of available water has been removed (from clay to sand respectively) at a tension of 500 kPa (Singer and Munns 1987).

Interestingly, **coarse-textured soils have more water readily available below 50 kPa than do fine-textured soils.** From Table 6–7, it appears that even below 200 kPa, a loamy sand has as much water readily available as the clay. This would be true if water were withdrawn slowly, but under normal transpiration, there would be such thin films of water in some sands that at even 50 kPa enough water might not flow fast enough to prevent water stress. In order **to have water equally avail-**

TABLE 6-7 ESTIMATED *AVAILABLE WATER* (AW) (*FIELD CAPACITY* [FC; 10–35 kPa] AND THE *PERMANENT WILTING PERCENTAGE* [PWP; 1500 kPa]); READILY AW (LESS THAN 30 TO 70 kPa DEPENDING ON SOIL TEXTURE); AND AW LESS THAN 200 kPa OF TENSION IN THE TOP 600 MM (2 FT) OF SOIL FOR SOILS OF VARIOUS TEXTURES (Adapted from Grant, Goldhamer and Gagnon 1986; Goldhamer and Snyder 1989)

	Available water[a]				
	FC–PWP	<30–70 kPa[b]		<200 kPa	
Soil texture	mm/600mm	%	mm/600mm	%	mm/600mm
Loamy sand	45 (1.8)[c]	55	25 (1.0)	90	40 (1.6)
Sandy loam	55 (2.2)	56	31 (1.2)	76	42 (1.7)
Fine sandy loam	62 (2.5)	55	34 (1.4)	82	51 (2.0)
Loam	70 (2.8)	40	28 (1.1)	71	50 (2.0)
Clay loam	85 (3.4)	23	20 (0.9)	45	38 (1.5)

[a] The depth of water available can be converted to liters per 100 m² by multiplying mm/m by 100, or gallons per 1000 ft² by multiplying in./3 ft by 650.

[b] Readily Available Water with soil moisture tensions ranging from 30 kPa for loamy sand to 70 kPa for clay loam. Water would be equally available to plant roots.

[c] Inches of water available in the top 2 feet of soil, which would include most of the absorbing roots.

able, a sandy soil usually has to be watered at a lower moisture tension than a finer-textured soil. It need not be a concern, however, if rains or irrigations are frequent.

Plants absorb moisture primarily from the upper 0.5 m (20 in.) of soil until moisture tension increases, then more water is absorbed from greater depths. The amount of available water may vary in different soil horizons and therefore influence rooting density and extent. As a result, a plant may be able to absorb enough water even though much of the root zone is at permanent wilting.

Movement of Water

Water movement in soil depends primarily on the height of the water column (head or potential), the texture and structure of the soil, and its initial water content. If water does not infiltrate a soil as quickly as it is applied by rain or irrigation, the runoff may cause soil erosion, uneven water distribution, and on a slope, less water than applied will accumulate in the root zone. Water will enter grass- or mulch-covered soil more rapidly than it will a bare soil (Table 6–8). The surfaces of bare soils often crust from *puddling* (dispersion of soil aggregates by drops of water) during rain or sprinkler irrigation (see Fig. 14–5); these crusts slow water intake.

Water penetrates a soil more rapidly before the surface 20 to 30 mm (1 in.) are wetted. Water moistens the surface soil to field capacity or above before it moves farther down in the soil. The soil below this level, however, will remain at its same moisture content.

Above field capacity, water moves down through the soil in response to gravity. The greater the head (potential) and the larger the pores, the faster the water moves. Below field capacity, water moves by capillarity from zones of higher moisture content (low tension or high potential) to those of lower content (high tension or low potential). Capillary movement can be in any direction. Water can move farther (though it will move more slowly) in soils with small pores than in soils with large pores. Unless it is near a water table, capillary flow is extremely slow and restricted in distance covered. Capillary movement decreases rapidly as soils dry below field capacity. A water table will supply the soil above it by capillarity for some distance, depending on soil pore size. In laboratory experiments, the upward

TABLE 6-8 TYPICAL WATER INFILTRATION RATES INTO THREE SOILS, EACH WITH A SURFACE AREA OF 148 SQUARE METERS (1600 SQUARE FEET) (Harris 1962)

Soil	Soil with cover				Soil without cover			
	mm/ hour	in./ hour	l/ hour	gal/ hour	mm/ hour	in./ hour	l/ hour	gal/ hour
Sandy loam	13	0.50	2000	530	8	0.33	1350	355
Silt loam	8	0.33	1350	355	6	0.25	1000	265
Clay loam	6	0.25	1000	265	5	0.20	850	210

movement of water was significant in a fine, sandy loam from water tables as deep as 8 to 9 m (27–30 ft) (Gardner and Fireman 1958). Rate of movement was fairly uniform until the water table reached to within 2 m (6 ft) of the surface; it increased by a factor of eight at 1 m (3 ft). If a water table remains at about the same level (within about 9 m [30 ft] of the surface) throughout the year, it can be a continuous source of water for plants.

As the permanent wilting point is approached in a soil, capillary water movement all but ceases and water moves only in the form of vapor. Although vapor movement in soil is extremely slow, it is more rapid than was once thought. Vapor can condense on one surface of water held between soil particles while water is evaporating from the other surface (Philip and DeVries 1957). Vapor movement is greatly increased by temperature gradients within the soil. Other things being equal, water vapor moves toward colder zones, where vapor pressure is lower than at higher temperatures. Surface soil is warmer in summer and colder in winter than soil at greater depths. A similar cycle can happen daily: Moisture vapor moves to the surface and condenses during the winter and during each night, particularly if strong radiation cooling occurs. The movement will be downward during the summer and during each day. Only in special circumstances, however, will the amount of water condensed in the root zone be sufficient to sustain plants. The more extensive the foliage canopy, the less radiation cooling of the soil surface will occur at night, and the less water vapor will rise. The water supply of small plants, however, may be augmented if a clear plastic sheet is used to condense water vapor from a 3 to 4 m^2 (30–40 ft^2) area of soil and concentrate it at the base of the plant (see Fig. 11–13).

Water movement in soils can be inhibited not only by compaction but also by differently textured strata. If a coarse stratum is on top of a finer-textured soil, water may accumulate above the lower layer until it infiltrates at the normal rate of the lower layer. When drainage is complete, the finer-textured soil will be at field capacity, but the coarser soil above will be somewhat below its field capacity because of the greater tension with which water is held by the finer-textured soil. Should the interface slope, water will flow to lower levels above the interface, resulting in uneven watering and a possible wet low spot.

A more serious situation occurs when a coarse-textured stratum is located below soil of finer texture. Water will not move from the fine-textured soil into the coarse soil below until it saturates the top layer and develops enough of a head (potential) to overcome the cohesive and adhesive forces holding it there. Even after drainage has been completed, the soil immediately above the interface will be saturated. This essentially constitutes a *perched water table*. Roots penetrate this interface with difficulty (Fig. 6–10). This phenomenon cannot be stressed too strongly because it comes into play in a number of landscape situations: Drainage pipes are sometimes lined with gravel to ensure lateral water movement to the drain; the bottoms of planters are usually covered with gravel; a container root ball may experience decreased moisture when planted in landscape soil; and gardeners will sometimes place gravel in the bottom of a planting hole with the fallacious notion that it will improve drainage.

Figure 6–10 An extreme example of stratification with abrupt boundaries between the top 300 mm (1 ft) of sandy loam, the second 300 mm (1 ft) of sand, and the underlying silty clay loam. Although soil density was not a problem, the roots of young pear trees would not grow through the boundary into the coarse white sand (left). Compare this root growth with that two years later when trees had been replanted in a soil mixed uniformly with the use of a trenching machine (right). (Wildman, Meyer, and Neja 1975)

LANDSCAPE SOILS

Many soils may adversely affect plant performance unless they are modified or precautions are taken. The original soil may be difficult to manage, and many landscapes are manipulated during site development by the removal of topsoil, addition of soil, compaction of soil, burial of construction materials, or construction of pavement. It is a wonder that plants grow at all in some soils. In addition to the following discussion, many soil problems are considered in the first parts of Chapters 11 and 20.

Soil Compaction

"The largest single killer of trees is soil compaction . . . by people and pigeons whose small feet exert greater pressure [per unit area at the surface] than heavy machines" (Perry 1982). Even so, vehicles and equipment compact soil more severely and to a deeper level. Vibration from traffic on paved streets seriously compact soils, particularly silty soils.

In preparing soils for structures, roads, and pavement, engineering specifications usually require soil to be excavated to within 150 mm (6 in.) of the specified depth to be compacted. The soil is loosened 150 mm deep, wetted to near field capacity, and compacted to 90 to 97 percent compaction.* The excavated soil is then replaced in 150 mm lifts (layers), and each lift is wetted and compacted similarly (see Chapter 7). For a street, the compacted soil is topped with 200 to 500 mm (8–20 in.) compacted-aggregate base and then 50 to 125 mm (2–5 in.) of asphalt. A soil at 95 percent compaction is almost, if not completely, impenetrable to roots. In such severely compacted soil, roots can grow only in any cracks that occur.

Minimizing Soil Compaction. Although some breakdown of structure within the surface 300 mm (1 ft) of soil may be inevitable when construction is taking place, an understanding of soil texture and structure can be applied to reduce structural breakdown. Compaction is easier to prevent than to remedy. The following practices should be observed:

Cultivate soil when it is dry or only moderately moist; avoid wet soils

Avoid recompaction of freshly plowed or loosened soil; the less tillage after loosening, the better

Schedule landscape maintenance work when the soil is dry

Keep travel over the site to a minimum; confine it to a few paths and keep it away from trees

Use lightweight vehicles with large, smooth, low-pressure tires

Spread thick, coarse mulch on the soil surface to disperse the load

Rejuvenating Soil of Poor Structure. In the effort to reduce the cost and time of construction, fill soil is often put on top of compacted soil without any prior preparation. An uneven and compacted surface overlaid with a shallow fill can lead to moisture and aeration problems. High areas of compacted soil under the fill will cause water to flow toward low areas under the fill, resulting in wet spots. Plantings in such an area will be watered unevenly and will consequently grow unevenly or poorly. If the fill is fairly deep, there will be less of a problem.

Before a compacted soil is brought to final grade, it should be broken up by ripping or deep plowing when it is relatively dry. The area should then be rough-graded. Should additional soil be necessary to bring the surface up to grade, its texture should be similar to the soil already there. After rough-grading or the addition of fill soil, the area should be thoroughly irrigated and allowed to dry in order to settle the soil before final grading.

Amending Soil

Amendments can influence the physical, chemical, and biological properties of soil but are primarily added to soil in attempts to improve its physical condition.

---

*The percent compaction for engineering purposes is the dry density of soil as compacted in the field compared to the maximum dry density of the same soil at optimum water content (near field capacity) as determined in a laboratory.

The chemical and biological properties can be improved less expensively by other materials. Amendments are added in efforts to improve the aeration and tilth of fine-textured or compacted soils, to increase somewhat the moisture and nutrient capacities of sandy soils, or to decrease bulk density (weight per unit volume) when excessive weight may be a problem. Mineral amendments (pumice, perlite, vermiculite) are fairly permanent in the soil, if they are not compacted. Organic matter (peat moss, bark, sawdust, manure, compost) is more commonly used, but it decomposes with time. To have much effect, an amendment must constitute at least 50 percent of soil volume.

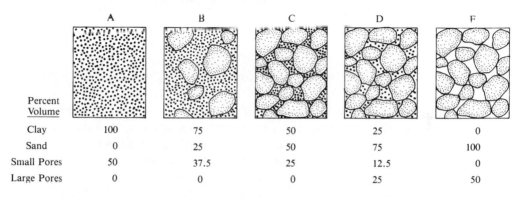

Percent Volume	A	B	C	D	E
Clay	100	75	50	25	0
Sand	0	25	50	75	100
Small Pores	50	37.5	25	12.5	0
Large Pores	0	0	0	25	50

Figure 6–11 The effect of mixing increasing proportions of a coarse sand to a fine clay, each having 50 percent pore space (A and E). In this example, even when equal volumes of sand and clay are mixed (C), large pore space is 0 and small pore space has been reduced to 25 percent. Soil mix C has the poorest aeration; aeration improves as a higher proportion of sand is present. Soil mixes A and B can be easily compacted while C, D, and E have better compaction resistance. (Adapted from Spomer 1983).

Amending an entire soil area for tree and shrub planting is seldom justified. It would probably be just as effective and less expensive to break up the compacted surface, bring it to rough grade, irrigate to settle the soil, and then, when the soil is dry enough, bring it to final grade, plant, and mulch. The first two steps should be performed when the soil is fairly dry. In most cases, amending the backfill with organic matter will not be of any particular benefit (see Chapter 9).

If a soil is to be amended, use a slowly decomposing organic material. It may include sphagnum peat, redwood sawdust, untreated fir bark, or certain wood products treated and fired so they are high in lignin and slow to break down in soil. Fresh redwood sawdust should be leached with water to remove a phytotoxic constituent before it is mixed with the soil. Rice hulls can be a source of weed seeds unless they are treated. Even though these materials decompose slowly, they can reduce the nitrogen level in the soil soon after they are added. Since peat and treated wood products usually have a low cellulose content, they should not deplete the nitrogen supply. Untreated sawdust should not be used to amend soil due to its rapid breakdown which ties up soil nitrogen. With other amendments, however, it might be well to spread nitrogen on the surface (5–10 g/m^2 or 1–2 lb/1000 ft^2) after the organic matter has been added. Spread the organic matter 75 to 100 mm (3–4 in.) deep or (2–4 m³/100 m² or 3–6 yd³/1000 ft²) and rototill into the surface 100 mm (4 in.) of the soil.

Amending a soil with organic matter or other materials essentially dilutes the soil so that some of the particles are spread farther apart. Unfortunately, sand still is occasionally suggested as an amendment (Flemer 1986). As sand is added, however, finer particles occupy the voids between the sand grains, producing a denser, less porous mixture (Fig. 6–11). Not until sand constitutes 45 percent or more of the volume will the mixture begin to have some of the properties attributed to sandy soils (Fig. 6–2). Sand is an unsatisfactory amendment.

FURTHER READING

HAUSENBUILLER, R. L. 1978. *Soil Science: Principles and Practices,* 2nd ed. Dubuque, IA: Macmillan.

PRITCHETT, W. L. 1979. *Properties and Management of Forest Soils.* New York: John Wiley,

SINGER, M. J. and D. N. MUNNS. 1987. *Soils: An Introduction.* New York: Macmillan.

UNITED STATES DEPARTMENT OF AGRICULTURE. 1957. *Soil: The 1957 Yearbook of Agriculture.* Washington, DC: U.S. Superintendent of Documents.

CHAPTER 7

Preserving
Exlsting Plants _____

When a site is being developed and landscaped, there may be existing plants that should be retained. Legal as well as horticultural measures may need to be taken to protect desired plants before and during construction. Efforts will also be necessary to create or maintain a continuing favorable growing environment for the plants. All too often action must be taken to remedy an unfavorable situation which could have more easily been prevented.

Considerable effort is frequently taken to save plants, particularly trees, in order to retain or create a mature appearing landscape. Wise planning, modest design adjustments, and simple precautions can greatly enhance success in preserving plants. Some considerations and techniques are presented in this chapter.

LEGAL MEASURES TO SAVE TREES

More city and county governments are using different ways to save existing trees, even on private property. Trees on private property are most vulnerable when the property is to be developed for more intensive use. Even though there are people who insist on saving almost every tree, all too often there are others who even want to remove trees that are on public property.

Tree Preservation Orders (TPOs)

Trees in Great Britain are protected under the Town and Country Planning Act of 1971, the Town and Country Amenities Act of 1974, and Criminal Law Act of 1977.

These acts empower a local planning authority to protect trees in the interest of amenity (British Standards Institute [BSI] 1989a). A *Tree Preservation Order (TPO)* can be issued for a specific tree or trees. The owner of such a tree is not allowed to "fell, top, lop or uproot" it without consent of the local authority unless the tree is dead, dying or dangerous, or is creating a nuisance. A TPO does not remove the owner's common law responsibilities for the tree or the cost of maintaining it. Administrative costs restrict some authorities to protecting only trees of high amenity value.

    Conservation areas can be set up in which no tree more than 75 mm (3 in.) in trunk diameter can be cut down or pruned without giving the local authority six weeks' notice. The authority then has time to issue a TPO, if appropriate.

Restriction on Tree Removal

A similar approach has been used in the United States by governmental entities to create an ordinance that states that no tree (private or public) above a certain trunk diameter, usually 100 mm (4 in.), can be removed (in some cases, not even pruned) without a permit. The tree is then inspected, the reason for removal considered, and a permit granted or denied. If a permit is denied, it is not clear what liability the governmental entity has assumed. Such ordinances have yet to be tested in a court of law as to their constitutionality or the liability involved. Such an ordinance requires considerable governmental expertise and expense to implement and can create considerable citizen animosity, even among those who favor saving trees.

Building Permit Approval

Another approach is for the agency that grants building permits to have a policy or regulation that when the use of a property is to be changed, every effort should be made to save as many trees and large shrubs as is reasonable. Development and building plans are not approved until the developer or builder and the planning and the tree-care agencies have agreed on a mutually acceptable tree-preservation plan. A skilled, concerned, and reasonable agency staff is essential for this to be successful.

    Cities that have had such programs have saved many fine mature trees that would have otherwise been lost. Most builders have found that property values are enhanced by saving trees and large shrubs. Unless property use changes, the owner is free to prune or remove trees. Seldom will a home owner remove a tree without good reason. Tree removal before a building permit is requested has led to frustrating and expensive delays in plan approval.

Compensatory Payment for Tree Removal

In 1980, the city of Cincinnati adopted the following ordinance (Sandfort and Runek 1986):

No person shall remove any public tree without replacing such trees with trees of equivalent (monetary) value in the vicinity of the removed trees. The value of trees shall be determined by the Urban Forest Manager in accordance with regulations considering the species, location, size, and condition of trees adopted by the Urban Forestry Board. If no suitable location exists in the vicinity of the tree removed or if the replacement tree is of lesser value, the person causing the tree to be removed shall make a compensatory payment to the City of Cincinnati equal to the difference in value between the tree removed and any replacement tree. Such payment shall be paid into a fund established by the director of finance for that purpose and used solely for the purpose of enhancing the urban forest.

The forestry division receives about $15,000 annually in compensatory payments. More important, many projects are not carried out or are relocated to save existing trees or minimize their loss. In addition, if a public tree is damaged due to an accident or improper pruning, the person responsible must compensate the forestry division for the loss in value of the tree. Funds so collected are used to plant trees in the vicinity of the loss.

Preserving and/or Replacing Trees

Fulton County, Georgia, a heavily wooded area adjacent to Atlanta, has a tree preservation ordinance which applies to property for which a "Land Disturbance Permit" is required, excluding single-family, detached, and duplex dwellings. Such a permit allows "defoliation or alteration of the site, or commencement of construction."

The ordinance states that a "site-tree density factor" of no less than 15 units per acre (37/ha) is required. Tables list the number of "units" allowed for existing and replacement trees by size; for example, an existing 200 mm (8 in.) tree is worth 0.5 units and a replacement tree of the same size is 1.2. Standards are prescribed for acceptable existing and replacement trees.

A tree protection plan is part of the permit process (Fig. 7–1). Adherence to strict and detailed protection specifications is required. The tree population must be sustained in perpetuity at or above the site-tree density factor.

In the five years it has been in effect, 40 percent more trees (based on the site-tree density factor) have been left or planted than was required. This was probably done initially to have younger trees planted to take the place of more mature trees that might not survive the redevelopment.

PROTECTING EXISTING TREES

Existing trees may warrant saving when land is developed for residential, industrial, business, park, roads, or other use. In many cases, the placement and design of structures and pavement can capitalize on some of the finer specimens and provide construction details to improve their future well-being. With early planning and

Figure 7–1 The tree preservation program in Fulton County, Georgia, saved this mature white oak in the planning and development of a shopping center near Atlanta. Little of the soil surface within the dripline of the tree has been disturbed.

certain precautions, some trees may be saved with modest effort and expense. Other trees in strategic locations may justify considerable effort on their behalf.

Choosing Trees to Save

Trees must be evaluated on the basis of their desirability in the new landscape and the effort necessary to save them. Factors to be considered include: (1) suitability of the species for the new use; (2) tree health, structural integrity, and expected longevity; (3) species tolerance to changes in the site; and (4) level of ongoing maintenance required (Matheny 1989).

Some species adapt better than others to changes in their environment. The USDA (1975) identifies elm, poplar, willow, plane tree, and locust as being tolerant to changes in their surroundings. Less adaptable trees include beech, birch, hickory, tulip tree, some oaks, most maples, and most conifers. Young, small trees adapt to change more easily than do older, larger trees. It may be easier and cheaper, however, to remove or stockpile trees that are below a certain size (100–150 mm or 4–6 in., for example) and to replace them later. Trees in groups, even if individually less desirable, may be more effective landscape features and easier to maintain than single trees. Fulton County, Georgia, encourages retaining small groves and groups of trees; tree health, safety, and survival are thereby improved.

An accurate survey of existing trees is needed to determine location, species, size, condition, and landscape value. This information should be part of the initial planning and design process in structure siting and design, grade changes, road and parking locations, utility routes, and major landscape features. Trees to be saved should be prominently marked so they will not be damaged or removed. Trees and shrubs that are not to be saved should be removed before construction begins if their removal would help to protect those that will be kept. Trees to be removed should be cut near the ground, and their stumps should be ground rather than pulled, so that the roots of the plants to be kept are not injured. Remaining trees may need pruning and even guying to reduce the likelihood of windthrow when the removal of adjacent trees exposes them to more wind. Previously shaded treetrunks exposed to the sun should be shaded or painted with white latex paint to minimize injury.

Construction specifications should include plans to protect the trees selected so that the contractors know their responsibilities and what the penalties would be if they damage those trees. Construction contracts are beginning to include penalties for damage to existing plants whose value is appraised according to the ISA tree valuation guide (CTLA 1992) before construction begins. Should damage occur, the loss in value is determined and assessed to the responsible contractor.

Protecting Trees During Construction

Unfortunately, *stripping* is the first operation to prepare most sites for development—all surface vegetation, organic laden topsoil, and debris are removed, and the area is essentially rough graded. Stripping near trees to be retained should not be allowed because most of the absorbing roots may be lost or seriously injured.

Even though the area around trees is not supposed to be disturbed during construction, all the trees to be saved need to be protected from compaction, mechanical injury, chemical damage, and drying out. **Sturdy temporary fencing should be installed usually at the outer dripline around each tree or group of trees to be saved.** A fence location in relation to a specific tree or group depends on species, soil structure, degree of surface rooting, and history of the site. Some species are more sensitive to soil and root disturbance than others and require more area to be fenced; others are more tolerant. Roots are usually closer to the surface in undisturbed areas than in cultivated or bare soils. Roots of a single irrigated tree will usually have less spread than one of the same size in an infertile, arid site.

Recommendations vary considerably as to placement of protective fences. Tattar (1989) and Bernatzky (1978) recommend that a fence protect the soil within the dripline of the tree. A USDA (1975) publication advises that an enclosed area be "at least 10 feet (3 m) square with the tree in the middle." Bernatzky also suggests that the fenced area can be as small as four times the trunk diameter if the root area outside the enclosure is covered with 200 mm (8 in.) of gravel with steel plates on top. To the detriment of many trees, protection at most construction sites falls short of even the minimal specifications.

Future planting areas outside of the fence-protected areas that may be subject to construction traffic, material storage, and equipment parking should be mulched 100 to 150 mm (4–6 in.) deep. In critical areas, interlocking metal sections on top of the mulch will give added protection for temporary roadways. Mulch is easy to handle after construction is finished; any that cannot be separated from the soil will tend to improve surface soil structure.

In addition to protective fencing, other construction site concerns include providing signs indicating protected trees, site clearing, utility trenching and tunneling, field office placement, construction parking, soil stockpiling, haul roads, material storage, chemical and fuel storage, concrete washout areas, tree-care procedures during construction, and other matters specific to the site. Tree pruning to make room for construction or equipment operation should be done by an arborist under the direction of the architect.

The construction site environment may differ greatly from the trees' previous surroundings. Trees and future planting areas should be examined at least once a month to check on their condition. Areas to be planted may be experiencing abuse. Water excess or deficiency commonly afflicts trees in building areas. These problems may be corrected easily or may require major design changes for the long-term well-being of the trees.

Grading and Construction Considerations

Almost every building and landscape development involves excavating, filling, compacting, and grading. These activities compact the soil not only under a structure or landscape feature but also adjacent to them (Table 7–1). Wind patterns, exposure to sun and reflected heat, and space limitations must be considered as well as the changes in the soil. Design details, engineering specifications, and construction operations in large measure determine the success of tree-preservation efforts and future landscape performance. If these are not well-conceived and well-executed, trees

TABLE 7-1 OCCURRENCE OF SOIL COMPACTION NORMALLY SPECIFIED FOR SITE DEVELOPMENT AND CONSTRUCTION[a] (Nelda P. Matheny, Pleasanton, CA, 1990 pers. comm.)

Within 1.5 to 3 m (5–10 ft) of a building

Within 1.5 m (5 ft) of a road edge

0.6 to 1 m (2–3 ft) beneath a road

0.2 to 0.3 m (8–12 in.) beneath a sidewalk

0.15 to 0.6 m (6–24 in.) beneath a footing

0.3 to 1 m (1–3 ft) beyond the toe of a slope
   If the toe of a slope is "keyed in" the compacted area will be the width of the equipment—2 to 4 m (6–12 ft)—and 0.6 to 1 m (2–3 ft) deep.

Throughout the depth of a fill plus 0.15 m (6 in.)

Backfill in utility trenches

Under mounds

Repaired slopes

[a]Compaction requirements vary with soil texture, shrink and swell characteristics, and other factors that affect potential of subsidence and the load the soil must bear.

will perform poorly, will require considerable ongoing care, and may soon die. Matheny (1989) has summarized the range of construction impacts to trees, probable causes, and ways to minimize the damage (Table 7-2).

Paving Around Trees

Installation of pavement around trees is probably the most frequent of potentially damaging construction activities. The common question is, "How close can the pavement be to the treetrunk?" or "How large an area must be left in the pavement for the tree?" And of course the answer is, "It all depends." It all depends on the species of tree and its present health, the conditions under which it has been growing, the soil porosity and drainage, and the measures taken to provide aeration and water. Some trees survive when paving is within 50 mm (2 in.) of their trunks, while others die quickly when only a small area within their driplines has been covered.

Pavement used by the public, including many walkways, is constructed to handle emergency equipment. Therefore, most paving is considered a structure so that soil preparation is similar to that for a building. Soil excavation and compaction practices for paving are discussed in Chapter 6. For asphalt, excavation ranges from a minimum of 300 mm (12 in.) for auto parking to as deep as 800 mm (30 in.) for trash loading pads. In highly expansive soils, excavations are often 500 mm (18 in.) deeper.

Excavation for cement paving for parking may need to be only 150 mm (6 in.). Cement paving near trees may be worth the additional cost over asphalt.

Streets and pavement obviously are not conducive to tree health. Most absorbing roots where pavement is installed and beyond are removed or severely injured. The compacted soil, aggregate base, and pavement are almost impenetrable except for cracks that occur due to expansion and contraction of the various materials. Even so, most trees in good health tolerate pavement to within half the distance from the dripline to the trunk, if the rest of their roots have not been unduly disturbed. In fact, growth of new roots under the pavement may need to be prevented, particularly if the roots could reach favorable soil beyond (see Managing Root Growth in Chapter 17).

Both cutting roots and installing street curbs and pavement close to tree trunks place mature trees in jeopardy, particularly if remaining roots have been injured. Cutting sinker roots increases the possibility of wind-throw, particularly when soil is wet. Every effort should be made to plan street and driveway pavement safe distances from mature trees.

Pavement to withstand vehicular traffic can be placed within the dripline of most trees with minimum detrimental effects (Fig. 7-2). For critical root zone areas, Steve Clark (Brentwood, TN 1990 pers. comm.) recommends doing little or no excavation; lightly grading the area to be paved, using *gradall* equipment to move soil and other material within the dripline of trees so that equipment weight on soil is not within the dripline; laying geotextile fabric on the area to be paved; placing aeration tubing or PVC aeration pipe on the fabric; covering the aeration passage with uniform gravel or coarse sand with no fines; laying geotextile fabric over the

TABLE 7–2 MAJOR CONSTRUCTION IMPACTS AND METHODS TO MINIMIZE DAMAGE
(Adapted from Matheny, 1989)

Impact to tree	Construction activity	Methods/treatments to minimize damage
Root loss	Stripping site of surface soil during mass grading	Restrict stripping of topsoil around trees. Woody vegetation to be removed adjacent to trees to remain should be cut at ground level and not pulled out by equipment, or root injury to remaining trees may result.
	Lowering grade, scarifying, preparing subgrade for fills, structures	Use retaining walls with discontinuous footings to maintain natural grade as far as possible from trees (Fig. 7–8). Excavate to finish grade by hand and cut exposed roots with a saw to avoid root wrenching and shattering by equipment. Spoil beyond cut face can be removed by equipment sitting outside the dripline of tree.
	Subgrade preparation for pavement	Use paving materials requiring a minimum amount of excavation (for example, concrete instead of asphalt). Design traffic patterns to avoid heavy loads adjacent to trees (heavy load bearing pavements require thicker base material and subgrade compaction). Specify minimum subgrade compaction under pavement within dripline.
	Excavation for footings, walls, foundations	Design walls/structures with discontinuous footings (Fig. 7–15), pier foundations (Fig. 7–14), and post and beam footings. Excavate by hand. Avoid slab foundations.
	Trenching for utilities, drainage	Coordinate utility trench locations with installation contractors. Consolidate utility trenches. Excavate trenches by hand in areas with roots larger than 25 mm (1 in.) diameter. Tunnel under woody roots larger than 50 mm diameter rather than cutting them (Fig. 7–16). If necessary, equipment should operate on double, overlapping, thick plywood sheets within the dripline.
Wounding top of tree	Injury from equipment	Fence trees to enclose low branches and protect trunk. Report all damage promptly so arborist can treat appropriately.
	Pruning for vertical clearance for building, traffic, and construction equipment	Prune to height requirements prior to construction. Consider maximum height requirements of construction equipment and emergency vehicles over roads. All pruning should be performed by an arborist, not by construction personnel.
Inadequate soil moisture	Rechannelization of stream flow; redirecting runoff; lowering water table; lowering grade	In some cases, it may be possible to design systems to allow low flows through normal stream alignments and provide bypass into storm drains for peak flow conditions. Usually flood control and engineering specifications are not flexible where the possibility of flooding occurs. Provide supplemental irrigation in similar volumes and seasonal distribution that would normally occur.

TABLE 7–2 *(Cont.)*

Impact to tree	Construction activity	Methods/treatments to minimize damage
Unfavorable conditions for root growth; chronic stress from reduced root systems	Compacted soils	Fence trees to keep traffic and storage from within dripline of trees. In areas of engineered fills specify minimum compaction (usually 85%) if fill is not to support a structure. Provide a storage yard and traffic areas for construction activity well away from trees. Protect soil surface from traffic compaction with thick mulch or double, overlapping, thick plywood sheets. Following construction, vertical mulch compacted areas or install an aeration system.
	Spills, waste disposal (for example, paint, oil, fuel)	Construction specifications clearly state disposal procedures. Post notices on fences prohibiting dumping and disposal of waste around trees. Require immediate cleanup of accidental spills.
	Soil sterilants applied under pavement	Use herbicides safe for use around existing vegetation according to label requirements.
	Impervious pavement over soil surface	Utilize pervious paving materials (for example, interlocking blocks set on sand). Install aeration systems under impervious paving (Fig. 7–2).
Excess soil moisture	Back-up of underground flow; raised water table	Fills placed across drainage courses must have culverts at the bottom of the low flow so that water does not back up before reaching the elevation of the culvert. Study the geotechnical report for ground water characteristics to see that walls and fills will not intercept underground flow.
	Lack of surface drainage away from tree	Where surface grades are to be modified make sure that water will flow away from the trunk, that is, that the trunk base is higher than surrounding soil. If the tree is placed in a well, provide drainage from the bottom of the well (Fig. 7–4).
	Compacted soils (many micropores but few macropores)	Auger or water-jet aeration holes to improve drainage (see Aeration/drainage section in Chapter 14).
	Irrigation of shallow-rooted plants requiring frequent irrigation	Avoid landscaping under trees sensitive to high moisture and poor aeration, or utilize plants that require little or no irrigation.
Increased exposure	Thinning stands, removal of undergrowth	Preserve in groups or clusters species that perform poorly when exposed. Maintain the natural undergrowth.
	Reflected heat from surrounding hard surfaces	Minimize use of hard surfaces around trees. Monitor soil moisture needs where water use is expected to increase.
	Pruning	Carefully thin tree stand; paint exposed bark with white latex to aid adjustment.

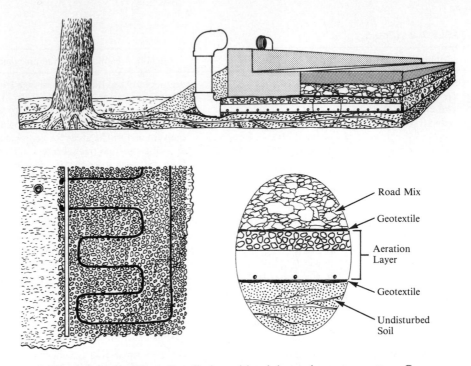

Figure 7-2 Pavement installation with minimum impact on a tree. Geotextile fabric is laid on carefully prepared soil surface. Aeration tubing or pipe and gravel are placed on the geotextile and covered with geotextile fabric. The aeration tubing loops can be 2 m (6 ft) apart since the gravel aids gas exchange. Each end of the tubing is attached to a metal riser (standpipe) with a protected top pointing into one of the two prevailing wind directions. The risers can be incorporated with landscape fixtures. Cement or rock base and asphalt can then be installed after the curb is in place.

aeration gravel and pipe; and covering the area with the desired depth of concrete or road base and asphalt. The geotextile fabric distributes compaction forces so that weight per unit soil surface is minimized.

The gravel protects the pipe in supporting the top geotextile fabric and pavement and aids air exchange to the soil beneath (Fig. 7-2). The main function of the aeration pipe is to ensure adequate air exchange; to do so, connect each of the two ends of the aeration pipe system (40–50 mm, 1.5–2 in. diameter) to strong vertical metal pipes (75–100 mm, 3–4 in. diameter) rising above the soil outside the paved area. The tops of the aeration standpipes have protected openings, one pointing in the direction of the prevailing wind and the other in the direction of the secondary prevailing wind (Fig. 7-2). The standpipes could be ballards to protect a tree, sign or lamp posts, or other landscape features.

This technique would minimize root disturbance and soil compaction and provide positive aeration to the covered roots; irrigation would also be possible.

Paving for sidewalks, driveways, and pavement from which heavy loads are restricted requires less extensive subbase preparation than for streets or parking areas. With little or no excavation or depth and degree of soil compaction, more absorbing roots would remain. Provision should be made to provide aeration and water for the roots under the pavement. With restricted loading, interlocking paving blocks on uniform sand might be a good pavement solution for trees.

Another solution would be to place base material and pavement on the natural grade with little excavation and compaction. The soil surface and pavement would need to be sloped to protect plants from excess runoff. If the area is fairly level, grade the soil surface to a 2 to 3 percent slope away from the treetrunk, without raising the soil around the trunk. Check and correct compacted layers that might interfere with water or air movement. Form a slight furrow (100 mm wide and 50 mm deep; 4 × 2 in.) at each 45° radius (Fig. 7-3). Apply gravel of uniform size (about 20 mm or $\frac{3}{4}$ in. in diameter) over the soil in a layer 100 to 150 mm (4–6 in.) deep. Firm it in place with a roller before putting down the pavement. Leave at least a 200-mm (8-in.) gap between the pavement and the treetrunk. Around the treetrunk, gravel can be filled to within 10 to 20 mm of the pavement surface. Near the drip line of the tree, cut holes 100 mm (4 in.) in diameter through the pavement

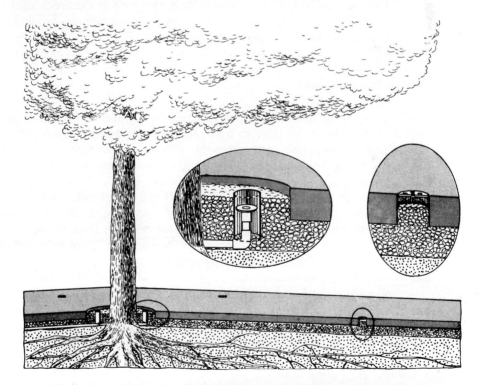

Figure 7–3 Most trees can be kept in good condition when paving for patios and courtyards (light traffic) is placed around them. The soil is graded to a slight slope away from the trunk, a layer of gravel is firmed in place, and the pavement installed. Irrigation bubblers at the tree trunk (left inset) and vents in the pavement at the dripline (right inset) should take care of irrigation and aeration. Aeration tubing (as in Fig. 7–2), vented at the trunk, could eliminate vents in the pavement.

approximately in line with the furrows at each 45° radius. Insert a pipe in the holes to hold them open and to seat a perforated cover. In asphalt pavement, each aeration hole should be set in a cement collar to provide protection to the opening and to minimize erosion around the aeration riser. The pavement should not drain into the tree basins or the aeration vents.

The trees can be irrigated by watering the gravel-filled basin by hose or by an underground bubbler system with two or more outlets at each treetrunk. Water should flow down the furrows to distribute the water more uniformly. As the water drains through and out of the gravel, air will be drawn in behind the water. The holes in the pavement at the drip line and trunk of the tree will provide some natural air circulation as well. Diffusion of the various gases in the gravel and soil will also actively keep oxygen near a safe level and prevent carbon dioxide and toxic gases from becoming too concentrated. When needed, nitrogen fertilizer (nitrate or urea) can be dissolved in the water used to irrigate the trees (see Chapter 12). The aeration-irrigation system under pavement must function for the life of the tree, but most trees will adjust within a few years to soil fills and cuts.

Pavement can be laid on cuts and fills under trees if the procedures followed are similar to those outlined above. On slopes, the surface must be shaped so that water will not drain to treetrunks but bypass them. The irrigation system should be adapted so that the water applied will flow fairly uniformly under the pavement.

Raising the Soil Level

If a fill is not to be load-bearing (no pavement, structure, or traffic), it may not be needed or need to be so close to trees. If, however, a non–load-bearing fill is to be put near existing trees, remember that fills create aeration, water movement, temperature, and root-growth problems in the existing soil. Most of these problems can be minimized if soil meeting the specifications given in Chapter 6 is added with care so as to reduce compaction of the original and added soil. Place soil and other material within the dripline of a tree using *gradall* equipment which has a retractable boom so that the equipment stays outside of the dripline. Even so, such operations can severely compact both the original and added soil. Keep in mind that slow warming of the root zone the following spring coupled with reduced aeration may delay root activity and could lead to early-season water deficits in trees (Fayle 1981).

Soil aeration is the critical factor. The roots must receive adequate oxygen and not be subjected to buildup of carbon dioxide or toxic gases. In most cases, measures that will ensure aeration need be employed only until new tree roots become established in the fill soil. Another concern is to protect trees from crown rot, which can occur if the trunks of certain species remain moist. Fills of 150 mm (6 in.) or less with fair to good drainage will harm few species (USDA 1975) and should not need aeration channels.

Numerous reports indicate rooting into fill soil by many species. Duling (1969) cites the following 1964 National Arborist Association Newsletter observation:

> Most everyone who has dug out stumps of trees around which there had been earth fills has occasionally found two root systems: the original roots, and another set that

had developed in the filled soil. In such instances the original roots are usually dead and the new upper story roots are keeping the tree alive and growing. These new roots developed from the original root collar at the base of the tree or from the trunk itself.

The phenomenon may be fairly common. Wilson (1984) states, "The small, short roots that grow lateral to the large roots grow at many angles to the vertical and a good many of them actually grow up into the forest floor." Zimmermann and Brown (1971) observe, "Each successive generation of lateral roots becomes less and less responsive to gravity, and other factors including available soil moisture and temperature often influence and control the direction of root growth. Moisture from above causes the roots to grow upward into the moist substrate instead of downward."

For the best evidence that roots do grow upward, lift a mulch that has been on the soil under a plant for several years. Where a bare soil exposed to the sun was previously devoid of surface roots, it will become alive with roots in and under the mulch. If soil conditions are favorable, roots of most plants will grow up into a fill.

Tree roots have a difficult time rooting into fill soil, however, if the fills are installed as has been commonly recommended: "Install a layer of gravel and a system of drain tiles over the roots of the tree" (USDA 1975). Variations of these recommendations include the use of stones under gravel and straw, burlap, or fiberglass matting to hold the 300 mm (1 ft) or so of fill soil (Duling 1969, Bernatzky 1978, Mayne 1982, Schoeneweiss 1982, Pirone and others 1988). Until some of the fill soil works down through the stone and gravel to the original soil or unless there are roots beyond the gravel layer, only roots of the most aggressive trees are likely to grow into fill soil installed this way. Although water can reach the original soil directly through the drain tile from the surface, the fill soil over the gravel layer must become saturated before water will move into the gravel and down to the original soil. After draining, the fill soil or at least the lower portion of it will remain saturated (see Chapter 6). Water will flow along the bottom of the fill soil (if it slopes away from the trunk) to the edge of the gravel layer and then into the original soil. In periods of excessive rain, the soil beyond the gravel layer will become waterlogged.

Aeration System. An effective system is even easier to install. Lay the smallest diameter aeration (drainage) tubing, usually 75 mm (3 in.) on the original soil in a meandering pattern within the dripline of the tree with each of its ends protruding into the tree well. **Fill soil between the loops of tubing will provide contact with the original soil,** so that water will drain from the fill soil into the soil below and roots can more easily grow into the fill. The tubing will allow oxygen to diffuse into adjacent soil and carbon dioxide and other gases to diffuse out.

To install this system, clear all plants, leaves, twigs, and debris from the area to receive fill soil. Keep equipment from going within the dripline of the tree. Use gradall equipment to do grading and soil placement under the tree. The original soil should slope away from the tree trunk. To ensure drainage from the tree well, install a drain line from the well to an outlet from which water can flow or be pumped as well as cleaned out and back flushed.

Protecting Existing Trees

Lay the aeration (drainage) tubing on the soil in a meandering pattern within the dripline, keeping the tubing loops 600 to 900 mm (2–3 ft) apart. Small stakes will help keep the tubing in place. The tubing ends should protrude into the tree well. The number of tubing units will depend on the size of the tree and whether the tubing ends connect into standpipes in the tree well as shown in Fig. 7–2. Two tubing units without standpipes should be sufficient for a medium-size tree or with standpipes for large trees; use four units without standpipes for large trees or with standpipes for extremely large and old trees.

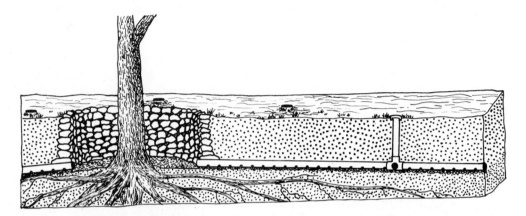

Figure 7-4   Aeration pipe laid on a ribbon of gravel radiating from a dry well 1 to 1.5 m (3–5 ft) in diameter provides gaseous exchange until new roots can grow into the fill soil. Another method would be to use aeration tubing similar to that in Fig. 7–2 with the aeration risers in the dry well. The flexible tubing loops or arcs should be about one meter (3 ft) apart; one aeration unit should serve the soil under half of the tree canopy. The aeration tubing would be simpler to install and would eliminate aeration vents in the fill surface. The aeration tubing is available with a cover of geotextile fabric or filter matting. If the aeration system is vented in the dry well, a separate drain line will be needed to keep the dry well free of water.

To keep roots from plugging the aeration tubing holes, use tubing with geotextile fabric or filter matting cover. Standpipes with their openings toward the prevailing winds will provide positive air exchange.

Construct the wall of the tree well at least 600 mm (2 ft) from the trunk. The wall can be of rock, cement, or wood. Tubing without standpipes should enter the tree-well wall 75 to 100 mm (3–4 in.) above the soil level in the well and the tubing openings covered with hardware cloth.

With the aeration system in place, scarify the original soil with hand tools to break up surface compaction. Broadcast 5 g/m² (1 lb/1000 ft²) of nitrogen. Place the fill soil with *gradall* equipment so as not to compact the soil unduly within the dripline of the tree to minimize tree and system damage, soil compaction, and the need to prune for equipment clearance. With the fill in place, water the entire area and allow the soil to settle and drain. More soil can be applied in a week or two to bring the fill to final grade. The new surface should drain away from the tree well.

Once the fill soil has been well watered to settle it, the usual irrigation practice (if any) should be sufficient to sustain the tree. It is neither necessary nor beneficial to apply water through the dry well–aeration channel system in order to irrigate the roots in the original soil. Water applied to the new surface soil will move down into the root zone where needed.

Sand Well. When grades are raised, a simple alternative to a dry well and aeration system may sometimes work. A number of reports indicate that many species form roots when soil is placed against or near their trunks. Duling (1969) reports that Charles Schmaltz of Rochester, New York, induced trees (including Norway spruce, black oak, and native varieties of maple and poplar) to root into fill soil. In May or June of 1936, he cut the bark of each tree to the cambium in several places about 300 mm (1 ft) below the final fill level. Even though the fill was as deep as 4 m (14 ft) on some trees, all were still alive 27 years later, when the report was made.

A number of species are known to develop new roots at places where the trunk is newly buried or just kept moist and dark. Wilson (1984) reports that poplar, willow, and black spruce root easily after the trunks had been buried during a flood. Coast redwood is known to root readily when the trunk is engulfed by river silt. Walter J. Barrows, Clovis, CA (1978 pers. comm.) observed that several large eucalyptus in a ravine in Whittier, California, were doing well after the soil level had been raised around them by 6 to 9 m (20–30 ft) more than 10 years before (Fig. 7–5). Pieces of broken sidewalk had been stacked around each trunk as the soil level was raised. A utility trench dug several years later revealed that roots had grown from the trunks through the concrete pieces into the new surface soil.

Figure 7–5 More than 10 years before this photograph was taken in Whittier, California, the soil level around these eucalyptus trees had been raised 6 to 9 m (20–30 ft).

Vigorous young trees that root easily and are tolerant of or resistant to crown rot will best survive raised soil levels. Preparation corresponds to that for a dry well. Water infiltration and penetration of the existing soil should be checked and corrected. The surface soil should be shaped so that water will not accumulate near the trunk and loosened so the interface with the fill soil will not interfere with water movement and root growth. Five g of nitrogen per m^2 (1 lb/1000 ft^2) should be broadcast on the surface of the soil.

In place of the dry well, put moist, uniform, medium sand (0.25–0.5 mm diameter) around the trunk so that it extends about 300 mm (1 ft) out from the trunk a little above the height of final grade. Firm, but do not compact, the moist sand so that its outer surface slopes away from the trunk but not more than 45° from the vertical (Fig. 7–6). Add fill soil of good structure and texture similar to the original soil. To avoid a hard-surfaced interface, loosen the surface of the moist sand slightly as the soil is filled against it. As water moves down near the trunk, it will follow the interface between sand and fill soil, thus keeping the soil near the trunk drier than it would be without the sand. The toe of the sand cone should be within 1 to 1.3 m (3–4 ft) of the trunk; otherwise, an excessively large volume of soil around the trunk may be inadequately watered. After the final grade is established to drain away from the trunk, be sure the sand collar is not covered with fill soil. Water will not enter the sand if its surface is covered with a finer-textured material, and aeration will be impeded.

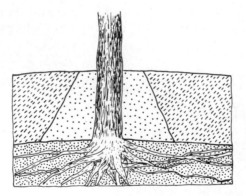

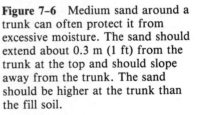

Figure 7–6 Medium sand around a trunk can often protect it from excessive moisture. The sand should extend about 0.3 m (1 ft) from the trunk at the top and should slope away from the trunk. The sand should be higher at the trunk than the fill soil.

Soil levels can probably be raised around trees with equal success at any season of the year but in cold-winter regions, late spring may be best (see previous section, Raising the Soil Level). Less soil compaction will result when the surface soil is dry or frozen. On the other hand, trunk rooting will be best encouraged if the bark is cut about 300 mm (1 ft) below the height of final grade in early spring.

To encourage rooting into the fill soil and to ease the transition after raising a grade, apply about 5 g of nitrogen per m^2 (1 lb/1000 ft^2) on the new soil surface. Rain or subsequent irrigation will move the nitrogen into the soil. This fertilizer recommendation differs markedly from those of others (Bernatzky 1978, Pirone

1978a, Tattar 1978, Pirone and others 1988) (see discussions on nitrogen, phosphorus, and complete fertilizers in Chapter 12).

Stabilizing Fill Slopes. *Brush layering* (or *buschlagenbau* in Schiechtl 1978) of woody branches can help stabilize fill slopes (see also Gray and Leiser 1982), both small and large, with grades of 3 : 1 and steeper. The technique involves placing alternate layers of fill and brush on the contour (Fig. 7–7). Depending on the size and steepness of the slope, the fill layer (lift) can be 0.6 to 1.5 m (2–5 ft) deep between two adjacent brush layers. The fill layer on which the brush is placed should slope into the hill at least 10 degrees from the horizontal. Branches of about 1 m (3 ft) long (for shallow fills) to 2 or 3 m long (for deeper fills) are placed more or less randomly, with some criss-crossing of stems, on top of each layer of fill soil. About one-fourth of the brush length should protrude beyond the fill when covered. Branches with butts 100 mm (4 in.) thick or thicker can be used on the deeper fills. Each layer of fill is compacted as it is placed. Species that have flexible stems and root easily are preferred because the resulting root and top growth gives more or less permanent protection to the slopes. The branches of species that do not root easily can provide interim protection until other vegetation is established.

Figure 7–7 If branches of easy-to-root trees are laid on the contour with their tips protruding as successive layers of fill are placed, they will stabilize steep slopes (left). Fill slope in New Zealand four years after willows planted by the brush layer method (right). (Photo courtesy R. L. Hathaway, Ministry of Works and Development, Palmerston North, N.Z.)

Brush layering can be used for original construction or as a remedial action on seriously eroding areas. For remedial work, one may have to cut into the slope in order to place the brush deep enough. Shorter brush can be used or the brush laid diagonally. In both remedial and original construction, one should start at the base of the slope and work up. Properly installed, brush layering can be an effective, relatively economical solution to a variety of problems concerning slope stability and erosion. Wattling is a variation of brush layering more commonly used for planting (see Chapter 11).

Retaining Walls. Developers favor steep slopes because they require less land area. It is difficult to establish vegetation on a steep, compacted slope, and the toe of such a slope may encroach on an existing tree and cause problems. A compacted slope is usually more impervious than a more level fill and can seriously interfere with aeration and rooting. Water from the slope may accumulate near its toe, further decreasing aeration and increasing the danger of crown rot for trees located there. Most serious of all, a 450- to 600-mm (18–24-in.) excavation is usually made to anchor the toe of a steep slope; this can sever a large proportion of active roots and jeopardize those below due to compaction.

If a tree near the slope is worth saving, several steps can be taken to protect it. Instead of excavating to anchor the slope toe close to the trunk, build a retaining wall up to 1.5 m (5 ft) high and support it with metal posts (2–2.4 m or 7–8 ft on center) set 3 m (10 ft) in the ground (Fig. 7–8). Such a retaining wall and the slope anchor excavation can be placed at least 2 m (7 ft) farther from the treetrunk than an excavation would be placed without a retaining wall. A retaining wall may mean that the slope above can be less steep, so that contour planting can be used in slope construction. A vertical perforated pipe or drain tile (50 mm or 2 in.) spaced 2 m (6 ft) along the slope can be used at or beyond the dripline of the tree. Connect each to a drain tile or gravel channel leading through the retaining wall to provide some aeration for the covered roots and encourage root growth into the fill. The gravel-covered tile or gravel channels should be covered with geotextile fabric or fiberglass filter matting in silt soil. They should emerge about 2 m (6 ft) apart along the retaining wall. More than one gravel channel can be served by each vertical vent pipe, but not more than 25 percent of the original ground surface should be covered with gravel. See Figs. 7–2 and 7–8 for construction details.

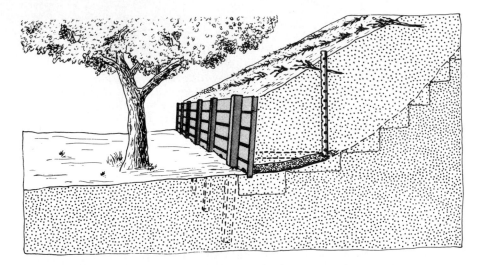

Figure 7-8 A well-anchored retaining wall can shorten the toe of a slope to protect a tree from encroachment. The fill is further stabilized with an anchor excavation (key-in) and step terraces on the slope. A gravel channel and vertical vent pipe provide aeration. Brush layers have been placed to stabilize the slope.

Lowering the Soil Level and Excavating

Removing soil from under a tree canopy can seriously damage roots and may even impair the stability of the tree. The root system of most trees in medium-textured soils exists primarily within a meter (3 ft) of the surface (Brown 1971). Most of the feeder roots occur in the upper 150 mm (6 in.) of most forest and many landscape soils. Each tree must be analyzed to determine the location of its major roots and the bulk of its absorbing roots. Isolated large trees may extend shallow roots two to three times the radius or height of the canopy. Horizontal roots tend to be close to the surface, particularly if the soil has been shaded or covered with a mulch of leaves. Even the long horizontal roots of oak, which are considered to be relatively deep, are within 300 mm (1 ft) of the surface (Wilson 1984). Most long horizontal roots of maples and birches are within 150 mm (6 in.) of the surface. On the other hand, many species have "sinkers," or large roots that grow down from large horizontal roots, most generally close to the trunk of the tree (Wilson 1984). There are seldom more than five or six sinkers per tree. Sinker roots provide anchorage and during droughts supply critically needed moisture.

At a fairly uniform site, even if 75 percent of the absorbing roots are beyond the dripline, lowering the grade on one side of a decurrent tree to a line tangent to the dripline would cut off about 20 percent of the roots to the depth excavated (Fig. 7–9). Excavating to within half the distance between the dripline and the trunk would remove about 35 percent of the roots. Most healthy trees should be able to withstand removal of one-half of their absorbing roots without serious effect (Sinclair, Lyon, and Johnson 1987), particularly if lightly pruned, fertilized with nitro-

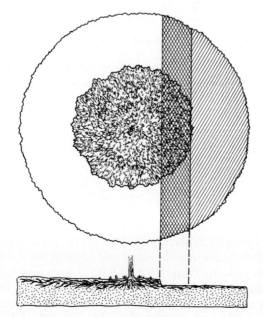

Figure 7–9 This plan sketch of a tree outlines the dripline, and the circle indicates the extent of most of the roots. Lowering the soil grade back to the dripline on one side would cut off about 20 percent of the roots; lowering the grade halfway from the dripline to the trunk on one side would eliminate about 35 percent of the shallow roots.

gen, and watered wisely. Reports indicate that most absorbing tree roots are in the upper 300 mm (12 in.) of soil (Zimmermann and Brown 1971, Perry 1982). If so, as long as sinker roots were not cut, a tree should be little affected whether an excavation similar to those in Fig. 7–9 were 0.6 or 6 m (2 or 20 ft) deep.

If you want to remove soil closer to the trunk than is shown in Figure 7–9, carefully fork the soil away from the roots, working toward the trunk. In many cases, you will encounter no large horizontal or sinker roots until within 2 to 3 m (6–10 ft) of the trunk. For tree stability and continued well-being, do not cut sinker roots unless they are a considerable distance from the trunk. Horizontal roots can usually be safely cut nearly to the point where their caliper begins to increase markedly toward the trunk.

A trench should carefully be dug to the desired depth by hand or a root-cutting or stone-cutting machine to establish the face of a cut. Roots should be cut with little disturbance of the root-soil interfaces in the soil on the trunk side of the cut. A backhoe or other equipment can be used to bring the area to be excavated to finish grade without injury to the roots that will be on the tree side of the wall.

Breast Walls. Cuts of 1 to 1.5 m (3–5 ft) in informal landscapes, for paths and driveways, and along minor roads can be stabilized with broken concrete or rock breast walls without footings (Untermann 1978). Breast walls differ from retaining walls in that they are built of loose, dry-laid rock placed against more or less undisturbed earth and receive little force from the soil behind them (Fig. 7–10).

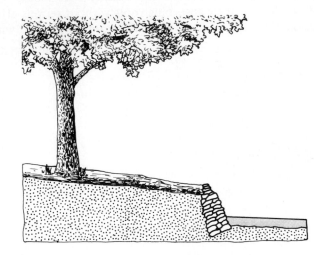

Figure 7–10 A breast wall of loose, dry-laid rock is used to retain the original grade near a tree trunk when the soil level is lowered near the tree. (USDA 1975)

For deeper cuts in more natural areas and along roads, as the rock is set in place, branches of woody species that root readily can be placed between the crevices to further stabilize and visually soften a wall (Fig. 7–11). Species, such as willow and poplar, are laid with their bases into the slope and covered with soil so that few voids remain. Cut branches no more than two days before placement and keep them from drying. Tamp the backfill firmly in place as the wall is built. Smaller rocks can be used in the larger interstices to reduce the amount of exposed soil. They must, of course, be held firmly in the wall. The branches, particularly if they root,

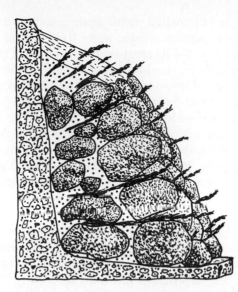

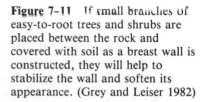

Figure 7-11 If small branches of easy-to-root trees and shrubs are placed between the rock and covered with soil as a breast wall is constructed, they will help to stabilize the wall and soften its appearance. (Grey and Leiser 1982)

increase stability, prevent or retard the sloughing of soil or rock from the slope. Plants that grow in the wall are particularly valuable for protection along stream banks.

Retaining Walls.   Concrete, or brick, or stone and mortar walls are usually required to protect cuts and fills adjacent to paved areas open to the public (Fig. 7-12). The cut is made 0.2 to 0.3 m (8-12 in.) closer to the tree than the tree-side of the wall will be. This allows room to set forms and pour a concrete base for the wall. A trench for the wall base is dug about 0.3 m (12 in.) deeper than the final excavation grade and is compacted. The wall base is at least 300 mm wide and 150 to 200 mm (6-8 in.) thick.

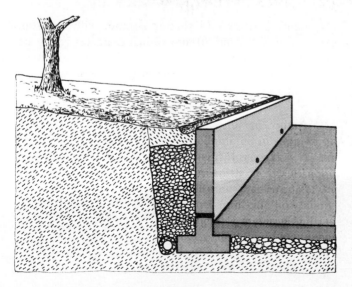

Figure 7-12   A concrete retaining wall protects the cut and the lower pavement from erosion. Weep holes in the wall allow excess soil moisture to drain; in wet areas a drain pipe in gravel behind the wall can keep the soil and pavement from becoming too wet. Using geotextile fabric, the width of the gravel behind the wall can be less than shown, but it is difficult to install. The drain pipe and the top of the gravel should be covered with geotextile to keep soil out of the drain and to keep the soil surface from settling.

A 50 to 100 mm (2–4 in.) perforated plastic drain pipe is laid in gravel along the tree-side of the wall base to remove seepage water from behind the wall. Weep holes should be placed about 2 m (5–6 ft) apart near the bottom of the wall to drain onto the pavement or other surface, should the drain pipe become plugged.

A vertical gravel layer to the tree-side of the retaining wall is needed to drain seepage and prevent pressure build-up behind the wall. Fill the space behind the retaining wall with gravel to within 150 to 200 mm (6–8 in.) of the top of the wall. Separate the gravel fill from the soil vertically and over the top with geotextile fabric. Cover the gravel and fabric with top soil from the site, forming a drainage way to carry surface water to a drain. Roots can grow in the humid drainage gravel.

Preventing the cut and exposed roots from drying out until the installation of the retaining wall is complete would seem desirable. Thick burlap or other porous, absorbent fabric hung from the top of the cut over the exposed roots and soil and misting are recommended to protect trees, particularly fleshy-rooted species such as magnolia. Unfortunately, such protection is seldom provided even for sensitive plants during hot, dry weather. The extent of tree damage by not protecting roots from drying is not known, but surprisingly, few trees are lost.

It is wise to consult an engineer whenever a design is created or altered because the person designing or altering a structure is responsible for its structural integrity.

Seldom should the entire surface beneath a tree be lowered. An excavation to accommodate a roadway, landscape feature, or building foundation can usually be accomplished by terracing to maintain the original level around the treetrunk and then building a retaining wall out from the trunk to lower the soil level the desired amount (Fig. 7–13). If the entire area around a tree is to be lowered much more than 150 mm (6 in.), the tree may not survive unless soil is retained within a fair distance of the trunk. Just how much of the original soil must be left intact will depend on the species and age of the tree, its particular rooting pattern, and soil and moisture conditions at the site. Obviously, the more soil that can be left undisturbed, the better. For maximum excavation, however, careful root exploration will be needed.

Equipment can seriously damage roots and shatter contact with the soil. Equipment excavation should be stopped when 50-mm (2-in.) diameter roots are encountered. If possible, place the retaining wall there; if not, fork the soil from the cut roots in toward the trunk, although no further than halfway between the

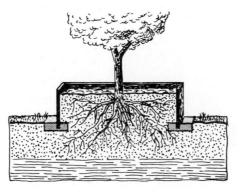

Figure 7–13 The largest possible area should be left at the original soil level when a grade is lowered on all sides of a tree. A retaining wall around the original soil will hold it in place and serve as a raised planter.

Chap. 7   Preserving Existing Plants

dripline and the trunk. Stop sooner if you encounter large roots. Cleanly cut the roots near the soil surface. The shape of the tree island can vary to retain the big roots, particularly those growing downward. Handle the roots, the retaining wall, and the rooting soil as described previously. Keep the soil surface and roots moist and shade the soil with plants or mulch. Aeration of the excavated area will be of little value even if roots are present unless care is taken to prevent soil compaction.

Excavating for Large Buildings and Basements. Steps similar to those described for installing retaining walls apply to excavating near trees for large buildings and basements. Such excavations are usually deeper and are not completely refilled until the building is completed, which may take weeks. Excavate first to the planned location of the building foundation. Fork the soil from the cut roots for 0.6 to 1 m (2-3 ft) toward the trunk to provide room for construction work. (See the previous section on Retaining Walls concerning handling cut roots.)

When the void between the cut and the building (sealed against moisture) is ready to be filled, install a plastic perforated pipe at the base of the wall to drain seepage. The void can be filled with gravel. Fill to within 200 mm (8 in.) of the final height. Cover the gravel with geotextile fabric and fill with soil to bring to grade. Shape the soil fill to drain surface water from the site. If desired, though difficult, the void can be filled with soil and gravel by separating the two vertically with geotextile fabric as gravel is placed next to the building and soil toward the tree. Compact the fill as it is being placed.

Piers, Pilings, and Posts. Using posts to hold retaining walls upright and constructing walks, porches, and buildings on piers minimizes their impact on fragile sites (Fig. 7–14). Compared to cement slabs on soil or peripheral foundations set below grade on compacted soil, construction with piers provides a favorable environment for continued root growth and function. To benefit most from this type of construction, take care to minimize soil compaction between piers.

The footing of a wall or rock fence can often be interrupted or modified to allow room for roots and their future expansion (Fig. 7–15). A similar technique can also be used for walks or paved areas near existing trees with large buttress roots.

Utility Trenching. Trenching for underground utility services causes serious root injury to trees that are restricted to the same easements. Morell (1984) reports that 12 years after trenching for water mains in Park Ridge, Illinois, 92 of the 262 mature street trees trenched were dead and 27 had significant top dieback.

The additional cost to have tunnelled (augered) under the trees on these projects would have been $150 to $215 per linear m ($45–65/ft), depending on pipe size; the value of the trees lost and the cost for removing and replanting dead trees was estimated to be four times what the additional cost of tunnelling would have been. The city now tunnels under trees. A 1991 estimate of the cost of tunnelling for 150 to 200 mm (6-8 in.) diameter pipe was $330/m ($100/ft) (John Morell, Park Ridge, IL, 1991 pers. comm.).

Several Illinois cities developed common tunnelling specifications (Table 7–3). Toronto has trenching and tunnelling specifications and the British Standards Insti-

Figure 7-14 In Maryland, an 18-unit, three-story condominium (right) was built on a foundation of 168 pilings (left). The pilings allowed trees to be preserved within 1.5 m (5 ft) of the building. The pilings also permitted a gravel retention system under the building to handle rain water, therefore trees did not have to be removed for a rain collection pond. Note the low lattice that screens the piling foundation (right). The wall construction was prefabricated so that no scaffolding was needed on the outside of the building during construction. The siting of and views from each unit were so enhanced by use of piers that even though the developer increased the price of each unit $30,000, more units were sold the first year than was originally projected. (Photos courtesy of Steve Clark, Brentwood, TN)

tute (BSI) (1989a) published trenching distances that were considered minimal (Table 7-3). Although no distances are given, the BSI recommends tunnelling directly under tree trunks (Fig. 7-16).

These specifications differ considerably from one another. To start tunnelling under a 50-mm (2-in.) tree 300 mm (1 ft) from its trunk seems too close; if the roots

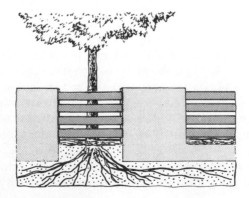

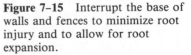

Figure 7-15 Interrupt the base of walls and fences to minimize root injury and to allow for root expansion.

Chap. 7   Preserving Existing Plants

TABLE 7-3 SPECIFICATIONS FOR TUNNELLING OR TRENCHING NEAR TREES FOR THE INSTALLATION OF UNDERGROUND UTILITY SERVICES

Distance of tunnel from each side of the treetrunk

Toronto[a]				Illinois[b]			
Trunk diameter		Distance		Trunk diameter		Distance	
mm	in.	m	ft	mm	in.	m	ft
50	2	0.6	2	50	2	0.3	1
75	3	0.9	3	75–100	3–4	0.6	2
150	6	1.5	5	125–225	5–9	1.5	5
300	12	1.8	6	250–350	10–14	3.0	10
450	18	2.1	7	375–475	15–19	3.6	12
600	24	2.4	8	>475	>19	4.5	15
750	30	2.7	9				
900	36	3.0	10				
1050	42	3.6	12				

Distance of open bypass trench from treetrunk

Toronto[a]				England[c]			
Trunk diameter		Distance		Trunk diameter		Distance	
mm	in.	m	ft	mm	in.	m	ft
50	2	0.9	3				
75	3	1.8	6				
150	6	3.0	10	200	8	1.0	3
300	12	3.6	12	250	10	1.5	5
450	18	4.2	14	375	13	2.0	7
600	24	4.8	16	500	20	2.5	8
750	30	5.5	18	750+	30+	3.0	10
900	36	6.0	20				
1050	42	6.6	22				

[a] Jack Kimmel, Toronto, Canada (1978 pers. comm.). Specifications being used by surrounding cities (Kimmel 1990 pers. comm.).

[b] Morell (1984). Developed by several Illinois cities. Distances approximated the dripline of the trees.

[c] Adapted from British Standards Institute 5837 (1989a). The information warned that excavation or root damage closer than given may render the tree dangerous. The distances should be doubled for trees of low vigor.

had grown 600 mm, a 450-mm (18-in.) wide trench would reduce the root system by 20 percent. On the other hand, the tunnelling distances for large trees may be more than necessary. A 450-mm wide trench on a radius directly toward and to within 1.5 m (5 ft) of a 300-mm (12-in.) trunk would seriously affect less than 15 percent of the absorbing roots (Fig. 7–17) and, if encountered, possibly 20 percent of the sinker roots. Moving the trench 600 mm (2 ft) so the tunnel would be 450 mm to one side of the trunk would affect about 35 percent more absorbing roots than going directly under the trunk (Fig. 7–17). Few mature trees have taproots. To impact the same proportion of absorbing roots as tunnelling directly under the trunk, a trench would have to be at least 4 m (13 ft) from the trunk (Fig. 7–17).

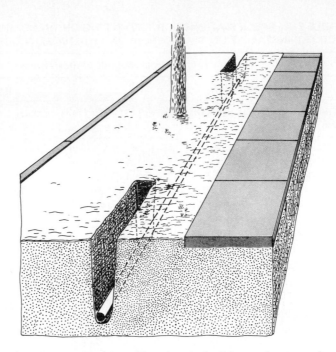

Figure 7–16 Trenching for utility pipes and cables can be stopped according to Table 7–3 or when major tree roots are encountered; a hole is then tunnelled under the major roots and the trunk.

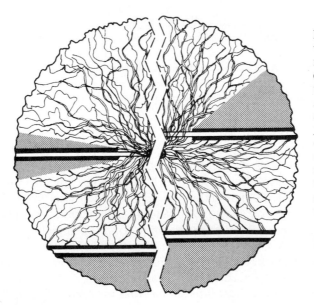

Figure 7–17 An overhead view showing that fewer large roots would probably be encountered by tunnelling under or close to a tree trunk. About 15 percent of the absorbing roots would be jeopardized (center left) compared to 20 percent (center right) for a tunnel 0.6 m (2 ft) from the trunk. A trench would need to be about 4 m (13 ft) from the trunk (lower left) or 3.3 m (11 ft) (lower right) in order to affect the same proportion of the absorbing roots as the tunnelling method for a tree with a 10-m (35-ft) diameter canopy.

A trench can be mechanically dug toward a tree to its dripline or a third the tree's height from the trunk, whichever is greater. The trench should be continued by hand until significantly large sinker roots (depending on tree size) are encountered or the distance indicated by the Toronto specifications is reached, whichever is less (Table 7–3). Tunnelling should continue under the central root system to reach the trench on the other side. A tunnel conduit may be advisable if more than one utility is to be accommodated, to minimize damage to the utilities and to ease future repair.

Tunnel-depth specifications vary as much or more than those for distance. The Toronto specifications vary from 0.9 to 1.5 m (3 to 5 ft), depending on tree size. The British recommend the tunnel be as deep as possible, while those from Illinois state a tunnel should be at least 0.6 m (2 ft) deep. The important thing is to tunnel below the major zone of absorbing roots, to be determined when digging the trench by hand; even for large trees it will probably not be much deeper than 0.6 m. Any roots encountered larger than 30 to 50 mm (1–2 in.), depending on tree size, should not be cut.

A governmental agency or developer can protect trees from trenching damage in utility easements, but trenches for buildings are usually located by the contractor, who will usually take the shortest and straightest route unless instructed to route trenches around trees. If possible, all utility services should be placed in the same trench. Protective fences around trees should not be violated if at all possible.

After any serious root injury, the tree should be wisely watered and, if necessary, pruned.

CHAPTER 8

Planting Site: Preparation

Landscapes range from intensive plantings in indoor and rooftop containers to natural areas that receive little or no maintenance; from deep, fertile agricultural soils to shallow, landfill soils of varied origin which are subject to toxic gases and high temperatures; from compacted soil in planting wells surrounded by asphalt and buildings to soil on an exposed seacoast. You name it; someone will want to grow plants there.

The performance of plants in the landscape depends on how well the species are adapted to the specific environment in which they are to grow, the quality of the planting stock, preparation of the site, planting methods, and later care. A site is usually determined in advance, so site preparation and species selection are quite important. A site that is less than ideal can often be modified to improve plant performance and ease of maintenance. Many of these modifications have already been covered; some others are dealt with now.

SOIL PREPARATION

Many soils become severely compacted during landscape construction, particularly if buildings are involved. As we have already discussed, soil compaction reduces water infiltration and movement, drainage, and aeration. These factors and the increased physical resistance to root penetration can greatly impair root growth and function and can increase the difficulty of later maintenance, particularly irrigation.

Irrigation or tillage will reveal if surface soil has been compacted. Former agricultural soils usually have a compacted layer (plowpan) 150 to 250 mm (6–

190

10 in.) below the surface, depending on the depth of earlier cultivations. Alluvial soils and soils graded during site preparation may have some layers that seriously interfere with root penetration and water movement. Fine textured soils may be naturally tight, restricting root growth and function as well as drainage and aeration. In many soils in the southwestern United States and other arid regions, hardpan of varying thicknesses appears at various depths below the soil surface. Bedrock beneath shallow soils also restricts rooting and drainage.

The interdependent factors of water infiltration, movement, and drainage are all influenced by one or more of these factors and in turn affect soil moisture, aeration, and root growth and function. Moisture infiltration can be taken care of most effectively after the surface soil has been altered during planting, although efforts should be made to keep soil compaction to a minimum.

Soil Water Percolation

At new landscape sites, it is wise to test the rate of water percolation in the soil; compaction, strata of different textures, clay soil, and plowpan are some problems that might be identified. A gross percolation rate can be estimated by digging a shallow (25–50 mm; 1–2 in.) hole into which a cylinder 150 mm (6 in.) or larger in diameter and height is pressed into the soil 25 to 35 mm (1–1.5 in.). Keep water in the cylinder for 8 to 10 hours. After the last filling, if the water does not lower at least 25 mm (1 in.) within two hours, you may have a percolation problem, but you cannot tell if the problem is throughout the profile or confined to one or more strata.

Percolation rates may need to be determined at different depths in order to locate compacted layers or lenses of different textures which can complicate irrigation, drainage, and rooting. Hardpan and bedrock near the surface can also be located. Even though a soil profile may appear to be uniform, portions of it may be tight and restrict water movement.

Water movement can be determined in different soil layers. Wildman (1969) devised simple infiltrometers by using 75-mm (3-in.) aluminum irrigation pipe, some glass jugs, rubber stoppers, and plexiglass tubing. He drove six sections of pipe, varying in length from about 0.3 to 1.1 m (12–44 in.) down about 70 or 80 mm (3 in.) into the bottom of auger holes spaced about 600 to 900 mm (2–3 ft) apart and drilled to various depths in the soil. The water reservoir for each infiltration pipe consisted of a glass jug fitted with a rubber stopper through which a plexiglass tube was inserted. When the jug was placed as shown in Fig. 8–1, water dribbled out of the tube until its level in the aluminum pipe rose above the end of the tube. No further flow took place until water moved into the soil and allowed some air to enter the tube. A scale taped to the side of the bottle was used to take readings at various intervals. The plexiglass tube arrangement provided a fairly constant head of water in the hole, and the jug held enough water to obtain sufficient readings without refilling.

The initial moisture content of the soil does not seem to greatly affect the usefulness of the readings. Water-level readings taken at 15, 30, 60, and 120 minutes will be sufficient to construct an informative infiltration curve. To obtain the depth

Soil Preparation

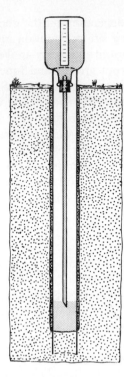

Figure 8-1 A set of inexpensive infiltrometers can be used to determine water movement at different depths in a soil and to locate any layers of slow penetration. An aluminum pipe 75 mm (3 in.) in diameter is inserted into a 90 mm (3.5 in.) augered hole in the soil and driven about 75 mm (3 in.) into the soil at the bottom of the hole. A 4-liter (1-gal) jug with a rubber stopper and a length of plexiglass tubing about 200 mm (8 in.) shorter than the length of the pipe is filled with water and inverted. The tubing is inserted into the pipe. The rate of water movement can then be measured by the rate of water loss from the jug. A series of infiltrometers at different depths will yield a profile of water penetration for a given soil.

of water infiltrated from the aluminum pipe, multiply the difference in water level readings by the radius of the jug squared divided by the radius of the pipe squared.

Strata with infiltration rates of 5 to 7 mm (0.2 to 0.3 in.) or less an hour from a 75-mm pipe are considered slow and in need of improvement. Since several infiltrometers at the same depth at the same site are likely to give different readings, at least two or three replications are desirable. This technique gives only an approximate water infiltration rate, one that is higher than the water percolation rate in the soil at large. It does, however, give a fairly accurate measure of the relative differences in water movement at different soil depths. Thus layers that restrict water penetration can be identified.

Improving Soil Moisture Conditions

Reducing Water Accumulation. Although water may move through a soil at a faster rate than the minimum given above, the soil may still be too wet to allow most plants to grow well. In such a case, water is entering the soil faster than it can move through the soil and drain away. Water may be accumulating in a low area on the surface, moving into the soil from an outside water source at a higher elevation, or collecting from pavement and building runoff. Initial grading should anticipate surface water flow, so that low spots do not occur or are provided with drainage. Where grading alone will not remove surface water, a *French drain* may remedy an existing wet spot or intercept underground flow if the amount of water is not great. A French drain is a slit trench, 50 to 100 mm (2–4 in.) wide, dug through the wet area toward an outlet, a lower area, or a more permeable soil. The bottom of the trench should slope (1 to 3 percent) toward the low area. Fill the trench with medium sand (0.25–0.5 mm in diameter) or fine gravel. Crown the fill above the surrounding soil level to delay the sealing that will occur when fine soil is washed into the fill (Fig. 8-2). In a turf area, the grass will soon grow and hide the trench; in shrub beds, plants or mulch will obscure the fill. Even if the trench does not lead to a drainage way, water will be able to move laterally from the trench into more permeable soil than that near the surface. Aeration will also be improved along the trench. To intercept water flowing underground into the planting area, locate the French drain along the upper side and essentially at right angles to the slope, though allowing for some fall.

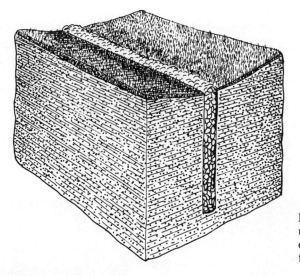

Figure 8-2 A French drain is used to remove surface water or to intercept underground flow.

Preventing or Reducing Soil Moisture Problems. Soil problems can often be avoided or at least reduced if certain precautions are taken before a site is graded or filled. If you plan much cut-and-fill grading, you must avoid locating impervious layers or infertile subsoil close to the surface (see Fig. 6-1) or creating new textural layers that could interfere with water movement and rooting. For these purposes, a knowledge of the soil profile is essential.

Compacted layers, particularly roadways, should be broken up and the area rough-graded and disked or rototilled before it is filled or graded. This will make grading easier, reduce the interface between the fill and original soil, and minimize impervious layers or pockets below fills that could impair water movement and drainage. Fill soil should be of a texture similar to or slightly coarser than the soil on which it is being placed. The site should be well watered to settle the soil and then graded when dry if the surface is uneven.

Correcting Slow Water Movement. If the tests described above reveal compacted soil only near the surface, disking or rototilling to a depth of 150 to 200 mm (6–8 in.) will loosen the soil for easier planting and improved aeration and water infiltration. A plowpan at 200 to 250 mm (8–10 in.) can usually be broken up by deep plowing or ripping (subsoiling) in two directions, intersecting at tree planting locations if possible. Ripping at directions that are diagonal to one another is somewhat more effective than 90° cross-ripping (Trouse and Humbert 1959). Subsoiling is most effectively done when the soil is dry, so that the compacted layers shatter. In Sweden, subsoiling compacted soil before planting trees increased tree growth (Rolf 1990). A good crop of weeds or other plants can dry a wet soil fairly effectively if there is no rain. Subsoiling naturally layered soils, however, has had only limited success (Aljibury, Meyer, and Wildman 1979), since the lenses often reseal in a short time.

Subsoiling may not be possible in many confined landscape situations. Although it takes longer, a backhoe may be the best equipment for preparing such a planting site. A backhoe can be used to dig a cubed hole of 1.2 to 1.5 m (4–5 ft) for each tree; it can break up the compacted layers, mix strata of different textures, and replace the soil in the hole (Fig. 8–3). A shrub bed can be similarly prepared. Chunks of hardpan, rock, or blue-black strata resulting from anaerobic decomposition should be removed and discarded. If the excavated soil is wet, it should be dried for several weeks in its original piles. When dry, the individual piles of soil should be mixed and returned to the holes. Some arborists add 25 to 30 percent by volume of a slowly decomposing organic matter. Studies, however, question the value of amending soil in shrub and tree planting (Pellett 1971, Whitcomb 1979b, Corley 1984, Davies 1987) and this doubt also applies to the preparation of some-

Figure 8–3 A tractor-mounted backhoe prepares a planting site by breaking through compacted layers and mixing strata of different textures.

what larger areas. Backhoe preparation has been successful in fruit orchards and landscape sites, even though organic matter has not been used (see Fig. 6–10). In addition, holes dug with a backhoe provide an opportunity to check the soil profile for compacted layers, changes in soil texture, structure, or color, and rooting from nearby plants. This information can be used to determine ongoing site preparation.

After deep cultivation or backhoe soil preparation, the disturbed soil should be wetted to settle it and remove air pockets. It may then take one to two weeks before the soil will be dry enough for planting. If the loose soil is not settled before planting, it will settle later, thereby creating a basin that will be slow to dry after rain or irrigation. This may lead to crown rot, poor growth, and even death. Soil preparation before settling should be done when the soil is not too wet; soil structure could otherwise be damaged.

Soil Drainage

If rain is frequent or irrigation available, general soil drainage can be estimated by a method similar to that used for checking water percolation in soil. Dig or bore a hole in the soil 1.2 to 1.5 m (4–5 ft) deep or to hardpan or bedrock. Leave the hole open to see whether water accumulates after a heavy rain or irrigation. If water stands in the hole more than five or six days, a drainage system or other measures may be necessary. If possible, dig the hole a meter or two (3–6 ft) deeper to see if a permeable layer can be located to provide drainage. Observe the drainage of water from the deeper hole after rain or irrigation to determine whether a permeable layer has been found.

Providing Drainage

Penetrating Semi-Impervious Layers. If the test holes provide drainage when deepened, as described above, bore a hole 100 to 150 mm (4–6 in.) in diameter at each tree location, or space holes about 2 m (6 ft) apart for shrub beds. Actual experience and experimentation in this area have been sparse and have provided almost no recommendations for the spacing and fill material to be used in such drainage holes, which are somewhat akin to dry wells although their purpose is to drain wet subsurface strata. After several holes are installed, an observation hole 1.2 to 1.5 m (4–5 ft) deep can be placed among them for observing water in it after the soil has been wetted. The spacing of remaining holes can be determined by this observation.

The drain holes should be filled with medium sand (0.25–0.50 mm in diameter) or fine gravel. If coarser fill is used and the surrounding soil is silty, saturated soil may slake into the fill material and plug the drain. If the soil is fluid when saturated, the outside of a perforated plastic pipe 70 to 100 mm (3–4 in.) in diameter should be covered with fiberglass matting about 10 mm (0.5 in.) thick or geotextile fabric and inserted in each drain hole. The covering will filter out the fine sand and larger particles so that they do not plug the pipe. The pipe should be capped at the surface to keep out soil and debris but does not need to be filled. Such a pipe drain can be back-flushed with water under pressure if the filter seems to be plugged.

Penetrating Hardpan.   Hardpan within 750 mm (30 in.) of the surface should be broken up or penetrated if possible. If the site is large enough and the hardpan not more than 600 mm (2 ft) thick, large subsoilers or slipplows can rip to depths of 1.5 to 2 m (5–7 ft). The hardpan must be broken to its full depth in order to provide drainage. Ripping is most effective in dry, brittle soils and hardpans and least effective in moist loams and clays. Compact sandy layers will shatter even when they are moderately moist (Wildman, Meyer, and Neja 1975). For best results the parallel ripping passes should not be more than 2.5 m (8 ft) apart, and passes in a second direction should be diagonal to the first direction.

Deeply tilled soil should be allowed to settle until several rains or sprinkler irrigations have occurred. If the soil is worked before it has settled, the final surfaces will usually be uneven. If the site is allowed to settle over a winter before it is extensively worked and planted, the plants will probably grow better and be more evenly wetted during rains or irrigations.

If the hardpan is too thick to rip or if ripping is impractical, holes should be dug or drilled through the hardpan to provide drainage and possibly some rooting into the soil below (Fig. 8–4) (Gowans and Hall 1971). Power augers of the sort used to set utility poles can drill through most hardpan. If the soil is quite permeable above the hardpan, a clump of trees can be drained by less than one drain hole per tree. The spacing of holes in shrub beds should vary from 2 to 3 m (6–10 ft), depending on soil texture and depth. The finer the texture and the shallower the soil, the closer the holes must be. For such drain holes to be effective, strata under the hardpan must be permeable.

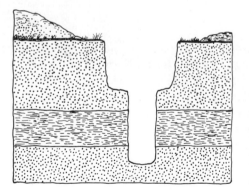

Figure 8–4 In shallow soils with underlying hardpan, it may be necessary to drill through the hardpan to provide adequate drainage and rooting volume. Fill with soil of the same or slightly coarser texture as the surface soil.

Unless the soil above the hardpan conducts water too slowly, the holes should be filled with soil of a texture similar to or slightly coarser than the surface soil. This will provide one less interface to impede water movement and will create a greater head (potential) at the bottom of the hole, which will aid drainage. After the soil has been repeatedly wetted, water in the hole should lower at least 2 mm (0.1 in.) per hour, or it will be of little value. If soil is well drained generally, roots may grow down the holes and into the pervious soil below the hardpan. This will increase the drought tolerance of the plant and lengthen the periods between irrigations.

196   Chap. 8   Planting Site: Preparation

Blasting hardpan or compacted layers with dynamite before planting trees and shrubs is usually not a good idea. It can form impervious basins where standing water can kill plant roots.

Providing an Internal Drainage System. If the downward movement of water is impeded by tight soil, rock, or hardpan, an internal drainage system may provide the only suitable rooting zone (Fig. 8–5). (Planters without natural drainage are discussed in Chapter 11 as a special case.) Landscape soils are often composed of fill material of various textures or are compacted during site preparation and may drain more slowly than similar soil nearby. The previously described methods can be used to determine the presence of soil layers that are permeated slowly.

The desirable depth and spacing of drain lines depend on soil texture and structure, depth of impervious soil or rock, depth of drainage necessary for the plants and the amount of water (from rain, irrigation, or seepage) that must be removed. To grow and stabilize adequately, most trees need deeper "drained" soil than do shrubs. Contrary to what many think, the more slowly water moves in a soil, the deeper the drain must be; the more water that must be drained, the deeper and closer together the drain lines must be. The drain line should be placed above any impervious layer.

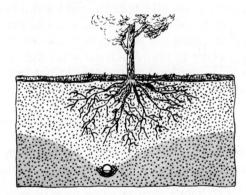

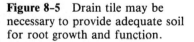
Figure 8–5 Drain tile may be necessary to provide adequate soil for root growth and function.

Drain lines should be at least 1 m (3 ft) below the soil surface, and plants will grow even better if the lines are at a depth of 1.5 m (5 ft) (Beauchamp 1955). If the lines cannot be placed this deep, they should be placed closer together. Table 8–1 lists recommended spacing between laterals for agricultural soils; in poorly drained landscape soils, the distances should be reduced by at least half.

To work properly, a drain system must have

An outlet from which the water can flow or be pumped
An even fall throughout its length, 50 to 300 mm per 10 m (6–16 in./100 ft)
A filter over the drain line to minimize clogging, particularly if soil is silty or sandy; clay particles remain aggregated or are suspended in the water

Drain lines are made of clay, concrete, or plastic. Clay and concrete tiles 100 mm (4 in.) in diameter and larger are commonly used for draining agricultural

TABLE 8-1 RECOMMENDED DRAIN SPACING[a] FOR AGRICULTURAL SOILS
OF DIFFERENT TEXTURES AND PERMEABILITY (Beauchamp 1955)[b]

Soil	Permeability	Spacing	
		Meters	Feet
Clay and clay loam	Very slow	9–20	30–70
Silt and silty clay loam	Slow to moderately slow	18–30	60–100
Sandy loam	Moderately slow to rapid	30–90	100–300
Muck and peat	Variable	15–60	50–200

[a] These spacings are quite similar to those given by Hillel (1980).

[b] In poorly drained landscape soils, reduce spacing by half. Drain lines should be 1 to 1.5 m (3–5 ft) below the soil surface.

soils. Because most landscape sites are less than 0.5 hectare (1.35 acres), plastic pipe 50 to 100 mm (2–4 in.) in diameter is large enough to carry water; easier to install; more stable in soils that vary in texture, compaction, and settling characteristics; and less likely to be invaded by tree roots. The larger the pipe diameter, the longer the drainage system will function without clogging. Plastic pipe should be used in important landscape areas, and provisions should be made so the drain lines can be reamed and cleaned out as well as backflushed. Tile (600–900 mm or 2–3 ft long) should be laid end to end on a thin layer of coarse sand or fine gravel, and a small crack (about 3 mm or 0.125 in. wide) left between each tile for entrance of water (Fig. 8–6). Plastic pipe with three to four 3-mm holes per 100 mm (10 0.125-in. holes/ft) can be laid in a similar manner, but the pipe sections should be cemented together. Flexible drain tubing also comes in rolls that can be cut to the desired length.

Tile joints should be covered with tar paper or plastic on top so that water must enter the tile from below; this allows some of the larger suspended particles to settle out of the water before they reach the drain. Some plastic pipe has holes in only one-third of its circumference; this portion is placed on the bottom. Pipe

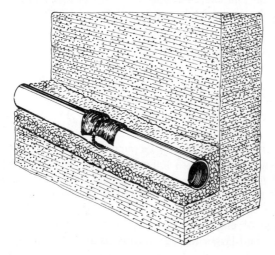

Figure 8-6 Drain tiles are usually laid on a bed of sand or fine gravel and covered with 50 mm (2 in.) of coarse material to allow water to move to the tile joints. The joints are covered to reduce the amount of sediment entering the drain.

with holes around the entire circumference should be covered on top with a strip of plastic if soil is silty; the holes on top should pose no problem in sandy or clay soils.

To allow water to flow to the drain openings and to keep soil particles from entering the drain, the drain line should be enveloped in 50 mm (2 in.) of coarse sand or fine gravel or a synthetic filter material. A synthetic filter is preferable in sandy soils, and coarse sand or fine gravel in clay soils. In the eastern United States, drain lines in clay soil are commonly not encased in any coarse material; most clay aggregates are fairly stable, and clay particles that do enter drain lines seem to be carried along with the drain water. Silty soils, on the contrary, can cause problems in drains: Silt particles slough easily and clog fine-mesh filters. Drain lines in silty soils should probably be encased on the top and sides in 50 mm (2 in.) of coarse sand or fine gravel and covered with synthetic filter material, although this method has not been tested. An irrigation or drainage engineer should be consulted for all but the smallest drainage systems.

Improving Percolation in an Existing Planting

Few landscapes have internal drainage systems, because most percolation problems are not recognized before a site is developed or are created during development. Also, the expense and delay of installing an adequate drainage system may be such that many people hope to minimize problems with proper maintenance alone.

Slow water infiltration can often be improved by aerification of turf or light surface cultivation of bare soil and the application of mulch (see Chapter 14). The procedures already discussed for site preparation will also apply to drainage problems in an existing landscape. Care should be taken to minimize damage to the root systems of important trees and large shrubs. Grading and French drains can improve surface drainage. Holes can be drilled near trees by power auger or water jet or fractured with an air blast (see Chapter 13) to penetrate layers that restrict water movement, whether they be lenses of different textures or compacted layers. Internal drains can be installed if needed. Drainage trenches can be dug by machine until root density and size indicate that digging should be continued with hand tools so that roots larger than 50 mm (2 in.) in diameter will not be cut.

IRRIGATION SYSTEMS

Landscape plantings can be irrigated in a number of ways: basin, furrow, sprinkler, soaker, and frequent–low-volume systems. The method used will depend on the type of plantings; amount, quality, and source of water; terrain; available funding; and sources of labor. In order to save water and labor, most intensive landscape plantings are being developed with sprinkler or frequent–low-volume, automatically controlled systems. Adequate design of the system is imperative. See Chapter 13 for a discussion of irrigation systems and water management.

Whatever methods of water application are chosen, the main delivery system to individual planters or beds should be planned and installed, along with the other utilities at the site, before paving and planting take place.

CHAPTER 9

Planting _____

Planting is the culmination of much preliminary work: analyzing the site, developing landscape plans, selecting appropriate and healthy plants, and preparing the site to preserve existing plants and accommodate new ones. Proper handling of stock, planting, and after-care can be fairly simple but are immensely important. Woody plants can be started in the landscape directly from seed (Fig. 9–1), as young plants easily handled by one person, or as larger specimens that must be transplanted with special equipment. Each method has its place.

Most direct-seeded plants develop in balance with their surroundings and require little or no maintenance once established. Direct seeding, liner plants, and cuttings can be economical ways to plant extensive areas that will receive minimum care and need not produce an immediate effect: highway and utility rights-of-way, reservoir areas, future recreation areas, urban forests, and open spaces (Chan, Harris, and Leiser 1977).

Young plants are most commonly used in home, commercial, industrial, and public landscapes because their cost is moderate and they have an immediate modest effect on the landscape. The landscaping industry is based almost entirely on designing, producing, marketing, planting, and caring for young plants.

Large shrubs and trees are key elements in new, intensively developed landscapes in buildings and shopping malls, on roof gardens, around office buildings, shopping centers, and public facilities. Large specimens are grown specifically for this purpose, become available when land use changes, or are harvested in woodlots, farms, and forests. Specialized equipment and techniques have been developed for handling large plants.

Figure 9–1 Spot-seeding technique perfected by Frank Chan using bur-lap-covered black plastic, a cylindrical carton with soil mix, and protective wire screen; the individual items (left), items in place with screen folded to keep out rodents, birds, and insects (center), and screen opened to allow seedling to increase in height (right). Rocks from the site help hold burlap in place and protect seed hole from large animals.

YOUNG PLANTS

Season to Plant

Unless the winters are too cold for young plants of a species, those planted in the fall will usually outperform those planted in late winter or spring (Whitcomb 1987, Pirone and others 1988). In the fall, the soil is warmer and may have better moisture and aeration conditions than in the spring. Even evergreen plants will have less transpiration due to shorter days and cooler temperatures; roots have more time for growth before there is an increasing demand for water by the top. Fall-planted plants get off to a good start the following spring; however, the season to plant young plants depends on how they have been grown and prepared for handling (see next section and Chapter 10).

Handling Plants Before Planting

Plants should be carefully inspected before or upon delivery to see that they meet specifications as to root quality and top conformation (Appendix 5). If the planting site is colder or hotter or more exposed than the location where plants have been kept, "harden off" or "acclimatize" them. This may take a few days to a few weeks. Keep plants moist and initially protected from temperature extremes.

Many deciduous trees and shrubs are received bare root during the dormant season. Though it is best to plant them soon after delivery, bare root plants can be held up to a few weeks if they are kept cool so that neither roots nor buds begin to grow. The roots must remain moist. The plants can be "heeled in" in the shade with their roots in moist sawdust, peat moss, or sand. If they must be kept in the open, plants can be "heeled in" by placing the roots in a trench running at right angles to the early-afternoon sun. The tops of the plants should point toward the early-afternoon sun so bud warming is minimized. Cover the roots with moist soil and work it in around the roots to avoid air pockets.

Plants can be kept in cold storage and bundled with moist packing material around their roots. The low temperature, by more completely satisfying the cold requirement that many plants have to overcome, will ensure more uniform bud break when the plant is put into the landscape. Cold temperatures (0°–4°C or 32°–40°F), by inhibiting bud break, will allow bare root plants to be planted two to four weeks later than would otherwise be the case.

Balled-and-burlapped (B & B) trees and container-grown plants can be planted almost year-round, although B & B plants are usually planted soon after they are dug. Containers for most plants are dark in color. In the afternoon sun, soil on the exposed side can reach 49°C (120°F) and remain above 38°C (100°F) for six to eight hours. In one experiment, the growth of black locust roots was reduced 75 percent by soil temperatures of 35°C (95°F) for six hours on four consecutive days (Wong, Harris, and Fissell 1971). Temperatures of 40°C (104°F) for only four hours will kill the root tips of most plants. In another study, no roots were found in the western third of 4-liter (1-gal) containers exposed daily to the afternoon sun (see Fig. 4–7) (Harris 1967). The tops of these container plants may show little or no effect because most are watered and fertilized daily. If the soil volume available to a plant, however, is only a portion of that in the container, it will have to be irrigated more frequently when transferred to the landscape than a plant whose roots had filled the entire container. Even these latter plants are seldom watered often enough.

While plants are being held for planting, place them close together, preferably in the shade. If they must be in the open, protect the outside cans from the sun. A 25 × 200 mm (1 × 8 in.) board, a strip of cardboard, aluminum foil, or a dike of mulch or soil will do the job. Keep the plants from wilting. Placing them close together will reduce transpiration somewhat, will reduce injury from excessive movement of the tops, and will help to keep the containers upright. In sunny weather, minimize root damage by placing the plants in the landscape just a short time before they are to be planted.

Preparing the Planting Hole

The following discussion assumes that soil at the planting site is of good tilth or that it has been prepared as described in the preceding chapter.

In soils that have good structure, the planting hole need only be deep enough to hold the root ball of the plant. **Plant "high" or "proud" in all but sandy soils.** The hole can be dug or augered 50 mm (2 in.) less than the depth of the soil in a 20-liter (5-gal) container plant (25 mm or 1 in. less for a 4-liter plant). If a deep hole is dug and loose soil returned, the plant usually settles after a few irrigations or rains. If the top of the root ball is below the level of the surrounding soil, water will collect around the trunk in all but very well-drained soils. Crown rot will frequently occur in such situations. In sandy soil, planting at the original depth or 25 to 50 mm (1–2 in.) deeper will keep soil around the roots from drying out so quickly.

Each hole should be about twice the diameter of the container or root ball, so that the backfill soil can be worked in easily around the plant. For bare root plants, the hole need only be large enough to take the roots without crowding. To minimize "glazing" of the sides when holes are dug with a power auger, digging should be done when soil is at or below field capacity. The glazed sides of many planting holes are almost impenetrable to roots. The sides and bottom of the hole should be scarified or roughened with a shovel to intermingle the backfill and field soil and to provide easier access for developing roots (Smith 1977).

When a planting hole is being dug, if the soil seems compacted or a plowpan is encountered, further measures should be taken. After planting, if there is space to create a more favorable rooting volume, loosen the soil at least 1 meter (3 ft) out from the planting hole to a depth of 200 to 300 mm (8–12 in.), or through the tight soil and plowpan. Adjust the depth (bottom) of the loosened soil so that excess water will not drain to the planting hole. Mulch the loosened soil and restrict traffic over it.

Unfortunately, a few publications still recommend placing about 100 mm (4 in.) of gravel in the bottom of planting holes if the soil is not well drained. The gravel layer is supposed to provide drainage for the soil above. In fact, **a gravel layer will have just the opposite effect, causing the soil above to become saturated when it otherwise would not** (Fig. 9–2). As the amount of water exceeds field capacity in the upper soil, it will begin to flow around the ends of the gravel layer, leaving

Figure 9–2 The fallacy of placing a layer of coarse sand or gravel in the bottom of a planting hole with no drain outlet is shown by the above laboratory soil profile. The soil above the sand layer must become saturated before water will move into the sand. Water flowed around the sand before penetrating it.

that layer essentially dry until the water rises from below or the head of water above becomes greater than the water potential in the soil. **If drainage is needed, there must be an outlet for the flow or pumping of excess water.**

Pruning Roots

Remove the dead, diseased, broken, and twisted roots of bare root plants by pruning to healthy tissue. Roots matted at the bottom or circling around the root ball of container-grown plants should be cut and removed or straightened. Some arborists cut the root ball vertically on opposite sides for at least half the distance to the trunk to decrease the chance that hidden circling roots will girdle later. When freeing the roots at the periphery of the root ball, break away some of the soil to provide better contact between the roughened root ball and the backfill soil. If the periphery roots are straightened so they extend 50 mm (2 in.) or more into the backfill soil, the soil available volume will be double that of a 20-liter (5-gal) root system. In addition, the roots will extend into the fill soil and will grow more easily than when they have to grow out of an undisturbed root ball. Removing one-quarter to one-half of the roots in the outer 25 mm (1 in.) of a root ball should set back none but the most sensitive plants. Most plants will in fact be stimulated.

Butterflying the bottom half of the root ball of container-grown plants by spreading the roots apart has been recommended to minimize root girdling and to hasten plant establishment (Gouin 1983, Feucht 1988). The root ball is removed from its container, laid on its side, and split with a sharp spade or cutting tool two-thirds of the way from the bottom to the top (Fig. 9–3). The two cut halves are spread and placed on a slight mound in a wider but more shallow hole. Backfill the hole so the top of the root ball is 25 to 50 mm (1–2 in.) above the surrounding soil; follow other planting recommendations on the following pages.

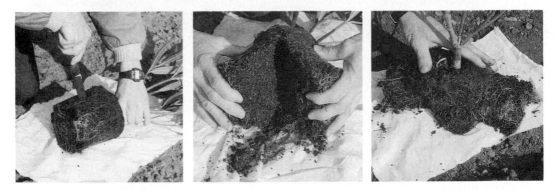

Figure 9–3 *Butterflying* (splitting) a root ball can help a plant adapt to a fine-textured or poorly-drained soil; more of the roots will be in the better aerated soil. The rootball is cut about two-thirds of its depth from the bottom with a knife (left) or a spade for large roots. The base of the root ball is spread open (center) and placed on top of a low mound in a planting hole (right) ready to be backfilled. The root ball is handled on a white cloth for better visibility. Splitting the rootball this way will not correct circling roots near the trunk surface (see Fig. 3–6).

The best reason for butterflying a root ball, as Gouin mentions, is to improve establishment, particularly in heavy or compacted soil. Gilman, Leone, and Flower (1987) found that 70 percent or more of root-length growth of thornless honey locust in the first three years after planting was in the upper 120 mm (5 in.) of a compacted soil. Even in a friable soil, at least 25 percent of root-length growth was in the top 120 mm. Planting a broader, more shallow root ball should be advantageous in shallow, heavy, or compacted soils.

Even though Gouin (1983) recommended cutting part way into the top half of a root ball to further "eliminate" girdling roots, he no longer recommends butterflying root balls because it does not eliminate girdling roots (Francis R. Gouin, University of Maryland, 1990, pers. comm.) That butterflying fails to eliminate girdling roots is not unexpected because a high percentage of girdling roots of plants started in seedflats are close to the trunk and the soil surface. Even so, **butterflying root balls may be wise in tight, poorly drained soils; just be sure the plants do not have circling roots close to the trunk** (see Chapter 3). Gouin now recommends cutting the sides of root balls from top to bottom in at least four areas.

Setting the Plant and Stakes

Just before setting the plant in the hole, loosen about 25 mm (1 in.) of the soil at the bottom of the hole and create a slight mound in the center on which to set the root ball or spread bare roots. Firm the mound by stepping on it lightly with one foot.

You may also want to consider a number of less-than-critical factors when setting a plant in the hole (you will soon observe that some recommendations may conflict with others in certain circumstances and will then have to judge their relative importance and appropriateness): Orient the plant to best advantage in the direction from which it will be viewed most often. Place the scion of budded (grafted) plants, particularly trees, toward the afternoon sun to reduce the possibility of sunburn in the crook just above the bud union (Fig. 9–4). Low foliage may shade this area, but if the trunk is exposed, it can be painted with white exterior latex paint.

Orient the side with lowest branches toward areas of least activity and higher branches toward areas needing greater headroom or clearance. This will reduce the amount of subsequent pruning. Aim the side with the most branches into strong prevailing winds; for deciduous plants, winds during the growing season are more important than winds of winter. On the other hand, Stribling (1966) recommends orienting a young, budded fruit tree, particularly a whip (a trunk 1.2 to 2 m or 4 to 6 ft high with no laterals), so that the scion faces downwind and the tree remains upright. "We expect a greater concentration of strong branches facing into the prevailing winds," he explains. If this is a valid observation, it may be that an uneven development of phloem above the bud union will influence the hormone balance so that more buds are released or stimulated into vigorous growth on the windward side.

If wind, sunburn, and appearance are not factors in plant orientation, place the largest branch or the side with the most branches away from the afternoon sun. The less developed side of the plant will then be favored with more light. With bare

Figure 9-4 When planting a budded tree, orient it so the bud faces toward the afternoon sun. If exposed to the sun, the bark on the inside of the curve just above the bud can be killed or seriously injured. Shading this area or painting it with white latex paint will minimize the possibilities of later injury.

root plants, place the largest root in the direction away from prevailing winds. This seems to be just the opposite of common sense. However, when the wind blows, the roots of young trees tend to be pulled out of the soil lengthwise on the windward side and forced downward on the leeward side. Since the latter encounters more resistance, particularly when the soil is wet, the largest roots should be on the downwind side to stabilize the tree. If the top of the tree is not vertical when the root ball or the trunk base is straight, tip the root ball in the hole to angle the trunk as you want it.

After the tree has been positioned in the planting hole, any stakes that are to be used should be placed close to the root ball or through the bare roots. Usually stakes are driven into undisturbed soil at the bottom of the planting hole to provide adequate support. Information presented later in this chapter will help you to determine whether staking is necessary, and what kind should be used.

Backfilling

In most cases, the soil dug from the planting hole is satisfactory for backfilling around the roots. This is particularly true for bare root plants, which would encounter no soil interfaces. If the soil has been previously loosened and mixed, it should be fairly uniform with depth and can be used directly for backfilling. If the soil has not been loosened and exhibits textural or compacted layers on the sides of the planting hole, adjacent topsoil or soil similar in texture to that of the coarsest layer should be used for backfill.

Although adding organic matter to backfill soil is often recommended, studies in England and different parts of the United States have shown the practice to be of no consistent benefit (Pellett 1971, Whitcomb 1979a, Corley 1984, Davies 1987, Hodge 1990). At five Oklahoma test sites with a range of soil types, Whitcomb found that fewer roots grew into the surrounding soil from amended than from unamended backfill soil. Whitcomb supplies no growth measurements for trees and shrubs planted in backfill soil amended with peat moss but states that the practice is detrimental. Wilson (1984), however, found that roots will often branch more profusely in friable, well-aerated, moist, fertile soil than in drier, infertile soil. The many small, short roots that form will not grow as far as roots in drier, less well-aerated soil. This may account for Whitcomb's conclusion: He correctly postulated that under drought conditions a plant whose roots were largely confined to the amended backfill soil would not fare as well as plants whose roots spread farther even though they were fewer in number. At any rate, in none of his trials did amended backfill soil produce a net benefit.

In England, Derek Patch (1980 pers. comm.) has observed poor results when backfill soil has been amended with organic matter. The main problem he identified was inadequate mixing: Large wet globs of organic matter interfered with root growth.

From the evidence to date, there seems to be little reason for amending backfill soil when planting trees and shrubs in the landscape. Commercial fruit tree growers, even those who plant subtropicals B & B, do not use organic matter. If organic matter is to be used, however, it should be added 30 to 40 percent by volume with the backfill soil and thoroughly mixed.

Placing fertilizer in the planting hole or mixing it with backfill soil is inconvenient and time-consuming and can injure plants. Even though most plants grow well for part or all of the first growing season without adding fertilizer, applying 25 to 50 g (1–2 oz) of nitrogen on the soil in the tree basin is good insurance (see Chapter 12). If nutrient deficiencies exist, nitrogen is usually the only element lacking. If nitrate or urea is applied to the surface after planting, nitrogen will move into the soil with the first rain or irrigation.

Work the soil around the roots so that they are not compressed into a tight mass but are spread and supported by soil beneath them. After each 75 or 100 mm (3 or 4 in.) of soil has been placed in the hole, firm the soil around the roots or root ball with your foot, taking care not to tear, bruise, or debark the roots.

In some regions, a wire basket with mesh openings of 25 to 100 mm (1–4 in.) is commonly placed around the root ball of a hand- or machine-dug tree after it has been wrapped with burlap. The purpose is to hold the root ball more firmly during handling from the field to the landscape. Seldom is any wire removed when the tree is planted. The roots of a number of cottonwood, ash, and pine trees blown down in tornado-like winds in Colorado broke at or just outside the wire baskets that had been left on when they were planted 7 to 10 years before (James Feucht, Colorado State University, 1990 pers. comm.).

Concern about the roots becoming physically and physiologically impaired by the wire has lead to several investigations (Fig. 9–5). In Ontario, Canada, Lumis and Struger (1988) and Lumis (1990) excavated the roots of 34 trees of seven species, including Norway and sugar maples. The trees had been planted in wire baskets 4 to 15 years before and ranged in caliper from 75 to 300 mm (3–12 in.). Growth of the trees appeared normal. Excavated roots were dried and sectioned. "In every instance the root tissues joined or bridged across the wire, . . . there was intact bark and wood over the wire" (Lumis 1990) (Fig. 9–5). Even though there was some

Figure 9–5 The roots of a Douglas fir tree eight years after planting with the wire basket left around the roots (left); the wire is embedded in a root but not completely enclosed (arrow). (Photo courtesy James Feucht and Eugene Eyerly, CO) A longitudinal section through a root of a columnar Norway maple, planted 13 years before, has enclosed a wire of the wire basket left on at planting (right); the structure of the xylem formed after enclosing the wire appears normal and dye has moved through similar tissue. (Photo courtesy Glen Lumis, University of Guelph).

initial infolding of bark of sugar maple just above a wire, root tissue bridged and resumed active growth. Injected dye moved through the root xylem bridges. In Colorado, Feucht (1990 pers. comm.) observed complete closure over the wire of some roots of cottonwood the third year after planting. Initially root tissue outside the wire appeared disorganized and similar to wound callus; later observations were not reported.

Even if movement of material in the phloem and xylem is not impaired, the stability of a tree planted with a wire basket around its roots may be a concern. However, if root tissue outside the wire resumes normal growth, it is this peripheral growth that gives strength against bending forces (Wagener 1963).

It may be cost effective not to remove wire from around the top of the root ball of the trees planted in informal groups where it would not be serious to lose a few trees. On the other hand, removing the wire from around the top 200 to 400 mm (8–16 in.) of the root ball (depending on size) would be wise for specimen trees, street trees, and those in formal landscape settings and in heavily used areas.

Basket wire remains intact in soil for many years, and wire strength diminishes slowly (Lumis 1990). If wire is to be removed, after setting the root ball in a hole but before the wire and rope or twine that holds the burlap is removed, add soil to support the lower root ball. If the root ball has broken apart or the tree is wobbly, do not plant it. Remove the wire from around the top of the root ball and fold back the burlap from around the trunk and the top of the root mass as long as the root ball stays firmly together. Cut off the loose burlap or fold it down to be buried when the rest of the fill is added. If the burlap is sturdy, particularly if it has been treated, carefully cut large gashes through it in the lower root ball area so that the roots will not be unduly confined. Complete the backfilling.

Papier-mâché containers are difficult to wet if they become dry and roots may have difficulty penetrating them even when wet. Be sure that the burlap or papier-mâché is not exposed above the soil, because either one can act like a wick, drying itself and the soil below.

The original ground level in the nursery (the soil line on bare root trunks or the top of the root ball) should be 25 to 50 mm (1–2 in.) above the finished ground level. Basins should be at least 750 mm (30 in.) in diameter. Fill the basin with water to further settle the soil and to provide the plants with water. For best results, attach a 1- to 1.2-m (3- to 4-ft) pipe to a hose and gently push the pipe, while water is running, through the soil around the roots. This will remove air pockets and improve contact between the fill soil and the roots. The plant can be rocked slightly to facilitate soil consolidation.

If the plant has settled so that the original soil line is below the soil surface, a bare root tree can be raised by lifting on the trunk while the soil is muddy (Fig. 9–6). It should be raised slightly higher than desired and then settled back into the soil. Use a shovel under the root ball to raise a container plant (Fig. 9–6). These movements should be kept to a minimum, since the more a plant is raised, the closer the roots are drawn together.

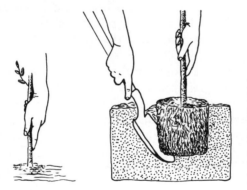

Figure 9–6 A bare root tree that has been planted too deeply can be raised to the proper height if you lift on the trunk when the soil is muddy (left). For B & B and container-grown plants, use a shovel to help lift the root ball while you lift on the trunk (right).

After the soil has drained, the final contour of the basin can be designed to make the base of the plant slightly higher than the bottom of the basin (Harris and Davis 1976). Unless some soil has fallen or been removed from the top of the root ball, do not add soil there to obtain the desired basin contour. This is particularly important if the backfill soil is finer in texture than that of the root ball. If so covered, the root ball will be wetted with difficulty or not at all (Fig. 9–7).

Mulch placed over the root ball in the basin and on the surrounding berm (ridge) will reduce moisture loss, moderate surface soil temperatures, and reduce settling and cracking of the soil and berm surface. Use 50 to 75 mm (2–3 in.) of coarse organic material (woodchips, bark, or forest litter), coarse gravel, small rocks, or built-up layers of dry lawn clippings. Around trees in a lawn area, a coarse mulch may interfere with mowing, but lawn clippings will usually work quite well (see Chapter 14).

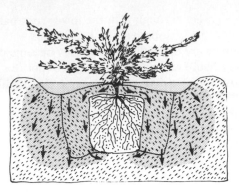

Figure 9-7 When a root ball is covered with a layer of soil whose texture is finer than that of the root ball soil, water will tend to flow around the root ball rather than entering it. The root ball will then be wetted only by water that rises from below.

Pruning

Newly planted trees and shrubs are often pruned with the thought that reducing top size will compensate for root loss of bare root plants or the relatively small root ball size of most container-grown plants. **If sufficient water is available, plants will usually grow better with little or no pruning than with severe pruning.** When water may be inadequate, however, prune severely enough to reduce the size or density of the plant's shadow on the ground to improve its chances for survival. Kozlowski and Davies (1975, citing Allen 1955) reported that the survival of newly planted unirrigated pine in the field was improved by shortening their needles. Fruit trees are commonly pruned severely at planting so that they will develop low-branching structures.

Thinning branches that are close together, crossing, or broken can remove considerable leaf area without greatly affecting overall plant size. One-fourth of the leaf area can usually be removed from most plants with only positive effects. If a plant's basic branch structure is not the one desired, some remedial pruning should be started; do not wait until the next season as is often recommended (see Chapter 15).

Staking

Before a tree or large shrub is planted, materials for any necessary staking should be at hand. Staking is used to protect or to anchor trees and shrubs and to support young trees (Harris, Leiser, and Davis 1976). The extent of staking for trees will depend on tree strength and conformation, expected wind conditions, the amount of vehicular and foot traffic, the type of landscape planting, and the level of follow-up maintenance. Many young trees can stand alone; others may need support to stand against the wind or to grow upright as desired. In order to decide whether staking is necessary and what type to use, you may want to review the consequences of supporting a tree by either staking or guying.

Consequences of Staking. Compared to a tree that stands alone and is free to move, a staked tree will

Grow taller

Grow less in trunk caliper near the ground but more near the top support tie

Produce a decreased or even a reverse trunk taper

Develop a smaller root system

Offer more wind resistance than trees of equal height (because the top is not free to bend)

Be subject to more strees per cross-sectional unit at the top support point (the ground for an unstaked tree)

Be more subject to rubbing and girdling from stakes and ties (Fig. 9–8)

Develop uneven xylem around the trunk if it is closely tied to one stake; the trunk will grow or bend away from the stake (see Fig. 2-11)

May be unable to stand upright when untied (Fig. 9–9)

All of these staking influences make a tree less able to stand without support (Fig. 9–9) and, if supported, more subject to injury, particularly if the ties or stakes should break. In addition, staking is expensive and time-consuming and often de-

Figure 9–8 Tree ties can injure. If ties are loose, the trunk or branches rub against the stake (left); if they are tight, a girdle may surround the trunk (center), restricting translocation in the bark and eventually causing breakage (right). Note the wire deeply embedded in the trunk; trunk caliper above the wire is greater than below it.

Figure 9–9 The cumulative influence of staking is shown by these two silver dollar gum trees grown for 11 months in 20-liter (5-gal) cans. One has grown unstaked with lower laterals on trunk headed back (left tree); the other has been tied to a stake with the lower laterals removed (right tree). The staked tree has been untied from the stake. (Harris, Leiser, and Davis 1976)

tracts from tree appearance, especially when improperly done. Even so, staking may be necessary to help young trees until they get root anchorage or their tops become strong enough to stand alone. Staking and other materials may be needed to protect young trees from vehicles, equipment, animals, and vandalism. Necessary and proper staking can overcome most of the problems associated with this procedure.

Types of Staking. *No staking* is necessary for most shrubs and for many conifers and other trees with limbs close to the ground (Fig. 9-10). Low branches keep people and equipment away from the trunk. These plants are usually short, with root systems or root balls adequate to hold the tops upright.

Figure 9-10 Most trees with laterals close to the ground need no staking (left). Many deciduous trees need only short stakes to protect their trunks from lawnmowers and other equipment (center). Other trees may have sturdy trunks but root development inadequate to maintain the tree upright, particularly when soil is wet. Such trees should be anchored by tying them between two low stakes (right). (Harris, Leiser, and Davis 1976)

Protective staking is used primarily to keep mowing equipment, vehicles, and vandals away from young trees that are able to stand alone: most conifers, small trees with upright growth habits, and most trees planted bare root. Even many trees with tops that are large in proportion to their roots can stand alone if about one-third of the branches in the crown are thinned out, reducing wind resistance and top weight.

If vandalism is not a problem, place two or three short stakes (about 50 × 50 mm or 2 × 2 in. thick and 1.2 m or 4 ft long) approximately 400 mm (15 in.) apart around the tree so that they protrude from the ground about 750 mm (30 in.) (Fig. 9-10). The stakes should be easily visible, so that people will not walk into them. To minimize vandalism, many public agencies are planting larger trees and using sturdy stakes encircled with heavy wire or metal grillwork (Fig. 9-11). More information is needed on staking that protects trees without unduly restricting top movement.

Anchor staking may be needed to hold the roots or root ball of an otherwise upright tree or shrub until the roots grow into the surrounding soil and can support

Figure 9–11 Trees in heavily used areas are frequently protected with metal grillwork (left) or more substantial barriers (right).

it. Spring rains or the frequent irrigations given to young plants add to their instability. If the roots are not well anchored, trunk movement could break new roots growing into the surrounding soil. Low anchor staking can hold the lower trunk to keep roots from moving while permitting freedom to the top.

The two or three short stakes suggested for protecting treetrunks usually provide enough anchorage for roots. Place one loop or figure-eight tie between each stake and the treetrunk (Fig. 9–10). Each tie should be secured near the top of its stake and should permit some movement at that level without allowing movement of the roots or rubbing of the trunk against the stake. The tops of some trees may benefit from thinning to decrease wind resistance and weight. Ties can usually be removed by the end of the first growing season, but leave the stakes in place to continue protecting the trunk.

Support staking is needed for trees whose trunks are not strong enough to stand without support or to return upright after a wind. Weak trunks often result when trees have been previously staked, grown close together, or had their lower branches shaded or removed (Fig. 9–9). Weak trunks are common on many container-grown trees. Top support for these trees should be about 150 mm (6 in.) above the lowest level at which the trunk can be held and still return upright after the top is deflected (Fig. 9–12). This location will give the top the greatest flexibility while providing adequate support.

Young trees are often tied to a stake higher on the trunk than is necessary.

Figure 9-12 Glossy privet supported by a stake to which it was tied during container production (top left); untied, it cannot stand upright (top right); to determine height of support tie, trunk is held at about 1.2 m (4 ft) and the top is bent (bottom left); this is the lowest height the trunk can be held and have the top return upright (bottom center); the support tie should be about 150 mm (6 in.) higher than the height determined (bottom right). Thinning out the top will reduce wind resistance. The top of the stakes should be cut to within 50 mm (2 in.) of the top ties to reduce the chance of rubbing.

Leiser and Kemper (1968) found on trees with tapered trunks that stress per unit of trunk area increased with increased height of staking. Just the reverse was true when trunks had little or no taper between the ground and the head of the tree. As the height of stake support increases in trees with untapered trunks, the lever arm of the trunk above the stake decreases and the stress at the tie decreases proportionately. On tapered trunks, cross-sectional area decreases proportionately more with increasing height than does the length of the trunk lever arm. The higher the support, the more upright and inflexibly the top is held, so that it must take the brunt of winds instead of taking them more obliquely. In addition, as the leader terminal is approached, the trunk is more succulent and less able to withstand wind stress. Leiser and Kemper determined that **the most critical stresses occur when staking height for a tapered treetrunk is above two-thirds the height of the tree.** Stress per cross-sectional unit increases rapidly as staking height increases above this point. When trees were staked at points closer than 0.75 m (30 in.) of the terminal, Leiser and Kemper found that stress per unit was 3 to 4.5 times as great as it was near the ground. They did not obtain stress figures within 10 percent of the height near the tip, but extreme stress would probably continue. The stresses found in the top portion of the trunk exceeded those that most species can endure (Fig. 9–13). It is a wonder that the tops of more staked trees are not deformed or broken by the wind.

Trees with little or no taper might be best suited for staking if they were not to grow after being staked. However, as previously stated, a staked tree grows taller and with less basal caliper than an unstaked tree and is less able to become self-supporting. Also, the stake and tie must withstand the stress of an increasingly larger tree.

Staking stresses vary within the head of a tree, depending on distances from the terminal and on the position and size of individual branches. Tying a leader closer than within 0.75 m (30 in.) of the tip, particularly of previously unstaked trees, will subject the leader to maximum stress and maximum likelihood of wind deformation. The onset of damage could be delayed by staking at ever-increasing heights, but practical limits and the weak trunk that would result make this solution unattractive. Sometimes small trees are tied to stakes that are taller than the trees. It should be apparent that, as the terminal grows beyond the top tie or the top of the stake, the leader will be subjected to extreme stress from wind unless it is in a protected area. It would be best in such cases to drop the height of the top tie and stake top to at least 1.2 m (4 ft) below the leader tip. A flexible auxiliary stake may be needed to strengthen the leader.

Only a single support stake is commonly used. Two stakes, however, can minimize the difficulties encountered in support staking; some arborists and landscape architects even recommend three stakes. In any case, care must be taken to tie or hold the trunk so that it can flex without rubbing against a stake. The ties must not be so tight or inflexible that they are likely to damage tender bark or girdle the expanding trunk. When you decide how many stakes to use per tree, consider the foregoing criteria, along with material and installation costs, expected vandalism, and expected frequency of follow-up maintenance. More latitude can be allowed in selecting the method of staking and tying if the trees will be inspected frequently.

When only one stake is used, the tree is often tied snugly against it at more than one level. Besides the direct effects of trunk immobility, three difficulties may result: (1) The ties may girdle the trunk, although frequent inspection can prevent this. (2) Greater stress will occur at the top tie during a wind if the trunk below is not able to flex in the opposite direction as the top is blown to and fro. The trunk is therefore more likely to break at the top tie or to be seriously deformed (Fig. 9–13). (3) The stake also will shade the trunk, causing the xylem cells to elongate more on the shaded side so that the tree actually tries to grow away from the stake (see Fig. 2–11).

Figure 9–13 Bending stress at the top tie of staked trees can cause serious deformation or breakage. *Melaleuca linariifolia* tied to an iron pipe (50 mm or 2 in. diameter) (left); a young liquidambar tied to a rigid stake broke at the top of the stake.

In addition to the many clever methods developed by commercial and public arborists, a number of commercial tree staking and tying devices are available to hold trees properly to single stakes. Black (1978) reports that street trees in Seattle have suffered less than 1 percent loss from vandalism when trees of 50-mm (2-in.) caliper are planted and a 16-mm ($\frac{5}{8}$-in.) rebar (metal reinforcing rod) 3 m (10 ft) long is driven 1.5 m (5 ft) into the ground close to the trunk. Three ties closely spaced near the top of the stake hold the trunk securely. At that height (1.5 m) it is difficult to break the trunk. The stake is removed after one year. One stake is cheaper and easier to install than two stakes, although any stake is easier to install if it is "planted" with the tree instead of driven afterward.

Two support stakes with one flexible tie near the top of each will hold the tree upright, provide flexibility, and minimize trunk injury and deformation (Fig. 9–12). If wind is not a problem, the stakes should be placed so as to provide maximum protection from traffic and equipment. In moderate to strong wind situations, though, an imaginary line drawn between the support stakes should be at right angles to the most critical wind direction. The ability of stakes to withstand wind, particularly when the soil is wet, can be increased by connecting them with a 25 × 75 mm (1 × 3 in.) wood crosstie partially or entirely below the soil surface. Place the crosstie to the lee (downwind) of the stakes, being careful not to injure roots near the surface. The crosswind orientation of the two stakes lessens the risk that the trunk will suffer rubbing injury and that the stake assembly will jack itself out of the ground in strong winds. One low crosstie keeps the two stakes equidistant along their length but still provides some flexibility in a strong wind.

Some arborists prefer three stakes for greater protection against equipment, vandals, and strong winds. Three stakes will more effectively support protective cylinders of wire or rod which may be desired or necessary in certain situations.

Stake and Tie Materials. Support stakes come in a variety of materials and cross-sectional shapes. Wooden stakes are usually sawn wood or small-diameter poles 50 mm (2 in.) in diameter (lodgepole pine is commonly marketed). Poles do not split or break as easily as sawn stakes while in use or while being driven (John Sue, Oakland, CA, 1973, pers. comm.). Metal pipe, "T" fence stakes, and reinforcing rebars (Fig. 9–14) of various sizes are also used. Metal stakes usually need a flange or plate just below the ground surface for extra stability. A metal flange entangled in roots can make stake removal difficult, however. Availability, cost, strength, durability, ease of installation, and appearance are all factors in selecting staking material.

Figure 9–14 A single tree stake provides support without shading or rubbing the trunk, as long as the tie holds the trunk secure.

Several different materials and methods are also used in tying trees to the stakes. Any material used should contact the trunk with a broad, smooth surface and have enough elasticity to minimize trunk abrasion and girdling. Common tie material includes elastic webbing, belting, polyethylene tape, tire cording with wire ties, and wire covered with hose or tubing. A number of patented ties and support devices are available for single and double staking. The use of rope, baling wire, covered electrical wire, string, and fishing line to hold tree and stake together is not advisable. Even commonly used hose-covered wire is inadequate to prevent injury (Feucht and Butler 1988). Brown (1987) observed that even careful once-a-week measurement of year-old birch stems with a hand-held micrometer interfered with normal stem formation even though there were no external signs of injury.

Especially on a single stake, wire ties break from flexing in winds. The break is usually between the trunk and stake, so that a circle of wire remains around the trunk. If the trunk is not immediately damaged by losing its support, it may be damaged later by rubbing against the stake and the jagged broken wire or by encountering the wire girdle as it grows. Such a girdle may be obscured by low foliage until it is too late (Fig. 9–8). Polyethylene or fabric ties that may last only a year will minimize the problem of girdling, although they may allow stake rubbing.

Except for the patented ties and devices, most methods of tying employ a figure-eight loop between the trunk and each stake. The figure-eight tie can hold the trunk secure and yet allow some flexibility; also, two layers of the tie material cushion the trunk against the stake. For two or more stakes, a plain loop between tree and each stake is often used, although these lack the cushioning capacity of the figure-eight if one tie breaks.

Trees with spindly trunks and little or no taper may need extra support along most of their length. A rigid support, however, will delay developing the ability to stand alone. A flexible auxiliary stake tied snugly to the trunk will usually give the needed strength. Its diameter and height should permit the trunk to flex and return upright. It should be slender enough to shade the trunk little or not at all. A 25×25 mm (1×1 in.) stake can deform treetrunks with diameters of 6 to 12 mm ($\frac{1}{4}-\frac{1}{2}$ in.) so that they bend away from the stake when untied (see Fig. 2–11) (Neel 1967). Spring-steel wire, fiberglass rods, bamboo, and split wood can be used for trunk support. Steel wire or rods are easily available and uniform. Harris, Leiser, and Davis (1976) have suggested that a rod 3 mm ($\frac{1}{8}$ in.) in diameter be used for 20-liter (5-gal) trees. If more support is needed lower on the trunk, a shorter section can reinforce the first rod, which will still provide flexibility at the top. A number of arborists use 6-mm ($\frac{1}{4}$-in.) steel rods for 20-liter (5-gal) trees and 9-mm ($\frac{3}{8}$-in.) rods for 60-liter (15-gal) trees. These larger diameters (the cross-section of a 6-mm rod has twice the area that two 3-mm rods have) give more support low on the tree but often lack flexibility at the top.

The rod should not extend within 1 m (40 in.) of the terminal of the leader, or deformation of the top may result, as already discussed. The auxiliary stake need not extend to the ground, though it may look more professional if it does. The tip of each rod should be rounded and wrapped with friction or duct tape to lessen the chance of injury to the trunk and to people. Tape on the tip will also keep the top tie from slipping off in a wind. Tie the trunk to the auxiliary stake with polyethylene tape at 200- to 250-mm (8- to 10-in.) intervals; the trunk and rod are then treated as a single unit and are tied to the support stake at one level.

The auxiliary stake may be needed through the first season. Remove it as soon as possible. Support stakes will probably be needed through the second year, but the height at which they are tied might be lowered before the second season and the stakes shortened.

Alternatives to Support Staking

Almost every response to support staking delays a tree from becoming strong. Support staking represents a last effort to hold a tree upright until it can finally stand alone. For the most part, the sooner a tree can stand alone, the sooner it will become strong.

As an alternative to support staking, a tree smaller than 50 mm (2 in.) in trunk caliper may be cut off about 300 mm (1 ft) above the ground or graft union, painted with white exterior latex paint, and one shoot allowed to grow a new top without support. Similar pruning before growth began was used on newly planted eucalyptus of 35-mm (1.5-in.) trunk diameter on the Santa Barbara campus of the University of California when these trees could not stand upright if untied from their stakes. One watersprout from each stub was developed into a main trunk. Straight, strong trunks grew, unsupported, into trees that at the end of the first growing season were almost as tall as when they were planted (William Smith, University of California, Santa Barbara, 1971 pers. comm.). Many Oregon nurseries cut deciduous landscape and some fruit trees close to the ground at the end of the first year in the field so that the next year's growth will be straight and vigorous and ready for sale. A few trees may die, but those that grow will be fine, strong specimens requiring less care. Such severe pruning should be done before growth begins.

Treeshelters

Treeshelters increase the growth and survival of most young tree species; protect against animals, mowers, and string trimmers; and permit easier herbicide application (Evans and Shanks 1987). Developed in England in 1979, treeshelters are vertical, light-colored, translucent, double-walled plastic tubes about 100 mm (4 in.) in diameter and 0.6 to 2 m (2–6 ft) tall (Fig. 9–15). Shelter height is determined primarily by the size of the animals that may browse and the desired height of the lowest permanent limb. The most common height is 1.2 m (4 ft).

Preformed shelters slide over a tree and fasten to a stake with ties or staples. Evans and Shanks recommend well-rooted transplants 150 to 400 mm (6–16 in.) tall with a good terminal bud and a stem diameter of at least 6 mm (0.25 in.). Trees taller than 1.2 m are not recommended.

Treeshelter plantings are best suited for woodlots and landscapes that have limited public access, such as highway rights-of-way, private gardens, and developing parks, particularly if animals are a problem. No vandalism was experienced, however, to 500 shelters the first year after being planted along streets and other public places in Pekin, IL (Fig. 9–15) (Sassman 1990 pers. comm.). An intensive informational program preceded the planting of the trees in treeshelters.

Shelters act like small glasshouses. Temperatures inside shelters are warmer than outside—important during windy, cold weather. On a clear summer day in England when the ambient temperature was 28°C (82°F), temperatures in double-walled treeshelters with small plants reached 38°C (100°F); those with larger plants reached 32°C (90°F) (Potter 1989). Concerned that in hot, arid summer regions high temperatures could be lethal, Harris (1989) advised that treeshelters should be

Figure 9–15 Some of the 466 trees of nine species planted in 1.2-m (4-ft) tall treeshelters along one of the boulevards in Pekin, IL, in the spring of 1990 (top). Many trees reached the top of the shelters by autumn. Coast live oak after one growing season in 13-l (3.5-gal) containers with and without a 0.9-m (3-ft) treeshelter in a San Francisco Bay nursery (bottom left); the plants were 250 mm (10 in.) tall when planted into the 13-l containers as indicated by the vertical measure between the plants. The plants are shown with the treeshelter removed (bottom right); the confined leaves were in good condition. Deodar cedar trees in the same experiment exhibited similar responses to the treatments but the deodar cedar trunks in treeshelters were spindly and would not remain vertical without a stake. (Top photo courtesy Tom Sassman, Pekin, IL; bottom photos from Burger, Svihra, and Harris, 1991)

perforated in order to dissipate the heat. However, perforations create a chimney effect and desiccate the plants. **Holes should not be put in shelters and the shelters should be firmly pushed into the soil.**

In hot, arid regions of Spain (Joe Lais, St. Paul, MN, 1990 pers. comm.), aleppo pine and holly oak in shelters without irrigation had 55 percent survival while only 3 percent survived in the open. In another trial, two pine species had similar survival rates when irrigated, but those in shelters grew 47 percent more than trees in the open. Bainbridge (1990) reports that similar treeshelter protection and minimal irrigation worked well in the southwestern United States for young honey mesquite, mulberry, oak, and other woody species.

High air temperatures in treeshelters are accompanied by high humidity (Evans and Shanks 1987). In mid-afternoon in England, the vapor pressure deficit (VPD, a measure of evaporative demand) within shelters with small plants was 20 percent greater than outside; however, the VPD in shelters with larger plants was 75 percent less than that outside. Some of the transpired water condenses on the shelter wall, flows downward, and rewets the soil.

Carbon dioxide levels in the lower half of a shelter can be two to three times greater than ambient air (Burger, Svihra, and Harris 1991). Soil microorganisms are probably the major source of elevated carbon dioxide levels.

Evans and Shanks (1987) summarize more than 130 experiments in England on 28 broadleaved species and 50 trials on 11 conifer species. The most extensively tested species, Sessile oak, in three years made five times the height growth in shelters as those not in shelters. Roots and tops significantly increased in size. Even though the sheltered oaks were kept relatively immobile, their stems were 45 percent larger in diameter and three times greater in volume than unsheltered stems. In spite of the sheltered trees having greater diameter for three years, they had to be staked for the next two. Evans and Shanks state, "Ideally a treeshelter should provide a greenhouse effect for the first two or three years and then continue to give support and protection for another two or three years while the stem thickens." They think a shelter life of five years is essential for maximum benefit.

At a coastal valley nursery in northern California, young holly oak, deodar cedar, and southern magnolia growing one season in 1-m (3-ft) tall treeshelters in 12-l (3.5-gal) containers increased in height 174, 74, and 84 percent, respectively, more than trees not in shelters (Burger, Svihra, and Harris 1991) (Fig. 9–15). The trunks of the sheltered plants were quite flexible; in fact, the deodar cedar bent to the ground when the shelter was removed. In addition, the roots of the trees not in shelters weighed 20 to >200 percent more than those in shelters. Most of the magnolia leaves inside the shelters dropped; leaves of the other two species in shelters were healthy.

The only reported pest problems using shelters has been infestation of aphid (*Aphididae*) on beech trees in England (Potter 1987) and on holly oak in California (Burger, Svihra, and Harris 1991). A small number of lady bird beetles placed in the one shelter on the holly oak controlled the aphids. On the other hand, oak mildew was observed to be less in shelters (Evans and Shanks 1987). Until trees grow out of the shelters, metal or plastic screen covers are sometimes put on the top of shelters to keep birds from becoming trapped.

Staking and adequate water may be problems for some species if the treeshelters are removed after one year. A flexible auxiliary staking procedure as described in this chapter should allow the most rapid development of a strong trunk. The limited root systems of the treesheltered trees in the California study indicate that wise irrigation the first growing season may be critical for continued good growth.

Treeshelters hold great promise for beginning with smaller trees and getting them off to a good start. Finding which species will benefit the most from treeshelters and how to handle the trees will be interesting and rewarding.

CARE FOLLOWING PLANTING

In areas where summer irrigation is usually thought to be necessary, deciduous trees and shrubs that are planted bare root and thoroughly watered at planting or afterward should not need irrigation until two to four weeks after growth begins. Plants in other areas may need mid-season irrigation if rainfall has been below normal. Early in the growing season, roots will actively grow into moist soil while the top has few leaves. Overwatering during this time can endanger root growth and function.

On the other hand, container-grown and B & B plants, both deciduous and evergreen, may require rather frequent watering, depending on the soil texture, the weather, and the relative sizes of leaf areas and root balls. Even though most container soils are coarse-textured, their shallow depth while in containers will keep moisture content near the saturation point following irrigation (Davis and others 1974). Depending on plant size and the weather, moisture in the root ball is used fairly quickly by the plant, so aeration is not a problem. **When container or B & B plants are transplanted, much of the water previously retained in the root ball drains into the soil below;** if the root ball soil is coarser than the surrounding soil, its moisture content can be reduced below field capacity by capillary forces. The soil of a small root ball can be brought close to the wilting point, even when the plant is using little or no water and the surrounding soil is near field capacity. Costello and Paul (1975) examined a plant newly planted in a loam soil from a 4-liter (1-gal) can and found that 80 percent of the available moisture in the soil mix was lost within 24 to 36 hours after flood irrigation. If a plant in a 4-liter container can last two days between irrigations before wilting, it would thus need irrigation at least every day when first planted in the landscape. The effects of capillarity will be less dramatic in larger root balls, but drainage into the soil below and capillarity will cause equally great losses down to field capacity and below.

Daily irrigations may cause the surrounding soil to remain extremely wet. For the first few weeks, water need only be adequate to rewet the root ball and a little of the surrounding soil. A small berm can be established just outside the root ball and high enough to hold water that will rewet the roots. The larger basin beyond need be watered only every two or three weeks. After the first few days, the watering interval for the inner basin can be lengthened until the first signs of wilting appear. An interval one day less than the wilting interval can then be used for irrigation for two or three weeks. Repeat the wilting experiment to adjust the irrigation schedule later. Rain must be taken into account, of course. A drip irrigation system will simplify application, but attention must be paid to the amount applied. See Chapter 13 for more irrigation information.

CHAPTER 10

Transplanting Large Plants

Large trees and shrubs transplanted into a newly developed landscape give a mature beauty that can be surpassed only by mature plants that were already on the site. There are almost as many ways of moving large plants as there are people who move them. They may be moved bare root or with a root ball that is completely exposed, covered with burlap and laced, wrapped with wire netting, boxed with lumber and banded or bolted, encased in cement, frozen, or dug and moved by special equipment. Care varies from the extremely meticulous (three years of root pruning prior to the move, digging a large root ball, careful wrapping and lacing, loading and moving with large equipment, preparing the site completely, and overseeing follow-up maintenance) to the extremely casual (simply scooping out the plant and moving it with a front-end skip loader or soaking the root area for several days and pulling the plant out with a crane). Any given method has its own advantages, disadvantages, and degrees of success. Transplanting success depends on the plant species and its condition, characteristics of the original and final planting sites, the season of the year, and follow-up care, as well as on the transplanting method itself.

SELECTION OF PLANTS

Species

Experience has shown that some species can be transplanted more easily than others. Plants that have primarily shallow, fibrous roots close to the trunk can usually be moved with greater success than plants with fewer and larger roots. Rooting patterns are determined by soil characteristics and growing practices as well as by species. Nursery-grown plants usually have a more fibrous, compact root system and more attractive tops than plants grown in the wild. Transplanting success is usually higher with shrubs than with trees and with deciduous plants than with evergreens.

Among the 10 genera of trees listed by Himelick (1991) as easiest to move some of the more common include: alder, elm, hackberry, linden, plane tree, poplar, willow, and yew. Also noted for ease of transplanting are honey locust, pin oak, and common pear. The trees of milder climates are not included in Himelick's list; among them, the olive and most palm species can be easily transplanted. Himelick listed hickory and pecan species, walnut and butternut species, sassafras, tupelo, and white oak as most difficult to move.

A survey by Rae (1969) of 24 commercial tree movers in the northeastern and midwestern United States produced the following list of "good risks" when trees are being moved by frozen root ball: crabapple, elm, honey locust, linden, maple, as well as red, Scotch, and white pine. When moved by frozen root ball, the following are "poor risks": birch, dogwood, hemlock, magnolia, oak, liquidambar, and tulip tree. Soft-rooted trees do not usually survive frozen root balls.

Plant Quality

Small plants transplant more successfully than do large ones of the same species. Small plants adapt more quickly to unfavorable surroundings and come into balance with their environment (Vanstone and Ronald 1981). When a 100-mm (4-in.) diameter tree (trunk measured 150 mm [6 in.] above the ground) is dug with a treespade, only 2 percent of the root system is removed with the tree; that is, 98 percent stays in the ground (Watson and Himelick 1982). Using a calculated rate of root growth, a transplanted tree with a 100-mm (4-in.) trunk diameter (measured 300 mm [12 in.] above the ground) will regain its original root volume in five years (Watson 1985). A 250-mm (10-in.) tree will take 13 years. Unless wisely watered, a transplant will be subjected to severe stress and be more susceptible to insect and disease attacks until the original root volume size is regained (Schoeneweiss 1981).

The cost and difficulty of moving increase rapidly with plant size. Large plants are often available for transplant when land use changes, roads are widened or built, or when plants become too large for their location. Many fine specimens can be obtained in these circumstances, though they may be difficult to dig because they are hemmed in by structures or other vegetation. Wherever large plants are obtained, their health and vigor are important to transplanting success.

SITE CHARACTERISTICS

The root systems of plants are usually more compact and fibrous in fertile, well-aerated soil than in sandy, infertile, droughty soil or in silt or clay soil with a poorly drained subsoil. In these latter situations, roots are more likely to be large, with few laterals or small roots near the trunk; those in sandy soil will tend to be deeper, while those in silt or clay will be more shallow and spreading. Plants are transplanted more easily from sites that are free of stones and other obstructions. It is particularly difficult to move trees from steep slopes and reset them vertically at a new site.

Planting sites with paved areas or utility wires and pipes may restrict the use of large tree-moving equipment, so that a boxed or B & B tree may have to be moved to the planting hole with great care. Large trees have sometimes been moved into place by helicopter.

A number of urban planting sites are extremely harsh for newly transplanted large plants. Paving and buildings can increase air temperature and radiation intensity. Buildings can create wind tunnels and extreme air turbulence. These conditions can make it difficult for the root system of an exposed plant to supply enough water and to support the top adequately. Frequent watering and sturdy anchor staking may be required.

SEASON OF TRANSPLANTING

The season of the year not only influences the stages of plant growth but in a gross way determines the weather. These factors, in turn, affect the ease with which a particular species can be transplanted. In most cases, however, the time of transplanting is determined by plant availability and the schedule of landscape installation.

In more temperate climates, deciduous plants are most easily transplanted in the fall after the leaves turn color or drop but before the soil freezes, or in the spring before growth begins. Winter planting may be desirable for species that can withstand being moved with a frozen soil ball. Conifers are planted most commonly in early fall or late spring, and broadleaved evergreens as growth begins in spring. In milder winter climates, plants can be moved throughout the winter with equal ease as long as the soil does not become too muddy for easy operation of equipment.

Late summer and fall have the advantage of warm soil to encourage root growth and their shorter and cooler days decrease transpiration. Winter planting takes advantage of cool or cold temperatures, reduced plant activity, and the relative ease of moving frozen root balls over frozen soil. Winter plantings, however, may desiccate or be injured by cold.

Spring planting before top growth begins will avoid most damaging cold weather, allow some root growth before top growth resumes, and ensure ample soil moisture. Most plants, except palms, should not be transplanted in late spring and summer, while they are still making rapid top growth. Most plants survive summer transplanting better after spring growth has matured. Palms, however, are to be moved in late spring and summer, when root growth is usually at a maximum (Muirhead 1961).

Whatever the season, the plant must be protected from freezing and desiccation. It may be wise to avoid moving plants on extremely cold, hot, dry, or windy days. Bernatzky (1978) reports that large trees in Europe are often transplanted at night to protect microorganisms that are thought to promote root growth after transplanting but to be injured or even killed by solar radiation or desiccation.

METHODS OF TRANSPLANTING

Methods of transplanting large plants depend on plant size and species, soil and site conditions, lead time before moving, time between digging and planting, distance between digging and planting sites, available equipment, personnel, and funds. Needless to say, the larger and more sensitive a plant is, the harsher the weather, and the greater the time and distance between digging and planting, the more protective the method must be.

Bare Root

Bare root transplanting is usually confined to deciduous plants up to 50 mm (2 in.) in trunk diameter. This method, however, has been successful with larger trees of many species in regions that have mild winter climates though not necessarily mild summers. Such handling has been confined primarily to plants that grow in sandy soils and are moved short distances. The tree is lifted by a crane with a cable attached to a hardened steel rod inserted through one or more holes bored in the upper middle portion of the trunk or in large branches. Since the holes are above the center of gravity when no soil is left on the roots, the tree will hang fairly vertical when free (Fig. 10–1). A trench is dug by machine or hand around the root system, similar to that dug for normal balling of roots. Soil is then forked from the roots beginning inside the trench and moving in toward the trunk. Exposed roots are covered or misted to prevent drying out. Sufficient upward strain is applied by the crane so that the tree will lift free when enough soil has been removed. The sandy soil that remains on the roots may be left if the tree is moved immediately to a new

Figure 10–1 An *Erythrina caffra* growing in sandy soil at Disneyland is moved bare root for the third time by lifting the tree with a crane. Cables are attached to large metal lifting pins inserted through three main branches above the tree's center of gravity. (Photo courtesy Morgan Evans, WED Enterprises, CA)

planting site; the soil should be hosed off if the tree is to be stored in a shaded area for a time. Well-decomposed organic matter should be worked in around the roots of stored trees, which are held upright and thinned out somewhat. The tops should be misted to reduce transpiration. During planting, moist, sandy soil from the site is worked in around the tree roots, which should be carefully spread. The tree can then be guyed upright, a vertical PVC pipe installed to mist the top, and the soil mulched.

Ball and Burlap

Moving a plant with a ball of soil wrapped in burlap or other material has been common for centuries, particularly for moving evergreens. The diameter of the root ball should usually be 10 to 12 times the trunk diameter measured 150 mm (6 in.) above the soil for trees up to 100 mm (4 in.) in diameter and at 300 mm (12 in.) for larger trees (Fig. 10–2). The higher ratio applies to trees 50 mm (2 in.) or less, the lower ratio to larger trees. Since most tree roots are in the upper 750 mm (30 in.) of soil, regardless of tree size, root ball depth need not increase in proportion to horizontal diameter (Fig. 10–2). Proper depth is best determined by root density: The ball has been dug deep enough when the root density decreases markedly; going deeper increases the size and weight of the root ball unnecessarily. To lighten the root ball after it is dug, remove surface soil until roots begin to be exposed, then protect them from drying out. The average relation of root ball size to weight is shown in Fig. 10–2. Root balls may be larger or smaller than listed in certain situations, depending on plant species and condition and expected weather.

If there is time, particularly with species that are difficult to transplant, prune back the roots one or two years before the move to promote fibrous rooting within the future root ball. Newman (1963) describes three root pruning procedures used in Europe. The one most commonly recommended involves digging a trench around the plant to leave a root ball slightly smaller than its final size will be. Arcs equal to half the root ball are dug one year and the remaining arcs the second year. These are backfilled with a mixture of sandy soil and organic matter. New, fibrous roots that grow out into the loose soil are thought to ease the moving process. Alternatively, the trench may be dug to future root ball size and backfilled with cinders or slag to encourage fibrous root growth within the soil ball. Finally, the larger root pruning trench may be left unfilled, although this entails a greater risk that the root ball will dry out. The advantage of the last two methods is that new root growth will occur within the root ball, not out at the periphery where new roots are easily damaged. As Newman (1963) points out, no experimental evidence favors one method or the other. In few cases, however, is time taken (or available) to prune roots before transplanting.

Sometimes leaves are stripped off or antitranspirant sprays are applied to plants in leaf to reduce transpiration during and after moving. The species sprayed and the concentration of antitranspirant applied must be carefully chosen. Antitranspirants seem to be more toxic to evergreens than to deciduous species. Used with caution, the sprays may reduce water loss not only during the growing season but also during the winter for conifers and other evergreens (see Chapter 13). A

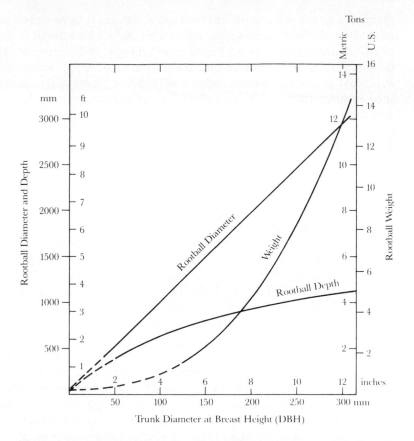

Figure 10–2 Estimated weight of root balls of recommended diameter and depth for standard shade trees of different trunk diameters. Weights are based on soil weighing 1500 kg per m³ (110 lbs/ft³). (Adapted from Thompson 1940)

compromise would be to spray the outside leaves that would be most exposed to water loss.

To minimize interference with the digging, lower branches should be pruned off or tied up. The root ball should be dug only when soil is moist. A trench is dug around the trunk 75 to 125 mm (a few inches) beyond the finished root ball size, straight down to just below most of the lateral roots and wide enough to be worked in comfortably, usually about 600 mm (2 ft) for large trees. Care must be taken to avoid damaging the roots or fracturing the root ball. The ball is finally shaped by shovel so that the base is 200 to 300 mm (8–12 in.) smaller than the top diameter. Some arborists begin with a root ball slightly larger than recommended and fork about 300 mm (12 in.) of soil from the periphery to form a slightly smaller ball than recommended, thus retaining more roots but decreasing root ball size and weight. The exposed roots will be appressed to the root ball when the burlap is in place and should be kept moist at all times. When planted, the appressed roots should be extended into the backfill soil as it is placed.

When the root ball is covered with burlap or other material, the covering is securely kept in place by rope lacing (Fig. 10–3). Root balls up to 750 mm (30 in.) in diameter have also been wrapped with 25-mm (1-in.) mesh, 20-gauge steel wire netting instead of burlap and lacing (*Amer. Nurseryman* 1970) or instead of the lacing (Himelick 1991). The ball periphery will dry more quickly in wire netting alone (see Chapter 9).

Figure 10–3 A large root ball has been covered with burlap, laced with rope, and then overlapped with chicken wire. (Photo courtesy W. Rae, Jr., Frost and Higgins, Burlington, MA; Himelick 1981)

After being wrapped and laced, a root ball up to a meter (3 ft) in diameter can be undercut on one side and tipped to break it free. Do not use the trunk to pry the root ball. After the plant is tipped, trim off excess bottom soil and cut projecting roots. Larger soil balls can be cut free if a steel cable is looped around the bottom of the ball and its two ends pulled with a winch. The cable will be pulled under the root ball, cutting the soil and roots.

Plants with root balls up to 600 mm (2 ft) in diameter can be moved by two people using shovels, a heavy canvas or burlap sheet, or a specially designed hand-cart. A front-end skip loader can be used to lift root balls a meter (3 ft) or more in diameter and to move them a short distance. If root balls more than a meter in diameter must be transported some distance, they are usually lifted onto a truck or trailer with a crane or lifted and transported by a truck with a hydraulic boom hoist or a self-loading trailer. The plant can be lifted with a chain or cable attached to the root ball and trunk base (Fig. 10–4). Some arborists insert a steel rod through the treetrunk a short distance from the root ball and use it as a major lifting point. In all of these operations, the trunk should be protected with thick padding where the lifting chain or cable might come in contact with it.

Methods of transporting B & B plants will depend on the plant size, the distance to be moved, and available equipment. Several plants with root balls up to 1 meter (3 ft) in diameter can be transported on one large truck. Special tree-moving trucks with hoists or self-loading trailers are most suitable for lifting and transporting large plants, one at a time, over reasonably short distances. Limbs should be tied up or in to reduce damage and to reduce the width of the load.

Figure 10-4 Lifting trees and root balls. The Newman Frame and Tree Mover lifts a linden 9 m (30 ft) tall and is ready to tilt it by truck winch for moving. The tree and root ball are stabilized by cables attached to a steel pin through the upper trunk (left). (Photo courtesy C. J. Newman, England). At the right, a heavily strapped root ball is lifted in a chain sling by the truck-mounted hoist. (Photo courtesy W. Rae, Jr., Frost and Higgins, Burlington, MA)

Follow the same general directions as given in Chapter 9 for digging the hole and setting the plant. Some arborists are quite particular about setting the plant with the same orientation it originally had. The main value in this is that the same sides of the trunk and main branches will be exposed to the afternoon sun. Previously protected bark can be killed or seriously injured if exposed to the sun's full intensity. A different orientation, however, may be desirable or necessary. Newly exposed bark can be protected with white exterior latex paint; this will weather so that the bark can withstand increased heat as it is progressively exposed. Selectively thin any branches that might be prone to wind damage because of the changed orientation.

Frozen Root Ball

The following discussion is based on information from Rae (1969) and Himelick (1991).

Only the outer 100 to 150 mm (4–6 in.) of the root ball periphery should freeze. Deeper freezing can cause serious root damage, but less freezing will not ensure a solid enough ball. In areas where the soil freezes to depths of 300 mm (1 ft) or more, trees can be moved with frozen root balls. This extends the transplanting season and can provide work during slow winter months. Frozen root balls require less wrapping and are less likely to be damaged in handling than other root balls. Root balls made of sand or gravel are more cohesive and easier to handle when frozen. Equipment can often be operated more easily and can cause less damage to lawns and thin roadbeds when the ground is frozen. Frederick L. Olmsted is credited with moving large trees with frozen root balls by barge to New York's Central Park in the 1860s (White 1987).

Serious disadvantages are that frozen root balls can increase costs and the chances of failure for the inexperienced. Work is more likely to be hampered by the weather in winter than at other seasons. Changes in the weather or delays in moving can lead to injurious freezing of the root ball or to a thaw that allows it to break apart. Winter working conditions are difficult for men and equipment; the days are shorter and breakdowns are likely to be more frequent. Many species cannot withstand being moved with a frozen root ball.

For trees that can be moved successfully with frozen root balls, certain procedures help ensure success: First, the trees to be moved and the planting sites must be selected in advance so they can be mulched before the ground freezes. When digging the root ball, do not break it free until the move is imminent. If a tree is to be left in a partially dug hole, mulch the excavation to prevent further freezing. The planting hole should be dug shortly before planting the root ball, so that the inside of the hole does not freeze. The soil from the hole may need to be covered or mulched before it is backfilled. If the backfill is not frozen and the temperature is near or above freezing, water only enough to help settle the soil around the root ball and mulch the tree well to prevent further freezing. In the spring, the tree should be watered thoroughly and soil added to the backfill if necessary.

Trees are usually dug and transported only when the daytime air temperature is at least $-7°\,C$ ($20°\,F$). During transport, the tree and root ball should be completely wrapped with canvas or heavy plastic to prevent drying, further freezing, and exposure to road salt. Other normal practices for plant moving and planting are applicable.

Bare Root Ball

A few hardy species of plants growing in silt or clay soils are moved with no covering on the root ball. These are usually shrubs and small trees that are replanted soon after digging, although some trees (olive, for example) are transported several hundred kilometers (200–300 mi) in the summer with little or no protection of the root ball or top. Although plants handled in this way may survive, they usually are slow to resume growth and regain a healthy appearance. With skill, an operator using a front-end skip loader can successfully move shrubs and fair-size trees up to a kilometer (0.6 mi) or so in a short time (Fig. 10–5).

Box

Boxing a square-shaped root ball can substitute for wrapping the root ball with burlap and lacing or wire. Boxing is commonly used on field-grown trees that may be held several months before replanting. Some nurseries grow trees to a large size in boxes. Some tree movers prefer to box trees, particularly large ones, because the ball is held more securely than in burlap and less experienced workers can successfully handle a boxed tree. Box materials, precut to a particular size, are quickly assembled around a root ball and can be reused.

Methods of Transplanting   **231**

Figure 10-5 A blue oak 8 m (28 ft) tall dug about 2 km (1 mile) away from the planting site in the Sierra foothills in early winter awaits the digging of the hole in which it is to be planted (top); note the small unwrapped root ball. The oak is positioned in the hole by the skip-loader (bottom left) that had lifted and transported it from its hillside site. The same tree is shown six years later (bottom right). About forty blue oak trees were similarly moved with less than a 10 percent mortality. (Top and bottom left photos courtesy California Dept. of Forestry)

For boxing, the root ball is dug as if it were going to be balled and burlapped, but it is shaped and trimmed to fit snugly in a square box with sloping sides. Box sides are made of one or two layers of 20-mm (0.75-in.) plywood or planks that are 25 to 50 mm (1–2 in.) thick and are reinforced with exterior bracing (Fig. 10–6). Two of the sides are wider than the others and have cleats along their vertical edges to hold the other two sides in place. Rods are inserted through and between these cleats to hold the four sides securely against the root ball. The ball can then be undercut and boards inserted to form a bottom for the box. Two beams are placed under and at right angles to the bottom boards; two similar beams are placed across the top of the box and attached to the lower ones by rods. If the soil is loose and

Figure 10–6 Ficus microphylla in 1.2 m (48 in.) boxes being loaded onto a trailer. (Photo courtesy Keeline-Wilcox Nurseries, Irvine, CA)

the tree is to be transported on its side, plywood can be fitted beneath the top beams to hold the soil ball in place. Heavy metal bands can be used instead of rods to hold the box together. Specially designed slings are used to lift the boxed trees by crane or hoist for loading and transport.

Trees so large they must be moved upright are pruned, side-boxed, and kept in place for at least 90 days before the bottom roots are cut and the trees moved (John Mote, Sylmar, CA, 1990 pers. comm.). The root ball should be watered after being side-boxed.

Mechanical Tree Movers

Two types of tree-moving equipment are manufactured in a range of sizes to dig, transport, and plant trees up to 200 to 250 mm (8–10 in.), d-1.4 m (4.5 ft); tree size capabilities vary somewhat with the manufacturer. The most common mover, called a tree-spade, can be mounted on either a trailer or a truck. It encircles a tree and hydraulically forces four pointed blades into the soil, one at a time, forming a root ball container when the blades meet below the tree (Fig. 10–7). Roots up to 100 mm (4 in.) in diameter can be cut fairly cleanly, if the blades are sharp, except when roots get caught between two blades; mangled or torn roots extending beyond the blades can be cut back when the tree is lifted. The root ball, encircled by the four blades, is hydraulically lifted out of the ground. The tree can be transported in a vertical position if there are no overhead obstructions; or it can be tilted forward, tied in at the top, and wrapped for longer hauls; or the root ball can be wrapped for storing or if several trees are to be transported at once.

A variation of the tree-spade, called a Tree-Porter, digs a tree mechanically by hydraulically hammering individual blades into the soil to cut the root ball and form the container (Fig. 10–8). The blades are locked together and the root ball

Methods of Transplanting

Figure 10-7 A Vermeer Tree-Spade is in position to hydraulically force the back blades diagonally into the soil under the roots (left). Blades are inserted in sequence so as not to raise the frame. At the right, a honey locust is being lowered into a predug hole. When the tree is in place, the blades will be retracted. (Photos courtesy Vermeer Manufacturing Co., Pella, IA)

lifted with a crane or hoist for transport to the planting site. The equipment, made in England, positions each blade for hammering into the soil. The curved blades come in two sizes, which can be used in various combinations to dig a root ball with one of seven diameters, from 0.8 to 2 m (31–78 in.). The equipment weighs about 115 kg (250 lbs) and can be moved short distances by two people or a small vehicle. A tractor or a portable gas-powered compressor supplies hydraulic power. Depend-

Figure 10-8 Trees in various stages of being moved by a Tree-Porter. The man is hydraulically hammering one of several blades that will encase the root ball. Note in the right foreground how the blades are bolted together. Chains are attached to the bolted members for lifting, as is shown in the background. (Photo courtesy C. J. Newman, England)

ing on the number of blades available, any number of trees can be dug before they are lifted for transport. The planting hole is dug by hand, backhoe, or other appropriate equipment, so its size can vary with soil conditions. The root ball and the blades that contain it are planted as a single unit; after the backfill soil has been firmed in place, the blades are unlocked and pulled out individually by hand or leverage.

The Tree-Porter digs a hemispherically shaped root ball, while tree-spades dig root balls that are more conical in shape. As a result, for a given top diameter, a smaller root ball will be dug with a tree-spade; and in the larger sizes, the bottom 450 mm (18 in.) of the tree-spade root ball adds soil weight without many roots. In other words, for a given diameter, a tree-spade root ball will usually have fewer roots than a root ball dug with the Tree-Porter.

A tree-spade can move small- to medium-size trees a short distance more quickly than the other equipment. For many trees of varying sizes and for longer distances, the Tree-Porter would be more economical. If the root ball is firm, the tree-spade can set it on the soil surface or a simple rack so that the ball can be wrapped and laced. Wrapping takes time and skill and does not protect the root ball as well as the Tree-Porter container. The Tree-Porter, since it is smaller and lighter, can dig trees in more difficult terrain than can the tree-spade. Also, its hammering action digs well in stony and frozen ground where other machines may have difficulty.

If the equipment is used to move a tree larger than it is designed for, the root ball may be insufficient to keep the tree upright and supply it with adequate water and nutrients. It is highly unlikely that many bigger tree-moving equipment will be built, because the largest equipment available now is the maximum size allowed on most highways. In addition, too few large trees could be moved in this manner to justify the added equipment cost. Large trees should be boxed for moving.

Members of the Forestry Division in Lansing, Michigan, found they could plant a surviving 64 mm ($2\frac{1}{2}$ in.) deciduous tree along a street with a tree-spade for two-thirds of the cost of planting a surviving 40 to 50 mm ($1\frac{1}{2}$–2 in.) bare root tree of the same species (Cool 1976). Survival of bare root trees was 72 percent compared to 99 percent of plantings with the tree-spade. Less than 5 percent of the bare root trees were lost due to vandalism and autos. Loss of bare root trees might have been lower had the trees not been so large. Working with 370 trees of eight fruit species in California, Hendrickson (1918) found that trees 12 to 16 mm ($\frac{1}{2}$–$\frac{5}{8}$ in.) in trunk diameter grew more than smaller or larger trees. On the streets, however, smaller trees might be more often vandalized and might require more care than larger trees. An important consideration in Lansing was the favorable response of homeowners when larger trees were planted in front of their property.

PLANTING

Planting a large tree, either bare root or with a root ball, involves the same steps and considerations as planting small trees and shrubs. Ideally, the planting site and soil surrounding the hole will be friable (or made so) and moist, in which case the planting hole need only be large enough to allow backfill soil to be easily worked in around the roots.

Usually, a planting hole is mechanically dug before the tree for that hole is dug by the same equipment. Even though the slick-sided mechanically dug hole and root ball may interfere with water movement and root growth, survival is usually high and subsequent growth acceptable (Cool 1976, Whitcomb 1987). Survival and first-year growth of 60-mm (2.5-in.) diameter mechanically dug trees have been better than similar trees planted bare root (Cool 1976, Vanstone and Ronald 1981).

Increasing the diameter of loosened soil around a tree mechanically planted in compact, stratified (see Fig. 6–10), or clay soil usually improves growth regardless of how the tree is planted. Four out of five species of 50- to 90-mm (2- to 3.5-in.) trees transplanted with a tree-spade made more shoot growth and/or were more attractive when planted into hand- or backhoe-dug holes than those planted into holes dug by a tree-spade (Preaus and Whitcomb 1980, Birdel, Whitcomb, and Appleton 1983). Backfill soil is filled around the tree-spade blades holding the tree upright in the planting hole and then the blades are removed. Although identical comparisons were not possible, B & B–dug trees made more growth and/or had better appearance than tree-spade–dug trees when both were planted in hand-dug holes.

It is interesting that a mud slurry compared to no slurry in a tree-spade planting hole significantly decreased the growth of crabapple and had little effect on pine (Birdel, Whitcomb, and Appleton 1983). It is possible that in some soils even though the slurry fills voids between the root ball and the sides of the hole, the slurry particles shear and form an even more impervious interface.

Most trees should be guyed after being set in the planting hole, to keep them upright and avoid shifting that will damage the root balls. The guys can be used temporarily until more permanent support is provided.

After a root ball covered with biodegradable burlap is set in the hole, the lacing can be removed, unless the ball is likely to crumble. The burlap should be rolled back or cut so the outer surface of the ball can be examined. If the ball has dried or been glazed in digging, the hard surface should be slit carefully or scarified with a sharp shovel. Backfill the bottom one-third to one-half of the hole, firmly working the soil around and under the root ball to give it support. If this has not already been done, the lacing can now be loosened and much or all of it removed, particularly if the drum lacing was done with open links. Roll the burlap off the top of the ball and down the sides if the root ball is firm. Cut the burlap above the backfill and remove it. Some burlap can be left to hold the soil together because it will rot in time, but avoid excessive folds of burlap, which could interfere with rooting. If the burlap is left in place, continue backfilling to within several centimeters (2 in.) of the surface and then cut the burlap close to the top of the backfill. Fold the burlap down so that none will be exposed when backfilling is complete, because the exposed burlap can act as a wick, drying the soil interface between the root ball and backfill.

Soil of the same or slightly coarser texture as that in the root ball should be added to the top of the ball to replace soil removed earlier. The new surface can be feathered to the level of the surrounding soil. The resulting crown on top of the root ball will compensate for any settling.

Anchoring the Root Ball

Anchor staking or guying is needed when trees are so large that the root ball may shift in the soil under windy conditions. Smaller trees should need only protective stakes to keep equipment and people away from their trunks (see Fig. 9-10). Larger (50–75 mm or 2–3 in. square or round) and longer (up to 2 m or 6 ft) stakes driven in the soil with about 1 m (3 ft) above ground can be used to anchor trees up to 75 mm (3 in.). In Japan, many street trees are anchored to withstand strong winds (Fig. 10-9).

Figure 10-9 Most street trees, as this liquidambar, in Japan are securely anchor staked to withstand strong winds.

Guys should be used to anchor trees larger than 75 mm (3 in.) in diameter. Most commonly, the guys with hose buffers are looped in a branch crotch about one-third up the trunk. The trunk is less likely to be injured, however, if each guy is looped around a branch growing at right angles to the direction of pull. If no crotch will provide the attachment desired, the guys can be attached to the tree by 10-mm ($\frac{3}{8}$-in.) screw eyes on trees with up to 150 mm (6 in.) and 12-mm ($\frac{1}{2}$-in.) screw eyes on larger trees. The screw eyes are usually placed on the side of the trunk that faces the direction of pull with the eye face parallel to the trunk. Newman (1963) recommends placing the screw eye on the opposite side of the trunk, so that the pull will be one of compression. A smaller screw can be used in this case and will provide a "fail-safe" anchor. The trunk must be protected where it is in contact with the cable. The screw eyes should be spaced vertically about 200 to 250 mm (8–10 in.) apart on the trunk so no point of trunk weakness will develop.

The guys should form about a 45° angle with the trunk and the ground. For staking the guys, Himelick (1991) recommends that for trees up to 100 mm (4 in.), stakes 50 × 50 mm (2 × 2 in.) and 1 m (3 ft) long driven at the same angle as the guy will form when installed. A stake so driven will be more secure than if it is at right angles to the direction of pull. For trees to 200 mm (8 in.), use 50 × 100 mm (2 × 4 in.) and 1.2 m (4 ft) stakes. Newman (1963) states that "Good anchorage can be obtained with metal stakes 1.2 to 1.5 m (4–5 ft) long, made of 40-mm (1.5-in.) commercial steel angle sections 5 mm ($\frac{3}{16}$ in.) thick." Trees in large planters can be guyed in the container corners or sides.

For larger trees, the guy cables should be anchored to deadmen (large wooden beams) 100 × 150 mm (4 × 6 in.) and 1.2 m (4 ft) long placed 0.6 to 0.9 m (2–3 ft) below the soil surface. Trenches for deadmen should be at right angles to the line of pull, with a channel cut for the cable that will be securely attached to the middle of the deadman. In light or disturbed soil, wide, flat stakes driven on the tree side of a deadman will provide additional anchorage.

A commercially available steel anchor (Duckbill®) with an attached cable can be driven into position with only the cable coming from the ground and is more easily installed on larger trees than a deadman.

There seems to be little agreement on how tight the guys should be. The findings of Jacobs (1939), Fayle (1976), and Leiser and others (1972), however, indicate that some trunk movement, unless the root ball also moves, will increase the size of roots near the trunk and will possibly foster root growth into the surrounding soil. Newman (1963) recommends that the tree be allowed a few inches of freedom at the head of the guys. Because such movement can put considerable strain on a screw eye or a stake when the cable snaps taut in a wind, a compression spring can be inserted in each guy to provide flexibility and increasing restraint as the spring is compressed and closed (Fig. 10–10). The size of the spring and its compression span will vary with the size of the tree and the flexibility of the trunk. A smaller tree will take a lighter spring with a longer span. For an acceptable amount of trunk sway at the height of the cable attachment, the compression span should be 0.7 that amount when the guy is at a 45° angle to the trunk; it should be only 0.45 that amount when the guy angle is 30°. Should a particular compression spring have a longer span than is wanted, the guy can be tightened until the right amount of span is left in the spring. Turnbuckles (screw tensioners) are usually used to adjust the slack or tension of guys. Guy wires should be marked with streamers to alert people to their presence.

Underground guying is particularly useful for new trees in pedestrian malls and along downtown streets. The root ball is held securely while the trunk and top are free to move, and no unsightly guys will trip pedestrians. Newman (1963) describes a method in which two cables are stretched across the top of the root ball and downward at about 45° to deadmen placed at the bottom of the planting hole (Fig. 10–11); Duckbill® metal anchors could also be used. Two boards are laid across the root ball under and at right angles to the cables to distribute pressure on the root ball. The backfill must be well tamped to keep the root ball from breaking as

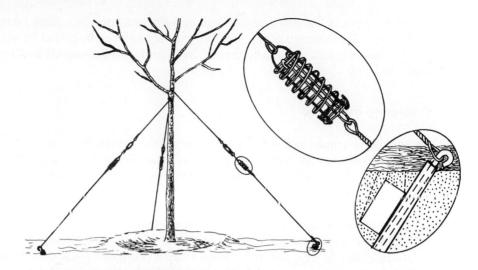

Figure 10-10 Large trees may need anchoring until roots grow into the surrounding soil. Compression springs (top inset) in guy wires provide limited flexibility and reduce the strain on the anchor stakes. In turf, guy wires can be attached to removable pins (lower inset) which are inserted into galvanized pipe permanently set flush with the soil surface. (Harris, Leiser, and Davis 1976)

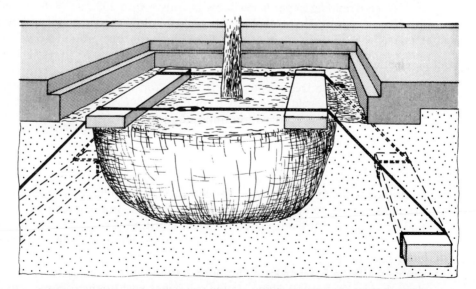

Figure 10-11 In heavily used public areas, underground guying may be preferable. Care must be taken to be sure the backfill soil is firmly in place before the cables are tightened so they do not break the root ball. (Adapted from Newman, C. J. 1963. Transplanting Semi-Mature Trees. In *Trees,* R. J. Morling, ed. London: Estates Gazette)

the cables are tightened. The cables can be tightened later by pulling them toward each other. In a variation of this method, the root ball can be cabled firmly to three or four stakes driven into the bottom of the planting hole alongside and around the ball.

Protecting the Trunk

Wrapping the trunks and sometimes the lower branches of newly planted trees with burlap or special paper has been thought to reduce water loss and protect the bark from sunburn, sun scald, and bark borers. Pellett (1981) found, however, that traditional wrapping did little to moderate the temperature of trunks of silver maple exposed to full sun; lack of temperature protection may account for Houser's (1937) observation that wrapped trunks of American elm and three maple species were not injured when uncovered and exposed to the sun. On the other hand, trunks painted with white latex or wrapped with Foylon I® were respectively 4.5° C (8° F) and 10° C (18° F) cooler than unprotected trunks when exposed to the sun. Foylon I® is an impregnated cloth fabric with a silver-metallic surface; although not as durable, aluminum foil should act similarly. Painting the trunks is easier and cheaper than wrapping them. As the paint weathers and cracks with trunk expansion, the bark is exposed to increasing radiation and becomes more temperature tolerant. Neither white paint nor Foylon I® may be particularly attractive in the landscape, however. The more complete protection by Foylon I® may sensitize the bark to sun exposure when the wrap is removed.

Tree wrapping paper is available in rolls 100 to 200 mm (4–8 in.) wide; Foylon I® comes in wide sheets. If used, each spiral of the wrap is overlapped about one-half the width of the material. Wrapping from the ground up will repel rain and sprinkler water, which may be desirable in rainy areas for trees such as pin oak that are susceptible to fungus cankers (Pirone and others 1988). In drier areas and for less susceptible trees, wrapping from the top down will keep the trunk more moist. The wrap is held in place by a cord or heavy string that is wound around the trunk and branches in a direction opposite to that of the wrapping.

Watering and Mulching

Even though some water may be applied during backfilling, more should be added after planting is complete to further settle the soil, improve contact between soil and roots, and ensure adequate initial soil moisture. Construct a berm around the trunk at the edge of the root ball so that water is initially confined above the root ball. The surface of the basin should slope away from the trunk. If the soil is well drained, fill the basin twice with about 100 mm (4 in.) of water, letting the basin drain before the second filling. If the soil is not well drained, more care should be taken to avoid overwatering.

After the basin has drained, it should be mulched 75 to 100 mm (3–4 in.) deep. See Chapter 14 for more details.

Misting. Large shrubs and trees that are already in leaf, or about to be, may not be able to get adequate water through their root systems. Misting such plants will cool and wet the foliage and decrease transpiration. One or more sprinkler heads can be attached to PVC pipe and placed in the upper part of the tree. The lower end can be joined to others from nearby trees or coupled directly to a standard hose or an automatically timed water source. A light sprinkling twice each afternoon will reduce the transpiration load on the new transplant. If the water is high in soluble salts, the sprinkling should not be continued for many days or the foliage will become gray with salt. Wet the foliage just to runoff, so that the soil below does not become too wet.

Fertilizing

Considerable differences of opinion exist as to the wisdom of fertilizing newly planted woody plants (see *Fertilizing Root-Damaged Mature Trees* in Chapter 12).

Pruning

Almost all plants will benefit from pruning after they are planted. Branches that have been broken or seriously damaged in moving should be removed or cut back to a good lateral. Pruning before transplanting should usually be limited to branches that will interfere with digging, loading, and transporting and that cannot be tied in. If branches destined for removal can be left on temporarily, they will protect other branches during transplanting. On the other hand, boxed trees whose branches are not likely to be damaged in moving are pruned before boxing (John Mote, Sylmar, CA, 1990 pers. comm.).

There are differences of opinion as to how much, if at all, a transplanted tree should be pruned; see Chapter 15 for more information on pruning root-damaged trees.

TRANSPLANTING PALMS

Much of the following discussion about palms is based on information from Henry Donselman (Rancho Santa Fe, CA, 1990 pers. comm.).

Palms are mostly tree- or shrub-like monocotyledons with one or more un-branched trunks, each with only a terminal bud. The vascular cambium, xylem, and phloem are in bundles within the trunk. A trunk increases little in diameter below its living fronds (leaves). Expansion that does take place is through the absorption of water in trunk cells. The cork cambium forms an outer psuedo-bark beneath frond bases but is unable to replace or repair trunk damage. Large palms flex more easily than do most dicotyledons of similar trunk size. Roots of most species are adventitious single strands growing out from the base of the trunk. These are continuously being replaced by new roots arising from the root initiation zone at the base of the trunk. Roots of many species live about three years; some must live longer since palm roots have been found up to 50 m (180 ft) from their trunks. Circling roots of container-grown plants are not a problem, since roots are continually renewed.

Even though some species can endure extreme conditions of salt water (coconut) and aridity (date), well-drained soil with adequate water, particularly when the trees are young, is crucial. On poorly drained sites, plant on a berm or install drain tile. Some palms can withstand $-6°C$ (20°F) and below.

Palms are considered low to medium maintenance, but in peopled landscapes pruning and fertilization are generally required. Most species have relatively few pests or diseases. Transplanting a palm with trunk sections of varying diameter ("hourglass") or small diameter below the terminal ("pencil point") should be avoided. A smaller diameter indicates previous stress of water or nutrient supplies and could restrict future growth and safety.

Palms for landscape planting in the United States are grown primarily in nurseries, though some are moved from older landscapes. In Florida, a few species are obtained from native plantings. Many species are nursery grown in containers. Field-grown palms may be boxed, particularly if they are to be held before replanting, or balled and burlapped for moving or storing. Sandy soil should be wetted before digging to provide a firm root ball for handling. Mechanical tree movers are well suited to move short- to medium-height palms short distances.

Preparing to Move

The misconception that cut palm roots always die back to the trunk has led to the belief that root ball size is not important. Root dieback will occur with the cabbage palm (*Sabal palpeitto*) but not with the coconut which regenerates new root tips or queen and royal palms which often regenerate new root tips if roots are cut 0.6 m (2 ft) or more from the trunk (Donselman and Broschat 1984).

A number of durable palms are dug with a small root ball and handled with little root protection. However, Donselman and Broschat found that the larger and the more protected the root ball the more successful the transplanting of coconut, queen, royal, and other palms known to regenerate new roots from cut roots. The minimum-size root ball for reasonable establishment of coconut should have a radius at least 150 mm (6 in.) greater than the trunk. Root pruning many palms six to eight weeks before transplanting stimulates new root initials; then taking a root ball large enough to include the new roots decreases transplant shock and increases survivability. Root pruning before transplanting is wise for large, valuable, or rare palm specimens.

The general practice is to remove most of the fronds before moving, leaving only six to eight per stem (Himelick 1991). Donselman and Broschat recommend removing $\frac{1}{3}$ to $\frac{1}{2}$ of the older fronds. The ring of fronds immediately below the bud is usually removed to prevent undue pressure on the bud when the fronts are tied up. The fronds are gathered around the bud and tied in place with biodegradable twine to protect the terminal bud and to reduce transpiration (Fig. 10–13). It is wise to keep the fronds tied up for at least eight months until new roots are formed.

In Florida, removing all the fronds when transplanting palms has been shown to improve survival and early growth compared to leaving fronds around the terminal. Broschat (1991) reports a 95 percent survival of mature Sabal palms when transplanted with complete frond removal compared to 64 percent when two-thirds of the fronds were removed. Virtually no cut roots of Sabal palm survive, regardless of length. All fronds left on the trees died and had to be removed. Trials with other species are needed. Removal of fronds is not necessary for palms that are container-grown or have been held in a box for a time after field digging.

Palms with long slender buds, such as Senegal date, should be supported with a beam attached to the trunk and extending beyond the bud. For transporting, single-trunk palms can be stacked or shingled on a trunk or flatbed trailer. Palms should be firmly secured to minimize vibration that can break the tender bud causing death six to eight weeks after transplanting. The individual stems of large, multi-stemmed palms should be braced and tied to maximize stability and to reduce the width of the load. A palm can be lifted with the same range of equipment that is used for dicotyledonous plants (Fig. 10–12).

Figure 10–12 A palm is usually moved with a small root ball and only a few fronds left on; these are generally tied together to protect the growing point.

The trunk surface must be protected to prevent damage from the lifting cables or chains. Nylon slings will provide both support and protection. When transported more than a few kilometers (miles), the entire palm should be covered to keep the roots moist, shade the trunk, and reduce water loss. Palms stored in a horizontal position should be shaded and the roots kept moist; the trunks of some palms can be damaged by direct sun exposure.

Planting

Planting palms is similar to planting other trees. A hole can be dug by hand or a back hoe. Himelick (1991) recommends that the hole be wide enough to work soil in around the roots; usually about 450 mm (18 in.) larger than the root ball radius. Along tropical coasts, holes are sometimes augered into coral rock. Pulverized rock from the drilling is used for the backfill. The palm should be oriented in the planting hole so that its most pleasing side faces the main viewing perspective; this is particularly important for palms with curved trunks.

Poor drainage and overwatering kill more palms than any other causes. Palms are sometimes planted deeper than they were originally grown in order to avoid staking or to have palms of different heights be the same height when planted. Unless drainage is excellent, deeply planted palms are usually doomed. They should be set at the level they originally grew or only deep enough to keep 50 mm (2 in.) of the root initiation zone above the soil. Some palms become manganese deficient if they are planted only 50 to 75 mm (2–3 in.) deeper than previously grown (Donselman and Broschat 1984).

Backfill with the soil that comes from the hole. Donselman and Broschat state that ''it is unnecessary and inadvisable to amend'' the backfill soil. After firming the backfill around the roots, water thoroughly. Most properly planted palms will need support. Place a band of burlap or padding with a fender of boards, each about 50-mm (2-in.) thick, on top around the trunk. Hold these in place with metal or nylon straps. The boards provide an anchor base for guy wires or, in Florida, the more commonly used wooden braces (Fig. 10–13). Do not nail or screw into palm trunks; infection may result. It is advisable to support the trees for eight months or so until new roots become established.

Figure 10–13 Cabbage or sabal palms recently planted in Miami, FL. The palms are braced with three or four wooden braces nailed to wooden fenders held in place against a fabric cushion with metal straps. The top five to seven fronds are tied together to protect the terminal bud.

Postplanting Care

Under normal conditions and drainage, newly planted palms should be watered once a week the first season by sprinkler or hand and monthly thereafter depending on the weather. If the weather is particularly hot and arid the first season, the foliage and trunk can be misted; palm trunks can absorb water.

A broad spectrum micronutrient spray following transplanting is recommended to prevent micronutrient deficiencies which are common on palms. Adding a spreader-sticker and 250 g of urea to 100 l (2 lb/100 gal) of micronutrient spray will enhance spray effectiveness and add nitrogen. Preventing micronutrient deficiencies is easier than trying to correct them.

Do not use a string trimmer or a machete around palms, young roots at the root initiation zone will be damaged or killed. Since existing roots are relatively short-lived, loss of new roots could greatly weaken a tree.

FURTHER READING

BERNATZKY, A. 1978. *Tree Ecology and Preservation.* New York: Elsevier Scientific Publishing Co.

HIMELICK, E. B. 1991. *Tree and Shrub Transplanting Manual.* Urbana, IL: Revision II–Intl. Soc. Arboriculture.

NEWMAN, C. J. 1963. Transplanting Semi-Mature Trees. In *Trees,* ed. R. J. Morling. London: Estates Gazette, pp. 29–39.

CHAPTER 11

Special
Planting
Situations _____

Landscape plants are often grown in extremely difficult sites. Success depends on correct analysis of site conditions, site preparations to minimize problems, selection of appropriate plants, and wise maintenance practices. We continue to learn how to improve our assessment of plant requirements in relation to site characteristics and attempts to adjust the microenvironment. Experiences must be shared to solve site problems more effectively. Certain landscape situations are reviewed here, and suggestions are offered or approaches described.

PLANTING TREES IN PAVED AREAS

Trees are commonly planted in paved areas: along downtown streets, in parking lots, in plazas, and in patios. Adequate provision and planning are seldom made for aeration, drainage, irrigation, root growth, and particularly for the time when the trees will be mature. This discussion will concern planting trees in existing or projected paved areas, where foot and vehicular traffic close to treetrunks usually limits soil exposure. Placing pavement around existing trees was covered in Chapter 7.

Research workers and municipal arborists in the Netherlands have been investigating street-tree problems since the late 1960s when natural gas caused as many as 20 percent of street trees to die in some cities (Kopinga 1985). Since much of the soil for street trees in the Netherlands is severely compacted material from the North Sea with a water table 1 to 2 m (3–6 ft) below the surface, some of the findings of their researchers are instructive for other regions as well (Kopinga 1985):

- When there is contiguous pavement around a tree, a ventilation system is highly recommended
- Wet and compacted soil surfaces also inhibit oxygen diffusion
- The oxygen supply through ventilation pipes in which air can circulate can be more than 10 times greater than the supply through vertical pipes containing stagnant air
- In smaller planting holes (1 to 4 m³ or 35–140 ft³), horizontal movement of oxygen from surrounding soil often exceeds vertical movement through overlaying pavement
- When vertical oxygen movement through pavement is possible, the tree planter should be rectangular with the long side parallel to the curb
- When no ventilation system is present, the planting hole should never be more than 1-m (40-in.) deep

Compacted soil and exclusion of air and water by pavement are universal problems for trees growing in pavement. More effort needs to be taken to reduce soil compaction during land shaping for landscapes and construction. Soil that has been compacted to the extent that root growth would be physically restricted should be replaced with quality soil to a depth of at least 0.5 m (18 in.), preferably 1 m (3 ft) (see Chapter 6). If artificial drainage is provided, replacement soil should be 1-m deep (see Chapter 8).

Openings in pavement, usually 1 to 1.5 m (3–5 ft) square, have been the standard method for planting trees in new and existing paved areas. In downtown sidewalks, the openings may be left with the soil exposed, covered with brick or stone on sand, or covered with a metal grate or precast concrete (Fig. 11-1). To try to minimize root damage to pavement, the soil level of some basins is as much as 0.5 m (20 in.) below the bottom of the planter covers. Small rock in the basin under the covers keeps soil erosion to a minimum if the tree is watered by hose or tank

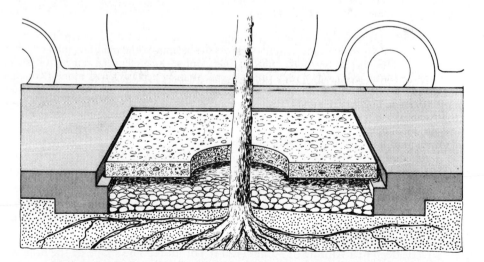

Figure 11-1 One method of planting trees in paved areas. An opening, about 1.2 m (4 ft) square, supports two half covers (one shown). The basin below provides for irrigation and aeration, while the covers provide a smooth surface for pedestrians.

Planting Trees in Paved Areas

truck. Rock up to the bottom of the covers reduces debris collecting in the planter. Hardware cloth (coarse metal screening) under the covers has been used to keep rats out; but the wire must be inspected frequently to prevent girdling.

Perforated PVC pipe (100- to 125-mm, 4- to 5-in. diameter and 0.6- to 1-m, 2- to 3-ft long) is commonly put in two or four corners of a planter opening for aeration and irrigation. As the people in Europe learned, dead-end aeration holes in the soil are seldom adequate. Connecting the bottoms of two or more of the vertical aeration pipes, however, allows air flow through the tubes, greatly increasing the oxygen concentration and aeration effectiveness (Kopinga 1985) (Fig. 11–2).

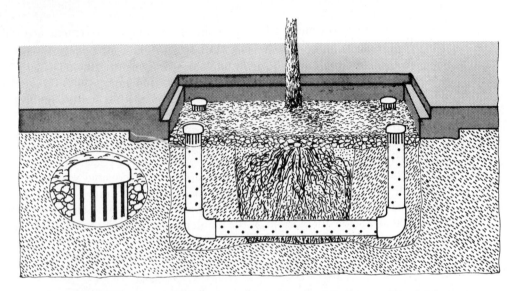

Figure 11–2 Aeration in pavement plantings can be increased by joining two vertical aeration pipes across the bottom of the planting hole, one pair each near opposite sides of the planting hole. Oxygen concentration in the pipes will be ten times greater than in vertical pipes alone (Kopinga 1985). Aeration could be further increased by having aeration risers as shown in Fig. 11–3 also serve as protective stanchions for the tree.

If the soil is compacted around the planter opening and under the pavement, aeration may also be increased and roots directed downward by drilling or water-jetting sloping irrigation and aeration holes under and out from the root ball and under the pavement. Within a year or two the holes are filled with roots.

Another pavement-planting scheme is also being used in the Netherlands. Essentially, a planting hole is established within a planting hole (Kopinga 1985, modified by Urban's [1989] survey) (Fig. 11–3). Excavate a hole approximately 0.6 m (2 ft) deep to a volume of 3.5 m³ (120 ft³) or more. The shape of the hole will depend on the space available. All but the finished planter opening (1 to 1.2-m, 3 to 4-ft diameter or square) will be paved over. The outer portion of the hole "is filled with a mixture of coarse lava slag (80–150 mm, [3–6 in.] diameter) and 'tree soil' in a ratio of about 2:1 (volume:volume, v/v)" (Kopinga 1985). A tree is planted in the

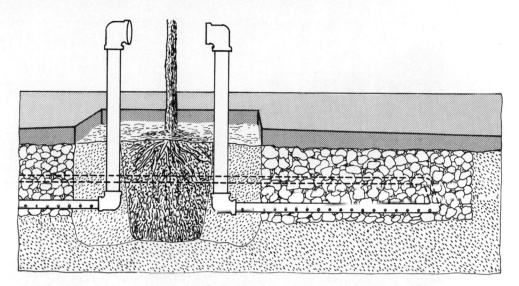

Figure 11-3 Favorable soil conditions for pavement plantings can be obtained by replacing 3.5 m³ (120 ft³) of soil 0.6 m (2 ft) deep with coarse lava slag and soil mix before installing pavement. The shape of an excavation depends on the space available. Aeration can be further enhanced by placing an aeration system as shown; the horizontal portion can be either PVC pipe or flexible tubing. The risers are of heavy metal pipe with additional support (not shown) to protect the tree and to withstand abuse.

center of where the planter will be. The rock-soil mixture is lightly compacted to give a solid surface of lava with soil filling the voids. The pavement can then be installed. This provides a firm base for the pavement with no further compaction and because of the porosity of the lava (about 48–50 percent v/v) an optimal supply of oxygen for tree roots (Kopinga 1985, citing Terlouw 1981). Tree roots should grow out into the soil among the lava. Rocks other than lava could also be used though the aeration would not be as good.

Urban (1989) examined thirteen 11- to 27-year-old plantings in tree pits surrounded by pavement in locations from Boston to Virginia. One of his conclusions is that "any planting site with less than 100 cubic feet (3 m³) cannot sustain long-term tree growth." Beyond a depth of 0.6 m (2 ft), increase in surface area is more beneficial than increase in soil depth for all but large-growing trees. Depending on priorities and adequacy of summer rain, an irrigation system may need to be installed.

Cluster and linear tree and shrub plantings are becoming more common along center-city streets and in plazas (Fig. 11-4). These provide more shared rooting space and exposed surface for aeration and rain deposition. In Europe and America, some planters are joined with channels of quality soil under the paving. For the soil channels to be successful, it would be wise to have at least one perforated aeration pipe in each channel which is connected to vertical risers in each of the planters. This is an extension of aeration system shown in Fig. 11-3.

Figure 11–4 A planting of Bradford pear trees at the J. F. Kennedy Center for the Performing Arts in Washington, DC. Note the difference in growth between the four trees in the openings in the paved area (right) and those in the linear planter with more soil surface exposed and shared rooting volume. (Photo courtesy James Patterson, National Park Service)

For planters at grade in new construction, the pavement around each planter should be sloped to direct runoff water to a drain or street gutter rather than into the planter. Such diversion will protect trees from excess water, from salt used to melt ice, from increases in alkalinity, and from oil and other toxicants on the pavement.

Trees that will be small to moderate in size when mature can usually be grown safely in pavement planters. The trees will grow according to existing conditions, so they may be healthy even when soil rooting volume is relatively small because of limited aeration and water distribution. Mature tree size may be smaller than typical for the species, but early growth should be good if the trees are given normal care. If full-size large trees are wanted, a gravel subbase should be considered for the surrounding pavement, as described in Chapter 7.

Enlargement of roots near the surface is a problem when trees are close to pavement, walks, and foundations. Roots can be especially destructive if they reach irrigated open areas beyond the pavement or other sources of air and water. If the soil level is below the pavement as shown in Fig. 11–1 and well removed from open areas, root damage to the pavement should be minimized. Damage can be further reduced if small or moderate-size trees are chosen that normally have few or no large surface or buttress roots. Managing root growth is discussed in Chapter 18. Robert Skiera (1990 pers. comm.) estimates that in Milwaukee the maintenance costs of a tree in a pavement planter, as in Fig. 11–1, is seven times more than a tree in an open landscape.

LANDSCAPING IN PLANTERS WITHOUT NATURAL DRAINAGE

Landscape plantings without natural drainage are common on roof gardens, in building foyers, on top of underground garages, in shopping malls, and in hotel lobbies (Fig. 11–5). Almost all of these situations require plant containers that have no contact with parent soil, yet can provide moisture, nutrients, aeration, drainage, and anchorage for successful plant growth. The weight of such plantings is an important consideration when they are placed on a structure, as are the weights of the soil mix, water, plants at maturity, other landscape features, and people occupying the area. Successful planters, whether individual and free-standing or built into a structure, must all be controlled for container size, waterproofing, irrigation, drainage, and soil mix (Davis and others 1974). These factors must be considered in the initial planning for a structure or project.

Figure 11–5 Landscape plantings can be successfully grown in many situations with the use of well-designed planters such as these shown in Constitution Plaza, Hartford, CT.

Container Size

The container must be deep enough to hold the drainage material and adequate soil mix to provide anchorage and drainage. Minimum container depth should be 150 to 300 mm (6–12 in.) for turf or turf-substitutes, 0.5 to 1.0 m (1.5–3 ft) for large ground covers and most medium-size shrubs, and 1 to 1.5 m (3–5 ft) for large shrubs and trees (Davis and others 1974). Bernatzky (1978) recommends a soil depth of at least 450 mm (18 in.) for shrubs and small trees. The weight of roof garden plantings can be minimized by varying soil depth with plant size even within individual beds (Fig. 11–7). The planting areas within a bed must conform to the pattern of the structural beams.

Both the texture and the depth of the soil influence the amount of water that will remain in it when drainage is complete. After a container soil is irrigated, a zone of saturation extends up from the bottom. The height of the saturation zone depends on the pore-size distribution (texture) of the container mix. Above the zone of saturation, the soil becomes drier with increasing height until field capacity is reached, providing the soil is deep enough (see Fig. 6–9). Note that, regardless of container depth, about the same amount of water remains in the bottom 150 mm (6 in.) when drainage is complete.

It is important to note that even in a coarse, sandy mix (0.5–1 mm), soil moisture is well above field capacity in the bottom 300 mm (12 in.), particularly in the bottom 150 mm. In a finer-textured soil mix, moisture will be above field capacity for greater distances above the bottom.

Except for a slight slope (about 2 percent) of the container bottom to assist the movement of excess water to drain lines, the bottom of each bed with a common drainage system should be fairly level. Similarly, the soil surface in each bed should be fairly level, or the high areas will drain to lower moisture contents than the lower areas (see Fig. 6–4). An extreme case would be a sloping surface with uniformly shallow soil (about 450 mm or 18 in.) and one side a meter (3 ft) or more above the opposite side. The soil at the top would dry rather quickly, while the soil at the bottom would be excessively wet.

If trees are grown in individual containers that are small enough to move with a forklift, the flexibility of plantings will be greatly increased. Plants can be moved to areas of high visibility when they are blooming or in fall color, and they can be replaced with others when they lose their showiness.

Waterproofing

For roof gardens and other landscaping above areas of use, containers must be completely waterproofed. The waterproofing should be planned before containers are built. A double-membrane construction is recommended for most built-in landscape planters.

Irrigation System

A proper and efficient irrigation system will reduce maintenance costs and will help ensure healthy plants. A water line should serve each container or container area of any size. Automatic watering systems with low application rates improve water use. Shallow, coarse-textured soil leaves little leeway between adequate and insufficient watering for good plant growth.

The irrigation system should be an integral part of the structure supporting the planters. If garden lighting fixtures are placed inside containers for partial concealment, they must be watertight.

Water-Reservoir Planter. Free-standing planters with water reservoirs are available for interior and outdoor use. A good example is one designed by the Milwaukee (Wisconsin) Bureau of Forestry and manufactured by Wausau Tile (in Wausau, Wisconsin) (Fig. 11–6). It is an insulated planter with a water reservoir for tree and shrub planting on downtown sidewalks. The square planters range in size from 0.75 m (30 in.) on a side and 0.75 m high (outside dimensions) to 1.5 m (60 in.) on a side and 0.9 m (36 in.) high. The poured concrete planters are lined with 50-mm (2-in.) thick styrofoam for insulation. A 150-mm (6-in.) gravel-filled reservoir is covered with a geotextile fabric mat to keep soil out of the reservoir. The reservoir is filled through a pipe to the surface; the pipe has a locked cap. Geotextile fabric wicks expedite capillary action from the reservoir into the planter soil. Excess water drains from the top of the reservoir through a weep hole (Myers and Harrison 1988).

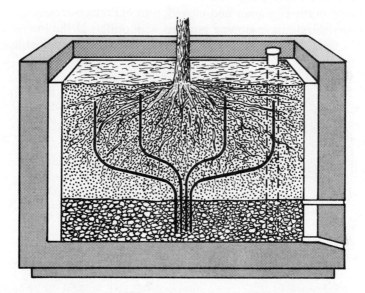

Figure 11–6 This free-standing poured concrete planter with a water reservoir for a tree needs to be watered only every 7 to 10 days during the summer, twice a week if bedding plants are added. Polypropylene wicks aid water movement from the reservoir to the soil above. Water is added through the pipe riser in the back right corner until water flows out of the hole (on the right) at the top of the reservoir. The reservoir can be drained by removing the plug in the lower hole. The walls are lined with styrofoam to protect from winter cold.

Forestry personnel use a 2000-l (500-g) tank truck to water the approximately 500 planters on downtown sidewalks. Water is added until excess runs out the weep hole. From early spring to midautumn, the reservoirs are filled twice a week, although they can go 7 to 10 days between refillings (Robert Skiera, Milwaukee, WI, 1990 pers. comm.) Similar containers without water reservoirs must be watered every other day during the summer. The time it takes to water and care for one tree and bedding plants in a planter with a water reservoir is estimated to be 6.5 hours during the 7- to 8-month maintenance period.

Drainage System

Few plants grow well if their roots must function for long without adequate oxygen. No matter how well a soil mix is designed for proper drainage, aeration, and water retention, excess water must be removed from the container. Drainage systems should be designed to handle maximum amounts of excess water, rather than average amounts. The water removal capability for any container surface area should be a minimum of 50 mm (2 in.) per hour.

The soil mix must allow the water to drain rapidly to the bottom of the container, and the drainage line and outlets must be large enough to collect water rapidly from the container and move it into building or street drainage systems. Containers should have sloping bottoms with outlets at the lowest point (Fig. 11–7).

Gravel in the bottom of the container will waste valuable space and be a poor

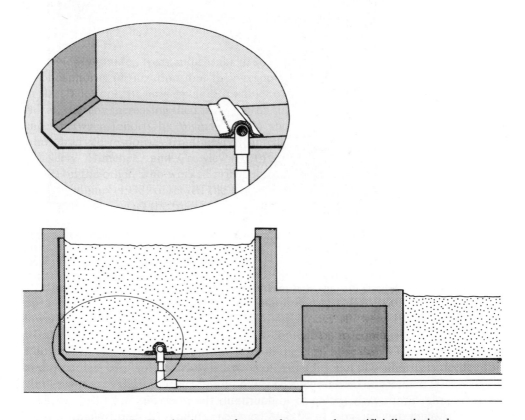

Figure 11–7 For landscape planters that must be artificially drained: Slope the bottom to a filter-protected drain line (inset), waterproof the container, and level each soil surface. (Adapted from Davis and others 1974)

medium for plant growth. When sandy soil overlies gravel, the extreme difference in pore sizes of the two materials causes drainage to cease when the sandy soil just above the gravel is still near saturation. Thus, instead of eliminating the water table, the gravel creates an artificial one nearer the surface.

Water moves into drainage tile when soil reaches saturation. In the past, 100-mm (4-in.) tile drain pipe was used in landscape planters. While this large size is needed for draining agricultural fields, much smaller plastic pipe can provide adequate capacity in planters at less cost and far less loss of container depth. A pipe 25 mm (1 in.) in diameter uses little more than 25 mm of container depth, whereas a 100-mm (4-in.) line, particularly one encased in coarse gravel, may use up to 150 mm (6 in.) of depth.

Plastic pipe has proven adequate for most containers or container areas when 4 rows of holes are spaced evenly around the pipe. Drill 3-mm ($\frac{1}{8}$-in.) holes about 75 mm (3 in.) apart in each row. Ten- to 25-mm ($\frac{1}{2}$- to 1-in.) lines, on 0.5- to 1.0-m (2- to 4-ft) centers connected to larger 25- to 50-mm (1- to 2-in.) pipe, will remove water rapidly from the bottoms of planters. Plastic lines are easy to install and add little to the weight load. Copper drainage lines are used between containers and as outlet lines that run to the street.

To keep drainage lines from silting up, wrap them with 12-mm ($\frac{1}{2}$-in.) fiberglass filter pads or strips of geotextile fabric laid directly over the drain pipe. Secure pads in place with mastic. After the drainage lines are installed, cover the entire bottom of the container to a depth of 25 mm (1 in.), with coarse sand or with sand used in the parent soil mix. Water will move laterally quite rapidly in this thin layer of sand, which will also help to protect the fiberglass filter pad from damage.

Even with properly installed drainage lines and filter pads, some silt may accumulate, so clean-out standpipes should be connected to each of the main collecting lines and capped. There should be one standpipe for each large container and several in a large container area. They can also be used to pump excess water from individual containers or container areas, if a major stoppage occurs in the system.

Soil Mix

Weight and aeration are the two most important characteristics of a soil mix to be used for roof gardens and other supported landscapes. Water and nutrients can be fairly easily supplied and large plants guyed if necessary. After the soil is in place, however, little can be done to decrease weight or improve aeration.

Davis and others (1974) examined the container soil mixes recommended by the University of California Manual 23 (Baker 1957) and found them to be quite satisfactory. These mixes combine uniform coarse sand (0.5–1 mm) and organic matter in volume ratios that range from 50:50 to 75:25. Commonly, a little silt or loam soil is added to the mix to increase water- and nutrient-holding capacities, but more than 5 percent of these materials will hamper drainage and aeration.

A coarse sand and organic mix offers a good combination of qualities. When it is saturated with water, a 50:50 sand and organic mix weighs about three to four times as much as when it is dry. The researchers found that the saturated weight was about 1.2 kg per liter (70 lbs/ft^3), but the bulk densities of planting mixes can vary greatly with slight changes in formulation. Therefore, the wet weight of a soil mix can be reliably estimated only by sampling the mix itself. In any planter landscapes of modest or large size, the soil mix should be specified and tested for compliance before it is put in place. The management of high-organic mixes is discussed in detail in the University of California Manual 23 (Baker 1957). The main feature of these practices is close attention to irrigation.

Plant Selection

Outdoor planters are usually focal points, yet the environment for such plantings may be extremely harsh. Wind, direct and reflected radiation, prolonged and varying intensities of light, low humidity, high day temperatures, low root temperatures in winter—all are common in outdoor container landscapes, and all jeopardize the well-being of plants. Generally, woody plants of arid climates will best survive such conditions (Bernatzky 1978). In most cases, trees and shrubs should reach a moderate mature size and grow slowly. Larger plants in windy locations must be supported with guys or anchor stakes. Evergreens are commonly used for their year-round foliage, but many are susceptible to air pollutants.

Determining which species will be suitable for specific microclimates and will also fulfill other design criteria can be a real challenge, but help is usually available: You can observe which plants grow well in similar situations and can consult local nurseries, botanical gardens, extension agents, public and commercial arborists, and state or regional publications.

Special Problems

In cold winter areas, roots growing in raised individual containers can freeze (see Chapter 4).

INTERIOR LANDSCAPES

Interior landscapes of more than 1000 m^2 (0.25 acre) are common in large buildings and shopping malls (Fig. 11–8). Critical to the success of such landscapes are the selection of plants and the control of irradiation (light intensity), fertilization, and irrigation. An indoor landscape must be planned as an integral part of the building to ensure proper access to electrical and plumbing utilities. Attempts to correct design and specification errors after construction can be costly and unsatisfactory.

Plant Selection

Plant selection depends largely on the irradiance (see Chapter 4) available within the interior site. In most cases, plants are not expected to grow much after installation, only enough to maintain attractive foliage. Tropical and semitropical foliage plants are most commonly used for interior plantings. Conklin (1978) has successfully grown tropical fruit trees: orange, lemon, guava, mango, and loquat. Though

Figure 11-8 Interior landscape plantings require good design, engineering, installation, and care if they are to be successful as this planting in the Ford Foundation building in New York City. (Photo courtesy Everett Conklin and Company, Montvale, NJ)

temperatures in most buildings will seldom go below 18°C (65°F), a number of temperate-zone trees and shrubs can also be used: southern magnolia, evergreen pear, and camellia, among others.

Radiation (Light)

If radiation is inadequate, most plants will grow satisfactorily for only a few months. Artificial light is generally necessary if less than 1600 lux (150 ft-c) of natural light are available. Plants of the dracaena and philodendron families will survive 550 to 800 lux (50–75 ft-c), but greater irradiance will allow for greater leaf retention and growth. Growing plants indoors is primarily a holding operation. Ideally, radiation will be great enough so that plants are slightly above their compensation point, growing slowly but not consuming reserves (Conover 1978). The large, thin, widely spaced leaves that develop in the shade have thin cell walls and usually only a single epidermal layer of cells at the surface. Such leaves are able to make maximum use of limited radiation. In the nursery, shade-grown plants usually grow more rapidly than plants in the sun but often have reduced trunk caliper and smaller root systems. The active absorbent area of the roots will not be substantially reduced, however.

Conover has found that the chloroplasts and grana of some sun leaves will reorient to shade conditions if the leaves are grown in radiation of intermediate intensity for four to eight weeks. If a sun-grown plant in a nursery is placed directly under low-radiation conditions, however, serious leaf drop usually occurs (Fig. 11–9) before existing leaves become acclimatized or new shade leaves form.

Figure 11–9 *Ficus benjamina* plants grown in full sun and then placed indoors under different radiation levels for ten weeks; left to right: full sun, 275 lux (25 ft-c), 800 lux (75 ft-c), and 1350 lux (125 ft-c). (Photo courtesy Charles Conover, Agricultural Research Center, Apopka, FL)

In 1975, about 85 percent of the plants grown for indoor plantings by producers in California, Florida, and Texas were shade-grown and therefore radiation-acclimatized (Conover, Poole, and Henley 1975). The plants that are grown in full sun include *Ficus, Brassaia, Dracaena marginata,* several palms, *Araucaria,* and *Sansevieria.* These plants should be checked to be sure they have been acclimatized before shipment. The youngest mature leaves of acclimatized plants will be larger, thinner, and farther apart than those of plants grown in the sun. Reliable dealers will correctly inform you about acclimatization.

Care must also be taken when transporting the plants to see that they are not exposed to ethylene or kept in boxes or other dark areas too long. Many plants start to lose foliage after seven days or less of darkness, and defoliate severely after 10 days.

Since windows may be shaded by nearby tall buildings or decrease radiation by means of tinted architectural glass, irradiance should be checked throughout the day, particularly in winter. A photographer's light meter will not measure irradiance accurately (see Chapter 4). When possible, new buildings should maximize the use of natural radiation for indoor landscapes. When artificial light is needed, it should be remembered that irradiance decreases by the square of the distance from its source, so that 100 lux (lumens/square meter) at one meter (3 ft) from the light source will be only 25 lux at 2 meters. Typical irradiances in most buildings are quite low (see Table 4–1). Side and bottom lighting is usually more efficient in conjunction with overhead radiation and can be more interesting as well.

Light quality is important for plant growth, but growth should be limited in most interior plantings. The wavelengths (artificial lights) usually recommended are blue (430–470 nanometers) and red (650–700 nm). Nevertheless, tropical foliage of excellent quality can be maintained with predominantly blue light, as from fluorescent fixtures (Conover, Poole, and Henley 1975). Biran and Kofranek (1976) found that cool white was the most effective of the fluorescent lights they used on nine foliage plant species. Fluorescent lamps are about three times more energy efficient than incandescent lamps, though some incandescent light (150 to 200 lux) may be needed to fill in the red and far red parts of the spectrum.

Fertilization

The proper nutrition regime for indoor plants depends on plant species and available irradiance (Conover, Poole, and Henley 1975). The level of nutrition must be higher under high radiation. Healthy plants can be maintained indoors on only 10 percent of the fertilizer needed in their nursery production. In fact, if their nutrition level is not reduced before plants are moved to low radiation, they may suffer chlorosis, necrosis, defoliation, general loss of healthy color, and, in severe cases, death. If you are not certain that soluble salts are at a low enough level, leach the containers with at least 150 mm (6 in.) of water before taking plants to the planting site.

Conover (1978) recommends using a slow-release fertilizer with a 3–1–2 ratio of nitrogen, phosphorus, and potassium. The rate of application depends on plant species and irradiance (Table 11–1). The low ratio of phosphorus and potassium is designed to keep soluble salts low. Periodic leaching may be needed to keep the soil solution at safe levels (see Chapter 12).

TABLE 11-1 FERTILIZATION RATES FOR INDOOR FOLIAGE PLANTS UNDER DIFFERENT LIGHT INTENSITIES FOR 8 TO 12 HOURS PER DAY (after Conover and Poole undated)

Plants	Lux[b] Footcandles	Grams of nitrogen per square meter of soil surface per year[a]			
		Low 550–800 50–75	Medium 800–1600 75–150	High 1600–2700 150–250	Very high 2700–5400 250–500
Ferns, *Peperomia*		3[c]	7	10	17
Calathea, Dracaena		5	10	17	23
Aglaonema, Cordyline, Dieffenbachia, Scindapsus, Syngonium		7	13	20	26
Palms, *Philodendron*		8	16	22	27
Ficus, Schefflera (*Brassaia*)		10	20	27	33

[a] The nitrogen was introduced as part of a 3–1–2 (N, P_2O_5,K_2O) fertilizer applied three times during the year.

[b] Fluorescent lamps (cool white) used: 100 lux = 1.5 $\mu E/m^2/sec$.
 Sunlight equivalent: 100 lux = 1.7 $\mu E/m^2/sec$.

[c] 10 g/m² is equal to about 2 lb/1000 ft². A teaspoon is equivalent to about 5 g or 5 cc.

Irrigation

Practically all of the factors that will influence the frequency and amount of irrigation are determined at the time of installation. These include the water-holding capacity and aeration of the rooting medium, the size of plants, and microclimate factors of temperature, radiation, humidity, and air movement. Most interior landscapes range from 18° to 27° C (65°–80° F), depending on the season. Humidity will usually be lower than that of nurseries and the native environments of plants, particularly in winter. Large, thin shade leaves will be particularly sensitive to moisture stress, which can be seriously aggravated by mite injury. Air circulation is determined largely by the air conditioning system. Sensitive, large-leaved plants should not be planted in the air stream of heating and cooling vents.

A moisture-sensing device will supplement careful observation to help determine an irrigation schedule. Indoor plantings usually evapotranspire less water than do plants outdoors, and overwatering interior plants is more common than underwatering. Most indoor landscapes, once established, will do well with a week or more between irrigations. Plants that wilt first when water is withheld can be a useful guide for irrigation schedules.

Some tropical plants are injured by air temperatures of 5° to 8° C (40°–45° F) or by cold irrigation water. It may be wise to monitor water temperature in the winter.

Other Concerns

Compounds used to shine leaves may reduce photosynthesis and can injure a number of species. If such compounds are used, they should not be used on a continuous basis. The foliage of shade-grown plants is particularly sensitive to some common pesticides and to air pollution.

PLANTING ON STEEP SLOPES

On slopes, watering can be difficult and uneven, erosion can expose roots or bury small plants, and increased exposure to sun can dry plants quickly. Three of many planting alternatives are suggested in the following sections.

Planting Pocket

Digging a planting pocket into a slope is the simplest and most common method of facilitating irrigation (Fig. 11–10). The plant is set well forward in the pocket, establishing a basin to the inside that will retain water and protect the plant from a certain amount of eroding or sloughing soil. An overflow spillway will prevent the pocket from being washed out by all but the heaviest rains. It should be cut into the undisturbed slope at one side of the pocket so that water will flow out of and away from the basin before the berm is breached. You may need to reform the

Figure 11-10 A plant is set toward the downhill edge of the basin near the original surface of the slope. The basin should be deepest near the back to accumulate water and any eroded soil. An overflow is cut into firm soil so the basin will not be easily washed out.

pocket occasionally in the first year or two, until the plant becomes established. For continued irrigation, you may want to consider a drip system (see Chapter 13).

Slope Serration

Slopes are sometimes cut into steps that measure about one meter (3–4 ft) in both the horizontal and the vertical. This method has proved particularly useful in certain highway embankments (Fig. 11–11). Instead of grading cuts to a smooth, steep, sloping surface, as has been so common, some contractors have stair-stepped certain rock or erosion-resistant earth slopes to establish plants and reduce erosion. The steps should slope toward hill so that water will drain into the soil. Small trees or shrubs can be planted in the middle of the steps and a slight basin dug to the inside of each plant in order to accumulate water. The lowest part of the basin, however, should be kept away from the trunk of the plant.

Figure 11-11 Slope serration provides growing sites on cut slopes. Woody plants or grass can be planted, or volunteer plants grown from local seed.

Wind, water, and sloughing will erode the edge of each step onto the inside of the step below. The speed of erosion will depend on the resistance of the slope material and on the weather. The loose soil and fine rock on each step are favorable germinating sites for seed from nearby plants. In many cases, the trees and shrubs planted or those that volunteer from local seed will become established before the serrations are obliterated. Even before the plants develop, the changing light-and-shadow patterns on a serrated slope can provide a more interesting landscape feature than a smooth slope.

Wattling

Wattling has successfully stabilized the surface of fill slopes, reduced erosion, and helped to establish plants on difficult sites (Kraebel 1936, Gray and Leiser 1982). Recently cut, long, slender branches are tied into elongated bundles that are partially buried in contoured trenches cut across the slopes (Fig. 11–12). The bundles, laid end to end, are staked down to hold them in place. Each row of wattling acts as a sediment trap for materials from up the slope and as an energy dissipater for soil and water movement. The area above a wattling bundle, with its reduced slope angle, is a favorable site for new plants. If the wattling species roots easily, as willows do, the wattling itself becomes part of the slope-stabilization system. Wattling and stuck cuttings root and grow particularly well in wet sites and can reduce slumping by drying up excessively wet areas. No amount of vegetation, however, will stabilize a basically unstable slope; such slopes must be laid back further or provided with drainage.

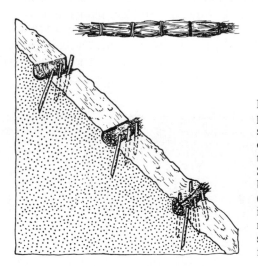

Figure 11–12 Wattling in the process of being installed on a slope: At the top, a shallow trench on the contour is ready to receive the bundles of wattling (center). Stakes are driven through the bundles for greater stability (bottom). While workers are installing the row above, the lower row of wattling will be covered by soil and firmed into place. (Adapted from Gray and Leiser, 1982)

Gray and Leiser (1982) have described the major features of successful wattling installation. Wattling bundles should be prepared from live thicket plants, preferably from species that root easily. The bundles may vary in length but should be tapered for 300 to 450 mm (1–1.5 ft) at either end. Stem butts should be less than 40 mm (1.5 in.) in diameter, with half of the stem butts at either end of the bundle. When compressed firmly and tied, each bundle should be 150 to 250 mm (6–10 in.) in diameter. Bundles should be prepared within two days of placement and protected from direct sun. If they are covered and kept moist, bundles can be stored up to a week before placement.

Trenches should be dug on the contour to about one-half the bundle diameter and about a vertical meter (3 ft) apart on the slope. Lay the bundles horizontally in the trenches so that their tapered ends overlap. Hold them in place on their downhill side with vertical stakes about 500 mm (20 in.) on center (distance between the center of two stakes). Drive stakes on 1-meter (40-in.) centers through the bundles into and perpendicular to the slope.

Stakes can be made of several materials, including: live wattling material greater than 40 mm (1.5 in.) in diameter, 50 by 100 mm (2 by 4 in.) lumber cut diagonally, or steel reinforcing bars. In most soils, 600 mm (2-ft) stakes are satisfactory; in loose soil the stakes should be a meter (3 ft) long. In compact or rocky soils where stakes cannot be driven 450 mm (18 in.) deep, steel reinforcing bars should be used.

Work from the bottom of the slope toward the top, covering and firmly packing the bundles with soil. If you walk on the wattling while working on the next row above, that alone may cover the bundles and firm the soil sufficiently. The downhill lip of the wattling bundles should be left exposed.

PLANTING UNIRRIGATED ARID SITES

Most of the considerations for successful low-maintenance landscapes—site analysis, species selection, and planting methods—have already been discussed. It may be wise, however, to review the more essential considerations for arid sites in particular—those that receive little or no rainfall during the growing season.

The site must be evaluated for its ability to accumulate and store water. Begin by observing uncared-for plants that grow on the actual or a similar site. Find out what has grown there in the past. Have the soil and native vegetation been disturbed? Have any nearby developments changed the natural surface or underground water flows? What are the average and extreme rainfalls, seasonal evapotranspiration, soil textures and depths, depth of the water table, and topography?

If the site is essentially unchanged from its natural state, species that have grown there before should be suitable, but the range of viable species may be larger than the range of indigenous species. In most cases, the smaller the plant when it is introduced into the landscape, the better the chance it will have of getting off to a good start. Certain locations within the site may be more favorable than others for plant survival and growth. Higher moisture supplies and lower evapotranspiration rates are most frequently found on easterly slopes facing away from the equator; near the bottom of canyons or slopes; in deep, well-drained soils; near streams; or on the shady side of large boulders. Individual planting sites can be modified to enhance water accumulation and conservation. One of the simplest methods, if the soil is well drained, is to slope the soil surface toward each plant or group of plants, but not so as to accumulate rainfall directly around the trunks.

Spacing plants farther apart than in irrigated landscapes will provide each with a larger soil-water reservoir. Mulching and weed control, discussed in Chapter 14, can materially increase water infiltration and conserve water.

For areas with clear night skies (high radiation heat loss), Jensen and Hodder

(1979) use condensation traps to condense soil water vapor from a 1.5 to 2 m² (16–20 ft²) area, concentrate it, and direct it toward plant roots. They designed a prefabricated condensation trap, which is easier to install and performs more consistently than earlier versions. A frame 1.2 m (4 ft) square, constructed of 25 × 250 mm (1 × 10 in.) lumber, supports a polyethylene sheet (Fig. 11–13). A 20-liter (5-gal) bucket is temporarily placed in the middle of the polyethylene sheet, whose edges are wrapped around 1.2-m (4-ft) long laths, which are then nailed to the outside of the frame. The plant is placed on a slight mound in a shallow basin and protrudes through a slit in the sheet. A few rocks pull the polyethylene taut, form a seal at the plant trunk, and prevent the condensate from dropping prematurely at a sag point away from the plant. Small slits beneath the rocks will drain entrapped precipitation into the basin. Although clear polyethylene collects about 30 percent more condensate than black, black polyethylene is used because it will last several years and will inhibit weed growth. A small berm of soil seals the outside of the frame. Available soil moisture is considerably greater beneath this type of condensation trap than beneath the basin and berm trap. Moreover, the frame trap does not silt up.

Figure 11-13 The water supply of a new plant is augmented by means of a condensation trap. Jensen and Hodder found that prefabricated moisture condensation traps were easy to install and performed consistently. (Adapted from Jensen and Hodder 1979)

Jensen and Hodder (1979) found that the amount of condensate accumulated daily ranged from 16 to more than 200 ml (0.5–7 oz) when a black polyethylene liner was used in midsummer at 1200 m (4000 ft) elevation in southcentral Montana, where annual rainfall was 260 mm (10.3 in.). During that period the maximum air temperatures ranged from 29° to 38°C (85°–101°F); those under the black polyethylene were 39° to 52°C (102°–126°F). Air temperatures under clear polyethylene were 42° to 57°C (108°–134°F).

PLANTING IN AREAS THAT ARE PERIODICALLY FLOODED

Many areas that are subjected to flooding are being planted to enhance the recreation potential of reservoirs, streams, and canals; to protect riverbanks and levees; to landscape marinas, drainage-water basins, and flood plains; and to improve wildlife and fishery habitats. Some reports identify the species, primarily trees, that have survived different periods of inundation (Whitlow and Harris 1979). Many trial plantings support and extend the flooding observations. As would be expected, species and site are the main determinants of plant survival and performance.

A number of species will survive water over the soil around them for more than three months a year (see Appendix 2). A few of these species have been completely submerged for that long with only reduced growth and trunk rooting of some willows. Some species that can withstand long periods of water over the soil will lose branches if submerged for more than two weeks. Plants have a higher survival potential on well-drained alluvial soils than on finer-textured shallow soils. Mature valley and blue oaks have grown on alluvial fans in areas with little or no summer rain and are able to survive 100 days of flooding annually. By contrast, interior live oak will die if the soil around their trunks is wet for two weeks. Whitlow and Harris (1979) report that trees that had been established for one growing season before flooding were no more tolerant of inundation than trees planted only two months before flooding. Although one would assume that plants are best able to withstand flooding during their dormant period, many are flooded in late spring and summer. In California, plants survive the shift from a flooded condition to an arid one for two to three months in midsummer.

Trees and shrubs on the leeward side of reservoirs may be broken off or seriously damaged by floating debris. These plantings are best located in coves or channels that can be buoyed off to prevent damage during high water. A number of easy-to-root species have been established successfully by pushing or driving stem cuttings, 450 mm (18 in.) long and 20 mm (0.75 in.) in diameter, into the muddy shoreline as the water recedes in spring and summer. An irrigation or two during the first growing season will enhance survival in areas of low summer rainfall. Heavy pruning before flooding can decrease the tolerance of young plants to flooding.

PLANTINGS AT LANDFILL SITES

Refuse-landfill sites have been developed into parks, golf courses, botanical sites, native-plant landscapes, wildlife management areas, and landscaped hillsides. Landfills are located in abandoned sand and gravel pits, bay-fill sites, canyons, excavations, or refuse mounds. When a landfill was full, the common practice was to cover the entire area with 0.6 m (2 ft) or more of soil and landscape or at least plant it.

The decomposition of organic matter, which usually makes up more than half of the refuse, causes the landfill to settle, produces about equal parts of methane and carbon dioxide, and generates heat. Each of these conditions, especially the production of methane, carbon dioxide, and heat, is detrimental to plant well-being. Plants adjacent to some landfills have been seriously damaged by gases escaping underground.

Research workers in New Jersey sought planting techniques to protect landfill plants and screened woody species for tolerance to landfill conditions (Appendix 3) (Leone and others 1977). Concern for leaching of toxic chemicals and escaping of toxic and flammable gases, however, has resulted in stricter landfill capping procedures. These procedures are minimizing some of the problems present in older land-fills.

Most landfills in the United States are now being lined with impervious geotextiles to minimize leaching of toxic chemicals. Some landfills have been capped with 0.45 m (18 in.) of a plastic-like clay that is quite impervious; unfortunately, much of the clay contains iron pyrite which oxidizes, producing toxic sulfuric acid. Some landfills are being capped with an impervious, fiber-coated (bonded, to prevent soil slippage on steep slopes) geotextile, such as *Terra-Tuff*® (a Hypalon/Geo-Textile composite by Stevens Elastomerics in North Hampton, Massachusetts) (Katy Weidel 1990 pers. comm.). The geotextile is covered with 0.6 to 0.9 m (2–3 ft) of soil for planting. Some landfills are only capped with soil that is compacted, 2 to 3 m (7–10 ft) deep.

Surface Settling

Even though the refuse is compacted and covered with soil or geotextile, decomposition of organic matter causes the fill to settle. Depending on the depth of the fill, organic-matter content, degree of compaction, and entrance of water, settling of 0.15 to 0.45 m (6–18 in.) is not unusual. Not only may the uneven surface hamper activities, but water accumulates in depressions, irrigation lines can break, and the soil or geotextile cap may rupture. Damage can be minimized by placing main irrigation lines, buildings, and large trees on firm soil around the edge of a fill.

Decomposition Gases

The rate of decomposition and the volume of gases produced by a landfill depend primarily on the organic-matter content and amount of moisture. Most deep land-fills produce between 3 to 6 liters of gas per kg (0.05–0.1 ft³/lb) of organic waste (Van Heuit 1979). The half-life of solid waste in relatively dry landfills is estimated to be about 15 years, but such estimates are difficult to make because it often takes more than 20 years to complete a large landfill.

Gas-collection wells can keep landfill-gas levels at safe levels protecting plants growing in the soil cap. Methane can be reclaimed and used for energy. The Hackensack Meadowlands (New Jersey) is expected to produce 190 million liters (6 million ft³) of methane a day for 10 to 20 years, enough to serve about 10,000 homes (Grant 1990).

A landfill cap may be permeable, crack, or rupture, and if the collection wells are not effective, plants can be seriously damaged or killed. Such areas are easier to see than they are to repair.

Landfill-Tolerant Species

With more impervious landfill caps, the selection of species tolerant of gases is not as important now as ones tolerant of shallow soil and wind. Trees that can be considered desirable for landfill sites are small-growing; shallow-rooted; open and flexible to withstand wind; and tolerant of flooding and poor drainage. Small, balled-and-burlapped and container-grown planting stock with mycorrhizal roots adapt more quickly than do large, bare root, nonmycorrhizal plants. Lists of woody plants tolerant and sensitive to landfill gases are listed in Appendix 3.

Landfill Temperatures

In landfills without a geotextile cap, soil temperature has reached 58°C (137°F) in the rooting zone of some landfill sites; this is at least 10°C above the temperature that will kill the roots of most plants. Soil temperatures in anaerobic soils may be 16°C (30°F) higher than in aerobic soils nearby (Leone and others 1977). In most cases, however, the temperature differences are insignificant.

High soil temperatures in a landfill are usually fairly localized. Reducing the amount of moisture that reaches these "hot spots" may slow decomposition and lower the soil temperature. The hot spots can be vented or supplied with extraction wells and suction blowers. On the other hand, these areas can be avoided for plantings.

Additional Comments

Mounded and even canyon landfills end up with much of the final surface as 3:1 and even 2:1 slopes. Soil is difficult to keep from slumping on geotextile slopes much steeper than 3:1. Grasses are hydroseeded to minimize erosion; woody plants are usually planted after the grass is established. Primarily because of settling, irrigation lines need to have flexible joints and connections or be laid on the surface to make adjustments and repairs easier.

Apparently **the secret to successful landfill planting is to have a good gas recovery system, good planting-hole soil, and wise watering.**

PLANTING IN SOILS THAT HAVE HERBICIDE RESIDUES

Where railroad rights of way, parking lots, driveways, vacant lots, roadsides, and canal banks were once treated with persistent preemergent herbicides to prevent weed growth, landscape plantings may now be desirable because aesthetics or land use has changed. Older residual chemicals, like the arsenites, the borates, and the chlorates, were used primarily along railroad rights of way and parking lots, usually in large quantities, and may be toxic many years after application. The more recently developed organic herbicides may persist for less than a week or more than two years, depending on the herbicide, soil texture, and moisture. Organic herbicides can usually be deactivated by adsorption on organic matter, but the inorganic soil sterilants cannot. Occasionally an herbicide may also be spilled, misapplied, or overapplied to a landscape planting (see Chapter 20).

Testing for the Presence of Herbicides

The arborist must first determine what chemical has been applied, in what quantities, and when. If no record or recollection of previous herbicide applications is available, the soil can be analyzed prior to a large or important planting.

In warm, sandy soils that are low in organic matter and exposed to moderate or heavy rainfall or irrigation, organic herbicides will break down or be leached from the surface soil fairly rapidly. Organic herbicides are much more persistent in cold, clay soils that are high in organic matter and exposed to little or no rain or irrigation. Such soils will allow the herbicide to remain effective for a considerable period but will be easy to reclaim because most of the herbicide will be concentrated close to the surface. Pre-emergent herbicides applied annually at recommended rates are less of a problem than are one or two excessive applications. Even after nine annual applications, simazine at 7.2 kg per hectare (6.4 lb/a; 33 percent above highest recommended rate) was herbicidal only in the top 150 mm (6 in.) of a sandy loam California vineyard soil (Leonard, McHenry, and Lider 1974).

A bioassay using annual plants can usually determine whether a soil is still toxic enough to injure new plantings (Lange, Elmore, and Saghir 1973). Toxic levels of herbicide in the soil will stunt, twist, give a characteristic chlorotic pattern, or kill test plants. To bioassay a soil for herbicide presence, collect soil in separate paper cups or small pots (about 200 ml or 6 oz) from depths of 0 to 50 mm, 50 to 100 mm, 100 to 150 mm (0–2, 2–4, 4–6 in.), and deeper if the herbicide is extremely soluble or if it has been applied repeatedly. Collect samples at several sites in both herbicide-treated and untreated areas. Plant the seeds of at least one grass and one broadleaved species and grow seedlings under good cultural conditions for at least 20 days; do not overwater or underwater. Compare the growth and foliage condition of plants grown in soil you suspect was treated with those in soil (from the same depth) you know to be untreated. If symptoms are moderate to severe, the soil in planting areas should be replaced or the herbicide deactivated.

Treatment

Soil containing a toxic level of herbicide can be removed, mixed with organic matter, leached, or planted with tolerant plants, depending on the herbicide, soil texture, available drainage, the area involved, the type of planting desired, and the budget available to support it. If the levels of inorganic soil sterilant are toxic or the amounts of organic herbicide excessive, the contaminated soil should be replaced with clean soil. A bioassay will determine the depth to which soil should be removed. If ground-cover plants, turf, or closely spaced shrubs are to be planted, all contaminated soil in the bed should be removed. For individual trees and shrubs, dig holes 1 to 1.2 m (3–4 ft) in diameter and deep enough to remove the contaminated soil. Line the sides of the hole with plastic, insert a 200-liter (50-gal) metal drum with top and bottom removed, or remove the bottom of a large plastic garbage can and invert the can in the planting hole. Fill the hole with clean soil, allow it to settle, and plant the tree or shrub. Do not allow surface water from contaminated soil to reach the clean soil in the planting hole.

Chap. 11   Special Planting Situations

If organic herbicides have not been used in large quantities and appear to be in the top 150 to 200 mm (6–8 in.) only, it may be easier to dig planting holes 300 to 450 mm (12–18 in.) deep and 0.75 to 1.0 m (30–40 in.) in diameter. Line the hole with plastic, leaving the bottom uncovered. Mix the soil from the planting hole with an organic amendment and return it to the lined hole. Activated charcoal (carbon) should be thoroughly mixed with the soil at a rate of about 40 g (1 oz) of charcoal for each 150 mm (6 in.) depth of soil in a hole 1 m (3 ft) in diameter. This approximates the recommended rate of 150 parts (weight) activated charcoal to one part active herbicide in the soil (Smith and Fretz 1979). Ottoson (1976) reports that a mixture of composted chicken guano and nitrogen-fortified sawdust will be as effective as charcoal, or more so. The amount to be added depends on the nitrogen level of the guano-sawdust mixture; actual nitrogen should not exceed 50 g (0.10 lb) per planting hole. Ottoson credits the improved growth of plants to the increased fertility this soil amendment provides. Any finely divided organic material should adsorb moderate amounts of organic herbicide; the organic matter, however, should not be allowed to deplete the nitrogen in the soil. Tolerance to herbicide-contaminated soil can be increased if roots or root balls are dipped into a charcoal slurry (110 g charcoal/liter or 1 lb/1 gal of water) (Smith and Fretz 1979).

Soluble herbicides in well-drained sandy soils can be leached by heavy rains and irrigation. Herbicides that break down under anaerobic conditions can be deactivated in poorly drained clay soils by several weeks of flooding (Lange, Elmore, and Saghir 1973). Field plantings of corn and grain sorghum are tolerant to simazine and atrazine and can degrade them in the soil (Clyde Elmore, University of California, Davis, 1981 pers. comm.). Certain landscape plants may have similar abilities to grow in contaminated soil and detoxify it. We need more information on such species.

MANAGEMENT FOR FIRE PROTECTION

Dramatic increases in losses of homes and other property to fire have been reported in wooded residential areas adjacent to cities and in second-home developments (Tokle and Marker 1987). Loss of life has also been involved. This woodland/urban interface refers to the geographical areas where two diverse systems meet and affect each other—where combustible homes meet combustible vegetation.

Wildfires have been commonly thought to happen only in remote forested regions subject to hot, dry summers. Many urban developments are spreading into adjacent wooded areas and retirement and vacation homes into forests. Few regions of the United States have escaped fire disasters. Wildfires in the United States in 1985 are reported to have killed 44 civilians and fire fighters and burned three million acres including 1400 homes and structures, at a cost of $500 million in estimated damages to property and natural resources and $400 million for fire fighting (Tokle and Marker 1987).

Three elements are essential for wildfire initiation: fuel, low rainfall, and ignition. The first two occur almost annually in the southwestern United States, but an unseasonably extreme drought has put almost every region in jeopardy at one time or another. Light, fast-burning fuels are highly receptive to ignition; heavy, dense fuels are less likely to ignite, and they burn slower but with greater intensity. Wind, low humidity, and slopes contribute to fire intensity and the rapidity with which it spreads.

Reducing Fire Hazard

Protecting lives and property from wildfires should be more than a one-household effort. To ensure acceptable protection at a reasonable cost, governmental regulations are necessary, not only to protect a given household from naturally caused fires but also to protect it from unsafe construction and practices of others. The Colorado State Forest Service published model regulations for protecting people and homes from wildfire in subdivisions and developments (Zeleny 1988).

Dry landscapes increase fire risk except in places where there are succulent ground covers and certain cleanly maintained trees. Healthy green plants have low fire potential, but when moisture is limited and plants begin to dry, they are more likely to burn. Hot winds, as are common in the Southwest, can turn shrubs tinder dry in a few hours—even after irrigation.

Individual homeowners can do much to reduce the hazards of wildfires to their families and properties. Some of the following recommendations have been codified by agencies:

> Do not build on or at the top of slopes that exceed 30 percent without special precautions
>
> Obtain and follow building standards to reduce fire risk, such as fire-resistant roof and fire-resistive siding
>
> Create a *safety zone* at least 10 m (30 ft) in all directions around major buildings
>> The safety zone can be a greenbelt around the building using low-growing, low-combustible, low-fuel plants (with adequate rain or irrigation during summer); or bare ground, rock, or other noncombustible material
>> Remove and do not install highly inflammable plants—those with high resin, high oil, or high litter content
>> Have no shrubs or trees (trunks) within 2 m (6 ft) of a house
>> Thin large trees to eliminate overlapping crowns
>> Remove branches which extend over building eaves or are within 5 m (15 ft) of a chimney
>> Maintain a distinct break (at least 3 m [10 ft]) between low-growing plants and high-growing trees to reduce the likelihood of ground fire reaching the treetops
>> Remove dead branches, leaves, and other litter from plants and the ground
>> Plant and encourage "heat shields" or "screens." Thick, well-watered hedges planted at least 6 m (20 ft) from buildings have been effective in preventing fire from reaching the buildings (Hamilton and Gowans 1970)
>> Install a high-pressure sprinkler system to keep plants well-watered during a dry, fire season
>
> The fire safety zone on a slope below a house must be greater than on the level (Fig. 11–14)

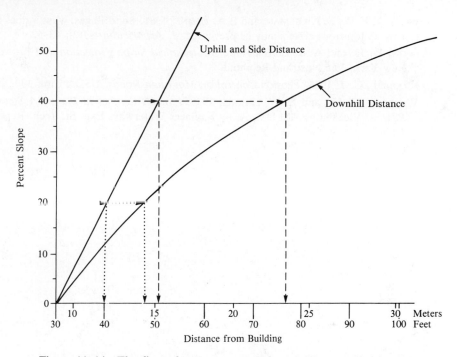

Figure 11-14 The fire safety zone on a slope should be greater not only below a house but also uphill and to the sides than when on the level (Colorado Springs Fire Dept. 1989). For example, on a 20 percent slope, the safety zone around a house should be extended to 12 m (40 ft) uphill and to the sides and to 14.5 m (48 ft) downhill.

Keep shrubs, ground covers, and grass down slope from the safety zone less than 0.5 m (18 in.) high

In wooded areas beyond the safety zone, thin nonirrigated shrubs so fire has difficulty spreading; remove brush

Clear a 2-m (6-ft) wide fire trail around the outer edge of the safety zone to slow down a running ground fire. Remove all burnable material exposing bare soil, unless erosion may be a problem.

Check with fire officials concerning local fire regulations and recommendations.

FURTHER READING

DAVIS, W. B., and others. 1974. *Landscaping in Containers without Natural Drainage*. Univ. of Calif. Agricultural Sciences Leaflet 2577.

ESTEVEZ, M. T. 1976. *From the Plants' Point of View*. Proc. Symposium on the Use of Living Plants in the Interior Environment. Alexandria, VA: Soc. of Amer. Florists.

FLOWER, F. B., E. F. GILMAN, and I. A. LEONE. 1981. Landfill gas, what it does to trees and how its injurious effects may be prevented. *J. Arboriculture* 7(2): 43–52.

GRAY, D. H. and A. T. LEISER. 1982. *Biotechnical Slope Protection and Erosion Control.* New York: Van Nostrand Reinhold.

KRAEBEL, C. J. 1936. *Erosion Control on Mountain Roads.* USDA Circ. 380.

WHITLOW, T. H., and R. W. HARRIS. 1979. *Flood Tolerance in Plants: A State-of-the-Art Review.* Vicksburg, MS: U.S. Army Engineer Waterways Exp. St. Tech. Report E–79–2.

CHAPTER 12

Nutrient Management

Fertilization is one of several cultural practices that can encourage the rapid development and continuing health of plants. Nutrient management can influence plant vigor, leaf size and color, susceptibility to certain pests and diseases, and tolerance to environmental stresses.

Fertilization practices have been cloaked in tradition and rules of thumb for decades. Until recently, woody landscape plants were seldom fertilized in urban and native settings. Off-color foliage and weak growth are typical nutrient-deficiency symptoms which can be brought about by root diseases, girdling roots, drought, compacted or water-logged soil, nematodes, salt injury, and so on. Yet, the use of fertilizers has been criticized by those involved in growing plants "naturally," without so-called "chemical poisons."

As older plantings deplete soil nutrients and as the danger increases that excess nutrients will pollute surface and ground waters, more attention must be given to wise nutrient management. An understanding of the inter-relationships among plant, soil, nutrients, and water is essential. This chapter reviews these relationships and offers suggestions for determining and meeting the nutrient needs of plants. Research on tree nutrition is also examined.

The next section presents information on nutrition that has fairly wide application.

Plant Adaptation to Low Nutrient Levels

The foliage of most trees is usually an acceptable green even though nutrient deficiencies may significantly limit growth. In a series of elegant experiments, Ingestad and Lund (1979), Ericsson (1981), Ericsson and Ingestad (1988), and others have used birch, willow, and other species to show that seedlings can adjust root and shoot growth to low levels of nitrogen, phosphorus, and other nutrients so that their foliage maintains a healthy appearance.

For example, when optimum nitrogen supplies to young birch seedlings are reduced to little or none, shoot growth is greatly reduced, leaves yellow, root growth is reduced though not as much as shoot growth, and roots grow long with little branching (Ingestad and Lund 1979). However, if the deficient nutrient(s) is then supplied at a rate proportional to the increase in plant weight (growth), even though below optimum, growth stabilizes at a slower rate, leaf color returns, and the root-shoot ratio stabilizes higher than when the seedlings were growing under more optimum conditions. The amount of chlorophyll and the rate of photosynthesis are not reduced in proportion to the reduction in the nitrogen supply or the overall growth of the plant. This is consistent with higher concentrations of starch found in plants with reduced nutrient availability (McDonald, Ericsson, and Lohammar 1986).

Ingestad and Lund (1979) postulate that leaf-color–deficiency symptoms occur primarily when a plant cannot adjust rapidly enough to a decreasing internal-nutrient level. The growth of a plant decreases (dilutes) its internal-nutrient level unless the plant has an increasing supply of nutrients. An increasing nutrient supply would be available if a plant: (1) increases its absorbing root volume (explores new soil), (2) has a high soil-nutrient level that can be absorbed in increasing amounts (a high cation-exchange capacity), or (3) has an increasing soil-nutrient availability (fertilization).

These findings have important implications for the care of young as well as old trees. For young trees to grow rapidly, they must have increasing amounts of nutrients available. In most situations, mature trees with little or no fertilization have good leaf color and low to moderate growth and are in balance with their surroundings.

There may be concern that even though leaves have good color, the reduced growth of a plant will curtail starch reserves vital for a plant's ability to begin growth in the spring and to withstand injury and stress. It is not clear whether greater care may be needed to minimize stress of such plants in order to maintain attractive and healthy plants. Trees with slow rates of growth have survived for centuries; it is the fast-growing trees that are short lived. Moderation is probably the message.

Nitrogen and Phosphorus and Root-Shoot Growth

An increase in soil fertility is commonly associated with a reduction in the root-shoot ratio; that is, root growth increases less (in weight) than shoot growth (Coutts and Philipson 1980).

Nitrogen is the nutrient that is almost universally deficient. The most visible response when nitrogen is added is increased shoot growth. On the other hand, phosphorus, also an essential nutrient, is not deficient for trees in most regions. Top growth is seldom affected when phosphorus is applied to trees. These observations have reinforced the notion that nitrogen stimulates top growth (at the expense of root growth) and that phosphorus stimulates root growth (Bernatzky 1978; Patch, Binns, and Fourt 1984; National Arborist Association 1987; Neely and Himelick 1987; Pirone and others 1988; the British Standards Institute 1989b; and Tattar 1989).

A common belief is that an increase (in weight) in top growth without a similar increase in root growth places a plant in jeopardy. Pirone and others (1988), Shigo (1989), and Tattar (1989) recommend or imply that injured trees should not be fertilized with nitrogen until they have recovered.

I have been unable to find convincing evidence to support these statements concerning the effects of nitrogen and phosphorus on root-shoot growth. Even if they were correct, comparing the weights of roots and tops is not a valid way to determine the well-being of a tree. As was pointed out in Chapter 2, the roots within the dripline of a tree are estimated to have 2.5 to 4.5 times more surface area than do the leaves (one side) of the same tree. In addition, mycorrhizae greatly amplify the effective surface of the finer roots.

Van der Meiden (1957), as cited by Coutts and Philipson (1980), reported proliferation of primary roots of poplar in the vicinity of phosphate. Roots, however, are known to branch more readily in soils favorable for growth (Wilson 1984).

When there is a deficiency, the addition of either nitrogen or phosphorus stimulates root and shoot growth. When nitrogen is optimum but phosphorus deficient, phosphorus addition stimulates shoot growth (weight) as well as root growth (Ingestad and Lund 1979, Ericsson and Ingestad 1988). Similarly, when phosphorus is optimum and nitrogen deficient, addition of nitrogen stimulates both root and shoot growth. In split-root experiments with spruce, Philipson and Coutts (1977) found that when deficient, addition of nitrogen stimulated more root growth than did the addition of phosphorus when phosphorus was deficient.

In England, at a difficult china-clay–waste site which severely limited growth, first-year nitrogen fertilization of birch and sycamore-maple resulted in 50 percent more root than shoot growth (Gilbertson, Kendle, and Bradshaw 1987). Also the root-shoot ratios were increased the greater the nitrogen application. They state, **"There seems little evidence for the common belief that N promotes shoot growth whilst P is for roots."** Also in England, Walmsley (1989) found no statistically significant differences when 576 *Fraxinus excelsior* 2-year-old nursery transplants were fertilized with one of 12 different combinations of N, P, and K. Although not significant, he stated that some trends were apparent: All treatments "which included nitrogen enhanced total, shoot, and root weights, and led to an increase in the root:shoot ratio, with the exception of the NP treatment (shoot and root growth was reduced)."

Although roots were not measured, experiments in mature unirrigated almond orchards in California are instructive. Lack of water was thought to curtail growth and yield. Nitrogen fertilization greatly increased growth and yield; other nutrients, including phosphorus, were without effect (Proebsting 1935). Root growth must have increased enough to supply adequate water for the increased foliage and crop.

Epstein (1972), Halfacre and Barden (1979), Kramer and Kozlowski (1979), and Mader and Cook (1982) make no mention of the root-promoting ability of phosphorus in their reviews of plant nutrients.

Even though addition of nutrients that are deficient in the soil may stimulate more shoot than root growth (weight, not surface area), it should not be a problem unless the root system is physically restricted (and becomes unstable) or excessive shoot growth results. Excessive shoot growth is usually kept in "balance" with the roots by increasing temporary afternoon water deficits as the leaf area begins to "outgrow" the ability of the roots (or conducting xylem) to supply enough water. A limited root volume of sandy soil, however, could dry so quickly, if not supplied with water, that a plant could not adjust fast enough.

Plants have evolved with this root-shoot–growth response to fertility, moisture, and other environmental variations—it must be in the best interest of most plants to do so or they would not have survived as well as they have.

Fertilizer Recommendations

Phosphorus and potassium are seldom deficient for trees and large shrubs in most regions. So-called "complete" or "balanced" fertilizers contain these two nutrients plus nitrogen and possibly others. Trials in the last century found these three nutrients deficient for field and vegetable crops in most soils. Even though hundreds of field trials have shown that most soils contain sufficient levels of phosphorus and potassium for trees and large shrubs, addition of these nutrients is still universally recommended for trees (National Arborist Association 1987, Neely and Himelick 1987, British Standards Institute 1989b, Swanson and Rosen 1989). Phosphorus and potassium are deficient for trees in fairly specific soils (see discussions of specific nutrients). Local extension agents and knowledgeable plantspeople should know if either is deficient in their regions and, if so, where.

Applying nutrients to be "safe" without knowing they are deficient wastes time and money and can lead to salt build-up in the soil and to water pollution. High amounts of phosphorus can cause iron and manganese deficiencies in broad-leaved trees (Whitcomb 1987) and palms (Henry Donselman, Rancho Santa Fe, CA, 1990 pers. comm.). High fertility inhibits mycorrhizae formation thereby jeopardizing disease protection (Sinclair, Lyon, and Johnson 1987) and absorption of nutrients in short supply.

Nitrogen is the nutrient to which plants most commonly respond.

RELATIONSHIPS AMONG SOIL, NUTRIENTS, AND PLANTS

Physical Properties of Soil

The physical properties of soil influence the amount of nutrients the soil holds and, to a certain extent, the availability of nutrients to plants.

Soil texture, the size distribution of soil particles, directly influences the amount of nutrients adsorbed by the soil. As a soil becomes finer in texture, it will hold more nutrients and water. To supply an equal amount of a nutrient, a sandy soil will require more frequent applications of smaller amounts than will a clay soil.

Soil depth can determine the nutrient- and water-holding reservoir available to a plant. As a plant is able to root more deeply and widely, it will reach more water and nutrients.

Soil structure, the arrangement of soil particles, influences root exploration and absorption by plants. A compacted soil, or one lacking an open, granular structure, may restrict root activity by limiting water and air movement and by physically impeding growth.

Chemical Properties of Soil

Sixteen elements have been found essential for woody-plant growth (see Table 6-3).

> In the absence of any one of the elements, a plant will fail to complete its life cycle
>
> Each element is specific; it cannot be replaced by another
>
> Each element has a direct effect on plants (as opposed to an indirect effect, such as repelling insects, which might prevent completion of the plant's life cycle)

As shown in Table 6-3, carbon and oxygen are available to plants from carbon dioxide in the air. Oxygen is directly available for respiration from either air or soil; hydrogen from water absorbed from soil. The remaining 13 elements are derived from soil and are called plant nutrients. Micronutrients, required in small amounts, are just as essential as macronutrients, required in larger amounts.

Nutrient availability is measured in terms of a soil's capacity to provide nutrients (amount available) and the intensity of these nutrients (concentration). When a soil is unable to meet the nutrient demand of the plant, it is deficient or infertile with respect to the particular nutrients it lacks.

With the exception of urea, which can be absorbed as a compound, **almost all elements—whether from organic or inorganic sources—are absorbed as inorganic ions.** Until organic forms are mineralized, that is, converted to inorganic ions, they are not commonly utilized by plants.

Plant roots absorb ions selectively. Some ions, such as K^+, NO_3^-, and NH_4^+, are absorbed rapidly and may accumulate in the plant at concentrations much higher than in the external solution. Other ions, such as $H_2PO_4^-$, $SO_4^=$, Ca^{++}, and Mg^{++}, are absorbed less readily. The selective uptake means that nutrient ions are not absorbed in the same proportion as they occur in a soil or soil solution. Absorption rates of specific ions are different for different species, and sometimes even for cultivars of the same species.

A plant does not distinguish between ions originating from inorganic and organic sources. This is not to say that all materials containing the same amount of a particular nutrient will be equally effective. Other factors are important: whether the material contains other essential elements, how it affects soil structure and soil pH, and how long it persists in a soil.

Compared with natural organic sources, inorganic fertilizers usually

Contain greater percentages of a given nutrient

Are easier to handle because they are more concentrated and compact

Are free from unpleasant odors

Are more uniformly available in the soil, not being dependent on the rate of organic decomposition (which is influenced by biological activity that varies with temperature, moisture, and substrate)

Cost less per unit of nutrient

Contrary to popular belief, plants will grow well on either nutrient source. When nutrients are the primary interest, inorganic forms are usually favored. The principal advantage of natural organic fertilizers is that they improve soil tilth (structure) and can meet the nitrogen requirement of plants if applied in sufficient amounts. Manures incorporated in surface soils, for example, reduce crusting and enhance seedling emergence but may contain weed seeds and/or excess salt.

Determining Nutrient Needs and Toxicity Problems

For woody plants, most soils supply adequate amounts of nutrients other than nitrogen, although there are some exceptions. Iron and manganese may be deficient in alkaline soils; phosphorus may be needed for extremely old, or acid sandy, or granitic soils. In semiarid regions currently or formerly used to corral livestock, plants may show severe zinc- and copper-deficiency symptoms. Young trees may not survive their first summer on these sites. Copper deficiency may also be a problem for woody plants on the sites of old Indian camps, even where annual plants grow well. If you suspect this type of nutrient problem, consult an extension agent, an agricultural commissioner, a consulting arborist, a farm supply employee, or a nursery person, all of whom are usually aware of such problems in a given area.

The presence of **toxic substances can seriously affect plant growth** and appearance; aluminum, boron, chloride, and sodium can be directly toxic. High salinity and high sodium (alkali) can be problems in areas of low rainfall, poor water quality, or poor drainage. Both nutrient deficiencies and toxic conditions may be diagnosed by soil analysis, plant analysis, and nutritional trials. If a governmental research station or the extension service does not have diagnostic services, commercial laboratories are usually available. It is best to use one specializing in agricultural crop analyses.

Soil Analysis.   Soil can be tested to appraise pH or the level of elements available before a plant is grown. In high rainfall areas, testing can indicate the amount of lime needed to adjust the soil pH to a more favorable level (see Chapter 6). Soil analysis can fairly accurately determine the availability of phosphorus and potassium to turf, herbaceous plants, and small shrubs, but it is of doubtful value with regard to trees and deep-rooted shrubs (Mader and Cook 1982). Only in extremely deficient soils, such as acid sands, have soil-test results correlated with woody-plant response, and then only as regards the level of phosphorus (Pritchett 1979). The amounts of other nutrients in the soil can also be determined, but, to date, are of limited value for determining the fertilizer needs of woody landscape species. Toxic levels of boron, chloride, sodium, and total salt can be determined by soil analysis. Water, too, can contribute toxic levels of these elements to the soil, and should be tested, if suspected.

A good analysis is dependent upon a careful, accurate, and representative sampling of soil. If you suspect soil problems or do not know the soil situation in the area, it may be wise to have the soil analyzed. Obtain sampling instructions from the laboratory that will analyze the soil; sampling procedures are given by Quick and Rible (1967). Several suggestions follow (Fig. 12-1):

Turf: Take samples at depths of 0 to 75 mm (0–3 in.) and 75 to 150 mm (3–6 in.); for special sites, such as golf greens, take samples at depths of 0 to 50 mm (0–2 in.), 50 to 100 mm (2–4 in.), and 100 to 150 mm (4–6 in.)

Flowers and shrubs: Take samples at depths of 0 to 150 mm (0–6 in.) and 150 to 450 mm (6–18 in.)

Trees: Take samples at depths of 0 to 300 mm (0–12 in.), 300 to 600 mm (1–2 ft), and 600 to 1200 mm (2–4 ft)

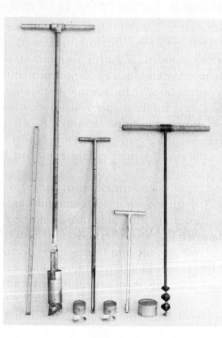

Figure 12-1 Soil augers (left and right), profile tubes (center two), and sampling containers can be used to collect soil samples for pH, nutrient, or moisture analysis. (Harris, Paul, and Leiser 1977)

To test saline or sodic conditions, take separate (not composite) samples to represent a range of conditions from no or poor plant growth to good plant growth. Depths should be at 0 to 150 mm (0–6 in.), 150 to 300 mm (6–12 in.), and 300 to 600 mm (1–2 ft), and, when needed, at 300-mm (1-ft) increments to lower depths. For sites you plan to seed, it is desirable to sample the surface 25 to 50 mm (1–2 in.) of soil separately. The results of soil and water tests must be interpreted by one who is experienced in evaluating laboratory data in relation to the plants and soils of the given region.

Plant Analysis.   *Visual symptoms* of nutrient deficiencies are seldom seen in most landscapes. Except for nitrogen and, occasionally in alkaline soil, iron and manganese, the other essential elements are seldom deficient. Plants fertilized with nitrogen will almost always make longer shoot growth, have larger leaves, and hold their leaves later in the autumn even though the leaves appear normally green in color before fertilization. Ingestad and Lund (1979) showed that if a tree is in balance with its supply of soil nutrients, it can have normally appearing green leaves even though growth is markedly reduced (see the discussion at the beginning of this chapter). Thus, if leaf-color symptoms do exist, they may indicate a marked reduction in the availability of a nutrient. This would be of greatest concern with young plants where rapid growth is desirable.

The length of shoot growth (see Figs. 2–3 and 12–2), leaf color, color pattern, and size, and the time of leaf fall can indicate the nutrient status of a perennial plant. The age of leaves affected and their position on the plant provide additional clues, because some nutrients are more mobile in certain plants. Symptoms for individual nutrient deficiencies and mineral toxicities are given in the next section. Some symptoms have been observed only in nutrient studies under controlled conditions. Because there may be multiple symptoms and symptoms reflecting disease or improper amounts of water, considerable experience is needed to visually determine the nutritional status of a plant.

Figure 12–2   These shoots of Chinese pistache exhibit differences in vigor. Leaves are present on current growth; bud scale scars (arrows) indicate starting and stopping points of growth (see Fig. 2–3). Annual shoot growth may be difficult to determine on species that may have several growth cycles in a year. (Harris, Paul, and Leiser 1977)

Tissue analysis can be used to detect certain mineral deficiencies and excesses (Chapman 1960). A tree in good health will usually be found to have higher concentrations of macronutrients when young than when mature (Binns, Insley, and Gardiner 1983). Leaves are most commonly analyzed because they are easy to collect and the correlation between their analysis and the nutrient status of the plant is relatively high. Collect recently matured leaves from the same relative position (the current season's growth) on both healthy and suspected nutrient-deficient branches or plants (Fig. 12–3). Foliar analysis for woody landscape plants is largely used in

Figure 12-3 Recently matured leaves near the shoot tip should be selected for nutrient analysis. For toxicity analysis, collect leaves that show symptoms as well as leaves that appear normal for comparison. (Harris, Paul, and Leiser 1977)

research studies and to confirm certain visual diagnoses. We must document more correlation among tissue analysis, plant symptoms, and fertilizer responses before we can develop meaningful standards for determining the nutrient status and needs of woody landscape plants.

Nutritional Trials. Nutritional experiments can provide valuable guidelines for fertilization, but they are not as practical for woody plants as may seem at first to be the case. To be accurate and worthwhile, fertilizer trials demand more skill and time than most arborists have or are willing to spend. Test plants and areas must be representative of the species and soils in question. Some plants must be fertilized and other similar plants left as unfertilized controls. Differences in the composition or quantities of nutrient must not be emphasized by the method of application. For example, if the nutrient is sprayed on foliage, control plants should be sprayed with the same amount of water; if the nutrient is injected into the soil, the same amount of water should be injected near control plants. Several comparisons (replicates) between fertilized and control plants must be included. Plant response (shoot growth, trunk increase, leaf size and color) must be measured and analyzed. Even in careful experiments, a difference of 15 to 20 percent may not be statistically significant.

CORRECTING NUTRIENT DEFICIENCIES

Nitrogen is the most commonly deficient soil nutrient. Other nutrients are likely to be deficient only under rather uncommon circumstances. The following text identifies those conditions under which individual deficiencies are most likely to occur. Typical visual symptoms are described and treatments recommended; both are summarized in tabular form, along with the relative mobility of each nutrient in plants and soil. Nutrient mobility within a plant will tend to determine where symptoms first appear and whether nutrients applied to the foliage will be translocated to foliage that develops later. **When nutrients are mobile in plants, deficiency symptoms are likely to appear first on older foliage; deficiency symptoms of less mobile nutrients tend to appear on newer leaves.**

To a large measure, the mobility of a nutrient in the soil determines how it should be applied and how vulnerable it is to leaching. Nutrients that move readily in soil can be applied to the surface in relatively small amounts, but those that are less mobile may need to be incorporated into soil and applied in larger quantities depending on depth of rooting or sprayed on foliage. Nutrient mobility increases loss from leaching. Mobility in soil is strongly influenced by soil texture and pH. As soil particles become coarser and the soil more acid, more nutrients are likely to be mobile. Organic matter and certain soil compounds may decrease nutrient mobility. The extent of nutrient mobility and its effect should be considered tendencies, not absolutes.

In the sections that follow, nitrogen, phosphorus, and potassium are considered first because they are among the first nutrients that were found lacking in soils, particularly for herbaceous plants. The rest of the macronutrients, and then the micronutrients, are presented in alphabetical order. Several nutrients may be present in toxic concentrations, which must be reduced if plants are to grow satisfactorily. Toxicities are also discussed individually.

Macronutrients

Nitrogen. **Plant response to nitrogen (N) fertilizer is almost universal.** Nitrogen becomes available to plants through

> Mineralization of organic matter
> Addition of fertilizers
> Fixation of atmospheric nitrogen by bacteria and actinomycetes

Nitrogen in soil is rendered unavailable to plants due to

> Absorption by weeds or other nontargeted plants and by organisms during the decomposition of organic matter low in nitrogen
> Denitrification by soil organisms
> Leaching
> Volatilization

These processes, except direct fertilizer application, are influenced by the environment (temperature, moisture, soil aeration, soil pH, and soil flora and fauna). Due to the transitory nature of soil nitrogen, soil and foliar analyses are not particularly useful for woody plants in the landscape.

Nitrogen fertilizers come in different chemical forms. Nitrate (NO_3^-) and ammonium (NH_4^+) ions are the most common of the inorganic forms. Nitrate ions, being negatively charged, are not adsorbed by soil colloids; they move with soil

Deficiency Symptoms: NITROGEN (Nitrogen is mobile in plants.)

Broadleaf: Leaves are uniformly yellowish-green, the color being more pronounced in older leaves; leaves are small and thin, have high fall color, and drop early; compound leaves have fewer leaflets. *Shoots* are short and small in diameter and may be reddish or reddish brown. *Flowers* bloom heavily but may be delayed. *Fruit* set is light; fruit is small, highly colored, and early to mature (Harris, Paul, and Leiser 1977).

Conifer: Needles are yellowish, short, and close together. Young seedlings may remain in the primary needle stage with little or no branching. Older plants exhibit poor needle retention. Lower crowns may be yellow, while upper crowns remain green (after Powers 1979).

Palm: Overall light-green foliage; decreased vigor.

Treatment: NITROGEN DEFICIENCY (Nitrogen mobility in soil depends on chemical form.)

Application	Chemical Form	Amount to Apply	How to Apply	Years Effective
		kg/100 m²ᵃ		
Soil	Several (N)	1–2	On surface	1–2
		kg/100 litersᵇ		
Spray	Urea	1	Foliageᶜ	1

[a] One kilogram per 100 m² is 1000 kg/ha (2 lb/1000 ft²). To convert kg/100 m² to g/cm diameter of treetrunk: Multiply by 20 when trunks are less than 150 mm in diameter; multiply by 40 when trunks are more than 150 mm in diameter. To convert lb/1000 ft² to lb/in. diameter of treetrunk: Multiply by 0.05 when trunks are less than 6 inches in diameter; multiply by 0.1 when trunks are more than 6 inches in diameter. When rainfall or irrigation is heavy, this amount might be split into two applications.
[b] One kilogram per 100 liters is about 8 lb/100 gal.
[c] Wet to runoff.

---

water, available for absorption by plant roots and other organisms. If excessive water is applied, some of the nitrate may leach below the root system and be lost to the plants.

Ammonium ions, being positively charged, are adsorbed on soil particles and do not move readily with soil water. Many plants have shallow roots and can absorb ammonium ions directly, particularly if the soil is protected by low branches or mulch. Ammonium ions, however, are commonly transformed to nitrate by soil microorganisms before they move down in the soil. Depending on the temperature and microbial activity of the soil, most ammonium ions will be converted to nitrate in about two weeks. Therefore plants will respond more slowly but for a longer period of time to ammonium nitrogen.

With the exception of urea, organic nitrogen is generally unavailable for plant use. Urea, though organic, is water soluble and moves with water in the soil, where it is converted first into ammonium nitrogen, then into nitrate. In practice, urea can be irrigated into the soil and made available to roots as urea, ammonium ions, or nitrate ions. If a second irrigation takes place when most of the nitrogen is still in ammonium form, it will not leach nitrogen out of the root zone, because ammonium ions do not move readily with soil water. Other organic forms of nitrogen, whether natural or synthetic, must be decomposed or transformed into more soluble forms before they move through the soil to plant roots. The rate of decomposition or transformation is usually dependent on the form of the material, temperature, soil moisture, and soil microorganisms.

In recent years, several nitrogen products that release nutrients over a longer period have become available. These "slow-release" fertilizers include

Slowly converted organic forms
Inorganic materials of low solubility, such as magnesium ammonium phosphate
Inorganic materials coated with plastic or sulfur or enclosed in plastic bags that govern their rate of release to the soil

The organic forms may release larger amounts of nutrients when the temperature is high. Inorganic forms, which release nutrients more uniformly, are therefore superior in most situations.

Slow-release fertilizers are usually more expensive than normal nitrogen sources, but their use may be justified where fertilizer application is difficult or required at frequent intervals, since they can be applied in larger quantities and less often. They provide a less concentrated but more continuous source of nitrogen with less likelihood of loss by leaching. This may better supply young plants with enough nitrogen to absorb in increasing amounts as reasoned by Ingestad and Lund (1979) (see the first part of this chapter). On the other hand, plant growth may not be controlled as easily by the slow-release fertilizers as by the more soluble forms. Slow-release nutrients will be available at a fairly uniform level for an entire season or more, while a soluble form applied at the correct rate in late summer or autumn will be available when growth begins the next spring and will drop to a low level by late summer. This may be desirable for winter hardiness or fall leaf color.

Even when the particular form of nitrogen supplied does not have a pronounced effect on plant response, it may influence soil reaction. For example, ammonium ions tend to acidify a soil (see Table 6–6). Also, as noted previously, forms vary in the ease with which they are leached from the soil.

Surface application is easiest and quickest and is just as effective as placement of nitrogen in the soil (Neely, Himelick, and Crowley 1970, Smith and Reisch 1975). If urea or an ammonium-containing fertilizer is applied on the surface of sandy alkaline soils, however, some nitrogen may be lost due to ammonia volatilization within the first two or three weeks. Laboratory tests estimate losses to be about 5 percent, but they may be as high as 50 percent on sandy alkaline soil that is alternately dried and wetted (Connell and others 1979). Volatilization loss is not a problem with nitrate or on acid soils. Ammonia loss from urea can be virtually eliminated if a good rain or irrigation dissolves and leaches the urea into the soil. Watering in ammonium fertilizers reduces volatilization by increasing adsorption of ammonium ions on soil particles. A mulch reduces ammonia loss by providing surface for adsorption and by maintaining a uniform and fairly high moisture level in the surface soil, which can absorb some of the ammonia that is volatilized. Incorporating ammonium-containing fertilizers or urea into the soil will practically eliminate the volatilization loss of nitrogen, though the savings may not be worth the effort.

Phosphorus. **In almost all soils, phosphorus (P) occurs in amounts that are adequate for trees and large shrubs.** Phosphorus fertilization research has focused much more extensively on fruit and forest trees than on landscape trees and shrubs. Trees and large shrubs are able to grow well even when soil phosphorus is available at relatively low levels. In certain California soils, satisfactory cover crops can be grown only when phosphorus is added; in these same soils, however, fruit trees have not responded to phosphorus applications (Proebsting 1958). Dickey (1977) reports similar response differences between woody species and field crops in Florida. Boynton and Oberly (1966) find no conclusive evidence that phosphorus applications affect the growth or fruiting response of apple trees under field conditions. By means of mycorrhizae, woody plants are apparently able to exploit larger volumes of soil than are annual crops and may use less available soil phosphorus (Pritchett 1979).

Deficiency Symptoms: PHOSPHORUS (Phosphorus is mobile in plants.)

Broadleaf: Leaves are green to dark green; veins, petioles, and lower surfaces may become reddish (dull bronze) to purple, especially when young; foliage may be sparse, slightly smaller than normal, and distorted; leaves drop early. *Shoots* are normal in length unless the deficiency is severe, but they are small in diameter. *Flowers* are few. *Fruit* is sparse and small (Harris, Paul, and Leiser 1977).

Conifer: Needles turn purple in young seedlings, starting at the tips of lower needles and progressing inward and upward. Few or no secondary needles may appear. Needles die, starting in the lower regions and spreading upward through the tree. Buds may set early or seedlings remain dormant longer than usual. Older trees take on a dull blue- or gray-green color. Roots are sparse with no evidence of mycorrhizae (after Powers 1979).

Treatment: PHOSPHORUS DEFICIENCY (Phosphorus is relatively immobile in most soils.)

Application	Chemical Form	Amount to Apply	How to Apply	Years Effective
		$kg/100\ m^{2\,a}$		
Soil	Several (P)	1–2 on sandy loam	Incorporate[b]	3–5
		2–4 on clay loam	Incorporate[b]	3–5

[a] One kilogram per 100 m^2 is 1000 kg/ha (2 lb/1000 ft²).
[b] Apply on surface if roots are near the soil surface.

Phosphorus availability to plants is low in most soils. Some clay soils, clay loam, red soils high in iron oxides, and highly organic soils have high phosphorus-fixing capacities. Very old soils (as in Australia and New Zealand) and some coastal sands and peat lands have extremely low levels of phosphorus. Availability is also influenced by soil pH: Phosphorus is most available to plants between pH 5 and 7 and becomes less available as the pH deviates further from this range. At pH 7, phosphorus combines with calcium and becomes increasingly unavailable to plants. Below 5, it is precipitated out of the soil solution, forming aluminum and iron compounds. Phosphorus deficiency occurs on many surface-mine spoils, which are often extremely acid.

Pritchett (1979) reports that young slash pine in the coastal plains of the southeastern United States respond more to phosphorus fertilization when soils are poorly drained. Conversely, in well-drained soils there is little or no response to phosphorus additions. Applying nitrogen to soil that is severely deficient in phosphorus may actually suppress plant growth until the deficiency is corrected (Maftoun and Pritchett 1970).

Phosphorus does not move readily in soil. It tends to concentrate near the surface and is often fixed in the humus layer there. Grading soils that are low in phosphorus may aggravate phosphorus deficiencies in young or shallow-rooted plants.

Newly planted trees and shrubs in soil that is low in phosphorus may respond to phosphorus fertilization, but their response usually decreases as plants increase in size. The roots of larger plants are able to explore more soil and become mycorrhizal, thereby increasing their phosphorus supply. This may account for the differences between reports by Wyman (1936) and Chadwick (1941) on the one hand, who observed responses from newly planted trees to phosphorus fertilization, and by Neely, Himelick, and Crowley (1970), Smith and Reisch (1975), and Dickey (1977) on the other, who worked primarily with more mature woody plants. In fact, when Pridham (1938) observed the trees originally fertilized by Wyman, he found that after seven years the trees not fertilized with phosphorus were as large as those that had been fertilized. In forest soils where phosphorus has been found to be deficient, it has been applied only to young stands of trees (Pritchett 1979).

It is reasonable, however, to apply phosphorus to young woody landscape plants if a deficiency is suspected. One should first ask appropriate authorities whether mature woody plants have responded in the past to phosphorus fertilization on a particular soil. Pioneer species in natural forest successions generally have lower nutrient requirements than succeeding species (Powers 1981a). It may be worthwhile to test leaves for phosphorus content or to conduct a fertilizer trial. Phosphorus should be applied if the level in the current season's foliage of conifers and broadleaved plants is below 0.1 percent. If additional phosphorus is needed for good plant growth, it needs to be applied only once every three to five years. When phosphorus is not needed, application will increase soil salinity; could tie up micronutrients, particularly copper and zinc (Bingham 1966); and can pollute surface water if soil to which phosphorus has been applied erodes.

Except in well-drained acid soils, phosphorus does not move well. If needed, the mineral should be placed among the roots or close to them. For young trees, mix 5 g of phosphorus for each 10 liters (0.3 oz/ft³) of backfill soil, or ring the bottom of the planting hole (500 mm or 20 in. in diameter) with 10 g (0.2 oz) of phosphorus. The hole should be 100 mm (4 in.) deeper if phosphorus is added, so it can be covered with unfertilized backfill to minimize toxicity.

In northern California forests, Powers (1981b) has obtained good response when he has sprayed phosphoric acid on the foliage of young white fir and sugar pine planted in phosphorus-fixing soil.

Potassium. **Most soils contain enough potassium for woody plants,** which absorb this nutrient in relatively large amounts, concentrating it in rapidly growing shoots. When leaves and brush are left on the ground, potassium is quite effectively replenished in surface soil, since it leaches readily out of organic matter.

Potassium deficiencies occur primarily in soils that are acid, sandy, low in organic matter, and low in total cation exchange capacity (Leaf 1968). Deficiencies are most likely to develop in soils of alluvial and blown-sand origin; are somewhat less common on heath lands, peat, and muck soils; and are rarer yet in heavier soils. On sandy soils in northwestern Florida, Dickey (1977) has identified potassium deficiency symptoms in wax-leaf privet, eastern dogwood, and red maple, but not in other plants. Potassium deficiencies have been identified on many agricultural and forest lands. Surface soil is usually higher in potassium than the soil underneath, so in areas where potassium is low, grading may leave cuts deficient and fill areas adequate in nutrient content. Similarly, cultural practices that encourage rooting in surface soil will decrease the likelihood of potassium deficiency symptoms (see Chapter 14).

If you suspect potassium deficiencies in a certain area, consult agricultural extension agents, extension foresters, or knowledgeable arborists. Leaf analysis may be unreliable because potassium is mobile in a plant and is readily leached from leaves during rain or sprinkler irrigation. Visual symptoms may be the best guide.

The amount of potassium required depends on the type of soil. In fine-textured soil that is neutral or alkaline and has little or no shallow rooting, apply potassium in larger amounts; less is needed in sandy soil. A positively charged ion, potassium tends to be adsorbed strongly on clay particles. Apply on the surface if rooting is shallow. On bare soil, apply it as a liquid by injection or dry in holes bored in the soil or in a trench dug near the dripline of the tree.

In sandy soils with an organic mulch cover to encourage shallow rooting, correct deficiencies by applying 2.5 kg of potassium per 100 m² (5 lb/1000 ft²) to a surface area that extends outward from the trunk for one and one-half times the radius of the plant canopy. In New York forests on sandy acid soils, White (1956) obtained response from conifers with only 1.2 kg/100 m² of potassium applied by air.

Fertilizer potassium comes from a number of sources. In areas where salinity (due to arid conditions or salting winter roads) is not a problem, potassium chloride (KCl), or muriate of potash, is generally used. Chloride, however, can be toxic, particularly in arid regions. Potassium sulfate (K_2SO_4) is a satisfactory source of potassium, particularly where salinity may be a problem. Potassium nitrate (KNO_3) is usually more expensive than other potassium sources and will apply an excess of

Treatment: POTASSIUM DEFICIENCY (Potassium is fairly immobile in soil.)

Application	Chemical Form	Amount to Apply	How to Apply	Years Effective
		kg/100 m² [a]		
Soil	Several (K)	2–8 on sandy loam	Incorporate[b]	5–10
		8–15 on clay loam	Incorporate[b]	5–10

[a] One kilogram per 100 m² is 1000 kg/ha (2 lb/1000 ft²).
[b] Apply on surface if roots are near the soil surface.

nitrogen if it is used to supply large amounts of potassium. At rates above 5 kg/100 m² (10 lb/1000 ft²), one application should last for five to 10 years.

Potassium deficiency is the most widespread and serious nutritional problem of Florida palms (Broschat and Meerow 1990); container-grown palms are also affected. In Florida soil, apply sulfur- or resin-coated potassium sulfate 1.5 to 4 kg (3–8 lb) per tree four times a year plus half as much magnesium sulfate to prevent magnesium deficiency. Affected fronds do not recover.

Wood ashes have been used as a source of potassium for a very long time. The term *potash* is derived from *pot ashes,* which were obtained by leaching wood ashes and evaporating the solution to dryness. Unleached hardwood ashes contain from 8 to 30 percent potassium, mostly in carbonate form (Branson 1980). Softwood ashes contain less potash, and ashes from trunk wood are poorer in potash than are those from twigs and small branches. Rain can leach about 90 percent of the potassium from ashes. Hardwood ashes also contain 15 to 25 percent calcium but less than 2 percent phosphorus. Potassium and calcium carbonates are quite alkaline: Although on acid soils they are beneficial for both pH adjustment and potassium, on alkaline soils they can raise the pH to critical levels (see Chapter 6).

Potassium is said to provide a number of benefits to plants. Bernatzky (1978) indicates that potassium increases root growth and drought resistance of trees in urban areas. In split-root experiments with conifers, however, roots did not respond to additions of potassium (Coutts and Philipson 1980). Potassium is also supposed to overcome succulence and brittleness (increase stem strength), hasten plant maturity, and enhance a tree's resistance to disease and cold (Murphy and Meyer 1969). I have been unable to find any experimental results to suggest that potassium affects these responses any more than other nutrients do. Again, Epstein (1972), Halfacre and Barden (1979), and Kramer and Kozlowski (1979) attribute no such responses to potassium in their reviews of the roles of plant nutrients.

All of these responses (except for the increased drought resistance and root growth) attributed to potassium are symptoms of low nitrogen levels. Early recommendations of fertilizers assessed quantities by total weight, not by the proportions of individual nutrients. By such a method of calculation, one nutrient increases in a fertilizer only when one or more of the other nutrients decrease. Because phosphorus and potassium are adequate for woody plants in most soils, the significant component in any fertilizer containing N, P, and K is nitrogen. As differing proportions of these three nutrients are used, varying amounts of nitrogen will be much more important to plant response than the presence or absence of potassium and phosphorus. Thus the responses attributed to increased amounts of potassium may in reality be due to the concomitant, and more significant, decrease in nitrogen.

Large amounts of potassium unfortunately reduce magnesium uptake, particularly in acid and sandy soils.

Complete Fertilizers. A *complete fertilizer* contains nitrogen (N), phosphorus (P), and potassium (K); sometimes other nutrients are added. Even though applications of complete fertilizers are still widely recommended, most soils contain phosphorus and potassium in sufficient amounts for fruit, forest, and landscape trees.

In fertilizer trials in Illinois (Neely, Himelick, and Crowley 1970), Ohio (Smith and Reisch 1975), and Florida (Dickey 1977), phosphorus and potassium fertilization neither increased growth nor improved the appearance of landscape trees and shrubs. In Tennessee, van de Werken (1981) obtained a response only to phosphorus on soils known to be phosphorus deficient. **It is the exception, not the rule, when woody plants respond to additions of phosphorus or potassium** (van de Werken 1984b).

Even when potassium is needed, a complete fertilizer will supply excessive amounts of nitrogen and phosphorus along with the required amounts of potassium. It is best, therefore, to apply phosphorus or potassium separately. Applications once every three to five years should be adequate even in soil originally deficient in one of these nutrients (Neely and Himelick 1987). In forest soils deficient in phosphorus or potassium, a single application of the deficient nutrient during or shortly after planting will usually produce sufficient growth (Pritchett 1979). Nitrogen, which is more soluble, must usually be added at least every two years if it is to continue to benefit plants. For certain shallow-rooted woody plants, a complete fertilizer may be most effective but ought to be applied only once every few years, with nitrogen applied in the other years.

Since complete fertilizers serve a genuine purpose for growing vegetables, herbaceous landscape plants, and turf, it is advisable to understand how they are labeled. In the United States and many other countries, a fertilizer container must have a nutrient analysis printed on its label. The analysis gives the percentage by weight of the three nutrients, always listed in the same order: nitrogen (N), phosphorus (P or P_2O_5), and potassium (K or K_2O). Percentages of the three nutrients are often prominently displayed on the label (as in 10–6–4 or 10–10–10), with the understanding that their order is N-P-K. Attempts have been made to standardize the elemental forms of all three nutrients, but the older oxide designations are still common. Be sure you know whether phosphorus and potassium recommendations and label analyses are given in terms of the elemental or oxide forms. Conversions can be made in this way:

1 gram* P = 2.294 grams P_2O_5	1 gram K = 1.205 grams K_2O
1 gram P_2O_5 = 0.436 grams P	1 gram K_2O = 0.830 grams K

*Other units of weight may be handled in the same manner.

Calcium. Most soils contain more than enough calcium for plant use. Some soils low in calcium become so acid that deficiencies of nutrients or excesses of chemicals other than calcium become more serious to plant health than the calcium deficiency alone. However, even when vegetative growth is normal, low calcium can result in fruit disorders such as bitter pit of apple. In soils that are about pH 7 (neutral), calcium represents 60 to 85 percent of the total exchange capacity (Chapman 1966a). In well-drained soils with an annual rainfall above 750 mm (30 in.), hydrogen and aluminum ions replace calcium and other bases. If the acidity becomes high enough, manganese, aluminum, copper, and other elements can reach toxic concentrations. Increasing sodium, on the other hand, displaces calcium, increases the soil pH, and reduces the solubility of remaining phosphorus, manganese, calcium, and, in certain cases, zinc, boron, and iron (Chapman 1966a). Soil structure deteriorates with high sodium content, leading to decreased water movement, permeability, and aeration. Low levels of calcium commonly occur in acid soils, in sandy soils where the annual rainfall is above 750 mm (30 in.), and in highly acid peat soils. Alkali or sodic (high sodium) soils are quite alkaline, so that any calcium present is extremely insoluble. Soils derived from serpentine rock are calcium deficient.

Serpentine soils, generally sterile and unproductive (Whittaker 1954), have been reported in isolated areas on all continents but South America. On these soils, plants are usually sparse or stunted and composed of many species restricted to the serpentine habitat. Molybdenum deficiency and heavy metal toxicities have been reported as the cause of serpentine infertility, but the basic cause is low calcium. In the United States, scattered serpentine outcroppings occur along the Appalachian Mountains from western Massachusetts to Georgia. The Pacific Coast states have even more extensive serpentine areas in their mountains and valleys.

Add lime to acid soils to increase calcium and pH (see Chapter 6). In alkaline soils, add gypsum or elemental sulfur to decrease pH and make more calcium avail-

Treatment: CALCIUM DEFICIENCY (Calcium is relatively immobile in soil.)

Application	Chemical Form	Amount to Apply	How to Apply	Years Effective
		kg/100 m²[a]		
Soil				
pH <6.0[b]				
pH >7.0	CaSO₄[c]	40–75 on sandy loam 75–150 on clay loam	Incorporate	5–10

[a] One kilogram per 100 m² is 1000 kg/ha (2 lb/1000 ft²).
[b] Apply lime to bring pH to about 6.0 (see Table 6–4).
[c] Twenty percent as much sulfur can also be used; it will act more slowly.

able. The amount of gypsum needed depends on the soil, its sodium and magnesium contents, and the change in pH desired. A simple soil test can provide this information (Chapman 1966a). Gypsum or sulfur should be worked to 100 or 150 mm (4–6 in.) below the soil surface. If heavy amounts are needed, apply them before you plant and incorporate the material to deeper levels.

Serpentine soils will require about the same amount of gypsum or lime to supply calcium, depending on the soil pH. It is important to incorporate material into the soil because movement of calcium can be slow, and root growth in soil that is extremely deficient in calcium is almost nil (Martin, Vlamis, and Stice 1953).

These drastic treatments should only be needed once if follow-up management maintains the new situation.

Magnesium. Although magnesium is sufficient in most soils, it is readily leached from sandy acid soils. In calcareous soils, it is tied up in relatively unavailable forms (Dickey 1977). Magnesium and potassium may be deficient in the same soils, and they can be antagonistic.

Magnesium deficiency in woody plants is often difficult to control. Raising the pH of acid soils to between 5.5 and 6.5 by liming (see Table 6–4) with dolomite limestone (magnesium and calcium carbonates) and supplementing the organic matter content will increase the availability of magnesium and its retention in the soil

(Dickey 1977). These treatments should be carried out before planting. Lowering the pH of calcareous soils by applying acidifying materials will also increase the availability of magnesium. Add magnesium sulphate to magnesium-deficient plants on acid soils in Florida in early spring and early summer, at rates of 5 to 10 kg per 100 m² (10–20 lb/1000 ft²). Substituting dolomite for up to half of the magnesium sulfate may improve response. Fertilizer should be spread evenly under the plant canopy and water used to soak it in; magnesium sulfate will move with the water into acid soil. In lawns, the material must be injected into the soil. For magnesium-deficient sandy soils, Dickey recommends that applications at this rate be repeated annually until symptoms disappear, then continued at a rate of 2 kg/100 m² (4 lb/1000 ft²) annually. For magnesium-deficient soils in California, Proebsting (1958) recommends magnesium sulfate for neutral or alkaline soils and dolomite for acid soils. Bernatzky (1978) cites Ruge as recommending that 2 percent MgO (about 1 kg $MgSO_4$/100 m² or 2 lb/1000 ft²) be regularly added to soils in Germany.

A foliage spray of magnesium sulfate alone or magnesium sulfate and calcium nitrate can be used to determine plant response to magnesium.

Palms in Florida are commonly magnesium deficient, especially the *Phoenix* species; date palms in California are also affected. Apply magnesium sulfate at 1 to 2 kg (2–4 lb) per tree four times a year plus resin-coated potassium sulfate at the same rate. Added potassium prevents an imbalance with magnesium. In Florida, higher rates may be needed for mature queen and Canary Island date palms.

Sulfur. Irrigation water, rainfall, decomposing organic matter, fertilizers, and some fungicides provide enough sulfur for normal plant growth in most soils. In the past, domesticated plants received sulfur from the decomposition of manure and other organic matter in the soil. Later, fertilizers began to contain inorganic materials such as ammonium sulfate, potassium sulfate, and superphosphates. Since the 1950s, the trend has been toward fertilizers that contain high concentrations of primary nutrients, but little or no sulfur. This trend could lead to sulfur deficiencies if it were not for the increased amounts of atmospheric sulfur dioxide (*acid rain*) produced by burning coal and oil. Maugh (1979) reports that Noggle of the Tennessee Valley Authority found that as much as 40 percent of a plant's accumulated sulfur had been absorbed directly as sulfur dioxide from the air. Compounds that are washed from the air by rain and settle as particulates supply more sulfur. Dickey

---

Deficiency Symptoms: SULFUR (Sulfur is mobile in plants.)

Broadleaf: Leaves are entirely pale yellow-green in both young and old plants; they are small on some species and exhibit other symptoms associated with nitrogen deficiency. *Shoots* are stunted (Eaton 1966; Harris, Paul, and Leiser 1977).

Conifer: Symptoms are similar to those associated with nitrogen deficiency; needle tips may be yellow, red, or mottled, particularly on older needles; necrosis may follow; needle retention is poor (after Powers 1979).

Treatment: SULFUR DEFICIENCY (Sulfur is somewhat immobile in soil.)

Application	Chemical Form	Amount to Apply	How to Apply	Years Effective
		kg/100 m²[a]		
Soil	$CaSO_4 \cdot 2H_2O$[b]	5–8 on sandy loam	Incorporate	5–7
		8–12 on clay loam	Incorporate	5–7

[a] One kilogram per 100 m² is 1000 kg/ha (2 lb/1000 ft²).
[b] Chemical treatment should not be needed if sulfur-containing materials are used to remedy other deficiencies or to control pests.

Correcting Nutrient Deficiencies

(1977) estimates that each centimeter of rain deposits 0.45 kg of sulfur per hectare (or 1 lb of sulfur per acre from 1 in. of rain). Although sulfur dioxide is a serious air pollutant that can damage plants and fish, at lower concentrations it may be an important source of sulfur for plant growth. In regions where sulfur is not added to the plant environment in one or more of the ways described, deficiency symptoms may appear.

Sulfur can be added by increasing the decomposition of organic matter or as part of a sulfate-containing fertilizer or other sulfur-containing compounds. Elemental sulfur is commonly applied to increase acidity in soil that is close to the neutral range (see Table 6-5) or as a corrective for alkaline soils (Proebsting 1958). Sulfur is readily available in gypsum, but it may take several weeks for microorganisms to oxidize elemental sulfur.

Micronutrients

Boron. Even though plants require only extremely small amounts of boron, that nutrient is deficient in many parts of the world. Boron excess, while not as widespread, is more serious than a deficiency, which can be easily corrected. Excess boron can be difficult to overcome.

Boron is a rather unusual nutrient because of the narrow range between deficiency and excess in a plant. For example, peach trees can be deficient at leaf concentrations less than 20 ppm, normal between 20 and 80 ppm, and in excess above 90 ppm (Bradford 1966). Roses growing in solution culture are deficient at leaf levels below 5 ppm and toxic at levels above 20 ppm. Most trees (primarily fruit) analyzed are deficient at leaf levels below 20 ppm, but toxic levels vary considerably with species, ranging from 80 ppm to 500 ppm. Woody perennials are more sensitive to boron excess than are most annuals, particularly field crops.

The boron content of soils is determined by geological origin and degree of weathering. Boron deficiencies occur most commonly in coarse sandy soil, in acid soils, soils derived from acid igneous rocks, soils from freshwater sediments, and alkaline soils, especially those containing free lime (Bradford 1966). As of 1968, Stone reported that all boron deficiencies in cultivated forests occurred in sandy or highly weathered soil. No native forest stands have been found to be boron deficient. In the United States, areas of known boron deficiency are located in the eastern third of the country and portions of the Pacific states. In Lake County, California, boron-deficient soils are within 10 km (6 miles) of soils with excess boron.

Boron deficiency can be aggravated by certain cultural practices. Irrigation water that is very low in boron and high in calcium will increase the need for boron. Symptoms of deficiency or excess can be intensified by drought. Adding lime to soils that are low in boron will inhibit boron uptake and utilization.

The correction of boron deficiency is relatively easy. Soil applications during summer and fall will usually produce growth response the following spring. Many species will respond if borax is sprayed on foliage during the early growing season. Borax sprays must be applied each year, while soil applications last several years. Sprays are usually more difficult in landscapes than are surface applications. Borax is fairly soluble and moves into the soil with rain or irrigation. Boron can be injected

Deficiency Symptoms: BORON (Boron is extremely immobile in plants.)

Broadleaf: Leaves are occasionally red, bronzed, or scorched, and young leaves are affected first. Leaves are small, thick, brittle, and, on some species, distorted. *Shoots* exhibit rosetting, discoloration, and dieback of new growth, which becomes zigzag, short, brushy, thick, and stiff. *Flowers* may be few. *Fruit* set is light and deformed, with cracked, necrotic, spotty, corky surfaces. Fruit may drop before it is mature (Bradford 1966; Harris, Paul, and Leiser 1977).

Conifer: Shoot tips are bent (J topping), and meristematic tissue of the main leader may split. Necrotic blotches are visible on magnified cross-sections of buds and cause the death of terminal and some lateral buds. Plants become more like shrubs than trees (after Powers 1979).

Treatment: BORON DEFICIENCY (Boron moves readily in soil.)

Application	Chemical Form	Amount to Apply	How to Apply	Years Effective
		$kg/100\ m^{2\,a}$		
Soil	Borax	0.2–0.5 on sandy loam	On surface	5–7
		0.5–1.0 on clay loam	On surface	5–7
		$kg/100\ liters$		
Spray	Boric acid[b]	0.125–0.250	Foliage	1

[a] One kilogram per 100 m^2 is 1000 kg/ha (2 lb/1000 ft^2).
[b] Boric acid is more soluble in cold water than is borax. One and one-half kg of borax equals about 1 kg of boric acid in boron content.

Toxicity Symptoms: BORON (see Fig. 20–6)

Broadleaf: Leaves are yellow along margins and tips, later turning brown or black; dark necrotic spots may appear interveinally and on undersides of the mid-rib and petiole. Symptoms worsen progressively from the tip to the base of shoots. *Shoots* exhibit dieback with swelling and cracking below lateral buds; shoots gum and have short internodes (Bradford 1966; Harris, Paul, and Leiser 1977).

Conifer: Shoot dieback is found in most pine (*Pinus patula* is an exception). The leaders of some pine may curl on drying. Resin flow is common in some pine species (Stone 1968).

---

into treetrunks, but soil application is much easier. **Care should be taken to avoid applications that are too large or too frequent because it is easy to turn a deficiency into an excess.**

Excess boron can be lethal to plants and is most likely to occur in soils that originate from marine sediments or geologically young deposits, or in parent material that is high in boron minerals or has developed in arid climates (Bradford 1966). Certain cultural practices can also increase boron toxicity: irrigating with water above 0.75 ppm boron, acidifying neutral or alkaline soils already high in boron, applying heavy amounts of boron fertilizers, and applying heavy amounts of potassium to high-boron soils.

High levels of boron may be reduced by leaching soil with water containing less than 0.5 ppm boron. The soil must be well drained (see Chapter 13). Some plants are more tolerant of high levels of boron than others (Table 12–1). The liberal application of nitrogen fertilizers, especially calcium nitrate, sometimes helps, as does moderate liming (Bradford 1966). Plants accumulate more boron when rates of transpiration are high. At the same boron level in the soil, a plant in a hot, dry

TABLE 12-1 BORON TOLERANCE IN SELECTED SPECIES OF WOODY PLANTS (François and Clark 1979)

Common name	Botanical name
Tolerant[a]	
Natal Plum	*Carissa grandiflora* 'Tuttlei'
Indian Hawthorn	*Raphiolepis indica* 'Enchantress'
Chinese Hibiscus	*Hibiscus rosa-sinensis*
Oleander	*Nerium oleander*
Japanese Boxwood	*Buxus microphylla* var. *japonica*
Bottlebrush	*Callistemon citrinus*
Ceniza	*Leucophyllum frutescens* 'Compactum'
Blue Dracaena	*Cordyline indivisa*
Semitolerant	
Brush Cherry	*Syzygium paniculatum*
Southern Yew	*Podocarpus macrophyllus* var. *Maki*
Oriental Arborvitae	*Platycladus orientalis*
Rosemary	*Rosmarinus officinalis*
Glossy Abelia	*Abelia* × *grandiflora*
Sensitive[b]	
Yellow Sage	*Lantana camara*
Juniper	*Juniperus chinensis* 'Armstrongii'
Chinese Holly	*Ilex cornuta* 'Burfordii'
Japanese Pittosporum	*Pittosporum tobira*
Spindle Tree	*Euonymus japonica* 'Grandifolia'
Pineapple Guava	*Feijoa sellowiana*
Wax-Leaf Privet	*Ligustrum japonicum*
Laurustinus	*Viburnum tinus* 'Robustum'
Thorny Elaeagnus	*Elaeagnus pungens* 'Fruitlandii'
Shiny Xylosma	*Xylosma congestum*
Photinia	*Photinia* × *fraseri*
Oregon Grape	*Mahonia aquifolium*

[a]Species are listed in order of decreasing tolerance. Tolerant species were affected little, if at all, by 7.5 mg of boron per liter of irrigation water.

[b]Sensitive species were severely damaged or killed by 7.5 mg of boron per liter and moderately damaged by 2.5 mg of boron per liter of irrigation water.

area may show toxicity symptoms, while the same species in a cool, humid area will be normal.

Chlorine. Although chlorine has been shown to be an essential nutrient, deficiencies of the element under field conditions for woody plants have not been reported. Chloride is highly soluble and widely distributed in soils and natural waters (Eaton 1966). If a chlorine deficiency occurs or is suspected, the equivalent of 0.25 kg of chloride per 100 m^2 (0.5 lb/1000 ft^2) can be included in a fertilizer mixture.

Chloride is much more likely to be toxic than deficient, particularly in irrigated arid regions, near seacoasts, and adjacent to roadways treated with salt during the winter. Symptoms include small leaves with marginal and tip scorch. Such leaves will yellow and drop early. Leaching is the most effective way to reduce the level of chloride and other salts in soil. The effects of excess chloride, sodium, boron, and total salts are discussed in Chapters 13 and 20.

Copper. Although not as common as boron or zinc deficiency, copper deficiency is fairly widespread on most continents. It is not likely to occur on soils that are sandy, organic, alkaline, or calcareous (Reuther and Labanauskas 1966). In the United States, old livestock corrals and Indian burial sites are commonly deficient in both copper and zinc; these elements are apparently tied up by the combination of certain organic matter and soil compaction. Copper deficiency can be aggravated by alkaline irrigation water and by nitrogen or phosphorus accumulation. The woody plants most likely to show deficiency symptoms are apple, apricot, camellia, citrus, jasmine, olive, peach, Japanese pittosporum, plum, wax-leaf privet, rose, and tung (Dickey 1977, Reuther and Labanauskas 1966).

Most soils deficient in copper respond readily to copper fertilization. The amount and frequency of fertilization should vary with soil texture, pH, and organic matter content. No additional copper should be added until and unless deficiency symptoms reappear. **Do not exceed recommended amounts; copper can become toxic at higher rates.** A 5 percent copper sulfate solution was one of the first sprays used to control weeds chemically (Reuther and Labanauskas 1966). In fine soil, where copper is slow to move into the root zone, you may have to drill or inject copper sulfate into soil (Proebsting 1958) or spray on foliage.

Copper-deficient plants usually respond quickly and satisfactorily to foliage sprays such as the Bordeaux mixture (copper sulfate and lime) (Reuther and Labanauskas 1966). In fact, many fruits and vegetables grow in copper-deficient soils without exhibiting deficiency symptoms because they have been sprayed with copper fungicides. Soil applications are easier and last longer than foliage sprays. Trunk injections are not recommended because copper is quite toxic to bark, cambium, and young sapwood. Proebsting (1958) reports that growing alfalfa in fruit orchards helps correct mild copper deficiency and greatly improves plant growth in severe cases. It is not known whether turf or other ground covers will improve copper availability.

Excess copper often occurs in soils derived from or influenced by copper ore deposits and those on which crops have been heavily fertilized with copper or sprayed with copper fungicides for many years (Reuther and Labanauskas 1966).

Treatment: COPPER DEFICIENCY (Copper mobility is greater in acid than in alkaline soil.)

Application	Chemical Form	Amount to Apply	How to Apply	Years Effective
		$kg/100\ m^{2\ a}$		
Soil	$CuSO_4 \cdot 5H_2O$	0.5–1.5[b] on sandy loam	Incorporate	5–7
		1.5–5.0 on clay loam	Incorporate	5–7
		$kg/100\ liters$		
Spray	$CuSO_4 \cdot 5H_2O$	0.4–0.8[c]	Foliage	1[d]

[a] One kilogram per 100 m^2 is 1000 kg/ha (2 lb/1000 ft^2).
[b] To reduce the possibility of copper toxicity in sandy soils, Reuther and Labanauskas (1966) recommend applying only 0.125 to 0.250 kg of copper sulfate per 100 m^2 (0.25–0.5 lb/1000 ft^2) annually for several years until the total of 0.5 to 1.5 kg/100 m^2 has been added.
[c] One-half kilogram of hydrated lime should supplement this amount.
[d] This is not a long-term solution, but will determine the plant's response to copper.

Copper excess commonly induces foliage symptoms similar to those associated with iron deficiency (Stone 1968). In solution cultures that receive excess copper, roots will be stunted and top growth noticeably reduced. A given level of copper will be most toxic in sandy soils of pH 5.0 or below.

Little can be done to reduce toxicity when an entire soil profile has excess copper. The soil will need to be replaced for planting pockets and a barrier created to prevent rooting into the surrounding toxic soil. If the excess copper is in the surface soil, as will result from heavy fertilization or spraying with copper, liming acid soils and spraying plants with iron chelate should reduce copper toxicity and encourage renewed rooting in surface soil (Reuther and Labanauskas 1966). If the soil is alkaline, large amounts of well-decomposed organic matter may be worked in to reduce copper toxicity. Replacing contaminated soil may be the only way to assure satisfactory plant growth, however.

Iron. **Iron deficiency is the most common micronutrient deficiency.** In many areas, its frequency is second only to that of nitrogen deficiency. Because symptoms occur most frequently in alkaline soils and those high in lime, the deficiency is sometimes called *lime-induced chlorosis*. Iron deficiency also occurs on particularly sensitive plants in acid soils. Plants growing in poorly drained soil or those high in salt often are chlorotic, especially young plants during cold, wet springs. Deficiency symptoms may be quite localized: A single limb may be affected, while others appear normal. Soil around new buildings made of block, brick, and particularly stucco is often contaminated with lime from mortar spilled from the walls or debris from foundations, interior plaster, or wallboard.

Iron deficiency usually occurs not because iron is lacking in soil, but because it is unavailable. The availability of soil iron to plants decreases rapidly as the soil pH rises above 7.0. In acid soils, an excess of heavy-metal nutrient elements (such as copper, zinc, or manganese) can also produce iron deficiency. The heavy use of copper fungicides has caused iron deficiency in acid sandy soil (Dickey 1977).

For more than 100 years, people have been trying to correct iron chlorosis with various iron compounds, soil amendments, and methods of application. These efforts have often been disappointing. Applying iron compounds to the soil or particularly to plants is usually only a short-term (1–3 year) solution. Longer-term solutions to correct iron deficiency include acidifying the soil, improving drainage, reducing high-salt content, or using plants tolerant of alkaline and high-calcium soils.

In Oklahoma, Whitcomb (1986) reduced or eliminated chlorosis of pin oaks growing on an alkaline (pH 7.9 to 8.2) heavy-clay soil by watering in surface-applied granular sulfur at rates of 30 to 50 kg per 100 m^2 (60–100 lbs/1000 ft^2), about half the rate suggested in Table 6–5. Even though soil acidity was markedly increased (pH 5.6 to 5.9) only in the surface 40 mm (1.5 in.), the availability of iron and manganese was increased within three to six months and in some cases, the nutrients were available for 10 years.

In Illinois, Messenger (1984) reported similar results on pin and white oaks and red maple by surface or subsoil application of granular sulfur or sulfuric acid to alkaline soil (pH 7.7 reduced to 4.0–6.2). Acid is hazardous to use, however.

Small planting areas can be acidified or more acid soil brought in. According to Dickey (1977), organic matter and organic mulch can reduce the possibility of iron deficiency. Dickey recommends that backfill soil low in available iron be amended with ample organic matter, probably for its acidifying effect.

Iron compounds can be applied on the soil surface, injected into the soil, sprayed on leaves and bark, and injected or placed in trunks and branches. Iron added as a simple salt is quickly made insoluble in alkaline soils, and little or none is available for plants. Heavy amounts of iron sulfate applied on the surface and worked into the soil have given inconsistent results. If iron sulfate or iron-sulfate solution is concentrated in holes or trenches, however, it will maintain localized zones of availability (Locke and Eck 1965). Dickey (1977) suggests that a mixture of three parts dusting sulfur and one part iron sulfate be applied once or twice a year at 5 kg per 100 m^2 (10 lb/1000 ft^2) and worked into the soil for 150 to 200 mm (6–8 in.) under the surface, if possible. Dickey has found this particularly successful for local areas that are alkaline or have been excessively limed, but not for extensive areas of calcareous coastal soil.

In the 1950s, the synthetic chelates improved the continued solubility of iron

Deficiency Symptoms: IRON (Iron is immobile in plants.)

Broadleaf: Young *leaves* are yellow with contrasting narrow green veins; older basal leaves remain darker green. Exposed leaves are bleached and will eventually exhibit apical or marginal scorch. Leaves may be small; symptoms will be severe in cold, wet springs. *Shoot* length is usually normal, but diameter will be small; twig dieback and defoliation will occur when the deficiency is severe. *Fruit* has poor color and some species, such as citrus, will drop fruit heavily (Wallihan 1966; Harris, Paul, and Leiser 1977).

Conifer: New growth will be very stunted and chlorotic; older needles and the lower crown will remain green. In seedlings, cotyledons remain green (after Powers 1979).

Treatment: IRON DEFICIENCY (Iron mobility in soil decreases with increasing pH.)

Application	Chemical Form	Amount to Apply	How to Apply	Years Effective
		$kg/100\ m^{2\,a}$		
Soil	Chelate[b]	0.2–0.5 on sandy loam	On surface	1–3
		0.5–1.0 on clay loam	On surface	1–3
	$FeSO_4 \cdot H_2O$[c]	12 on sandy loam	Incorporate	3–5
		18 on clay loam	Incorporate	3–5
		$kg/100\ liters$		
Spray	Chelate	0.12–0.2	Early foliage	1
	$FeSO_4 \cdot H_2O$	0.5	Early foliage	1
Implant and Injection	Ferric ammonium citrate or Chelates	Several products Follow directions	Early spring	2–3

[a] One kilogram per 100 m² is 1000 kg/ha (2 lb/1000 ft²).

[b] A number of iron chelates are available: for acid soil, FeEDTA (Sequestrene NAFE® or Versene Iron Chelate®); for neutral, alkaline, and most lime soils, FeHEEDTA (Perma Green Iron 135® or Versenol Iron Chelate®), FeDTPA (Chel 330®), or FeEDTAOH (Versenol®); and for difficult soils, FeEDDHA (Chel 138®) (Wallihan 1966; Dickey 1977). Manufacturer's directions should be followed for effective treatment.

[c] Apply one half as much during the growing season. For liquid incorporation in soil, dissolve 1 kg of ferrous sulfate in 10 liters (1 lb/gal) of water and inject into holes 250 to 300 mm (10–12 in.) deep and 750 mm (30 in.) apart around the dripline of the tree. In the dormant season, inject 5 liters in each hole (half that amount during the growing season). For upright trees, space the holes 600 mm (24 in.) apart. In a shrub area, apply 35 ml (2 oz) at sites 750 mm (30 in.) apart.

---

even in alkaline soils. Iron chelates are more mobile in soil and plants than are other forms of iron. The differences among the various iron chelates principally involve their relative stabilities at high pH values (Wallihan 1966). Iron chelates can be applied to the soil surface and watered in or, for more rapid response, can be dissolved and applied as a soil drench. For more concentrated iron sources in large trees, Wallihan suggests applying the solution at 10 to 50 spots under the crown and then irrigating to leach into the root zone. Neely (1976) recommends injecting the chelate solution into soil to a depth of 300 to 375 mm (12–15 in.). He obtained the best results by injecting chelate in the spring and found that treatments remained effective for two or more years.

Iron sulfate or chelate solutions can be sprayed on foliage. Response may be irregular among species and nonexistent in citrus and gardenia (Dickey 1977). Iron sprays have certain disadvantages, particularly for landscape plants, because

The sprays are difficult to apply
They stain brick and cement
They leave an undesirable residue on the foliage
Iron does not move to new foliage
Chelates can burn certain plants

Iron sprays have a place when symptoms must be corrected immediately or the user wants to determine whether chlorotic symptoms can be corrected by iron treatments.

When organic salts, such as iron citrate or iron tartrate, are placed in holes in trunks, they can correct iron deficiencies for up to three years (Bennett 1931). Gelatin capsules containing ferric ammonium citrate (sold as Medicaps®) or ferric citrate can be tapped into holes drilled into the trunk. The number of capsules and the amount of iron salt in each depends on tree size. Solutions of iron sulfate or one of the chelates can also be injected into holes bored into treetrunks.

Smith (1976) reports that trunk implants of ferric ammonium citrate are more effective in correcting chlorosis in Ohio pin oak than are either foliar sprays or soil applications of various compounds. In Illinois, Neely (1976) has also successfully treated pin oak with iron implants. Neither Smith nor Neely tried injections of iron solutions.

Nutrient implants and injections should only be used when other measures have failed. Such applications injure trees and are more expensive and time consuming. They may be warranted, however, for quick response until acidification or drainage takes effect.

When iron chlorosis symptoms are corrected, applications of iron can be discontinued until symptoms reappear. If soil acidification is part of the treatment, it can be continued at a more modest level or with the aid of acid-forming fertilizers (see Table 6–6).

Manganese.　Although it is deficient less often than iron, manganese is commonly deficient under similar conditions and in a wide variety of cultivated plants. Reports of manganese deficiency in forest trees are rare, but are quite common in shade trees and shelterbelts on calcareous soils and in orchards (Stone 1968). The solubility of manganese decreases as soil alkalinity increases, and it is not readily available to plants above pH 6.5 (Labanauskas 1966). At pH values above 6.5, manganese is converted from the manganous to the manganic form, which is less soluble. Manganese deficiency is more likely to occur in poorly drained soils that are high in organic matter. Calcareous soils are noted for their alkaline reaction, poor drainage, and their low availability of iron and manganese. Manganese deficiency can also occur in sandy, acid, mineral soils that are low in native manganese or have been heavily leached. Cultural practices that increase soil alkalinity—such as liming, irrigation (in most soils), and burning (in organic soils)—increase the possibility or severity of manganese deficiency (Labanauskas 1966). Manganese deficiency often occurs when soils are alternately well-drained and waterlogged. Deficiency symptoms are more likely to occur under drought conditions than when moisture is sufficient.

Labanauskas notes that apple, cherry, and citrus trees are especially sensitive to manganese deficiency. Dickey (1977) lists more than 30 woody species in Florida landscape plantings that have exhibited manganese deficiency. It is interesting to note that at least eight species of palm exhibit manganese deficiency symptoms, but no palms are thought to exhibit iron deficiency symptoms. Fruit trees may exhibit pronounced chlorotic leaf symptoms of manganese deficiency before there is any reduction in growth or fruiting (Boynton and Oberly 1966). More severe deficiencies cause necrotic leaf spotting and margins, twig dieback, and reduced growth and fruitfulness.

Mild leaf symptoms of manganese deficiency may not be worth correcting, because such correction is unlikely to increase growth or fruitfulness. To restore dark green foliage, however, corrective steps must be taken.

Acidification of alkaline and calcareous soils with sulfur or sulfuric acid will usually correct the deficiency, but the slow response and initial cost make it unattractive. The rates used by Whitcomb (1986) or Messenger (1984) might be more quick acting. **Soil acidification will improve availability of iron and manganese.**

Trunk injections, foliar sprays, and soil applications of manganese have corrected manganese deficiency in trees, primarily fruit trees, with varying success. In general, foliar applications of manganese sulfate have been more successful than soil applications (Labanauskas 1966), but soil application, where effective, is easier and lasts longer (Dickey 1977). Even so, Dickey had no success with soil applications of manganese sulfate to camphor trees, crape jasmine, and citrus on calcareous sandy soil. The landscape plants Dickey examined probably received minimum cultivation and had roots near the surface. Soil applications of manganese sulfate will be more effective for these plants than they would be in cultivated orchards with few or no roots in surface soil. Except in acid or very sandy soils, little or no manganese sulfate applied to the surface reaches the deeper roots of large woody plants. Manganese sulfate can be placed in holes or injected into the soil, however. Although it is more expensive, manganese chelate may be more successful where soil applications of manganese sulfate have failed.

Foliar sprays have been almost universally successful in correcting manganese deficiency symptoms, but they must be applied nearly every year. Lemon foliage has been injured by manganese sulfate sprays, while orange and grapefruit foliage was not (Parker and Southwick 1941). Parker and Southwick found that soda ash (anhydrous sodium carbonate) is preferable to hydrated lime for mixing with manganese sulfate because the resulting spray leaves less visible residue on foliage and fruit. Sprays of manganese chelate are also effective. A combination of foliage and soil application might provide both a short- and a long-term solution. Once manganese deficiency has been corrected, additional applications should be made only when symptoms reappear.

For palms, especially queen, royal, paurotis, and pygmy date palms, a broad-spectrum micronutrient spray is recommended plus 250 g urea per 100 l (2 lb/100 gal) of spray with a spreader sticker. Foliar sprays are not effective on *Phoenix dactylifera*.

Although trunk injection of manganese sulfate or chelate has been successful, it is time-consuming and causes wounds that are unsightly and can lead to decay. Trunk implants of manganese are also available and are easier to install than injections.

Manganese toxicity frequently causes a mottled chlorosis of leaves, and necrotic spots develop as the excess increases (Halfacre and Barden 1979). Manganese toxicity symptoms often resemble those associated with iron deficiency, and, in fact, the two minerals can be competitive. Of the fruit plants, apple, citrus, and pineapple are most seriously affected by manganese excess. Apples, particularly Red Delicious, develop corky measles.

Manganese toxicity can occur in strongly acid and poorly aerated soils (Labanauskas 1966). In conifers, toxicity may occur in high elevation fir sites with leached, exposed, sandy loam soil that is low in calcium and pH (Powers 1981b).

Sprays or soil amendments that acidify soils high in manganese will chemically reduce manganic manganese to the manganous form (which is much more soluble), thereby increasing toxicity. The effects of manganese toxicity can be reduced if lime is applied and drainage improved. Because the effects of toxicity appear to be more severe in young trees, if toxicity may be a problem, lime the soil to obtain a pH of 6.0 to 6.5 before new landscapes are planted (Boynton and Oberly 1966).

Molybdenum. Among the woody plants, molybdenum deficiency has been reported only in Chinese hibiscus and citrus in Florida (Dickey 1977), rose in Australia (Johnson 1966), and apple in New Zealand (Boynton and Oberly 1966). This deficiency is more common in vegetable and field crops.

Molybdenum deficiency has occurred in soils derived from a variety of parent materials. Its incidence depends more on soil formation and leaching. In contrast to most of the other micronutrients, molybdenum is less available at lower pH values. Deficiencies can occur in heavily leached, well-drained calcareous and serpentine-derived soils, highly podsolized soils, and old soils with extensive secondary mineral formation (Johnson 1966). Molybdenum deficiency commonly occurs in soils that

---

Deficiency Symptoms: MOLYBDENUM (Molybdenum is mobile in plants.)

Broadleaf: Leaves are similar in color to those deficient in nitrogen; they exhibit marginal scorching and rolling and reduced width (strapping). *Shoot* internodes are short when deficiency is severe. *Flowers* are few and small when deficiency is severe (Johnson 1966).

Conifer: No description is available.

Treatment: MOLYBDENUM DEFICIENCY (Molybdenum is mobile in soil.)

Application	Chemical Form	Amount to Apply	How to Apply	Years Effective
		$g/100\ m^{2\,a}$		
Soil pH 5.5[b] pH 5.5	$Na_2MoO_4 \cdot 2H_2O$ or $(NH_4)_2MoO_4 \cdot 2H_2O$	2–20[c]	On surface with other fertilizer[d]	3–7
		$g/100\ liters$		
Spray	Either of the above	10–100[c]	Foliage	1+

[a] One gram per 100 m² is 1000 g/ha (0.03 oz/1000 ft²).
[b] Apply lime to bring pH to 5.5 or above (see Table 6-4). If this does not work, apply molybdenum.
[c] Note that the amount specified is in grams not kg; excess molybdenum may poison forage.
[d] Molybdenum is usually added to phosphate fertilizer.

are extremely low in phosphorus and sulfur. Because of its role in the biochemistry of nitrogen fixation, molybdenum may be deficient when nitrogen-fixing species, such as acacia, alder, and locust, show symptoms of nitrogen deficiency (Johnson 1966).

Molybdenum deficiency in soils can be treated in one of three ways: soil application of sodium molybdate at low rates, usually in conjunction with superphosphate; foliage spray of sodium molybdate; or liming of acid soils to above pH 5.5 when total molybdenum is thought to be adequate but unavailable because of low pH (Johnson 1966). Dickey (1977) suggests affected plants be sprayed to runoff with 8 g sodium molybdate per 100 liters (1 oz/100 gal) of water and the soil drenched with the excess solution, up to a total (on plant and soil) of 10 to 20 g of sodium molybdenum per 100 m² (0.5–1 oz/1000 ft²). This treatment should be effective for at least one year and should be repeated when symptoms reappear.

Molybdenum excess or toxicity in plants rarely occurs in the landscape, but ruminant animals can be seriously affected by forage from soils that contain excess molybdenum (Johnson 1966). This danger must be kept in mind when fertilizers are applied or soils are limed.

Zinc. Zinc deficiency is fairly common among cultivated trees and large shrubs, and its effects on growth can be quite marked. However, when Viets (1966a) and Schütte (1966) published their papers on this subject, South Africa was the only country whose native vegetation had responded to applications of zinc (Stone 1968). The level of zinc available in surface soil is often double that in the soil below (Chapman 1966b). Grading that removes surface soil and other practices that reduce rooting in surface soil will aggravate zinc deficiency.

As with many other nutrients, zinc is more likely to be unavailable in the soil than low in total amount. As with several other micronutrients, **zinc compounds decrease in solubility as alkalinity increases.** It follows that zinc deficiency is more likely to occur in alkaline and calcareous soils like those of the western United States. Zinc availability may also be low in some organic soils, in clay soils with low silicon-to-magnesium ratios, and on the sites of old livestock corrals and Indian burial grounds (Chapman 1966b). Total zinc may be deficient in soils derived from granites and gneisses, and in acid, sandy soils that have been severely leached.

Zinc deficiency can be aggravated in high-phosphate soils, in those that have received heavy or prolonged applications of phosphate fertilizers, and in acid soils that have been limed. Zinc deficiency is more likely when topsoil is removed and when soil is cultivated frequently, an act that restricts rooting near the surface. Nitrogen fertilization has been said to increase zinc deficiency, but this may be due to an increase in pH or to cation effects rather than to nitrogen itself (Chapman 1966b). Sodium nitrate has been found to decrease zinc uptake, while ammonium nitrate and ammonium sulfate have increased it.

The symptoms of zinc deficiency were known long before the cause was determined in 1931. Disorders included: *rosette* (the formation of extremely short internodes) of pecan and walnut, *little-leaf* of peach and pear, *frenching* or *mottle* of citrus, and *bronzing* of tung. Dickey (1977) lists a number of Florida landscape plants commonly afflicted with zinc deficiency, including orange jessamine, loquat,

Broadleaf: Leaves are uniformly yellow, sometimes mottled with necrotic spots. Leaves are small (*little-leaf*), very narrow, and pointed; older leaves drop. *Shoots* of small diameter have tufts (*rosettes*) of leaves at their tips, which may die back. *Fruit* set is light with small, pointed, highly colored fruit (Harris, Paul, and Leiser 1977).

Conifer: Branches and needles are extremely stunted; foliage yellows. Trees lose all but their first- or second-year needles; terminals die back (Powers 1979).

Treatment: ZINC DEFICIENCY (Zinc mobility in soil decreases as pH increases.)

Application	Chemical Form	Amount to Apply	How to Apply	Years Effective
		kg/100 m²[a]		
Soil	Chelate	1 when soil pH <6.0	On surface	1–3
		2 when soil pH >6.0	On surface	1–3
		kg/100 liters		
Spray	Chelate	0.125–0.25[b]	Early foliage	1
	$ZnSO_4 \cdot 7H_2O$	1.25–6.0 during dormancy	Dormant twigs	1
		0.5–0.75 during growing season	Foliage[c]	1

[a] One kilogram per 100 m² is 1000 kg/ha (2 lb/1000 ft²).
[b] Also apply 0.125 liter of detergent.
[c] Also apply 0.4 kg of hydrated lime.

wax-leaf privet, surinam cherry, carambola, barbados cherry, and silk oak. Fruit trees are particularly susceptible to zinc deficiency; vegetable and field crops are less often affected (Halfacre and Barden 1979).

Depending on the plant species, soil, and climate, zinc deficiency can be corrected by spray and soil applications, by injections into treetrunks, by zinc-coated nails or pieces of galvanized iron driven into trunks and limbs of woody plants (Chapman 1966b), and by implants.

Soil application is usually preferred because of its ease and its longer-lasting effects. Surface applications of zinc sulfate are primarily successful in soils that are low in total zinc and have a pH lower than 6.0. At higher pH values, zinc sulfate can be more effectively applied in holes or injected into soil. Zinc chelate has been successful in a number of soils in which plants did not respond to zinc sulfate, but the variable fixation and movement of zinc in soils and the danger of toxicity have discouraged soil applications in some areas. In alkaline soils or when soil applications are ineffective, zinc sprays will correct deficiency symptoms in most plants; sweet cherry and walnut are two reported exceptions (Proebsting 1958). For many deciduous species, an annual dormant spray of zinc sulfate will control deficiency symptoms in all but young and vigorous plants. A foliage spray of zinc sulfate and lime, zinc oxide, or zinc chelate can be used on evergreen plants or when deficiency symptoms appear during the growing season. Zinc oxide, however, can injure thin-skinned fruit.

Metallic zinc points, or pieces of galvanized iron driven into the trunks and branches parallel to the grain, and implants will correct deficiency symptoms in most species for several years. This is the most satisfactory method for sweet cherry and walnut, and perhaps for other plants that do not respond to soil or foliage applications (Proebsting 1958). An area around each piece of metal will die, and if affected areas merge, the trunk or branch will be girdled. To prevent this, the metal pieces should be placed in a spiral at least 50 mm (2 in.) apart with 1 to 2 pieces per 100 mm (4-6 in.) of circumference. This placement is most effective when the trunk or branches are less than 250 mm (10 in.) in diameter. Zinc solutions injected into trunks or dry salts placed in bored holes in trunks have also been successful. Deficiency symptoms have been corrected for three years or more when one gram of zinc sulfate per hole is placed in holes about 100 mm (0.04 oz/4 in.) apart around the trunk. Because of the labor involved and the danger of decay, implants, metal, salt, or solutions should be placed in trunks only if soil or spray applications are ineffective or will cause toxicity or drift problems.

Alfalfa and cover crops can reduce or prevent zinc deficiency in fruit orchards (Chapman 1966b), and a build-up of organic matter on the surface will improve the growth and appearance of plants deficient in zinc. Both of these practices tend to accumulate zinc in surface soil. If soil is fumigated or sterilized before planting, zinc availability will frequently be improved.

Initial symptoms of zinc toxicity resemble those of iron deficiency and may, in fact, be a sign of iron deficiency, because the two nutrients compete with one another. Plants can be severely injured and killed by toxic levels of zinc in soil. This has occurred in some acid peat soils, soils contaminated by mine spoils or seepage, and soils derived from rock or materials high in zinc. Tree seedlings can be stunted and, in some cases, killed when they are grown in galvanized iron tubes (Chapman 1966b). Liming soil and applying phosphate fertilizer may help mitigate zinc toxicity.

Application Methods

Fertilizers can be applied in a variety of ways; they can be

> Broadcast on the soil surface
> Placed in holes in the soil
> Injected into the soil in solution under pressure
> Sprayed on foliage
> Injected into or placed in holes (implants) in treetrunks

The appropriate method will depend on the nutrients applied, equipment available, other plants in the area, nature and slope of the soil surface, and, in some cases, the species to be treated.

Surface Application. Surface application is the easiest and most effective method for applying nitrogen, most chelated micronutrients, and, where shallow roots are present, even phosphorus (Perry 1982, van de Werken 1984b) and potassium (White 1956). On a phosphorus-deficient soil, van de Werken compared two nitrogen and two complete (N-P-K) fertilizers applied either on the surface or in 450 mm (18 in.) deep holes. The trunks of five of the six landscape tree cultivars at the end of the eight-year experiment were 56 percent larger when a slow-release complete fertilizer was broadcast than when placed in holes in the soil. Except for the broadcast and subsurface applications of slow-release fertilizer, there were no consistent differences among the fertilizers or methods of application.

Lawn fertilizer spreaders or cyclone seeders can be used to apply nitrogen fertilizers and, when tree roots are shallow (as in turf, mulch, etc.), even phosphorus and potassium. For trees growing in a lawn, fertilizer should be applied in late fall or winter before the turf begins rapid growth. Grass blades should be free of moisture. After spreading the fertilizer, sprinkle the area thoroughly to wash fertilizer from the grass and into the soil. A second irrigation the following day will move the material further into the root zone, will minimize injury from high concentrations of fertilizer, and will reduce volatilization of nitrogen from urea and ammonium fertilizers.

Soil Incorporation. Phosphorus, potassium, and other nutrients of low solubility usually need to be incorporated in the soil only where tree roots are not near the surface, such as in bare or cultivated soil. Then they can be effectively applied by injection or placement in holes dug in the soil. Placing such nutrients in the root zone will make them more readily available. In addition, the holes will increase aeration and water penetration. In poorly drained silt and clay-loam soils in Ohio, Smith and Reisch (1975) found that young crabapple, linden, and maple trees benefited as much (they produced 20 percent more caliper growth) when holes were drilled 300 mm (1 ft) deep and no fertilizer added as they did when 3 kg (6 lb) each of nitrogen, phosphorus, and potassium per 100 m² (1000 ft²) were applied in holes or on the surface.

Holes can be cored or punched in soil with a bar (which may compact the soil) or drilled by hand with an auger, with an electric drill, or with the power takeoff on a tractor (Fig. 12-4). To minimize compaction around each hole, the moisture content of the soil should be below field capacity before holes are made. Drill holes 150 to 200 mm (6–8 in.) deep and 0.6 to 1 m (2–3 ft) apart, beginning away from the trunk to avoid injury to main roots and extending up to one-fourth of the radius beyond the dripline. The number of holes will range between 100 and 275 per 100 m² (110–250/1000 ft²) with such spacing. Apportion the fertilizer among the holes.

Another method of incorporating fertilizer is to dig a circular trench 150 to 200 mm (6–8 in.) deep at the dripline of the tree. Place the fertilizer in the trench and cover it with soil.

Fertilizer can be effectively injected into the soil with proper equipment (Fig. 12-5). Fertilizer materials must be soluble in water so they can be applied uniformly, even though the nutrients may later be adsorbed or reduced in solubility by the soil. Water-soluble fertilizers containing phosphorus, potassium, and other nutrients are

Figure 12-4 A two-man power auger is often used to drill holes to aerate the soil, to apply the more insoluble nutrients, and, more recently, to fracture tight soil (see Fig. 14-9). Its primary value is improving soil aeration. (Photo courtesy Michael Hutnick, Sta-Green Tree Service, Carmichael, CA)

Figure 12-5 Needle probes are used to inject into the soil those solutions of nutrients which are strongly adsorbed and do not move readily in the soil. Since most of a tree's absorbing roots are in the top 150 mm (6 in.) or so of the soil, injection of nutrients below the surface for most landscape trees is not as important as once thought. The holes created, however, improve soil aeration. The nutrient solution under pressure washes a hole for the probe to penetrate and forces solution laterally into the soil. (Photo courtesy Michael Hutnick, Sta-Green Tree Service, Carmichael, CA)

usually more expensive per unit of nutrient than are fertilizers not readily soluble in water. Ammonium phosphate, potassium phosphate, and potassium nitrate are water-soluble. Potassium chloride is water-soluble but should not be used where salinity may be a problem.

A high-pressure hydraulic sprayer is probably the most economical piece of equipment to use for injecting fertilizer, because it can be used for other maintenance practices. Fertilizer solutions are corrosive, however, and the equipment must

Correcting Nutrient Deficiencies

be cleaned thoroughly after being used for fertilizer application. Stainless steel or plastic-lined tanks reduce maintenance, but pump and lines must be cleaned carefully.

Because the fertilizer is in solution, the injections can be spaced somewhat farther apart than can holes for dry fertilizer. The probe should penetrate the soil about 200 mm (8 in.), applying 100 to 150 kg/cm^2 (150–200 lb/in.^{2}[psi]) of pressure to force the solution into the soil. Inject at least 800 liters (200 gal) of solution for each 100 m^2 (1000 ft^2) treated (5 liters or 1.2 gal/injection). The concentration of the solution will depend on the nutrients and the desired rate of application. Equipment that is supposed to pneumatically loosen soil can also force nutrients into the soil (see Chapter 14), but the nutrients may be placed below most of the absorbing roots.

Foliage Sprays. Sprays can quickly overcome deficiency symptoms involving micronutrients like iron, zinc, and manganese. Small-scale spray applications of a single micronutrient will quickly determine a plant's response; this can be a very useful preliminary to a longer-term corrective program (Fig. 12–6). A detergent spreader (a dishwashing detergent, for example) added at a rate of 60 to 120 ml/100 liters (0.5–1 pint/100 gal) will increase the effectiveness of sprays.

Figure 12–6 Applying foliar sprays of an individual nutrient may be a quick way to determine plant response. (Harris, Paul, and Leiser 1977)

Spraying to alleviate deficiency symptoms can be expensive unless included with pesticides. A number of micronutrients have been incorporated into the pesticide spray programs of many horticultural crops, but few landscape plants are sprayed so routinely. Spraying any material on large shrubs and trees is difficult in most landscapes and may involve damage to nontargeted plants, to other property, and to people. Sprays may leave unattractive whitish film or spots on foliage. Fertilizers and pesticides applied together must be compatible.

Trunk Implants and Injections. **Trunk implants and injections of nutrients should be used only when other methods prove to be ineffective or too difficult.** Boring the necessary holes, implanting or injecting fertilizer, and sealing holes takes a great deal of time, disfigures trunks, and can lead to decay around insertion sites. On the other hand, these methods may be the only effective way for plants to obtain adequate nutrition.

Fertilizer implants and injections are almost entirely restricted to micronutrients whose availability is reduced by soil conditions. Implants contain small amounts of a soluble nutrient in a gelatin capsule, and they are placed in holes drilled into trunks or branches. Holes should usually be 10 to 13 mm ($\frac{3}{8}$–$\frac{1}{2}$ in.) in diameter and deep enough to allow the implant and a plug to be placed entirely in the xylem. Shurtleff and Jacobsen (1983) recommend plugging the holes with grafting wax, putty, glazing compound, or asphalt but not wood dowels or corks. Place the plug deep enough so the cambium can form callus over the wound. Some commercially available capsules, such as Medicaps®, seal the hole when inserted properly, so no plug is needed. Follow the manufacturer's directions for the number of implants. If several implants are to be placed in a tree, they should spiral around the trunk to minimize injury and should start low on the trunk.

Trunk injection holes can usually be smaller than those necessary for implants. A hollow, tapered 6-mm ($\frac{1}{4}$-in.) wood screw with a quick coupler attached can be used effectively for injecting orchard and landscape trees with pesticides and nutrients (Reil 1979). Nutrient solutions can be injected at pressures of 7 to 14 kg/cm^2 (100–200 psi). Some species may gum severely after injections at more than 7 kg/cm^2, and pressures higher than 14 kg/cm^2 will not appreciably decrease injection time.

With regard to implanting iron in trees, Neely (1976) observes that although injection and implantation have consistently provided the most prompt and thorough correction of iron chlorosis, some authorities are reluctant to recommend the method. Objections center on the numerous holes that must be routinely drilled (every three to four years) into the trunk, sap leakage from holes, toxicity to cambium where salts are inserted, and toxicity to leaves when too much nutrient is applied at the wrong time to sensitive plants. In his experiments on pin oak, however, Neely (1976) observed no injury to treated trees regardless of the time of treatment, source of iron, or amount implanted. Neely reports that implantation holes were closed within one year.

Working with red maple in West Virginia, Shigo, Money, and Dodds (1977) found negligible amounts of discolored wood and cambial dieback (8 mm or 0.3 in.) above and below injection wounds made one year before; no nutrients had been injected into these holes. Similar holes that had received Mauget® injections of iron, magnesium, manganese, or zinc were more seriously affected. The cambium died back 20 mm (0.75 in.) above and below the holes injected with magnesium, while 50 to 60 mm (2–2.3 in.) of the cambium was killed around holes injected with any one of the other nutrients. New wood formed after treatment was healthy around all wounds. One year after red maple trees were injected with Bidrin, the dead cambium extended 350 mm (14 in.) above and below wounds; two years after injection, it extended only 110 mm (4.3 in.). The wounds apparently close fairly rapidly, but there is always the possibility of decay, especially in trees of low vigor. The researchers did not try implants or other fertilizer methods.

Time of Application

If needed, one annual application of nitrogen is sufficient in most soils. Late summer or early autumn is usually the most effective application time. Conditions are favorable for nutrient absorption and accumulation then: Top growth has ceased, the soil is warm, carbohydrate supply is high, the weather is cooling, and moisture is usually available. In regions of warm autumns and/or sandy soils, applications should be later than regions where early winters and/or clay soils prevail.

Numerous studies involving many species show that **early shoot growth (first four to six weeks) depends almost entirely on the level of dormant-stored nutrients, even though soil nutrients are abundant** (Kozlowski 1971). At moderate to high dormant-nutrient levels within a tree, 70 to 90 percent of the nutrients present in current shoots come from stored reserves (Meyer and Tukey 1965, Weinbaum, Merwin, and Muraoka 1978). Therefore, nitrogen applied later than about a month before growth begins seldom affects early season tree growth and may not be effective until the following year. The dormant-nutrient status of a tree is closely related to the soil-nutrient level the previous summer and autumn (Weinbaum and others 1984).

Even when deciduous oak and many conifers, whose growth is limited by the number of preformed initials in their overwintering buds, are fertilized in late summer or autumn, they may show improved leaf color the next season but little or no increase in shoot growth until the following year.

In sandy, well-drained soils, particularly in high rainfall areas or where irrigation must be heavy to control salt build-up, plants may respond best to split applications in late spring and autumn. Split applications or "slow-release" fertilizers can be used to encourage rapid growth of young trees. If leaching is not a problem, however, little or no benefit results from applying the same total amount of fertilizer in more than one application (Whitcomb 1987).

Fertilizing semitropical plants in USDA hardiness zones 8 and 9 (see inside front cover) in the autumn may increase their susceptibility to winter cold, but increased cold susceptibility of temperate-zone plants from autumn fertilization is unlikely (Whitcomb 1987). Heavy amounts of dormant-applied nitrogen or moderate rates during the growing season, however, can extend late-season growth making plants liable to injury by fall and winter cold.

Shigo (1989) recommends timing application to control the level of nitrogen (especially in a stressed tree) in relation to one or more of five phenological periods (periodic biological phenomena, such as shoot growth or wood formation). With our present knowledge, however, species, soils, fertilizers, weather, and their interactions are too complex for us to be able to time applications in order to vary the levels of nitrogen for specific phenological periods.

For example, in apple trees, two phenological periods overlap: Wood formation often begins before and continues later than shoot growth (Head 1968). Young prune trees absorb six to eight times as much nitrogen during rapid shoot growth as when dormant or during bud swell (Weinbaum, Merwin, and Muraoka 1978). On the other hand, euonymous plants absorbed 10 to 15 times as much nitrogen between growth flushes as during periods of rapid growth (Hershey and Paul 1983). Other species exhibit different nutrient-absorption/growth patterns. In addition, fertilizer formulations and various soil properties affect nutrient movement and availability. It is possible to have relatively high nitrogen levels in the spring and have them decrease during the growing season. I know of no study, however, which verifies the feasibility of being able to time fertilizer applications to have high or low levels of nitrogen at specific phenological periods of plant activity.

Shigo (1989) states that healthy, mature trees can be fertilized to have nitrogen and other elements available from the "onset of growth" until the beginning of "energy storage" (usually late summer). For most species, however, to have sufficient applied-nitrogen in a tree at those times, the applied-nitrogen would also have to be in the tree during much of the dormant period. **To have nitrogen available for growth in the spring, apply nitrogen fertilizers in late summer or autumn.**

For macronutrients other than nitrogen, unless the nutrient is deficient, the time of application is not as important for perennial plants. If needed, apply when convenient.

Boron and chelates of iron, manganese, and zinc, which are water soluble, should be applied to the soil in autumn. Spray application of these nutrients, however, should begin in spring, after the first few leaves attain full size. Additional sprays may be needed during the growing season, if deficiency symptoms reappear (refer to earlier discussions of the specific nutrient). Injections are most effectively administered after the first leaves have reached full size. Implants should be in place two to four weeks before growth begins or when deficiency symptoms are seen.

Fertilizing Young Plants

Recommendations for fertilizing newly planted trees and shrubs vary considerably: Some arborists suggest that a complete fertilizer be mixed in backfill soil; others advise no fertilizer during the first growing season. Experimental results vary with some trees responding to nitrogen fertilizer the first season; many not responding until the second season; and, in one case (van de Werken 1981), there was no response until the fourth season after planting. As already discussed, current shoot growth depends primarily on the nutrients in a tree before growth begins, so little or no growth response the first season might be expected. Weinbaum, Merwin, and Muraoka (1978) did find isotopic nitrogen in small amounts in the tops of young prune trees within 20 days after fertilizing two months before bud swell. If applied at or soon after planting, nitrogen should be absorbed to enhance photosynthesis, keep leaves on the tree longer, and be stored for growth the following spring. **Fertilizing soon after planting is good insurance.**

Nitrogen is the only nutrient needed in most situations and can be applied on the surface after planting more easily and with less likelihood of toxicity. Slow-release fertilizers can supply low concentrations of nutrients for two to 18 months. These fertilizers are particularly useful for plantings that are inconvenient or expensive to fertilize regularly, or for those in sandy soils. Slow-release fertilizers can be mixed with the backfill at planting according to the manufacturer's directions.

The quantity recommendations for nitrogen application also vary considerably: from 10 to 50 g (0.02–0.10 lb) of nitrogen in the planting hole or backfill soil, and from 10 to 75 g (0.02–0.15 lb) of nitrogen per surface application in the planting basin. In Florida, Meskimen (1970) made monthly surface applications of 80 g (0.18 lb) of nitrogen per square meter (as 6-6-6) to newly planted red gum in a fine sandy soil. In California, Harris (1966) made two applications of 125 g (0.25 lb) of nitrogen per square meter (as NH_4NO_3) to newly planted southern magnolia 'Saint Marys' and sawleaf zelkova in a silt loam soil. Both field experiments produced significant growth responses and no injury to trees.

Nitrogen (25–100 g or 0.05–0.20 lb) applied to an area one-meter square (3 ft × 3 ft) around each tree after planting will ensure an adequate supply. Do not apply close to the trunk. A second application in early summer may enhance growth the following spring.

Fertilizing Root-Damaged Mature Trees

Contrary to several recommendations, modest amounts of nitrogen (0.5 kg/100 m²; 1 lb/1000 ft²) applied to a newly transplanted tree or one with injured or pruned roots should benefit both root and top growth. It is thought that fertilizing with nitrogen, particularly having it available in the spring, will stimulate not only top growth at the expense of root growth and energy reserves but also disease organisms (Shigo 1989). As pointed out at the beginning of this chapter, most woody plants studied absorb nutrients primarily when roots are growing. Nitrogen has been shown to overcome moisture stress of unirrigated almond trees in California; root growth must have been adequate to supply moisture for the increased top growth (see the discussion of nitrogen and phosphorus). Stimulation of disease organisms should be minimized by applying the fertilizer only above uninjured roots during late autumn when microbe activity is low. Response to nitrogen may not occur the season of application, but nutrients will be available the following spring for root and top growth.

Fertilizing Palms

Palms can suffer quickly and conspicuously from improper nutrition, whether due to insufficient or incorrect fertilization, particularly in Florida and other areas with acid, sandy soils. Some nutritional problems in palms are difficult to diagnose because symptoms of different deficiencies may overlap. A high level of potassium or magnesium can cause a deficiency of the other nutrient. Iron deficiency symptoms seldom occur on palms in Florida unless the palms are growing in poorly aerated soil or are planted too deep.

Symptoms of and treatments for deficiencies of nitrogen, potassium, magnesium, and manganese have been described. Most of the nutrition information is from Henry Donselman (Rancho Santa Fe, CA, 1990 pers. comm.) and Broschat and Meerow (1990).

Estimating Fertilizer Costs

The cost of fertilizer nutrients is important not only as you budget for landscape maintenance but as you determine which fertilizer to buy. Fertilizers are available in a number of formulations and mixtures; they vary in the amount of nutrient as a percentage of weight (Table 12–2) and can vary greatly in price. Fertilizer is usually applied by area or individual plant. Be careful to consider what amounts of a nutrient are to be applied and how large an area or how many plants this amount will cover. The following equation may be helpful:

$$\frac{\text{Area to be}}{\text{fertilized}} \times \frac{\text{Rate of}}{\text{application}} \times \frac{100}{\frac{\text{Analysis of}}{\text{fertilizer}}} \times \frac{\text{Price of}}{\text{fertilizer}} = \text{Cost}$$

$$\text{Area} \times \frac{\text{Weight}}{\text{Area}} \times \frac{100}{\frac{\text{Nutrient analysis on}}{\text{a percentage basis}}} \times \frac{\text{Price}}{\frac{\text{Fertilizer}}{\text{weight}}} = \text{Cost}$$

Example 1: What is the cost of applying 2 kg of nitrogen per 100 m² to a grove of trees in an area 300 by 400 m? Ammonium nitrate ($33\frac{1}{3}$ percent nitrogen for ease of calculation) costs $9.00 per 40-kg bag.

$$300 \text{ m} \times 400 \text{ m} \times \frac{2 \text{ kg N}}{100 \text{ m}^2} \times \frac{100\% \text{ NH}_4\text{NO}_3}{33.3\% \text{ N}} \times \frac{\$9.00}{40 \text{ kg NH}_4\text{NO}_3} = \text{Cost}$$

$$120,000 \times \frac{2}{100} \times \frac{100}{33.3} \times \frac{\$9.00}{40} = \$1620.00$$

Example 2: What is the cost of applying 2 lbs N/1000 ft² (about 1 kg N/100 m²) to an area 3000 ft by 4000 ft? Ammonium nitrate ($33\frac{1}{3}$ percent nitrogen for ease of calculation) costs $9.00 per 80-lb bag.

$$3000 \text{ ft} \times 4000 \text{ ft} \times \frac{2 \text{ lb N}}{1000 \text{ ft}^2} \times \frac{100\% \text{ NH}_4\text{NO}_3}{33.3\% \text{ N}} \times \frac{\$9.00}{80 \text{ lb NH}_4\text{NO}_3} = \text{Cost}$$

$$12,000,000 \times \frac{2}{1000} \times \frac{100}{33.3} \times \frac{\$9.00}{80} = \$810$$

Note that all of the units except the price cancel out to give the cost.

Pollution Caused by Fertilizers

As concentrations of nitrogen and phosphorus increase, plant life, particularly algae, increase in water. Lakes with high concentrations of these nutrients may bloom

TABLE 12-2 ANALYSIS AND RELATIVE COST OF SOME NITROGEN-CONTAINING FERTILIZERS FOR WOODY PLANTS

Fertilizer	Analysis on percentage basis			Cost $	/	Weight kg (lb)	Cost per kg (lb) of N
	N	P_2O_5	K_2O				
Calcium nitrate	15.5			9.00	/	36 (80)[a]	1.61 (0.73)
Ammonium sulfate	21			5.30	/	22.5 (50)	1.12 (0.50)
Ammonium nitrate	33.3			13.95	/	36 (80)	1.16 (0.52)
Urea	46			14.00	/	36 (80)	0.84 (0.38)
Ammonium phosphate	16	20		8.95	/	22.5 (50)	2.49 (1.12)
Complete	6	20	20	11.25	/	22.5 (50)	8.33 (3.75)
Complete	15	15	15	17.65	/	36 (80)	3.27 (1.47)
Slow-release fertilizers							
Urea formaldehyde	38			36.50	/	22.5 (50)	4.27 (1.92)
Complete	14	14	14	56.00	/	22.5 (50)	17.78 (8.00)
Special fertilizers							
Spikes-trees and shrubs	16	8	8	14.98	/	3.2 (6.5)	29.26 (14.40)
Water soluble[b]	20	20	20	29.95	/	11.4 (25)	13.16 (5.99)

[a] The weight of a bag of fertilizer in lbs and the cost per lb of nitrogen are given in parentheses. These are 1990 retail costs from a farm supply store. Commercial and public accounts would get about a 20% discount; buying by the ton would save about 10%.

[b] These fertilizers should be diluted for application to the soil.

with algae in summer, resulting in low oxygen levels that may be fatal to fish. Swimming and boating under such conditions are unpleasant.

Nitrogen leaves soil and enters bodies of water primarily through leaching. Phosphorus loss is almost entirely due to surface soil erosion, though it can also be leached from sandy, coarse-textured soils.

The following measures, designed to minimize pollution, will also increase the efficiency of fertilization programs (Gowans 1970):

Do not over-irrigate; large amounts of water seeping below the root zone leach out nitrates

Take care to apply only the amount of nitrogen fertilizer needed; apply at optimum times and in such a manner that the fertilizer is most readily available to roots

Keep erosion to a minimum, particularly near streams and lakes

Avoid using phosphate to fertilize trees or large shrubs on sites that are susceptible to excessive leaching, water runoff, or erosion; if plants require phosphate for satisfactory growth, carefully control the amount used and the method of application to minimize transport of phosphate from the site.

Use plants that are suited to conditions in the area and will require only minimal fertilization

FURTHER READING

CALIFORNIA FERTILIZER ASSOCIATION. 1985. *Western Fertilizer Handbook,* 7th ed. Danville, IL: The Interstate.

CHAPMAN, H. D. 1966. *Diagnostic Criteria for Plants and Soils.* ed. H. D. Chapman. Riverside, CA: Chapman.

HAUSENBUILLER, R. L. 1985. *Soil Science: Principles and Practices,* 3rd ed. Dubuque, IA: W. C. Brown.

NEELY, D., E. B. HIMELICK, and W. R. CROWLEY, JR. 1970. Fertilization of Established Trees: A Report of Field Studies. *Ill. Natural History Survey Bull.,* 30(4):235–66.

SINGER, M. J., and D. N. MUNNS. 1987. *Soils: An Introduction.* New York: Macmillan.

TENNESSEE VALLEY AUTHORITY. 1968. *Forest Fertilization: Theory and Practice.* Symposium on Forest Fertilization. Muscle Shoals, AL: Tenn. Valley Authority.

TISDALE, S. L., and W. L. NELSON. 1975. *Soil Fertility and Fertilizers,* 3rd ed. New York: Macmillan.

CHAPTER 13

Water Management _____

Landscape plants probably suffer more from moisture-related problems than from any other cause. For them, it is either feast or famine, flood or drought, air or suffocation, acceptable or saline water. Life as we know it is not possible without water. Water is a primary constituent in the photosynthetic production of organic matter. It is the solvent for nutrient and food transport within plants. Transpiration cools plants. Roots extend into soil and shoot tips grow only by absorption of water and the turgor it provides. But although water is vital to plants, excessive water is often responsible for their decline and death.

As a source of water for plants, irrigation is more reliable than rain. Irrigation is no longer restricted to arid-summer regions. Even where summer rain is common, plants can be seriously injured by a short dry period. Supplemental water may be needed. Irrigation, however, has drawbacks with which horticulturists must reckon. Irrigation without adequate drainage can create aeration, disease, and salinity problems that are almost as serious as drought. It is essential to understand the relationship among soil, plant, and water in order to grow healthy, attractive landscape plants. The retention and movement of moisture in soil are discussed in Chapter 6. Some methods of improving soil-moisture conditions in the landscape are presented in Chapter 8. This chapter discusses factors that affect water use by plants: the timing and quantity of irrigation, soil salinity, drainage, antitranspirants, hydrophilic gels, and wetting agents.

WATER USE BY PLANTS

When managing irrigation, you may want to determine the rates at which water is lost by transpiration from plants and evaporation from the soil (these losses are called *evapotranspiration* or ET).

Environmental factors that influence ET are sunlight, temperature, humidity, and wind. The combination of factors that creates the lowest vapor pressure around leaves and the highest vapor pressure within leaves results in the highest ET. Intense solar radiation invariably produces higher temperatures, particularly in exposed leaves. High air temperature is often accompanied by low humidity, which tends to increase transpiration. Transpiration is usually higher on windy days, but it does not increase in direct proportion to wind velocity. The wilted, dull appearance of some leaves after strong winds is probably due to a combination of low humidity, high temperature, intense sunlight, and mechanical injury. These factors determine the rate at which plants will use water under different conditions and in different seasons. Of course, ET varies considerably from seacoast to desert (Fig. 13–1), but it can also vary greatly within a single landscape.

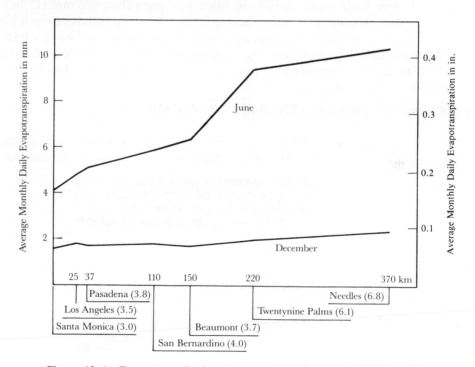

Figure 13–1 Evapotranspiration (grass surface) from the Pacific Coast (Santa Monica) to the California-Arizona border (Needles) for June and December. The "Average monthly daily ET" is the "average ET for each month expressed on a daily basis." The numbers in parentheses indicate the average annual daily ET in mm per day. (Data courtesy W.O. Pruitt, University of California, Davis)

Determining When and How Much to Irrigate

ET is usually expressed as water loss in mm or inches (similar to rainfall) per unit time. The area needs to be considered when you want to know the volume of water or if plants shade less than 60 percent of the ground. ET varies greatly during the year, particularly for deciduous plants. In summer, deciduous trees may use 40 percent more water per day than broadleaved evergreen trees of similar size. Annually, however, deciduous and evergreen trees use about the same amounts of water.

ET figures are usually based on a solid cover of vegetation (a closed canopy). The relationship of the ET of small trees and the percent of ground shaded by tree canopy to the ET of the same species in a closed canopy provides an estimate of ET for solitary plants (Table 13-1). For example, a solitary tree whose canopy shades 100 m² (1076 ft²) is surrounded by more than 500 m² of bare ground (< 20 percent of area is shaded). If the ET for the day and species were 5 mm (0.2 in.), the ET of the solitary tree would be 250 percent greater or 12.5 mm (0.5 in.) if confined to the shaded area. To replace the water transpired and to wet most of the absorbing roots, it would be wise to wet an area about twice the area shaded.

The estimated amount of water needed to replace the ET of the solitary mature tree in a 2 × 2 m (6 × 6 ft) planting basin in a paved parking lot is mind boggling. The tree basin would have to be filled with more than 300 mm (12 in.) of water daily, even more for smaller planting basins. Trees that survive in such paved areas must have access to other sources of water or can survive periods of minimum water availability without serious problems. Plants' ability to survive on minimum water is discussed in sections on Minimum Irrigation and Drought later in this chapter.

DETERMINING WHEN AND HOW MUCH TO IRRIGATE

Variations in the weather and length of day, adequacy of previous irrigation or rainfall, depth and spread of rooting, and size of top affect the moisture require-

TABLE 13-1 ESTIMATES OF THE INCREASE IN ET FOR A SOLITARY TREE OR SHRUB IN BARE SOIL IN MIDSUMMER IN RELATION TO ET FOR THE SAME SPECIES IN A CLOSED CANOPY (Based on Fig. 4.8, Fereres 1981)

Percent of ground area shaded	Percent increase in ET for an area equal to that shaded
< 20	250
30	220
40	200
50	180
60[a]	170
70	140
80	125

[a] For plantings with 60 percent or more of the ground shaded, ET is the same as that of a closed canopy; therefore, 100 percent ET for 60 percent of the area shaded would equal (100%/60%) 170 percent of ET for the area shaded.

ments of plants. If you know when and how much to irrigate, you will be able to balance plant performance and water use.

Observe the Plants

Most plants wilt noticeably when too little water is available. Leaves that were once shiny become dull, and bright-green leaves fade or turn gray-green. Leaves fall early and young leaves sometimes die. Plants must be watched carefully as the time nears for the first irrigation of the season, when the amount of moisture in the soil and the rate of use are difficult to assess. A few plants wilt before others because of their species or location and can be used as indicators.

Feel the Soil

With experience, you can roughly estimate moisture adequacy by the feel of the soil (Harris and Coppock 1977). The soil sample must be representative of the site. Collecting a representative sample with a soil probe (see Fig. 12–1) can be difficult and time-consuming, particularly when the roots of large shrubs and trees range far and deep. To estimate moisture adequacy, roll or squeeze a small sample of soil into a ball. If the soil will not mold into a ball, it is too dry to supply adequate water to plants. If the ball formed will not crumble when rubbed, the soil is too wet. If the soil can be molded into a ball that will crumble when rubbed, the moisture is probably about right. Sandy soils, however, will crumble even when wet.

Use Soil-Moisture Sensors

In both production agriculture and landscape maintenance, soil-moisture sensors are being used more widely to determine irrigation needs, particularly for turf in the landscape. *Tensiometers, electrical resistance blocks,* and *digital capacitance sensors* measure soil moisture in different ways.

Moisture availability depends on the tension with which water is held by the soil; this tension is measured in kiloPascals (kPa) (see Chapter 6). Even though the permanent wilting point is 1500 kPa and many plants can withstand lower tensions, most plants should be watered before soil in the root zone exceeds tensions of 70 kPa, in order to minimize moisture stress. With observation and some experimentation, irrigations can be scheduled according to soil-moisture tensions.

A *tensiometer* measures soil moisture by determining the tension with which moisture is held by a soil. It is a closed tube filled with water. A hollow ceramic tip is sealed to the bottom, and a gauge or device for measuring vacuum is attached to the other end (Fig. 13–2). The tube is placed in the soil so that the ceramic tip is located at the spot to be measured. As the soil dries, water is sucked out through the porous wall of the ceramic tip, creating a partial vacuum inside the tensiometer. This power of the soil (soil suction) to withdraw water from the tensiometer increases as the soil dries, raising the gauge reading. When soil is irrigated, soil suction

Figure 13–2 Two tensiometers, one (dial visible) installed in the soil with its ceramic tip at the depth soil moisture is to be measured, can be used to determine when to irrigate.

is reduced and water is drawn back into the tensiometer, lowering the gauge reading. In practice, the operation range of most tensiometers is from 0 to 85 kPa.

An *electrical-resistance block* (still called a *gypsum block*) evaluates soil moisture indirectly by measuring the electrical resistance between two electrodes embedded in a small block of gypsum, fiberglass, or nylon buried in the soil. The electrical resistance is measured by a meter and converted to soil-moisture–tension values.

The blocks operate more effectively in the drier range of soil moisture (>0.33 kPa). Therefore, blocks are more useful in medium- to heavy-textured soils, which tend to retain more available water as soil tension increases. Gypsum blocks slowly dissolve; a block's usual life of two to three years can be extended by adding a small amount of gypsum to the backfill soil. Irrometer, a Riverside, California, company, now markets an electrical-resistance sensor, the Watermark, said to register at lower tensions useful for a wider range of soil textures. The stated Watermark range is from 10 to 125 kPa.

The *capacitance sensor,* introduced in 1988, uses resistive capacitance (impedance [resistance] to the flow of an alternating current) to measure water in the soil. The electrodes of the sensor are thin metal plates embedded in a plane of Teflon® about the size of a credit card. The sensor and wires, to which it will be connected, are encased in inert material impervious to water and chemicals. The design of the sensor is said to minimize the effects of solutes in water and ensure that it responds only to changes in the water content of the soil. The sensor is reported to be sensitive to moisture levels from saturation to below permanent wilting.

The sensor is connected to a "moisture adjustor" at a controller. It does not give a tension or electrical reading; but by trial and error, the setting is adjusted to the observed soil moisture level at which you want irrigation to begin. Aquametrics, Inc., in San Diego, California, manufactures this sensor, and it looks promising.

Tensiometers and electrical resistance blocks can be attached to controllers to apply water at selected moisture tensions. The present capacitance sensor is adjusted relative to the observed moisture level wanted. Controllers can override time clocks so that if sufficient moisture is available, irrigation will be delayed 24 hours.

Estimate Rate of Evapotranspiration (ET)

The rate of evapotranspiration (ET) is increasingly being used to schedule irrigations and to indicate how much water to apply for a number of agricultural crops and turf. ET usually refers to potential ET (*Reference ET,* ET_o) which is considered equivalent to the ET of a large area of 100- to 180-mm (4- to 7-in.) tall cool-season grass with adequate water. ET_o is most commonly calculated using several weather parameters which affect ET. ET data are available from weather or extension agencies in most irrigated agricultural areas. Urban areas, where summer irrigation is common, in the future may be supplied ET information as water becomes more expensive and conservation more imperative.

The actual ET of a species varies from ET_o depending on the characteristics of the species and amount of foliage, particularly with deciduous plants. These relationships, called *crop coefficients* (K_c), have been determined for a few agricultural tree crops. Except for turf, crop coefficients are not available for landscape plants. In fact, the problem is even more complicated. In most landscapes, ET is influenced not only by plant species and sizes, but also by degrees of exposure to sun, reflection, and wind; and the range and sizes of different microclimates within a site.

For most landscapes, an educated estimate of ET may be most useful. As a first approximation, the ET of a landscape of trees, shrubs, and ground cover or turf might be 110 to 140 percent of ET_o during the middle and later parts of a growing season (Table 13–1). This approach is covered in the section on Establishing a Water-Management Program later in this chapter.

Measure Foliage Temperature

As soil moisture becomes limiting, transpiration slows, and leaves become warmer. The midday temperature of leaves exposed to the sun compared to the moisture content of the air (Idso and others 1981) or to the temperature of shaded leaves (Sachs, Kretchun, and Mock 1975) has been related to the availability of soil moisture. Hand-held infrared thermometers (Fig. 13–3) are used to sense radiant energy of foliage which is then related to temperature.

Differences of opinion prevail as to the accuracy with which the moisture status of one plant in a field or of the entire field can be determined from the radiant energy they emit. Many factors influence the measurement and the relationship of radiant energy emission, leaf temperature, and moisture status. Until more versatile equipment and procedures are available, infrared thermometers will have limited value in most landscapes.

Figure 13-3 By indicating temperature differences between exposed leaves and those not exposed to the sun, infrared sensing instruments may be useful in evaluating a plant's need for water.

APPLYING WATER

The method of application, the system design, and the skill of the irrigator determine the uniformity and the major costs of applying water to a given site. The way in which water is applied greatly affects its distribution; soil topography and the rate of application greatly affect water infiltration and runoff.

Water can be applied by basin (flood), furrow, sprinkler, and low-volume/high frequency (also called drip, mini-sprinkler, or soaker) systems. In most irrigated landscapes, water is applied by sprinkler, drip, or mini-sprinkler systems. Low-volume systems are commonly used in raised planters and basins in pavement. Additional information about basin, furrow, and soaker irrigation can be found elsewhere (Reuther and others 1981).

High-pressure sprinkler systems are most commonly used for turf and mixed plantings of trees, turf, and other landscape plants. The sprinkler heads either pop up when operating in turf and low-growing plantings or are on permanent risers above the ground to apply water to the tops of taller plantings. Drip emitters and mini-sprinklers apply water on the surface and are usually used with plantings that are not subject to much traffic.

Sprinkler Application

Properly designed, maintained, and operated sprinkler systems provide fairly uniform water distribution even on hilly terrain. The rate of application may need to be slow or interrupted on uneven areas to allow for penetration; otherwise, high areas will be underwatered and low areas overwatered. Most plants do well with overhead watering, which keeps humidity higher and washes off dust and insects. Sprinkler irrigation moistens mulches but does not wash or float them away. Sprinklers can reduce maximum air temperatures by 6° to 9°C (10°–15°F) (Hamilton 1978).

Uniformity of water sprinkling depends on the sprinkler system and its design, water pressure, and wind conditions. Penetration of water into the soil depends on application and infiltration rates, topography, and soil cover. Most sprinklers apply water faster than the soil can absorb it. Mulch, turf, and ground cover slow water movement on sloping surfaces so that more will penetrate. Sprinkler heads with slower application rates can be installed, but they may adversely affect the distribution pattern.

Sprinklers are best used in the early morning, when water pressure is usually greater, there is little or no wind, irrigation will not interfere with other activities, and foliage will dry more quickly. This last factor may be important for plants that are susceptible to water-related diseases. With sprinkler irrigation of full canopy plantings, time of day makes little difference in the total water loss by evaporation and transpiration under relatively calm conditions. Even on a hot, dry afternoon, due to foliage wetting, the increase in ET during sprinkler application to a large planting will be less than 3 percent (Heermann and Kohl 1980), because evaporation during sprinkling and from the wet foliage reduces transpiration by an almost equal amount.

Sprinkler irrigation can pack the surface of bare soil and reduce infiltration. Wind can result in uneven distribution. Water can damage flowers and knock them down or bend tall plants. Frequent light sprinklings with water with high salinity may cause an unsightly and even toxic build-up of salt on foliage; longer, less frequent applications usually correct or reduce this. Some believe that sprinkler application favors mildew and rust diseases on susceptible plants, but observations do not bear this out. Mildew and rust are particularly unlikely in arid regions, or when water is applied in early morning so that foliage dries quickly. **Trees susceptible to Phytophthora and other root rots will be in jeopardy if sprinklers wet their trunks.**

To determine the rate of application and the distribution pattern of a sprinkler system, distribute several cans evenly between two adjacent sprinkler heads in the same row and diagonally between two sprinklers in adjacent rows (Walker and Kah 1989). After running the sprinklers for a specified length of time under the pressure at which they will usually operate, compare the depth of water in different cans. Many systems in use today distribute more than twice as much water to some areas as to others, but the timing of irrigation is usually based on the area receiving the least water. Many inadequate sprinkler systems can be redesigned or heads can be adjusted or changed to improve distribution, plant performance, and water conservation. If a planting is irrigated by a movable sprinkler, variations in the distribution pattern can be minimized if the distances between sprinkler settings are equal to the radius of the area covered by the sprinkler.

Drip and Mini-Sprinkler Application

The goal of drip and mini-sprinkler irrigation is to apply water slowly and to wet 50 to 75 percent of the soil within the dripline or a diameter equal to the height of a plant, whichever is greater. Water is usually applied every two or three days. When plants are young or farther apart, water can be applied so that little of it wets soil without roots. The amount of soil wetted depends on soil characteristics, irrigation operation time, and the number of emitters or sprinklers used.

Drip emitters apply water more slowly and to a smaller area than do mini-sprinklers. Drip irrigation is better suited to smaller, low-growing, or widely spaced plants. Drip systems operate under lower pressures than sprinklers so smaller-diameter tubing can be used. The low pressure, however, necessitates better filtration (Fig. 13–4) or expensive emitters and more frequent cleaning (Fereres 1981).

The higher flow rates of mini-sprinklers overcome some of the problems of drip irrigation: Plants can be irrigated in a shorter time; a mini-sprinkler wets more surface so fewer mini-sprinklers are needed; and the larger orifices and higher pressures minimize clogging. However, larger pipe diameters are needed, and the greater area wetted may restrict activity longer after an irrigation.

Drip emitters, mini-sprinklers, or bubblers are often used to water planters and basins in pavement (Fig. 13–5).

Follow the manufacturer's instructions for the installation, operation, and maintenance of drip and mini-sprinkler systems. Drip emitters must be checked visually each week for correct flow. Manufacturers also describe how to clean the systems and control for bacteria and algae.

Low-volume irrigation systems only wet a portion of the root system, adding small amounts of water frequently. **There is little reserve water; should the system fail due to clogged emitters or filters, the plants would quickly become stressed.** Regular and proper maintenance is essential.

Concern has been expressed that wetting only a small volume of soil will limit root development and lead to instability of large trees. This does not seem to be a serious problem. If the entire root-system soil is wetted sometime during the year,

Figure 13–4 Underground installation of a control station for a low-volume irrigation system: (from left to right) back-flow valve, manual and electric control valves, filter (also inset), pressure regulator, and pressure gauge.

Figure 13–5 An irrigation bubbler (near corner) at each tree greatly reduces the labor needed to water plantings in paved areas. Light and electrical outlets are also present.

roots in soil not wetted by drip emitters should be able to grow and function enough to develop and maintain a satisfactory supporting root system.

Some Common Irrigation Problems

Table 13–2 summarizes solutions to some common irrigation problems.

ESTIMATING HOW MUCH WATER IS APPLIED

The amounts of water in the soil that evapotranspire or are applied as rainfall or irrigation are most easily expressed as the equivalent to the depth (in millimeters or inches) of an equal amount of water on the surface of the ground. That measure is useful even when the area involved is not known. The depth of rainfall and sprinkler application is measured easily. The amount applied by other means must be measured by volume (in liters or gallons), although some irrigation distribution systems have a meter to record the amount of water applied. You can estimate the rate of water application by measuring how long it takes to fill a container of known volume when the rate of application is similar to the irrigation rate. For a drip system, catch the flow of each of several emitters in a calibrated cylinder for one minute. The rate of discharge will vary with changes in water pressure and in the length and diameter of the distribution line beyond the point of measurement. The amount of water applied can be determined by the time and rate of discharge. The amount of

TABLE 13-2 SOLUTIONS TO COMMON IRRIGATION PROBLEMS

Problem	Solutions	
	Sprinkler systems	Drip and mini-sprinkler systems

1. Surface runoff	Decrease application by
	  a. replacing worn nozzles
	  b. reducing nozzle size
	  c. reducing pressure
	Decrease set time
	Repair system leaks
	Increase intake rate by
	  a. installing ground cover or turf
	  b. surface mulching
	  c. applying amendments
		Repair system leaks

2. Ponding at lower end	See (1) above	See (1) above

3. Uneven distribution	Check sprinklers for proper operation
	Repair leaks
	Check operating pressure for pump and system
	Shut down during high wind
	Change
	  a. sprinkler spacing
	  b. sprinkler head
	  c. nozzle size
		Check proper operating pressure
		Check for clogged emitters or sprinklers
		Change emitter or sprinkler spacing and/or location

4. Erosion	See (1) and (3) above	See (1) above
5. Saline water supply	Maintain high soil moisture content by irrigating more frequently Apply adequate water for leaching	
	Replace system with a more corrosion resistant material Irrigate at night to reduce leaf burn	
6. Salt deposits on plants	Irrigate at night to reduce leaf burn	Not applicable
7. Salt deposits on soil surface		Apply adequate water for leaching using other irrigation methods
8. High water table	Improve water management Improve subsurface drainage	
9. Uneven plant growth	See (2), (3), (7), and (8) above	
10. Distribution system losses	Repair pipeline leaks	Repair pipeline leaks

Adapted from Interagency Agricultural Information Task Force. n.d. *Drought Tips: Common Irrigation Problems: Some Solutions.*

water needed can be calculated by multiplying the area to be irrigated by the depth of water wanted. **Few people know how much water they apply or how uniformly.**

MANAGING WATER FOR YOUNG TREES

Plant health and vigorous growth are important so that young trees can quickly fulfill their landscape purposes. The amount of water that might be saved by being frugal is not worth the possible result of reduced growth. Experiments on young fruit trees summarized by Reuther and others (1981) show that frequent irrigations (3 to 7 vs. 12 to 16 days) and at lower moisture tensions (40 to 80 kPa) increased shoot and root growth markedly.

During the growing season, a tree in a nonturf area should receive rain and/ or irrigation within its dripline equal to twice ET_o. Young trees can be irrigated most efficiently by drip or mini-sprinkler systems. In areas where summer irrigation may be needed, basins are often used. A basin berm just outside the planting hole is fine for bare root trees and for B & B and container trees if the planting-hole soil is similar to that of the root ball. If the root ball soil is coarser than the backfill, it may be best to place a berm at the edge of the root ball and a second berm outside the backfill. Until roots grow into the backfill, the root ball can be watered more frequently than the area between the two berms. Less water will be used and the backfill soil will be less likely to remain too wet. The berms concentrate water around the tree which is good in low rainfall areas, but they should be broken before the rainy season or if summer rains are heavy. Berms around young trees are not necessary if summer rain will be adequate or the trees will be watered by a drip, mini-sprinkler, or sprinkler system.

For a tree in turf, keep the turf at least 0.3 m (1 ft) away from the trunk. Be sure the tree receives enough water if rain or the turf irrigation is the only source of water.

ESTABLISHING A WATER-MANAGEMENT PROGRAM

Control of water-application uniformity and amount in relation to evapotranspiration is the key to efficient, effective water management. In order to take full advantage of an irrigation system, it is recommended that

The plants that are watered by a single controller station have similar water requirements (a *hydrozone*)

The irrigation system be well designed and maintained in order to apply water uniformly and efficiently

Current ET_os be available in order to calculate the amount of water necessary to replace that evapotranspired

The application efficiency (AE) of the irrigation system be used to figure the amount of water to apply to replace that evapotranspired or the amount desired

Adjustments to ET_o

To obtain an estimate of the ET for a landscape planting (ET_L) irrigated by one controller or valve, the generalized or the current (real time) ET_o for the period involved would be adjusted (multiplied) by the

K_c (crop [species] coefficient) for the type of planting
K_h (hydrozone [microclimate] coefficient) for the planting and
K_d (density coefficient) depending on the planting density (Table 13–3)

The planting density can vary considerably and depends on the amount of ground surface covered and the vertical distribution of plant foliage.

When the accumulated ET_L reaches a predetermined level, the plants would be irrigated, or, if on a schedule, the accumulated ET_L would indicate the amount of water that needed to be replaced.

The K_c for turf or an evergreen planting does not vary much during the year. The K_c of a deciduous plant, on the other hand, begins from near zero at bud break and increases to a plateau as its canopy fills out. In late summer, its K_c decreases as water movement through leaves slows. A planting of trees and turf has a K_c less than the sum of the K_c's of each (Fig. 13–6).

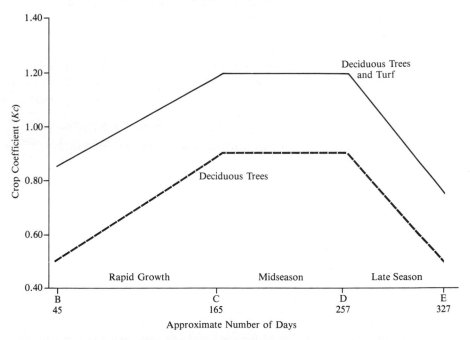

Figure 13–6 Relationships between crop coefficients (K_c) and growth and development periods in central California for two hypothetical landscape plantings of deciduous trees with and without turf. B: Trees begin to leaf out; C: Trees in full leaf; D: Leaf aging begins to reduce transpiration; E: Leaf fall (B is 45 days after January 1 in the northern hemisphere, July 1 in the southern hemisphere). An appropriate K_c can be selected from the graph to calculate when to irrigate. (Goldhamer and Snyder 1989)

Application Efficiency (AE).   In order to rewet the soil by the amount evapo-transpired, more water than that evapotranspired must be added. Few irrigation systems apply water uniformly. A well-designed sprinkler or drip system kept in good repair and operated properly will have an application efficiency between 75 and 90 percent (Goldhamer and Snyder 1989).

For a given landscape, the K_c, K_h, and K_d would be the same for the same period of the year. The coefficients could be multiplied together to give K_L for a specific landscape. To figure the amount of water to apply for the soil to be wetted to field capacity, divide the product of the accumulated ET_o and K_L by the AE of the system.

An example:

$$\text{Average } ET_o = 35 \text{ mm (1.4 in.)/week}$$
$$K_c = 0.8$$
$$K_h = 1.3$$
$$K_d = 0.9$$
$$\text{AE} = 75\% \ (0.75)$$
$$K_L = K_c \times K_h \times K_d$$
$$K_L = 0.8 \times 1.3 \times 0.9$$
$$K_L = 0.94$$
$$\text{Amount of water to apply} = ET_o \times K_L \ / \ \text{AE}$$
$$ET_o \times K_L \ / \ \text{AE} = 35 \text{ mm} \times 0.94 \ / \ 0.75 = 44 \text{ mm/week}^a$$
$$1.4 \text{ in.} \times 0.94 \ / \ 0.75 = 1.75 \text{ in./week}^a$$

[a] This can be converted to liters/100 m^2/week by multiplying depth in mm by 100 (or to gallons/1000 ft^2/week by multiplying depth in inches by 625).

Even though the K_c for deciduous trees increases in the spring as the trees refoliate (Fig. 13–6) and the K_h of a given landscape changes with the seasons, their values only need to be changed a few times during the year. AE remains much the same unless the system is repaired, develops leaks, or has other problems. Since these factors are difficult to assess accurately, field verification of soil and plant conditions must be part of setting up a program. A verification procedure is described to tell how well the program is working and what adjustments would be wise.

Calculating when and how much to irrigate a specific landscape can be simplified by preparing a table listing the K_L values for the different periods of the year. Then the accumulated ET_o would only have to be multiplied by K_L. This could be done by a hand-held calculator or with a computer by entering the code for the specific landscape, the date, and the accumulated ET_o.

Soil texture need be of little concern if the irrigation schedule is on a two- or three-day cycle. Turf is usually sprinkler irrigated twice a week and other plants, low-volume irrigated, three times a week. On such a frequent schedule, plants would be subjected to little stress (50 kPa of tension or less) even in hot weather (see Table 6–7). Interestingly, **coarse-textured soils have more water available below 50 kPa than do fine-textured soils.** From Table 6–7, it even appears that below 200 kPa, the loamy sand has as much water readily available as the clay. This would be true if water were withdrawn slowly, but under normal ET there would be such thin films of water in some sands that at even 50 pKa enough water might not flow fast enough to prevent water stress. In order **to have water equally available, a sandy soil usually would be watered at a lower moisture tension than a finer-textured soil.** Just what those tensions would be would have to be determined for each soil. It need not be a concern, however, when irrigating frequently.

Putting It All Together to Start a Management Program

Essentially, all that needs to be done is to irrigate to replace the equivalent of ET_L since the last irrigation.

1. Obtain current ET_o data or use historical average ET_L values for the same dates, if conditions appear normal
2. From Table 13–3, obtain the K_c, K_h, and K_d values for the type of planting

TABLE 13-3 ESTIMATED COEFFICIENTS FOR SPECIES (K_c), HYDROZONE (MICROCLIMATE) (K_h), AND PLANTING DENSITY (K_d), USED TO DETERMINE THE LANDSCAPE COEFFICIENT(K_L) FOR SELECTED VEGETATION TYPES (adapted from Costello, Matheny, and Clark 1991)

Vegetation type	Species factor (K_c)[a]			Hydrozone factor (K_h)[b]			Density factor (K_d)[c]		
	high	avg	low	high	avg	low	high	avg	low
Trees[d]	0.9	0.6	0.2	1.4	1.0	0.5	1.2	1.0	0.5
Shrubs	0.7	0.5	0.2	1.3	1.0	0.5	1.1	1.0	0.5
Groundcovers	0.7	0.5	0.2	1.2	1.0	0.5	1.1	1.0	0.5
Mixed[e] (trees, shrubs, and groundcovers)	1.0	0.6	0.2	1.4	1.0	0.5	1.3	1.1	0.6
Turfgrass	0.8	0.7	0.6	1.2	1.0	0.8	1.4	1.0	0.6

[a] K_c estimates for trees are based on research from agricultural tree crops, while ground cover estimates are based on preliminary research.

The high K_c values are for high-water-use species (cherry, birch, alder, etc., and include cool-season turfgrasses). The low values are for water-conserving species (olive, oleander, juniper, etc., and include warm-season grasses).

[b] K_h considers the microclimate. A high value would be for sites with full exposure to sun and wind and surrounded by heat-absorbing and reflecting surfaces. Low values would be for sites shaded much of the day, protected from wind, with low-reflective and radiating surfaces. Average values would be for sites in full sun.

[c] K_d estimates are based on 1.0 being equal to 60 percent or more of nonbare ground shaded or covered with plants (see Table 13-1). The high values are for dense foliage and/or a mixed planting of tall and/or tiered foliage (for example, trees and shrubs). The low values are for less than 60 percent ground coverage with no bare soil exposed (mulch or pavement) in the planting. If bare soil is exposed, K_d should be increased 10 to 20 percent, especially for trees and shrubs, due to soil surface evaporation.

[d] The tree, shrub, and groundcover categories listed are for landscapes that are composed solely or primarily of one of the three vegetation types.

[e] Mixed plantings are composed of two or all three vegetation types; no single type predominates.

3. From tests or estimates, determine water application efficiency (Walker and Kah 1989); the AE of properly designed and adjusted sprinkler and drip systems ranges from 0.75 to 0.9 (Goldhamer and Snyder 1989)
4. Multiply the four figures in steps 1 and 2 together and divide by the AE to obtain the amount of water to apply to return the soil to field capacity or the level desired
5. Apply that amount of water according to the regular irrigation schedule
6. As the next irrigation approaches, observe the plants carefully, particularly those most likely to become stressed, for signs of wilting

The first year or two, it is wise to check each time the coefficients change whether or not the program is delivering the amount of water desired. The best way to do this is to use moisture sensors in a few representative locations (at least one should be near the most sensitive plants or driest area) within each of the different types of hydrozones. At each selected location, place one sensor at 0.5 m (18 in.), a second at 1 m (3 ft), and if turf is present, a third 0.15 m (6 in.) below the surface. Be sure the sensors are in good contact with the soil.

Moisture Tensions Before Irrigating. The upper-sensor readings may be at or slightly above the tension desired without cause for alarm as long as the lower-sensor readings remain below the tension desired (more than enough water). If one or more sensor readings are much above the tension desired, either not enough water is being applied or the interval between irrigations is too long. On the other hand, if the tensions are much below the desired level, too much water is being applied.

The sensor readings should be recorded along with ET_o data each time readings are made so that moisture-level trends can be spotted before problems occur and program schedules adjusted for the next year.

Moisture Tensions After Irrigating. The upper-sensor readings should be less than 20 kPa of tension indicating that the water added had penetrated that far. If lower-sensor readings have not decreased to 20 kPa or less by the next day, more water may need to be applied at the next irrigation or irrigation may need to occur sooner than scheduled.

By comparing accumulated ET_o data with depths of moisture penetration before and after each irrigation along with the amount of water applied, more accurate values for K_L can be obtained. The depth to which water has penetrated the soil can be easily determined using a meter-long, 3- to 4-mm diameter (40-in., $\frac{1}{8}$ to $\frac{5}{32}$ in.) spring steel rod, sharpened on one end and bent into a handle at the other.

Adjust Water Use. If the plants are growing vigorously, you may want to reduce water use to conserve water, save money, and possibly improve plant well-being. To do so, reduce the amount of water applied by 15 to 20 percent of the amount previously applied but still irrigate on the same schedule. Check the sensors to see that the before-irrigating readings stabilize about where you want them— possibly about 100 kPa instead of 60 kPa. Remember that the permanent wilting point (PWP) is 1500 kPa. Be sure to observe the most sensitive plants for wilting. With such observations, the amount of water and the irrigation schedule can be fine-tuned.

By the end of the first year, the amount of water to apply at each irrigation should be able to be determined. K_L should have been fine-tuned to more accurately adjust future accumulated ET_o readings. From then on, adjusted accumulated ET_o values can be used to determine when to irrigate, if not on a schedule, and how much water to apply.

Additional Comments

Rain. Accumulative ET_o values should be adjusted for rain by applying less water or increasing the time to the next irrigation. Less than 5 mm (0.2 in.) of rain will do little more than settle the dust. On the other hand, more than 20 mm (0.75 in.) of rain in a short time may result in runoff. A reasonable estimate of the influence of rain should be made and subtracted from the amount of water to be applied at the next irrigation.

Sensor Override. Moisture sensors can be integrated into an automated-controller system so that if there is adequate water in the soil, the controller will delay activating a scheduled irrigation for 24 hours. Such setups have improved plant health and significantly reduced water use (Morgan 1965).

Adjust Sprinklers. Areas within a hydrozone may need more or less water than the hydrozone as a whole. Individual sprinklers or emitters can be adjusted, changed, or installed to solve the problem. Removing the most sensitive plants and those susceptible to problems associated with frequent irrigation may be wise.

It Will Be Worth Doing. Establishing an efficient, workable irrigation program based on ET may appear tedious and complicated. It will take time and skill, but **the benefits in balancing plant well-being and water conservation should be well worth the effort.**

MINIMUM IRRIGATION

As the demand for and the cost of water increases, interest is turning to landscapes that use less water. Since 1981, such landscapes have often been called *xeriscapes*. They are developed through wise plant selection, planning and design, and maintenance. They can include those species that have adapted so that they transpire less than others or species that can survive periods of drought because they have extensive root systems but transpire freely when water is available. Many people incorrectly think that both of these types of species always use less water.

Landscapes can be designed to have small focal areas that may require considerable water, with the rest requiring little or no supplemental watering. Mulching, weed control, efficient irrigation, and other practices can help conserve water.

Many woody plants, even some fruit and nut species, function satisfactorily when supplied with less water than would be transpired under adequate soil-moisture conditions. In Arizona, irrigation of mature pecan trees was scheduled by infrared sensing so that the average amounts of water were 100, 70, and 55 percent of the amounts that gave maximum yields. For the first three years of the experiment, the yields for the trees receiving the least water were reduced 8 to 16 percent and trunk growth 23 to 37 percent (Garrot and others 1990). In spite of the marked reduction in tree growth, Donald Garrot (Tucson, AZ, 1990 pers. comm.) reported that the shoots and leaves were smaller but otherwise the trees appeared normal.

By scheduling irrigations in relation to tree and fruit growth, peaches in California have been grown with no loss in yield, fruit size, or return bloom even though the amount of water applied was 33 percent less than ET_c (Goldhamer 1989). Shoot growth and tree size were reduced, but the trees appeared normal.

On deep California soils near San Jose and Santa Ana, one irrigation in midsummer adequately maintained the appearance of six out of eight mature shrub and ground-cover species (Sachs, Kretchun, and Mock 1975). *Cotoneaster punnosa* and mirror plant required two irrigations to maintain health and appearance. English ivy, two ice-plant species, and shiny xylosma maintained adequate or excellent appearance with no irrigation. Even though moisture in the top six feet (2 m) of soil was at the permanent wilting point for several weeks and leaf temperatures were 6°–15° C (40°–59° F) above ambient, few species were seriously injured. Growth was reduced by the less frequent irrigations, but appearance was quite acceptable for most species (Fig. 13–7).

Figure 13–7 The growth of established Monterey pines was influenced by the frequency of irrigations during the growing season at San Jose, California. All pines were sheared to a height of about 1200 mm (4 ft) each winter. Tree heights at the end of a growing season reflect the irrigation frequency: no irrigations (left), bimonthly irrigations (center), and biweekly irrigations (right). (Photos courtesy Roy Sachs, University of California)

Goldhamer (1989) points to two adaptations that many plants are able to make to better survive during drought periods. As soil water becomes limiting, solutes within cells, including leaf cells, become more concentrated which reduces transpiration but also delays stomatal closure (this is called *osmoregulation*). Photosynthesis can be carried on at the same time that transpiration is decreased.

Deep-water absorption is minor under normal soil-moisture conditions, since trees preferentially absorb water from the upper layers of soil. Only after the upper soil is largely depleted does absorption increase from the deeper soil.

Most landscapes in California were in satisfactory condition after two years of drought (1975–1977), when winter rainfall was less than the winter ET and only half the normal amounts of water were available for summer irrigation. Many trees and landscapes had no supplemental water. Some plantings did not show stress until two years after normal watering was resumed. It is obvious that the amount of water applied to most irrigated landscapes could be reduced to the benefit of plants and owners.

Trees of 39 species in southern California that had been irrigated in home landscapes for at least 25 years showed little or no effect from eight years without irrigation (Hodel 1986). The trees are in a proposed freeway site that has yet to be built. All structures were demolished. The area has warm, virtually rainless summers and relatively cool, moist winters. The rainfall average for the previous 10 years was 450 mm (18 in.). Several years were "exceedingly wet," but three of the winters were the driest on record.

In general, the trees were slightly smaller and more compact than would be expected of irrigated trees of the same age. Damage, if any, was insignificant and the trees were aesthetically acceptable. Of the 39 species, there were 21 broadleaved evergreens, 6 deciduous, 8 conifers, and 4 palms.

Research and/or experience should eventually provide easier ways to assess soil moisture reserves and how plants respond to moisture stress at different stages of growth. It may then be possible to provide plants with enough water to make modest spring growth before becoming stressed enough to check shoot growth. Deficit irrigation for the rest of the summer will keep the plants under moderate stress. Water use should be able to be reduced 30 percent or more below ET for many woody plants with little or no adverse effects on plant appearance or performance.

Several factors should be considered in minimum irrigation situations.

The transition from a regime of adequate soil moisture (and even excess or too frequent watering) to minimum irrigation may require care to make sure that plants are not severely injured because they have an inadequate root system or moisture reservoir.

Some plants tolerate infrequent irrigations because they have extensive root systems. In drought years, the water reservoir may be low, thus reducing a plant's supply of water. It may be necessary to apply more water than normal to keep the plant alive.

Minimum irrigation regimes are designed for maintaining plants, not for promoting vigorous growth. New plantings require frequent irrigation because of their limited root systems. Even after they become established, the plants usually need frequent watering until they reach optimal size.

Under saline conditions, irrigations must be more frequent to avoid salt toxicity.

Plants differ in their water requirements, so irrigation schedules may have to differ accordingly. In mixed plantings, irrigation schedules should be determined according to the needs of species that are the least tolerant of heat or desiccation. Alternatively, plants with similar water requirements can be grouped in the same areas, so that each group can be watered as needed.

Minimum irrigation regimes are not restricted to so-called "drought-tolerant" species; most plants can survive on less water than they receive in the landscape. A minimum irrigation regime merits serious consideration in the design and maintenance of landscape plantings, particularly areas such as freeways, regional parks, and open spaces.

DROUGHT

Summer-Rain Regions

Droughts occasionally occur in regions where summer rain is usual. When a drought occurs, some plants are already stressed before the problem is realized. Although plants may not die, they may be severely weakened and succumb two or three years later from a secondary infection, infestation, or stress (Holmes 1986).

Many soils in summer-rainfall regions are shallow or poorly drained so that few plants root deeply. In deep well-drained soils, however, plants root deeper even though moisture in the surface soil is usually adequate. Under low soil moisture, such plants are able to exploit moisture from depths not available to most plants. It is wise to know how deep your plants root.

Arid-Summer Regions

Lack of winter and spring rains and a shortage of water for summer irrigating are the ingredients for droughts in regions having Mediterranean climates. The seriousness of the situation can be fairly well assessed and treated several weeks before most plants are affected unless the plants go into a dry winter with low soil-moisture reserves; then the situation may become serious.

Managing Landscapes During a Drought

Wisely watering the more valuable plants during a drought can mean the difference between survival and death. In order to have the best chance of success,

Identify the most important plants; drought-sensitive plants, turf, and those weakened from other causes may be abandoned, removed, or killed. You may not be able to save them all.

Control weeds and apply mulch around plants.

For priority plants, estimate their effective rooting volume and the amount of available water (AW) in the soil. The effect of the drought can be appraised more realistically.

Film-forming antitranspirants have reduced transpiration of some trees (see the section on Antitranspirants later in this chapter).

If water is available for landscape use, wait to apply it until the first wilting is observed. Depending on the extent of plantings, water only those plants beginning to show stress. Apply about 50 mm (2 in.) of water to half the area within the dripline of a plant (about 500 l/m^2 or 1.25 gal/ft^2). This much water should last 7 to 10 sunny, summer days.

Household wastewater can also be used. If the water is softened, its use would reduce soil permeability and cause sodium-sensitive plants to burn. These hazards can be minimized by adding gypsum (calcium sulfate) to the water. The easiest way is to put 100 ml (0.5 cup) of gypsum in a bucket in which wastewater is accumulated. When about full, stir the gypsum and water, let the gypsum settle, and then water with the liquid. Enough gypsum will have dissolved to reduce the sodium problem and enough will be left over for other buckets of wastewater. It takes less than 15 ml per 10 l of water (1 tsp/2 gal) (Branson 1977). Long-term use increases salinity.

Pruning is often suggested to reduce water use by woody plants. At least half of the leaves would have to be removed to have much effect and then trunk and branches might be exposed to sunburn injury. From the information to date, pruning would be a last resort.

SOIL SALINITY

Soil salinity is a problem in many arid and coastal regions, where annual rainfall is commonly less than 750 mm (30 in.). Soils contain a mixture of salts. A white crust on the soil surface is usually a mixture of sodium, calcium, and magnesium salts. Most moist, dark, oily spots on the soil surface indicate an excess of calcium chloride. These salts dissolve easily in water, thereby increasing salinity. Other salts, such as lime and gypsum, are only slightly soluble and do not appreciably increase soil salinity.

Excess salt can also be a problem along roads and streets where sodium or calcium chloride is used to melt ice in winter (Carpenter 1970). Sodium chloride has been shown to be more toxic than calcium chloride, whether applied to the soil or sprayed on plants. When salts are applied to roads, the salt in the spray generated by traffic may be more damaging to plants than the salt that drains into the soil. Plants that are the most tolerant to soil salinity may not necessarily be the most tolerant to salt spray or mist on foliage, and vice versa (see Chapter 20).

Saline soils are not desirable for several reasons. Salinity decreases growth of sensitive plants, kills leaf margins, and may cause death. Sodium and chloride can cause specific injury to certain plants, including leaf burn, leaf drop, and stem dieback. High sodium content also adversely affects soil structure, resulting in poor aeration and extremely low rates of water infiltration. The availability of water to plants is also decreased in saline soils because of increased osmotic tension, by which water is held in the soil. In such circumstances, irrigation must be more frequent. If salinity is suspected, soil samples should be analyzed. A list of commercial laboratories is usually available from the local cooperative extension or soil conservation service.

Salinity has usually been expressed in millimhos per centimeter of soil extract (mmhos/cm), a unit for measuring electrical conductivity (EC). The metric unit for electrical conductivity is decisiemens per meter (dS/m). One dS/m equals one mmho/cm. The higher the soil salinity, the greater will be the conductivity of the extracted soil solution (EC_e). A soil is saline when its total soluble salt content is high enough to affect plant growth adversely. This could be as low as 3 dS/m for sensitive plants. A few plants will grow satisfactorily in soils with levels up to 10 dS/m. Most plants tolerate salinity of 5 to 7 dS/m (Bernstein 1964).

Water Quality

Salt may accumulate in the soil from irrigation water, high ground-water tables, de-icing salt, or existing salt deposits. If irrigation water does not penetrate the soil deeply, most of the salt brought in by that water will remain near the surface. Ground water generally contains more salt than does surface irrigation water. When the ground-water level is near the surface, some water moves upward because of evaporation and plant use. The salt content of the surface soil increases as this happens and plant growth can decrease. A good soil can become saline in one season.

There is increasing use of secondary-treated sewage effluent for landscape irrigation. The main concern is salinity and the presence of heavy metals. Check the range of effluent quality before beginning effluent irrigation. In addition, monitor effluent quality regularly because incoming sewage can vary considerably. Close cooperation with the treatment facility manager is essential to alert you to changes in effluent composition that might adversely affect plants.

Assessment of water quality is based upon three main considerations: the total salt content, the proportion of sodium in salt, and the boron concentration (Table 13-4). Water from the Colorado River contains up to 75 kg of salt per hectare cm (2000 lb/acre ft) of water (800 ppm, or an EC of 1.25 dS/m). With reasonable irrigation practices, however, water with an EC of 0.75 dS/m or less should present few or no salinity problems. An EC greater than 3.0 will cause severe problems except for a few salt-tolerant species (Stromberg 1975). Salt-sensitive species can tolerate an EC no greater than 2.0 dS/m.

Even though the total content of irrigation water may be low to moderate, if the sodium constitutes more than 60 percent of the total of calcium, magnesium, and sodium, soil permeability to water and air is usually adversely affected. It is worthwhile to remember the expression: **"Hard water (high in calcium and magnesium) makes soft soil, and soft water (high in sodium or very low in total salt) makes hard soil."** Before you can assess the salinity hazard of any irrigation water, you

TABLE 13-4 GUIDELINES FOR INTERPRETATION OF WATER QUALITY FOR IRRIGATION (Ayers 1977)

	Water quality guidelines		
	No problem	Moderate problems	Severe problems
Salinity			
EC_w in decisiemens[a] per meter	0.75	0.75–3.0	3.0
Boron			
Parts per million (ppm)	0.5	0.5–2.0	2.0

[a]EC_w = electrical conductivity of water. EC expressed as decisiemens per meter (dS/m); one decisiemen per meter equals one millimho per centimeter (mmho/cm).

must know how much salt the plants can tolerate and how much leaching is needed to maintain the desired salt level in the soil water.

Removing Salt by Heavy Irrigation

Each area or basin to be irrigated should be fairly level to obtain uniform leaching with the least water. Adequate drainage is also important. If natural drainage is inadequate, you may have to install tile or open drains. If poor soil drainage is caused by excess sodium, you can first reduce sodium by adding gypsum ($CaSO_4$) and then leach salt out of the root zone.

Planters, containers, and beds with shallow soil are quite vulnerable to salt build-up. Soils of limited volume and depth need frequent and excess watering to keep salinity from increasing to injurious levels.

Ordinary irrigation methods should produce some leaching, which reduces but does not eliminate the accumulation of salts in the soil water. The amount of water that must be applied to reduce soil salinity to a safe level depends primarily on the salinity of the water and the sensitivity of plants. Generally, for each 100 mm (4 in.) of medium-textured soil

50 mm (2 in.) of irrigation water leaches out $\frac{1}{2}$ of the salt
100 mm (4 in.) of irrigation water leaches out $\frac{4}{5}$ of the salt
200 mm (8 in.) of irrigation water leaches out $\frac{9}{10}$ of the salt

After soil salinity has been reduced to a level plants can tolerate, water must be applied in excess of the quantities used by the plants in order to maintain soil salinity at safe levels. The saltier the irrigation water and the more sensitive the plants, the more water must be applied (Table 13-5). For instance, a moderately salt-tolerant plant should be irrigated with 125 to 166 percent of the water that

TABLE 13-5 THE DEPTH OF WATER REQUIRED FOR MAINTAINING A SAFE LEVEL OF SOIL SALINITY (Bernstein 1964)

Salinity of irrigation water EC_w (dS/m)[a]	Amount of water (percentage of evapotranspired water) required to maintain safe salinity levels for plants of different salt tolerances		
	Tolerant (10 dS/m max)	Moderately tolerant (5–8 dS/m max)	Sensitive (3 dS/m max)
0 (salt free)	100	100	100
0.5	105	106–111	120
1.0	111	115–125	150
2.0	125	125–166	300
4.0	166	200–500	

[a] Electrical conductivity expressed as decisiemens per meter (dS/m); one decisiemen per meter equals one millimho per centimeter (mmho/cm).

transpires if the salinity of the water is 2.0 dS/m. If 30 mm of water has transpired, the plant should be irrigated with 38 to 50 mm of water (30 mm times 1.25–1.66). This would maintain soil salinity at a safe level (below 5–8 dS/m).

It has generally been assumed that the effects of saline water can be offset by increasing the amount of leaching, so that the *average* salt content of the root zone will not be increased. The U.S. Salinity Laboratory has demonstrated that yields of alfalfa (and probably other crops) are governed not by the average soil salinity but primarily by the salinity of irrigation water (Bernstein and François 1973). Experimental yields were reduced as the salinity of the water increased, no matter how much leaching was done. With this in mind, the practice of blending saline drain waters with low-salt irrigation water must be reassessed. In native landscapes where yield is not a factor, reduced growth due to saline water may be desirable as long as plant appearance is satisfactory.

Irrigation water may be more saline than sensitive plants can tolerate. Irrigation water with a conductivity of 4 dS/m can never lower the soil salinity below a lethal level for certain trees, such as the Lombardy poplar. Additionally, if the irrigation water is strongly saline, the amount of water needed by sensitive plants may make leaching impractical. In fact, the high amount of water necessary for adequate leaching may be as harmful as the salinity.

Other practices may reduce salt build-up in soil. High-analysis fertilizers provide adequate nutrition with the least increase in soil salinity. For example, an application of 2 kg of nitrogen per 100 m^2 would require 2000 kg of 10-6-6 fertilizer per hectare but only 600 kg of urea (45 percent N). Surface mulch reduces evaporation and salt build-up in the surface soil. If the mulch is of wood chips, bark, or other organic matter, it should be at least 50 mm (2 in.) deep. Sprinkler irrigation may be necessary to keep the mulch from floating to low areas.

Salt build-up on surface soil may be encouraged to minimize salt concentration in the root zone of shallow-rooted plants. Since salts usually accumulate at the high spots in a planting, berms and ridges used in basin and furrow irrigation can be formed so that salt accumulates away from the root zone (Fig. 13–8). However, since light rains may carry this salt into the root zone, heavy irrigation may have to be used if heavier rains do not follow. In some landscapes, it may be necessary

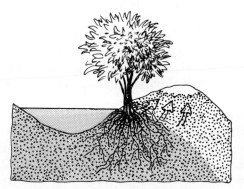

Figure 13–8 Establish small shrubs at the bottom or side of a furrow to minimize salt build-up in the root zone since much of the excess salt will accumulate at the top of the ridge.

to select relatively salt-tolerant plants for satisfactory growth and appearance with reasonable maintenance (Bernstein 1964, Maire and Branson 1972).

ANTITRANSPIRANTS

Antitranspirants are chemicals capable of reducing transpiration. They have been used for a number of years by arborists and nursery persons to reduce water loss in plants, particularly during and after transplanting. Antitranspirants have also been used to reduce the water requirement of plants during drought, to reduce the frequency of irrigation in arid regions, or to protect conifers against windburn in winter. Since plants lose most water through their stomates, antitranspirants are primarily foliar sprays.

When a plant wilts, growth is retarded. By slowing down the rate of water loss, antitranspirants can help prevent or delay wilting. Measurements have shown that the trunks of trees usually shrink and branches rise during daylight hours, when transpiration exceeds water uptake. A film-forming antitranspirant sprayed on five-year-old almond trees can reduce daytime trunk shrinkage by more than 50 percent, indicating that the water balance of an entire tree can be improved if water loss from the leaves is curtailed (Davenport, Hagan, and Martin 1969).

Foliar antitranspirants may reduce transpiration in one of three ways. They

Use reflecting materials that reduce absorption of radiant energy, thereby lowering leaf temperatures and reducing transpiration

Apply emulsions of wax, latex, or plastics that dry on foliage to form thin, transparent films, which hinder escape of water vapor

Apply certain chemical compounds that affect the guard cells around stomatal pores and prevent stomata from opening fully, thus decreasing the loss of water vapor from leaves

To be effective, reflecting materials must coat the exposed leaves with a light-colored film, but a plant so colored may not be acceptable in most landscapes, except when needed to keep the plant alive. Another drawback is that the film restricts light penetration into leaves and reduces photosynthesis. Spray emulsions that form transparent films not only reduce water loss but reduce the carbon dioxide intake needed for photosynthesis. Chemicals that inhibit stomata from opening fully appear promising because the smaller openings restrict water vapor movement more than they restrict the passage of carbon dioxide (Kramer and Kozlowski 1979). Three such chemicals are phenylmercuric acetate, abscisic acid, and silicone. Although phenylmercuric acetate has been more widely tested than the others, its use will probably be limited because it poses the danger of phytotoxicity and mercury contamination of the environment.

While antitranspirants of the reflecting type cause a reduction in leaf temperature, the film-forming and stomata-closing types tend to increase leaf temperature by curtailing transpiration and thus reducing evaporative cooling. Under normal conditions, however, the increase in leaf temperature induced by antitranspirants is usually not serious, because thermal emission rather than evaporative cooling becomes the more important means of heat dissipation.

The effectiveness of antitranspirants seems to depend on species, the plant's stage of development, and atmospheric conditions (Gale and Hagan 1966). Even when antitranspirants are effective, the duration of effectiveness will vary with the efficiency and durability of the material, the accuracy of application, environmental conditions, and the amount of new foliage produced after spraying. Cracking of the film and new growth limit the duration of reduced transpiration. Repeated application can retard growth by seriously reducing photosynthesis. Even so, film-forming materials are the most commonly used antitranspirants and have a place when reduced transpiration for a few days to two weeks is crucial. Antitranspirants have been helpful in transplanting plants that are in leaf, particularly during the growing season. Though they may reduce water loss initially by 40 percent, the film materials have limited application for longer-term reduction of water loss. In tests by Turner and DeRoo (1974), antitranspirants did not reduce winter-burn injury to conifers when air temperatures were below freezing, but they seemed to be helpful in the spring when air temperatures were warmer but the soil was still cold or frozen.

Antitranspirants can be used occasionally on small plantings or container plants to extend the interval between irrigations when it is not convenient to water on schedule. Cut foliage, including Christmas trees, will stay fresh longer if they are sprayed with an antitranspirant before cutting. Antitranspirants can also reduce salt damage to conifers along highways that are salted during the winter months (Gouin 1979).

In general, the use of antitranspirants will be limited until more effective materials are found and benefits are more specifically documented.

HYDROGELS

Hydrophilic polymers have enticed horticulturists for more than 20 years with the possibility of increasing the water-holding capacity of soils and soil mixes. Hydrogels can absorb up to 1500 times their weight in pure water (Johnson 1984). Increased water-holding capacity would be particularly helpful for plants through all stages from production to landscape planting, since the interval between waterings could be extended.

The first hydrogels were short-lived, expensive, and variable in effectiveness. Now, several general types of hydrogels are available. Starch- and urea-based polymers are subject to microbial degradation whereas the preferred cross-linked polyacrylamides remain active longer (Walmsley 1989). Hydrogels, as a dry powder, can be mixed with soil or, as a water suspension, can be injected into soil.

Results from the use of hydrogels have been mixed, both for plants grown in containers and field soil (Walmsley 1989, Bowman, Evans, and Paul 1990). Walmsley cites reports that polymers increased the planting survival and shoot growth of a number of amenity and orchard apple trees. He also cites Gilbertson's (1987) studies in urban Liverpool in which polymers had no effect on the growth of *Sorbus aria* or *Acer pseudoplatanus*. Sandell and Kube (n.d.) obtained no benefit from two hydrogels (type unspecified) mixed in the backfill on the survival or growth of *Ecalyptus torgyata* planted in central Australia.

In Liverpool, Walmsley (1989) transplanted 2-year-old bare root *Acer pseudo-platanus* into coarse sand or peat:coarse sand in 15-l (4-gal) polypots or in field soil backfilled with soil. Powdered polymer was mixed 0.004:1 (volume basis) with the container and field soils. Trees in the polymer-amended soils made more total dry weight and shoot length gains although the gains were not always significant. Most interesting was that in all three experiments, root dry weight of the polymer-treated trees was double that of those with no polymer. The weight of the tops was increased 5 to 45 percent with the addition of the polymer; however, the increases were not significant. Walmsley is the first to report such increases of root growth in polymer-treated soil.

Mixing the polymer with the peat-sand mix doubled the amount of available water for the container-grown trees and the number of days (14 to 28) before wilting (Walmsley 1989).

Soil variation, the influence of solutes, and the quality of irrigation water may account for the divergent reports on the success of hydrogels in increasing available moisture for plants. The more concentrated a soil solution, the less water a hydrogel is able to absorb (Woodhouse 1989, Bowman, Evans, and Paul 1990). Hydrogels have the greatest increase in available water when mixed with an infertile sand. They have the least benefit in a fertile, saline soil that is irrigated with hard water (high in Ca and Mg). Bowman, Evans, and Paul found that gel hydration was not affected by urea at twice the equilibrated amount of potassium nitrate that reduced hydration 75 percent.

The high fertility that is usually desired in the production and early growth of landscape trees and shrubs may negate hydration of hydrogels. If there is interest in the use of polymers, it would be wise to conduct a trial under the conditions in which the polymer would be used. If nitrogen is needed, add urea.

SOIL CONDITIONERS

Water infiltration into and through soil and root growth and activity are greatly influenced by soil factors such as texture, structure, degree of compaction, and surface crusting. Soil conditioners have long been studied in efforts to favorably alter the physical conditions of soil. The early synthetic polymers (Krilium® was the best known) were effective stabilizers of soil structure. But the soil had to be brought to the desired physical condition before or while mixing it with the polymer. The polymers proved difficult to use and too costly for commercial agriculture (Oster and Singer 1984).

More recently, water soluble polymers have become available that show promise in improving the condition of soils using relatively little conditioner (Helalia and Letey 1988). The new polymers are most efficiently applied as a soil drench or in irrigation water (Wallace and Nelson 1986). Some of the possible benefits suggested for these chemicals sound too good to be true. More research and field experience will determine if and where they can be used to advantage. Landscape soils can benefit from the new polymers if they live up to early findings. If you are interested in using these polymers, be sure to test them under your conditions before treating large areas.

WETTING AGENTS

Wetting agents, which reduce the surface tension of water, are thought by some to improve water infiltration into soil. **There are few soil situations in which water infiltration might be improved by adding a wetting agent to irrigation water.** There is more than one solution to slow infiltration of water. Infiltration rates vary due to differences in soil texture, structure, cover, and alkalinity; subsoil restricting layers; and maintenance practices (Oster and Singer 1984).

Infiltration problems that may respond to wetting agents include soils that have been subject to hot fires or are highly organic (James Oster, University of California, Riverside, 1990 pers. comm.). Organic soils, particularly in containers, and turf thatch that are allowed to dry become extremely difficult to rewet without adding a wetting agent. On the other hand, it is doubtful that wetting agents would improve water infiltration of mineral soils.

If you think a wetting agent will improve infiltration, experiment to determine its effectiveness before treating a large area. Most household dishwashing detergents (such as Joy® or Ivory® Liquid) are effective wetting agents.

CHAPTER 14

Soil
Management _____

A number of cultural operations (other than fertilization and irrigation) may be employed to protect the soil, maintain or improve its physical condition, and provide an attractive surface for landscape use and enjoyment. Most of these operations involve management of the surface soil, but some involve the tops of plants as well as deeper soil and roots.

Development and management of the soil surface should be integral to any landscape plan. Paved surfaces are used for thoroughfare and recreational activities. Lawns provide attractive surfaces for leisure and for soil protection. Areas not covered by pavement or lawn are usually treated in one of the following three ways:

    With ground covers
    By clean cultivation or chemical weed control
    With organic or inorganic mulches

Proper handling of the soil surface will depend on the function of the landscape, the type of soil, and the kinds of plants that are grown.

TURF AND GROUND COVERS AROUND TREES

Turf and ground-cover plants present few problems for existing trees. When turf or other low plants are irrigated, however, care must be taken so that trees and shrubs are not over- or underwatered. Overwatering encourages surface rooting, may suffocate roots, and even allows crown rot fungi to infect treetrunks. On the other hand, light, frequent irrigation may keep turf healthy but may fail to supply enough water to trees and shrubs.

Turf

Turfgrass can greatly reduce the growth of young plants if it is planted too close. Adequate irrigation and fertilization will help overcome some of the growth inhibition, but not all of it. A number of turf species and some broadleaved plants have allelopathic effects on young trees and shrubs (see Chapter 2). Keeping turf and other plants at least 300 mm (12 in.) away from the trunks of young plants will eliminate almost all such retarding effects (Fig. 14–1) (Mookimen 1970, Harris, Paul, and Leiser 1977, Whitcomb 1987).

Clear areas near treetrunks or protective stakes will also minimize trunk damage caused by mowing equipment (see Fig. 20–9). After turf growth has slowed in

Figure 14–1 The influence of alta fescue turf on young southern magnolia trees (left). The photograph shows growth after two years: 0, 0.36, and 17.6 m² (0, 4, and 196 ft²) of soil (front to back) were kept free of turf around each tree (Harris, Paul, and Leiser 1977). The three 42-year-old ginkgo trees show the influence of turf on tree growth: For the first 25 years the one on the right was surrounded by turf while the other two grew at the edge of the turf in a shrub bed. The smallest was so low in vigor that the bark exposed to the afternoon sun was killed; transpiration did not "pull" enough water up to keep the cambium cool.

autumn, nitrate or urea nitrogen fertilizers applied on the lawn surface under trees will favor tree uptake. Tree roots are in the surface soil. Turf will be less stimulated at this time.

Ground-Cover Plants

Where foot traffic is infrequent or not wanted, low-growing shrubs, vines, and herbaceous perennials can be used and will require less maintenance than turf. Dense plantings will quickly minimize weeds, but physical or pre-emergent weed control will usually be needed until the ground cover blankets the soil. Irrigation may be needed until the plants become established. Most ground covers, particularly vines, must be pruned to keep them in bounds and attractive (see Chapter 15).

Edging. The spread of turfgrass, weeds, and other ground-cover plants can be controlled with herbicides, growth regulators, mechanical edgers, and mulches. The development of string trimming equipment has greatly facilitated edging around plants and isolated objects and along walks and fences (Fig. 14–2). A short length of flexible cord rotating at high speed cuts or breaks leaves and small-diameter stems. The flexible cord allows trimming in difficult-to-reach places. Caution must be exercised, however, when trimming around the base of plants, particularly in the spring when the bark of even mature plants is easily separated at the cambium. Some trimmers have a guard to more safely trim near plants.

Figure 14–2 A string trimmer is effective in edging turf around shrub beds and trimming around trees. In the spring when the cambium is active, bark can be damaged or torn loose by the impact of the flexible cord.

BARE SOIL AROUND PLANTS

Cultivation

Seedbed preparation, weed control, and incorporation of organic matter are the primary purposes of cultivation. There is reason, however, to minimize cultivation in landscape plantings. The process increases costs, cuts and injures shallow roots, and makes soil on slopes more subject to erosion. Even though cultivation can destroy weeds, it brings buried seeds to the surface to perpetuate the species. In plantings, weeds can be controlled with chemicals, mulch, or surface hoeing, which disturbs little of the soil.

Some people would insist on the necessity of stirring the soil around existing plantings in order to improve aeration and water infiltration. Cultivation may appear to improve tilth, but in fact it breaks down soil structure; the surface of cultivated soil can quickly become sealed from splashing rain or sprinkler drops. Cultivation can, however, break up a tight soil, so that wetting and drying of the fractured soil (large clods) from rain (irrigation) and plant extraction can bring about an internal cultivation that will improve soil structure.

Chemical Weed Control

Chemical control of weeds came into its own with the development of 2,4-D (2,4-dichlorophenoxyacetic acid) in the early 1940s, and later with the introduction of pre-emergence herbicides. Wise use of herbicides can minimize weed problems and greatly reduce the cost of weed control. Herbicides are classified in a number of ways. *Contact* herbicides kill only those plant parts they contact, while *translocated* herbicides can affect the entire plant even though only a portion of it has been treated. A translocated herbicide that is particularly effective against perennial broadleaved weeds, 2,4-D is translocated from the leaves throughout the plant, including the roots. Contact sprays are primarily used against annual weeds; once their tops are killed, few will sprout again. Contact herbicides are usually not *selective* and kill everything they strike. At the right concentrations, 2,4-D can be a selective herbicide, one that can kill broadleaved plants without damage to surrounding grass.

Fumigants can be used before planting to rid confined planting areas of established and seedling weeds. This *preplanting* treatment will greatly reduce the need for future weed-control measures, particularly if the treated planter is isolated from other weed-seed sources.

Other chemicals can be applied to the soil and landscape plants to kill germinating weed seedlings. These *pre-emergent* herbicides, such as oryzalin (Surflan®), simazin (Princep®), and trifluralin (Treflan®), are valuable in keeping ground cover and shrub plantings weed free. The soil must be free of weeds before these materials are applied. A selective *postemergent* herbicide can be sprayed on landscape plants while leaving desirable ones uninjured. Sethoxydim (Poast®) or fluazifop (Fusilade®) can kill grass in broadleaved plants; 2,4-D kills broadleaved plants in turf.

As with any chemical, herbicides must be used with care. Nonselective, post-emergent herbicides like glyphosate (Roundup®) or the short-chain fatty acids (Sharp Shooter®) can be used under trees and shrubs, but the spray should not touch leaves or young trunks of desirable plants. The margin between killing weeds and seriously injuring or killing landscape plants can be narrow. Careful attention must be given to correct concentration, timing, and rate of application. Avoid spraying desirable plants or applying herbicides too heavily, because they may be further concentrated by being washed to the bottom of the basin around a tree or shrub. **Applying sprays with large droplets at low pressure will minimize drift,** which can cause plant injury, particularly under windy conditions. Weed oils, which have a disagreeable odor, will stain concrete, but kerosene and most other contact sprays will not.

In Scandinavia, weeds growing in walkways in parks, churchyards, and sports areas are killed with a propane burner to avoid using chemicals (Niels Hvass, Ballerup, Denmark, 1990 pers. comm.).

Effect of Bare Soil Around Plants

Whether weeds are controlled by hand pulling, cultivation, or herbicides, the soil surface that is cleared of them is left unprotected from water, wind erosion, and compaction. The soil surface on slopes can be shaped to trap water so it will soak in near the site of application, however. Bare soil under plant canopies can protect plants, particularly trees, against grass fires in unirrigated, arid landscapes.

MULCHES

Material placed on soil to cover and protect it is called *mulch.* Organic mulches include bark, wood chips, leaves, conifer needles, lawn clippings, straw, corn cobs, peanut hulls, and numerous other organic byproducts. Inorganic or synthetic mulches include crushed stone, black polyethylene, spunbonded-polypropylene or polyester fabric, asphalt, pavement, and aluminum foil. Although not recommended now, in the early 1900s, loose, dry surface soil, called a *dust mulch,* was thought by many to reduce water loss from deeper soil.

Benefits of Mulching

The plant microenvironment is so extensively modified by mulching that it is seldom possible to isolate specific causes for each of the benefits of a mulch. It is helpful, however, to consider each of the possible benefits so as to better understand how mulches can be used or modified to be most effective.

Soil moisture is conserved by mulches because they reduce evaporation from the soil surface and reduce weeds that use water. In 1939, Russel conducted a laboratory study with soils at field capacity and found that a layer of straw mulch 40 mm (1.5 in.) thick reduced evaporation primarily by obstructing solar radiation (about 35 percent), protecting against wind (about 25 percent), reducing soil temperature, and obstructing vapor escape (about 40 percent) (Fig. 14–3). Mulches are most ef-

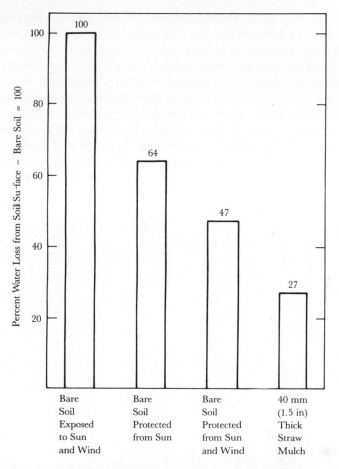

Figure 14–3   Water loss from four soil columns, initially at field capacity, as influenced by soil surface protection during four summer days. The straw mulch did more to reduce evaporation than mere protection from sun and wind would account for. (Adapted from Russel 1939)

fective in reducing evaporation when the surface soil is near field capacity. Once the water content of surface soil is much below field capacity, evaporation is slow. Weeds can use as much or more water than landscape plants, depending on their size and the area they occupy. A cover crop or a dense stand of weeds in a mature deciduous orchard will increase evapotranspiration about 25 percent, or 1.5 mm (0.07 in.), per day in the summer. A thick mulch (100–150 mm or 4–6 in.) can suppress all but the most persistent perennial weeds, such as bindweed and quackgrass.

A mulch will also hold water near where it falls, so that more of it soaks into the soil. In addition, moisture can accumulate under a mulch because water vapor in the soil condenses on the cold mulch at night. This moisture may be equivalent to 0.1 mm (0.004 in.) of rain per day. Under high humidity and radiation cooling conditions, moisture from the air also condenses on cold mulch surfaces. For a peat-

Mulches

mulched soil, Jacks, Brind, and Smith (1955) estimate that dew equivalent to 1.25 mm (0.05 in.) of rain per day condenses on the mulch and wets the root zone. The mulch surface must be exposed to clear sky if that much condensation is to occur.

Soil erosion and water loss are reduced because mulch breaks the impact of rain and sprinkler drops, slows down the movement of water, and keeps water in contact with soil that has a faster infiltration rate than bare soil. Borst and Woodburn (1942) have shown that raindrop impact is a much more important erosion-causing factor than is flowing water (Fig. 14-4). On a soil that had been puddled and sealed by long exposure to rain, they found that during a heavy rain, a straw mulch 15 mm (0.6 in.) thick reduced water runoff by 43 percent but reduced soil erosion by 86 percent. When a similar mulch was placed on wire mesh 25 mm (1 in.) above the soil, water runoff increased 25 percent, but soil erosion did not

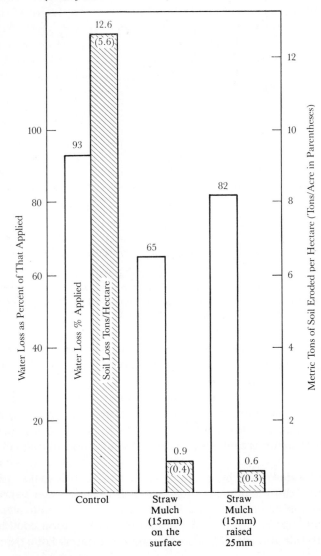

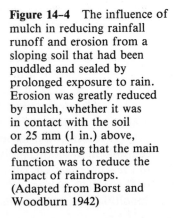

Figure 14-4 The influence of mulch in reducing rainfall runoff and erosion from a sloping soil that had been puddled and sealed by prolonged exposure to rain. Erosion was greatly reduced by mulch, whether it was in contact with the soil or 25 mm (1 in.) above, demonstrating that the main function was to reduce the impact of raindrops. (Adapted from Borst and Woodburn 1942)

Chap. 14 Soil Management

increase. More water would probably have infiltrated the soil if the mulches had been left in place long enough to improve soil structure.

Soil fertility is usually increased by nutrients from mulch, either by direct leaching or by decomposition. The surface soil under a mulch is favorable for microorganisms, which increase the availability of nutrients. Mulch protection also allows increased rooting in the more fertile, better aerated surface soil. Jacks, Brind, and Smith (1955) cite results from both farm and forest studies in which almost all soils showed increased nutrients under organic mulches. In short-term studies, nitrogen may be depressed by the use of mulches, while in studies of more than five years, nitrogen levels are usually enhanced after the early years. The more easily a mulch decomposes, the lower the initial nitrogen levels in a soil. A fine-textured mulch that is high in cellulose can cause a nitrogen deficiency because of the increased number and activity of microorganisms.

Weed competition is reduced by mulches, so more water and nutrients are available for the landscape plants and any allelopathic effects on young plants are minimized. A thick, impervious, or special fabric mulch will reduce weed-seed germination and keep weeds that do germinate from growing through the mulch.

Soil structure is improved when soil is covered with a thick mulch, particularly if the mulch is organic: The size of soil aggregates and voids and the total porosity of soils increase (Jacks, Brind, and Smith 1955). Improved aeration, temperature, and moisture conditions near the surface encourage rooting and other biological activities that enhance soil structure. Just the absence of cultivation and the low amount of compaction that even a thin inorganic mulch provides will allow soil structure to improve through wetting and drying cycles and through biological activity. Improved soil structure increases the infiltration rate and allows more uniform water distribution and less soil erosion, all of which favor plant growth.

Soil compaction is reduced when a protective covering reduces rain and sprinkler impact on the soil surface and disperses the weight of vehicles, people, and animals. Bare soils can be seriously compacted so that water and air movement is extremely slow and root growth impeded. If surface soil becomes puddled from the impact of rain and sprinkler drops (Fig. 14–5), water infiltration and air exchange are limited.

Mulches provide an improved surface for traffic during wet weather or after irrigation. A coarse organic mulch reduces dust when soil is dry and mud when soil is wet. Many mulches improve the appearance of landscape plantings and are also good sound absorbers.

Clay soil is less likely to crack when protected by mulch. Some clay soils shrink upon drying and form large cracks in the surface that increase moisture loss from the soil below, break small roots, and make furrow and basin irrigation difficult.

Mulches reduce salt build-up in surface soil by reducing evaporation of water that may be high in soluble salts. This is particularly important when summer rainfall is low and water tables are shallow and saline.

Mulches moderate soil temperatures so that surface soil is cooler in summer and warmer in winter than it would be otherwise. Moderate soil temperatures and good aeration encourage rooting in the fertile surface soil; this can be quite important in shallow soils. In areas with extremely cold winters, soil under a mulch is less

Figure 14-5 The impact of a raindrop on wet soil splashes water drops and soil particles in all directions and leads to soil crusting and slow water infiltration. (Photo courtesy U.S. Soil Conservation Service)

likely to freeze; if soil does freeze, the mulch will minimize the alternate freezing and thawing that can injure the roots of young plants. Snow can serve as a mulch to effectively insulate the soil from cold and changing air temperatures. Mulch that has highly concentrated, tiny air spaces provides the most effective insulation. When they become soggy, compressed, or thin, mulches lose much of their insulating properties. Thin, dark mulches or those that are made of dirty paper or polyethylene increase soil temperatures because they absorb more heat than they dissipate. In fact, summer temperatures under clear plastic can be lethal to plant roots to a depth of 300 mm (12 in.). On the other hand, aluminum and white paper mulches reflect radiation and keep the soil cooler in summer than if it were bare or covered with clear or dark films.

Coarse mulches reduce the reflection and reradiation of heat, thus increasing human comfort and plant well-being in summer. Rough pieces of mulch diffuse heat in many directions instead of concentrating it, as a smooth, bare surface or pavement may do. The mulch surface, however, can become quite hot and damage emerging plants.

A preliminary USDA report indicates that reflective colored mulches increase the yields of certain vegetables (American Horticultural Society 1988). This might be of interest in plant production, but white, red, or black mulches would be acceptable in few landscapes.

The incidence of some diseases is reduced when mulch minimizes the splashing of soil and pathogenic organisms onto plants during rain or sprinkler irrigation. Avoid mulch materials, of course, that may harbor diseases to which the landscape plants are susceptible. Though it is probably not appropriate for landscape plantings, aluminum foil has been used to reduce thrips and aphids on certain vegetables.

Problems Caused by Mulches

Most problems associated with mulches can be avoided or greatly minimized. The potential problems are discussed in this section; the methods of avoiding or minimizing them are discussed here and also more generally in the later section on Mulch Management. Mulches may not be advisable in some situations.

Nitrogen deficiency may develop when mulch materials, such as sawdust and straw, decompose rapidly, particularly if materials are partially incorporated in the soil. This problem can be overcome fairly easily if you use mulch that decomposes more slowly, add nitrogen fertilizer, or minimize incorporation of the mulch in the soil.

Excessive moisture may occur in fine-textured and poorly drained soils, resulting in poor aeration, particularly below a compact organic mulch. Rickman (1979) found that mulches can aggravate anaerobic conditions in slowly draining soils, leading to soil denitrification or loss of nitrogen fertilizer. Even in better-drained soils, studies (e.g., Jacks, Brind, and Smith 1955) report slower nitrification in mulched soils than in bare soils. This may be due to the higher moisture content and lower temperature of mulched soil during the warmer part of the year. When such a problem occurs, the use of mulch should probably be discontinued and drainage improved (see Chapter 8).

Insufficient water may be available under plastic mulch because large sheets of unbroken plastic greatly reduce the amount of rain or sprinkler irrigation reaching the root zone. Geotextile fabric or openings around the plants, at the edges, and along the overlapped sheets usually allow enough water to flow through or under the plastic and wet the soil. Pricking holes in sheet plastic will also overcome this problem without encouraging weeds. The problem would not arise, of course, under flood irrigation.

Air temperature extremes are accentuated above a mulch. Because mulches insulate soil from radiation and air temperatures, temperatures are more extreme just at and above a mulch. As mentioned, a mulch surface can become so hot in summer that young plants growing through it will be injured. A more frequent threat to most plants is the increased cold above a mulch under radiation frost conditions (see Chapter 4). A mulch has low heat absorption and storage characteristics, and its surface cools quickly under night radiation. In Maryland, Creech and Hawley (1960) found that minimum air temperatures in midfall (Oct. 4 to Nov. 12) averaged 3°C (5.5°F) colder 50 mm (2 in.) above a hay mulch than above bare soil; temperatures above the mulch were as much as 4°C (7°F) colder on clear nights. They attributed the winter injury of buds and bark-splitting of two-year-old azaleas during the first fall frost to the prolonged growth and the abnormally low air temperatures brought about by continuous mulching. If cold is a hazard to plant tops, mulch should be removed before cold weather arrives so that the heat capacity of the soil can moderate air temperatures.

Mulched plants usually grow more vigorously and longer in the summer, thereby delaying development of hardiness. Whitcomb (1980) examined winter dieback of mulched one-year-old Chinese pistache in Oklahoma. Of trees mulched with black plastic and 50 mm (2 in.) of bark mulch, an increasing number experienced top dieback as the fertilizer level increased from zero to 3.5 to 7 kg of actual nitrogen per 100 m² (0 to 7.5 to 15 lbs/1000 ft²). Trees mulched with only 50 mm (2 in.) of bark were killed back only at the highest nitrogen level. Unmulched pistache trees were not injured, nor were any of the other three species tested (sawtooth oak, dwarf burford holly, and pfitzer juniper), regardless of the mulch and fertilizer treatments. Around the pistache trees mulched with black plastic and bark, soil moisture had been higher than around the other pistache during a dry period the previous summer. Where bark or plastic and bark were present, growth was 30 percent and 21 percent higher, respectively, than that of trees in bare soil. The most vigorous late top growth was most susceptible to winter injury. Therefore, you can avoid injury by not mulching, by mulching only in the summer, by protecting the plants from the cold (see Chapter 4), or, as the three uninjured species testify, by planting hardier species.

Some diseases and pests may be more serious on mulched plants. A mulched soil favors the increase and development of most microorganisms, some of which may be infectious. Armillaria root rot (*Armillaria mellea*) (see Chapter 21), a soil-borne disease, is encouraged by the moist soil under a mulch. Crown rot is more likely if the trunks of susceptible plants are kept moist by mulch piled against them. Proper mulch selection and placement will greatly reduce the likelihood of infection. In certain situations favoring disease, a mulch should not be used.

Because rodents often burrow and live in loose mulches, particularly if snow is common, mulch should be kept away from the trunks of plants. Trapping or poison may be necessary if the problem becomes serious. Mulches also harbor slugs and some insects that can occasionally be troublesome. These pests can be eliminated by specific control measures. Some mulches, such as straw and manure, contain bothersome weed seeds and should be avoided.

Some mulches increase fire hazard, but most can be used with little or no danger. Except for the surface centimeter (0.5 in.) or so, most mulches are quite moist and resistant to fire. Many federal and state agencies use wood chips as ground cover to control dust and reduce soil compaction in picnic areas and campgrounds, even in the arid Southwest. That type of coarse mulch works down into the soil and stays fairly moist. Even mulches around stoves and campfires have not proved hazardous. Vapor from the soil below condenses in the colder surface mulch at night and keeps it moist even during dry periods. Rain and sprinkler irrigation, of course, are even more effective in moistening mulches.

On the other hand, a loose mulch of straw or other dry, lightweight organic matter can be a fire hazard even in irrigated areas. Straw dries quickly and is highly inflammable. Such materials are usually unattractive in landscapes, although they are frequently used in conjunction with vegetable and fruit production where summer rains are frequent.

A dry, compact organic mulch can catch fire from a discarded match or cigarette and smolder for days. In a peat-moss mulch plot in California, two days of continuous sprinkler irrigation wetted only the mulch surface and failed to extinguish a smoldering fire under the surface. The mulch had to be broken apart so that a fine spray of water or fire retardant could be applied directly to the burning material. Wyman (1957) reports a similar experience. Although a smoldering mulch fire may not endanger most plants, wind or nearby inflammable material can cause it to grow to dangerous proportions.

Nearly all fire problems can be eliminated by proper mulch selection and management.

Some mulches may be toxic, particularly to young plants. The bark, sawdust, and foliage of a number of tree species have been reported to be toxic to other plants (Jacks, Brind, and Smith 1955, Del Moral and Muller 1970, Green 1978). Toxic tissues include eucalyptus sawdust and leaves, redwood and cedar sawdust, Douglas fir, larch, and spruce bark. Reported toxicities occurred with fresh mulch on young plants in almost all cases. Toxic substances can be removed or inactivated by leaching or composting (Hoitink and Poole 1977).

Mulches may affect depth of rooting. The belief that mulched plants develop shallower root systems than unmulched plants is not borne out by several studies. There is no doubt that more roots grow near the soil surface in mulched than in unmulched soil. However, the roots of mulched trees usually penetrate as deep or deeper than those of unmulched trees. Jacks, Brind, and Smith (1955) cite a Japanese study in which mulched peach seedlings developed almost 50 percent more total root weight than did unmulched seedlings and double the weight between 0.6 and 1.2 m (2 and 4 ft) deep. In a mature Michigan apple orchard, Kenworthy (1953) found that moisture depletion was uniform at all depths to 1.06 m (42 in.) under a hay mulch on an orchard grass sod, while there was no appreciable depletion under an unmulched sod below 0.6 m (24 in.), indicating that mulch promoted root penetration. Beckenbach and Gourley (1932) examined the root distribution of an apple orchard that had been mulched for 35 years and found that mulched trees in the lower soil horizons had root systems at least as dense as those of unmulched trees. Whitcomb (1980) modified his earlier (1979a) observation to say that although fine, fibrous roots concentrate just under a plastic/bark mulch, both mulched and unmulched plants have larger roots penetrating to a depth of 0.3 m (12 in.) or more.

Mulches usually encourage roots to grow deep as well as close to the surface. Mulches should not jeopardize plant stability, though increased surface rooting may later damage nearby curbs and sidewalks.

Mulch Materials

The range of mulch materials is great. Most are organic and are usually byproducts of industry and agriculture. The following considerations should guide selection of a mulch.

Availability, including price and delivery, is probably the most important factor in determining which mulch to use. Several mulch materials may be free, but transportation and handling make delivery expensive. Large quantities of potential organic mulch materials are now being used as alternate sources for energy, building products, and nursery soil amendments. Plant litter obtained near the site and used directly or after composting may be the most economical source.

Lawn clippings and the leaves of some plants are considered poor mulch materials because they compress and mat together, restricting air and water movement. They can also become slimy and unattractive. This often happens with fresh leaves that are applied so thickly that they are slow to dry. If applied in thin layers, they dry more quickly; shredding leaves with a mulcher before application will quicken drying and reduce matting. Composting will also reduce the problem.

In Nebraska mulch trials, dried turf clippings proved as effective as dried alfalfa in moderating soil temperatures, conserving soil moisture, and improving growth and nitrogen content of plants (Shearman and others 1979). Of six turfgrass species tested, five were about equally effective and all were much better than creeping bentgrass. Bentgrass clippings were little better than no mulch at all, because they were difficult to dry, tended to compact, and had a crusted surface. Alta fescue and buffalograss should form a more open mulch, since they are coarser and usually heavier when dry than other species.

Ease of application and maintenance are important considerations, particularly in existing plantings. Wood chips are easy to apply and maintain, while plastic sheeting or fabric can be difficult and expensive.

Mulch appearance may be a key element in the design of a landscape (Fig. 14–6). A mulch should enhance the appearance of a garden and conform to its design. Many organic materials are quite versatile. Gravel, crushed colored rock, cobblestones, and sand are often used, usually underlaid with fabric or black plastic for weed control. The scale of the landscape should suggest an appropriate mulch texture.

Mulch stability under windy and wet conditions is essential. A number of materials, such as rice hulls, blow easily and are unsatisfactory for mulching in most locations. Organic mulches may be washed away during heavy rains, just when they are most needed. Coarse or heavy materials in close contact with the soil are less likely to be blown or washed away. Surface water flow may have to be diverted to protect the landscape as well as the mulch.

Freedom from contamination in a mulch will protect plants and ease maintenance. Mulch material should be free from weed seeds, harmful insects and diseases, and toxic chemicals. Any one of these can create maintenance problems and inhibit the success of a planting. Weed seeds are most likely to occur in mulch material made from the entire tops of mature plants, such as hay, straw, and weeds, or from leaves of small-seeded fall-fruiting plants, such as hackberry and privet. As mentioned, mulch material made from plants that are susceptible to serious diseases such as Armillaria root rot (*Armillaria mellea*) and fire blight (*Erwinia amylovora*) should be selected with caution. Some organic mulch materials, when fresh, may be toxic to plants, particularly young plants. These materials should be leached or composted (see next section) before being applied.

Slow decomposition will prolong the effectiveness of a mulch and reduce the chance of nitrogen deficiency. Organic materials that are high in lignin, are partially decomposed (composted), or have large-sized particles are slower to decompose.

Low fire hazard characterizes many coarse organic materials that work into the soil. Rock and plastic pose no problem, but straw and peat moss can be hazardous when dry.

Permeability is necessary so that water and air can penetrate the mulch and enter the soil. Spunbonded polypropylene fabric, such as Typar® #3301 or #3401, is pervious. It effectively controlled weeds and root suckers of flowering crabapples

Figure 14-6 A wood-chip mulch is being renewed in an area of a park that had originally been planted to turfgrass (upper left). Two colors of crushed rock on black plastic edge a shrub and tree border along a sidewalk (upper right). An entire front yard is mulched with gravel on black plastic; specimen roses and other shrubs are planted through the plastic (lower left). Weeds are growing (left foreground) in an old gravel mulch on plastic (lower right).

in Iowa (Bickelhaupt 1983). Bickelhaupt prefers it to black plastic, even though it is more expensive.

Peat moss does not rewet easily after it dries. Thick thatches of some turf-grasses may repel water so that little gets into the soil.

Mulch Management

Mulches can be most effective and most easily maintained if appropriate guidelines are followed. A mulch may be used primarily for appearance, weed control, water conservation, erosion control, improvement of soil structure, protection of soil from cold or heat, provision of a firm, clean surface for pedestrians, or for disposing of garden litter. These purposes are not mutually exclusive, but the choice of a mulch and maintenance methods should depend on the primary purpose.

Soil Preparation. Installing a mulch is usually one of the last steps in developing a planting bed or landscape. The site should first be brought to grade to provide surface drainage. Weeds should be killed and large ones removed, especially if they have gone to seed. If surface soil is not in good physical condition, loosen it to increase water and air infiltration. Unless fabric or plastic is used, apply a pre-emergent herbicide before the mulch is placed to reduce weed growth. In small planting areas isolated by buildings and pavement, it may be wise to fumigate soil to destroy most weeds and their seeds, as well as any disease organisms and other pests.

Mulch Decontamination. The bark, wood, and foliage of some plants contain substances toxic to other, particularly young, plants. Leaching will remove water-soluble toxicants, and composting will remove or inactivate toxicants and kill most disease organisms. The toxic substances can be leached from redwood and cedar sawdust by heavy sprinkling or several months of rain. Water should drain into a sewer or special settling basin so that surface and underground water and nearby plants are not polluted. Pollution-tolerant fish have been killed by as little as 10 ppm of leachate extracts from western red cedar (Scroggins 1971).

Even though they may contain substances toxic to other plants, pieces of freshly chipped wood and bark (with an average diameter of about 20 mm or 0.75 in.) are commonly used to mulch mature plants with no apparent ill effects. A mulch containing toxins, however, is most likely to cause injury if the mulch particles are small (as in sawdust), if the mulch is particularly deep (over 100 mm or 4 in.), if heavy rains or irrigation follows application, if the mulched plant is young, or if a high proportion of the plant roots are in the surface soil. When there is concern about the toxicity of fresh chips, they should be spread thinly (in layers 20–30 mm or 1 in. thick) under the plants. The toxins will be adsorbed on the soil particles and inactivated by other soil constituents. If injury should appear, the mulch can be removed from underneath affected plants or the soil heavily leached to move toxins below the major part of the root zone, or both.

Compost Preparation. **Bark and sawdust can be safely used as soil amendments, as well as mulch, provided they are composted first to minimize later nitrogen draft and to eliminate toxins.** Leaves, grass clippings, and other plant refuse can be composted along with sawdust and chips of wood and bark. Decomposing microorganisms need nitrogen to increase the speed of decomposition. Wood decomposes more quickly than bark, softwoods more quickly than hardwoods, fine particles more quickly than coarse, succulent material more quickly than woody, fresh tissue more quickly than dry. For each cubic meter of compost composed primarily of woody material, 0.5 to 1.5 kg of nitrogen (1–3 lbs/yd³) should be used, the larger amounts for more rapid decomposition. The total carbon to nitrogen ratio of the materials in a compost pile should be about 25 to 1 (Raabe 1974).

Methods of handling compost depend on the amount of material and the availability of space or facilities, equipment, and labor (Fig. 14–7). Bins can be used to store compost for small landscape operations, while large stacks or long windrows can be used for larger quantities of compost. A compost bin is usually 1 to 2 m (3–6 ft) wide and about 1 meter deep, with a variable length. The sides of the bin can be built of wire, wood, or open brick or cement block; if the compost will be turned frequently, the sides should be solid. Large amounts of compost should be handled with equipment and formed into stacks or windrows about 5 m wide × 2.5 m high (15 × 8 ft); the amount of material and space will determine length.

Figure 14–7   After the 21-day composting period, a pile of wood-chip and sewage-sludge compost is moved to a holding area ready for use (left); 100-mm (4-in.) diameter perforated flexible hose was used to force air through the compost pile. (Photo courtesy N.H. Experiment Station, University of New Hampshire) Compost material is forked from the right to the left bin (left bin cover visible above the person's hand) (right). Each bin (about 1 m^3 or 1 yd^3) has a cover to confine heat and maintain optimum moisture. (Photo courtesy Robert Raabe, University of California, Berkeley)

Those who use bin composting are commonly advised to fill the bin with alternating layers of organic matter 150 to 300 mm (6–12 in.) thick and of topsoil about 25 mm (1 in.) thick. To each layer of organic material, add an appropriate amount of nitrogen fertilizer. Recommendations for composting, however, vary a great deal: Some people use organic matter and nitrogen alone, while others add alternating layers of soil and manure, more exotic fertilizer combinations, and lime. Each compost enthusiast seems to have a favorite recipe.

Garden wastes can be converted to compost in three weeks by turning the pile every second or third day if the compost material has a carbon-to-nitrogen ratio of about 25 to 1 and enough moisture to favor decomposition (Raabe 1974). A desirable carbon to nitrogen ratio is not difficult to obtain. Fresh plant material (green leaves, grass clippings, and weeds) has a ratio of about 12 to 1; dried plant material (dried leaves and straw) has a ratio between 50 to 1 and 100 to 1; woody plant materials range from a low ratio in small, soft material to 200 to 1 in sawdust. Mixing equal volumes of green and dried (straw or leaves) material will give approximately the correct ratio; no nitrogen needs to be added (Raabe 1974).

Compost materials should be placed in solid-walled bins to hold the heat. Compost turning is most easily and completely accomplished by transferring the compost from one bin to another. Material toward the outside of a bin should be placed near the center of the second bin to equalize temperatures because temperatures are higher in the center. Frequent turning aerates the compost to aid decomposition and prevents the compost from becoming so hot that most of the decomposing organisms are killed. At the beginning, the pile should not be turned until it has become very warm (several days). If the pile is slow to heat, nitrogen fertilizer can be sprinkled on the material as it is moved to the second bin. If the pile smells of ammonia, sawdust (carbon) should be added during the next turning to absorb the excess nitrogen.

Compost piles in the open should be packed more at the edges than at the center so that the center will settle lower than the sides. Rain or water applied will then wet the compost rather than run off. Compost windrows are built in about 300-mm (12-in.) layers with nitrogen fertilizer spread between the layers. The top of the windrow should be fairly level to retain water.

Frequent turning and mixing of the compost of woody material (once every two to three weeks) hastens decomposition and raises compost temperatures to between 50° and 80°C (120° and 160°F) (Hoitink and Poole 1977). Compost mixing can be a slow, tedious procedure with a pitchfork, but a skiploader can turn large quantities of compost quickly. In warm weather, frequently turned composts high in hardwood bark should be ready to use for mulching in 10 to 12 weeks, and those high in softwood bark within six weeks.

Many cities are composting wood chips and sewage sludge as an alternative to incineration or landfill disposal of sludge. A system of windrowing and forced aeration is simple, easy to operate, and adaptable to different situations (Fig. 14–7) (Leighton, Harter, and Crombie 1978). Wet sewage sludge (about 20 percent solids) is mixed with wood chips at a ratio of three parts chips to one part sludge by volume, then the pile is covered with about 300 mm (1 ft) of composted sludge for insulation purposes. Fans draw air through the pile for the first 10 to 14 days and exhaust it into a small pile of composted sludge for odor absorption. After this period, the fans are reversed, and air is blown through the pile for another 10 days. The resulting compost is stable, practically odorless, and essentially devoid of pathogens.

Leighton, Harter, and Crombie (1978) found that composting costs were competitive with other methods of sludge disposal. Composting appears to be a viable and satisfactory method of sludge disposal and compost production for communities with a low concentration of heavy metals in the sludge.

Plant toxins and pathogens will normally be eradicated during composting. During cold weather, detoxification and composting will be slower. Even though the material may not be completely composted, it will be safe to use around plants.

When compost is to be used as a soil amendment, composting should continue until the desired degree of decomposition has occurred.

Mulch Application. Loose organic mulches are usually applied in a layer 100 to 150 mm (4–6 in.) deep, but even a thin layer of these mulches (10–20 mm or 0.5 in. thick) can quite effectively reduce surface puddling, erosion, and mud splashing from rain. It can also moderate summer soil temperatures and reduce evaporation from soil. Mulches, however, must be thicker to effectively control weeds, minimize soil compaction from traffic, and provide firm footing under wet conditions. The insulation value of an organic mulch, particularly against winter cold, depends on its thickness.

Fresh leaves and grass clippings should be spread in thin layers so they can dry before more are added. Coarse leaves are less likely to mat, but wind can often blow them away. Composting or shredding large leaves might be the best solution.

Geotextile fabric and black plastic covered with bark or rock can be used to mulch shrub plantings and other areas free of traffic. They effectively control weeds and surface evaporation, but the plastic is easily torn. The soil should slope away from plants for drainage. The fabric can be cut in 1250-mm (50-in.) wide strips for ease of application. The plastic can be slit and a hole cut to fit around a plant trunk. Pricking the plastic will provide holes for air and water movement without encouraging weeds. Adjacent sheets of either material should be overlapped about 100 mm (4 in.).

After the fabric or plastic is in place, apply bark, gravel, crushed rock, cobblestones, or other material thickly enough to cover the plastic. The border of the planting should be high enough to contain the mulch. Some geotextile fabric is guaranteed against damage from ultraviolet light for five years. The plastic should last several years if it is not disturbed by traffic. Problems may include dust, plant litter, and small pieces of trash that accumulate in the mulch. After a few years, seeds will also germinate in the mulch and lower its effectiveness (Fig. 14–6). It is tedious to remove leaves and litter from a mulch; most litter would just blend in with organic mulches. To replace the sheeting and bark or rock mulches, you must remove the mulch covering, install new sheeting, and replace the mulch cover. Most other mulches can be simply upgraded by adding more mulch material.

Fertilization. Improved growth from mulching may decrease or even eliminate the need for fertilizing. Whitcomb (1980) reports winter injury to young Chinese pistaches that were invigorated after being mulched with plastic; the number of injured plants increased with increasing nitrogen application. In some instances, however, nitrogen deficiency may occur after installation of an organic mulch; most other nutrients increase under mulching. A fine-textured, rapidly decomposing material in close contact with a low-nitrogen soil will be most likely to cause nitrogen-deficiency symptoms in mulched plants. **Observation of plant growth is the best guide to fertilizer needs.**

If growth is vigorous and leaf color dark, no fertilizer is needed. On the other hand, if growth is less than desired and leaves are yellowish, nitrogen should be applied. If the nitrogen-deficient plants are small, the mulch can be raked from beneath them, 25 to 100 g of nitrogen per square meter (0.05–0.20 lb/yd^2) spread on the surface around each plant, and the mulch replaced. Fertilizer should not be concentrated in basins. If the plant canopy is more extensive, apply a nitrate or urea fertilizer on the mulch (assuming no plastic) at a rate of about 1.5 kg N/100 m^2 (3 lbs/1000 ft^2) and sprinkler irrigate the fertilizer (with about 25 mm or 1 in. of water) through the mulch into the soil below. If experience indicates that mulching may cause nitrogen deficiency, nitrogen can be applied before the mulch is put in place (see Chapter 12). Alternatively, fertilizer can be mixed with the mulch, at a rate of 1 kg N/100 kg or 600 liters of mulch (1 lb/100 lb or 10 ft^3). The fertilizer, whether mixed with the mulch or applied on top and watered through, will quicken decomposition of the mulch but should also correct or prevent nitrogen deficiency.

If deficiency symptoms persist two months after fertilization, the soil may be poorly aerated because of high moisture under the mulch. The mulch may have to be thinned in such a case.

Irrigation. If irrigation is necessary, mulching can reduce its frequency. Plantings treated with organic mulches respond best when they are drip-, soaker-, or sprinkler-irrigated. Flood and furrow irrigation will carry light mulch to low areas or the downwind portions of a basin. Sprinkler irrigation keeps a mulch moist and thus reduces fire danger. Plantings treated with plastic mulch can usually be irrigated satisfactorily by flood, sprinkler, or drip irrigation.

Fire Protection. Straw and peat moss are highly inflammable and should normally not be used for landscape mulches. Wood-chip mulches are used extensively in campgrounds without posing serious fire problems. Wyman (1957) has also found that spent hops burn slowly. Sprinkler irrigation and rain will keep most organic mulches resistant to fire. Straw used on top of ground-cover plants to protect them from winter cold should be removed in spring to allow for the best growth and appearance of plants, as well as to reduce the fire hazard.

In arid regions, where organic mulches are more liable to catch fire, they can be sprayed with a solution of ammonium sulfate or diammonium phosphate (1–1.5 kg/40 liters or 2–3 lb/10 gal of water per 100 m^2 or 1000 ft^2). These effective fire retardants will also enhance soil fertility and mulch decomposition as they soak in below the mulch surface. They are not used in irrigated landscapes or areas with summer rain, where mulches are already slow to burn, and chemicals would wash off the mulch surface.

Inorganic mulches offer most of the benefits of organic mulches and in addition will not burn or decompose.

Disease and Pest Control. Do not use mulch materials that might be infected with disease organisms to which your plants are susceptible. Even though composting would probably kill any pathogens, it is better not to compost diseased material. Do not mulch plantings where *Armillaria mellea* is present, because the higher soil moisture under a mulch will encourage root rot.

If rodents damage the trunks of trees and shrubs during the winter, place gravel or crushed rock 100 to 150 mm (4–6 in.) deep around the trunks. Rodents find it difficult to burrow in gravel. Keeping mulch away from the base of plants will also decrease rodent damage, even without the gravel barrier. If roots are also being damaged, the rodents must be trapped or poisoned; see a pest control officer in the local agricultural commissioner's office for specific recommendations.

   Protection from Cold.   Mulches protect against heat loss by reducing radiation and conduction from below. Low-growing plants can be protected from cold if covered with a loose mulch, such as straw, from soon after the first frost until spring growth. On a calm, clear night, temperatures above a mulch will be colder than those above bare soil. This difference in temperature can sometimes be the difference between severe injury or death and safety to the upper parts of plants. In these situations, the mulch should be removed so the soil is warmed by the sun during the day and in turn protects the plants at night. Under a leafy canopy, however, a mulch will make little difference, because exposure to the sun and the cold night sky will be limited.

   Mulch Incorporation.   Although most organic mulches will improve the physical and chemical properties of soil, that is usually not their primary purpose. In most situations, mulches should remain on the surface and should not be incorporated into the soil.

AERATION/DRAINAGE

Aeration of soil is closely related to the adequacy of its drainage. If a soil is well drained, it is usually well aerated; if a soil has poor drainage, it is poorly aerated. A continuous, adequate supply of oxygen is required for roots to supply plants with sufficient water, nutrients, and certain hormones (Kozlowski, Kramer, and Pallardy 1991). All too often compacted soil, fine-textured soil, surface crusting, pavement, flooding, or a high water table interfere with gas exchange between the soil and air.

   Major efforts (deep ripping, drainage systems, pavement installation, and full-aeration systems) usually made during landscape development to improve drainage and aeration were discussed in Chapters 8 and 11. Practices that might be used in an established landscape are presented here.

   Surface crusting and future compaction can usually be minimized by mulching. A number of practices have been recommended to overcome aeration and drainage problems due to compaction and fine-textured soil.

   Keep traffic to a minimum on landscape soils, particularly when they are wet. Restrict traffic to paths or certain areas. If the planting is irrigated, sprinkler and drip systems should ideally apply water at or slower than the infiltration rate. Allow the soil to be dried to near the critical level for the plants before irrigating; this fosters soil shrinking and swelling which improves aeration.

   Aerifying turf at least 100 mm (4 in.) deep (150 mm is better) under trees should improve water penetration, drainage, and aeration for both the turf and trees. Fertilizer and aeration holes are occasionally drilled 0.3- to 0.45-mm (12- to 18-in.) deep, spaced 0.6 to 0.75 m (24–30 in.) on center. Deeper holes affecting a wider volume of soil can be washed with a water jet (Fig. 14–8) made from a 20-mm (0.75-in.) galvanized pipe 1.5 to 2 meters (5–7 ft) long with a beveled hard-

Figure 14-8 A water jet is used to drill (wash) a hole to improve future water penetration and aeration. Jetting holes at an angle will have a greater effect than a vertical hole.

steel tip on one end, and a valve and hose connection on the other. Water is best supplied by a high-pressure sprayer at 10 to 20 kg/cm^2 (150–300 psi) pressure. A garden hose will work, but its low pressure makes the job a slow one. The water jet is to be used for making holes, not for irrigating.

To use the water jet, push its tip into the soil at an angle, turn the water on, and push the pipe into the soil. The water washes the soil ahead of the tip and carries it to the surface around the outside of the pipe. A channel about 50 mm (2 in.) in diameter can be washed fairly easily in all but the tightest soils. Holes are usually started near the trunk of a large shrub or tree and slanted down and away from it. Three or four holes per plant have been thought to be sufficient; for large trees, additional holes near the dripline may be desirable. Holes may have to be plugged while the remaining ones are bored so mud and water do not flow into them. After all holes have been dug, they are usually left open or filled with gravel.

Roots usually grow into the channels and eventually impair their effectiveness. After two to three years, new channels can be made in between the previous holes. Water-jetting young trees in pavement planters (which are open to the soil around) and along streets should encourage deeper, wider rooting, which improves tree growth, increases the interval between irrigations, and minimizes later damage to sidewalk and pavement.

The Bartlett Tree Research Laboratory (Fraedrich, Booth, and Smiley 1989) offers an aeration alternative: Dig trenches, 150-mm wide × 300-mm deep (6 × 12 in.), radiating out from the trunk at 45° intervals. They should start close to the trunk, without injuring roots, and extend to the dripline of small trees or halfway to the dripline of large trees. Additional trenches could be dug between the primary ones keeping 0.6 to 1.2 m (2–4 ft) from them. Fill the trenches with topsoil of good structure.

Woodtli (1989) reported on equipment and techniques to liquify surface soil with water and pump the slurry from around the trunk and surface roots. The soil is replaced with "new vital" soil (soil/compost/fertilizer and small balls of specially burned clay).

These various methods of "aerating" soil under trees should improve drainage and aeration, but few studies have been done to quantify the improvement or compare different methods or variations of them. A fertilizer trial in Ohio found that drilling holes under trees in a poorly drained soil improved tree growth whether fertilizer was put in the holes or not (Smith and Reish 1975). When aerating is done, the trees are usually also fertilized, pruned, and possibly irrigated, with no checks (controls) left. Growth is probably improved, but what was the contribution of each factor? It would be useful to have more reliable information on which to base a practice that most people believe to be worthwhile.

Fracturing Soil. Dynamite was one of the early treatments for tight soil and for successfully breaking up hardpan and plowpan. A dynamite blast will not decrease the bulk density (compactness) of a soil; in fact, in all but the driest soils, it will increase the bulk density, particularly below the blast. In addition, roots may be seriously damaged.

Fracturing soil with compressed air began in the late 1920s by several tree-care companies using equipment which they developed (Smiley and others 1990). In Ohio, Charles Irish patented and sold the *Aero-Fertil Gun* for aerifying and fertiliz-

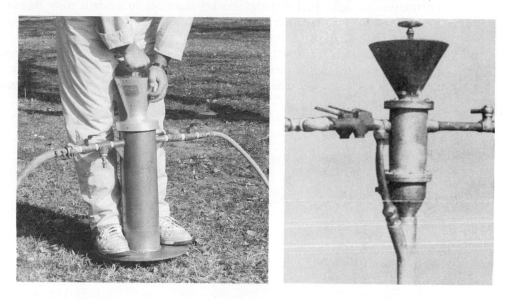

Figure 14–9 Air can be injected under pressure (7 to 10 kg/cm^2; 100 to 150 psi [lbs/in.^{2}] as is being done with this Grow Gun® to fracture the soil (left). After the soil has been fractured, water, dry or liquid fertilizer, or amendments can be forced into the soil under pressure. Charles Irish in Cleveland, Ohio, patented the Irish Aero-Fertil Gun of similar design in the 1930s (right). Trees have responded to the use of this equipment, but it is not clear whether it is due to the soil being fractured, aerated, fertilized, irrigated, amended, or some combination; research now underway should clarify the situation. (Left photo courtesy Michael Hutnick, Sta-Green Tree Service, Carmichael, CA; right photo courtesy Edwin Irish, Warren, MI)

ing trees. Its purpose was to inject air under pressure into the soil to fracture compacted layers; fertilizer, amendments, and/or water could be added after fracturing. Use of the equipment was essentially discontinued with the beginning of liquid fertilizer injection in the 1940s (Smiley and others 1990).

In the last decade, new equipment has renewed interest in fracturing and aerating soil. Two are the Grow Gun® sold by the Grow Gun Corp. in Arvada, Colorado, (Fig. 14–9) and the Terralift® sold by the R. E. Jarvis Co. in Fayville, Massachusetts. The Grow Gun® is similar to the design and operation of the Irish Aero-Fertil. The injection hole must be drilled before the injection tube is inserted; a separate compressor supplies the air. The Terralift® is a self-propelled, self-contained unit. These two pieces of equipment were described, evaluated, and tested by Smiley and others (1990) at two sites in North Carolina, one having a sandy clay loam over clay and the other a clay loam over clay. The fracture radii produced by the Grow Gun® were 0.9 and 0.6 m (34 and 25 in.) and by the Terralift® were 1.1 and 1.2 m (45 and 47 in.) for the two soils respectively. There were no significant decreases in bulk density (soil compaction) following any of the treatments. The oxygen diffusion rate (ODR, soil aeration) was higher only at the fracture layer. No increase in ODR was found within or outside the fracture. At the end of two growing seasons following treatment, slight (but not significant) increases in growth were measured on the willow oak and Bradford pear trees growing in the test areas (Thomas Smiley, Charlotte, NC, 1990 pers. comm.).

It may be that fracturing soil with compressed air may be little different than with dynamite; though it certainly would be difficult to break up hardpan with compressed air. More experimental results are needed to properly assess such aeration practices.

CHAPTER 15

Pruning

Pruning is the removal of plant parts—usually shoots and branches, but sometimes buds, roots, and even flowers and fruit. By pruning, one can control the growth of plants to enhance their performance or function in the landscape. Pruning can increase the structural strength of trees, the productive capacity of fruit trees, the trunk quality of lumber trees, the quality and size of flowers and fruit, and the aesthetic appeal of many plants. **Most importantly, pruning as part of the training of young trees can ensure structurally strong trees which will be safer and require less corrective pruning when mature.**

GENERAL PRINCIPLES OF PRUNING

Purposes of Pruning

Plants grow in many shapes and sizes. Some have central leaders with tall, straight trunks (an *excurrent growth* habit) (see Fig. 2–14). Others have several main branches with spreading crowns (a *deliquescent, diffuse,* or *decurrent* growth habit) (see Fig. 2–15). Between these extremes, intermediate forms occur. The natural characteristics of different plants can be exploited through landscape use and maintenance practices. Pruning can enhance plant appearance and safeguard plant health and well-being. Depending on its extent, pruning can affect a plant from root to crown. Growers should therefore be familiar with pruning techniques and plant responses to pruning. You should be able to determine the growth habit of a plant by observing its growth and its response to previous pruning and thereby be able to properly prune even unfamiliar species. This discussion will cover pruning techniques and plant responses in general but will offer little advice on pruning particular species.

Plants are pruned for a number of reasons.

Training Young Plants. The arrangement, attachment, and size of scaffold branches can be controlled to produce vigorous and mechanically strong plants. Pruning should take advantage of the plant's growth habit, accentuating its natural tendencies, seldom modifying them greatly. You can create unusual plant forms through pruning, including topiary, espalier, bonsai, pleach, and pollard forms (discussed later), but to do so you must be familiar with the plant's responses to pruning.

Maintenance of Health and Appearance. Pruning can remove dead, diseased, injured, broken, rubbing, and crowded limbs. A dense top may be thinned to allow for the passage of light and air. Light is needed by the interior foliage of a plant and by other plants beneath it; air circulation reduces the incidence of certain diseases and allows sprays to penetrate more effectively. Proper thinning reduces wind resistance, which can create deformities or uproot a tree; however, an occasional branch lacking taper may break due to greater exposure.

Control of Plant Size. Pruning can reduce shade, the danger of windthrow, and interference with utility wires; can simplify pest-control spraying; and can prevent the obstruction of views and traffic. Choosing plants that will be an appropriate size at maturity minimizes the need for pruning. If a plant must be pruned more than every five to seven years to control its size, it is the wrong plant for the particular location or use. Withholding nitrogen fertilizer and growing lawn under trees and shrubs will slow growth and reduce the frequency of pruning as a plant reaches the desired size.

Influencing Flowering, Fruiting, and Vigor. Pruning influences the balance between vegetative growth and flower bud formation. If young plants that flower on one-year-old wood are pruned, the development of flower and fruit may be delayed. On mature plants, pruning helps maintain vigor, minimizes overcropping (which results in small blossoms and fruit and broken limbs) and encourages annual flowering and fruiting throughout the plant. Pruning plants that flower on current season's growth stimulates shoots, producing fewer and larger flowers—particularly those that flower only from terminal buds.

Compensation for Root Loss. This is often given as a reason for pruning back the tops of newly planted plants. Considerable root loss occurs between the nursery field and landscape planting. A 100-mm (4-in.) tree may lose 95 percent of its absorbing roots when mechanically transplanted (Watson and Himelick 1982). But, if water is not limiting, removing a large portion of the top may not necessarily improve a plant's ability to survive and grow (Whitcomb 1987). Such pruning may instead delay the initiation of root growth (see next section).

Invigoration of Stagnating Plants. When plants are doing poorly but show no symptoms other than extreme lack of vigor, they may be pruned in a "kill or cure" operation.

Increasing the Value of Conifers. Nursery stock and Christmas trees may be sheared to improve their shape, density of foliage, and appearance.

Pruning Responses

Young Plants. Pruning removes leaves and buds that would develop into leaves. Two seemingly opposite effects occur when young plants and those that do not have a heavy flower and fruit load are pruned.

Invigoration of individual shoots is the universal response to pruning (Fig. 15-1). Pruning off foliage and buds that would develop into leaves allows the root system (which is not immediately affected) to supply each remaining shoot, leaf, and bud with more water and nutrients than before. Individual shoots grow more rapidly and later into the season; leaves become larger and darker in color. Even with larger leaves, though, total leaf area will be reduced on more severely pruned plants because there will be fewer shoots. A pruned plant will transpire less water than will an unpruned plant if its shadow is reduced in size or density, but the reduction will not be in proportion to the reduction of foliage.

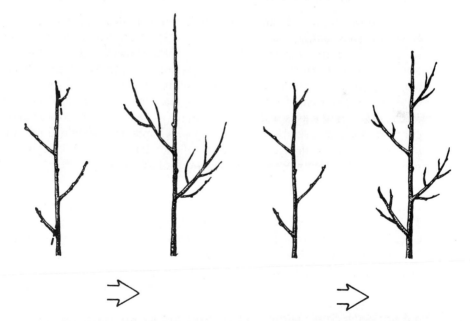

Figure 15-1 The effects of pruning. The trees to the left of each arrow were exactly the same before two of four lateral branches were pruned off the tree on the left. The two remaining laterals and the terminal of the pruned tree grew more than the corresponding laterals on the unpruned tree, but by the next dormant season the unpruned tree (extreme right) will have made more total growth.

General Principles of Pruning

Dwarfing **results from pruning young plants and those that do not flower and fruit heavily.** Fewer leaves and buds are left. Though individual leaves may be larger, total leaf area will be less. Shoots will grow later in the season, using foods produced by the leaves. After shoot growth stops, the pruned plant has less time to produce food for other growth and to store food for the next season. Less total growth is the result. This can be observed and even measured by the relative sizes of trunks on plants that have been pruned more and less severely.

A young plant that has been pruned will usually exhibit the following characteristics at the end of the growing season:

The top and root systems are in balance

The top and roots will be smaller than if the plant had not been pruned

There will be less stored food because the plant had a smaller leaf area for photosynthesis, and its leaves were active for a shorter time

The extent of invigoration and dwarfing will depend on the severity of pruning. A branch that needs subduing should be pruned more severely to reduce its relative growth. A branch that needs encouraging, however, should be pruned lightly or not at all; branches that shade or compete with the branch to be encouraged should be pruned more severely. Removing dead, weak, and heavily shaded branches has little or no effect compared to removing healthy, well-exposed branches.

Mature Plants. **Pruning does not always result in dwarfing.** Plants, such as peach and pyracantha, that produce flowers and fruit heavily on one-year-old wood can attain a greater leaf area by the end of the growing season if they are pruned when dormant. Both leaf and flower buds are removed by pruning. No new flower buds will form and bloom in the spring following pruning, but shoots from the leaf buds that remain can form more leaves than they would have without pruning. Thus, shoots and leaves will form in greater quantities than flowers and fruit. It is possible to prune severely enough that the leaf area developed will not only increase fruit size but also enhance total growth of the plant. Pruning a stagnated plant can stimulate it to grow more vigorously and also more in volume.

Top Pruning. Removing branches of a young plant reduces not only the food supply to the roots but also the flow of auxin formed in developing buds and leaves. The spring root growth of deciduous plants may be delayed by the removal of shoot terminals, which provide the first auxin needed to stimulate root initiation. Richardson (1958) found a close correlation between the initiation of spring root growth in sugar maple and the presence of at least one physiologically active bud. The roots of sugar maple whips began to grow just before the terminal bud began to expand; when the terminal bud was removed, root growth slowed to zero in three days and did not resume for another five days, when the two uppermost lateral buds began to swell. This response may explain Whitcomb's (1987) finding that pruning bare root trees at planting (designed to balance the top with a reduced root system) improved neither survival nor growth. On unpruned trees, expanding buds and new leaves may stimulate enough additional root growth to more than compensate for any increased transpiration due to a larger leaf surface. **The best established reasons** for **pruning plants at planting are to remove damaged branches and to begin to develop tree structure.** It may be desirable, however, if there will be sufficient water, to leave as many terminals and leaves as possible in order to stimulate root growth.

The Influence of Pruning on Flowering and Fruiting. In most young plants, particularly those that produce flowers on one-year-old wood (Fig. 15–2a), pruning

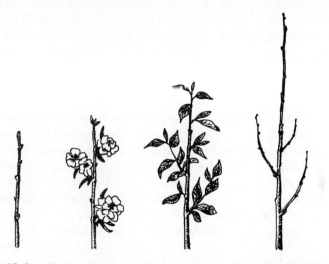

Figure 15–2a Spring-flowering plants bloom from buds formed the year before (extreme left). Flowers usually open before new shoot growth is extensive (left center). New shoots grow after the setting of fruit and fading of bloom (right center). Dormant buds on newly formed shoots overwinter to bloom the next spring (extreme right).

Figure 15–2b Summer-flowering plants bloom from buds on the current season's shoots. Overwintering twigs have dormant vegetative buds and often remnants of the previous summer's blossoms (extreme left). New shoots grow from lateral buds in the spring (left center), and flowers or flower clusters form on current shoots, often terminally (right center). The following winter, flower stalks may remain on the previous season's shoots (extreme right).

General Principles of Pruning

will delay flowering. It will invigorate the growth of individual shoots, so less food and fewer hormones will be available during the early summer, when the buds for the next season are differentiating into leaf or flower buds. If you keep pruning to a minimum and withhold nitrogen fertilization, plants will be more floriferous and will flower at a younger age. On the other hand, plants of low vigor may produce few flowers because of a lack of nitrogen and associated compounds; pruning channels available nutrients into fewer shoots, thereby increasing the nitrogen supply of each. Plants that flower laterally on growth of the current season will produce more flower buds along invigorated shoots. Plants that flower terminally (Fig. 15–2b) will have larger flowers or flower clusters. Measures that increase flowering will usually increase fruiting.

Root Pruning. As would be expected, root pruning has essentially the opposite effect of top pruning. It decreases the supply of nutrients and water to the top, though usually not in proportion to the roots pruned off. Thus, less vigorous growth of individual shoots and less total growth are produced. On a young, vigorous plant which flowers on one-year-old wood, root pruning will encourage flowering and fruiting at a younger age. The roots of established trees are sometimes pruned to obtain more fibrous rooting within the root ball (see Chapter 10); this is thought to enhance transplanting success. When surface roots threaten to damage walks, streets, paving, and foundations, they are commonly pruned back and efforts are made to stop or divert their growth. Root pruning can be used to decrease the vigor of severely top-pruned, vigorous shrubs. Root pruning is hard, dirty work, and the results are not always obvious, so it is usually discussed more than it is practiced.

The Time of Pruning. The appropriate time to prune will depend on the type of plant, its condition, and the results desired. Light pruning can usually be done any time. Unwanted growth is most easily removed while it is small, and early removal will have less of a dwarfing effect. Broken, dead, weak, or heavily shaded branches can be removed with little or no effect on a plant, no matter what the timing.

Most deciduous plants can be pruned any time during the dormant period between leaf-fall and spring growth with similar results. Evergreens will be set back the least if they are pruned just before spring growth starts. A few broadleaved evergreen plants grow most rapidly after the weather warms up later in the season. If these plants are pruned just before the period of most rapid growth, leaves will be kept productive for the longest time and pruning cuts will be concealed more quickly by new growth. The growth of young plants can be directed during the growing season itself. Branches in desired positions can be encouraged if you pinch back or remove competing shoots.

Plant development can be slowed and plant size maintained if pruning takes place soon after growth is complete for the season. Such pruning should not be so severe or so early as to encourage new shoot growth. If maximum dwarfing is desired, most plants should be pruned in the period from early to midsummer. This will reduce leaf area for the longest period. Pruning cuts should be discretely placed for minimum visibility.

Corrective pruning may be easier during the growing season. Branches that hang too low from the weight of leaves or fruit can be thinned. Dead and weak limbs can be easily spotted for removal.

If you wish to maximize flowering, timing will depend on the flowering habit of the plant (Fig. 15–2). Those that flower in summer or fall on the current season's growth, such as crape myrtle, jacaranda, and rose, should be pruned during winter, before growth begins. Moderate-to-severe pruning will favor the growth of fewer but larger blossoms or blossom clusters. Plants that flower in the spring from buds on one-year-old wood, particularly flowering fruit trees, should be pruned near the end of their blooming period. You can enjoy the blossoms and then remove them before many set fruit that competes with new shoot growth. Pruning at the end of spring bloom will have little or no debilitating effect on growth. Blossoming utilizes food in the bud and its shoot and makes little or no demand on other food reserves. This is well demonstrated by the branches that will bloom and initiate shoot growth after they are cut and brought indoors to flower.

Callusing and woundwood formation should be more rapid if a wound is made a few weeks before or after growth begins, assuming that bleeding is not a problem. In Illinois, Neely (1970) found that wounds on ash, honey locust, and pin oak closed as rapidly in the next growing season whether cuts were made in the spring, summer, or winter, but about 20 percent more slowly when they were made in the autumn. Summer wounds, however, closed much less than spring wounds during the current growing season, because they had less time to do so. This difference in closing did not carry over into the following season.

Shigo (1989) recommends that pruners avoid the spring flush of growth, when the cambium is active and bark is particularly vulnerable to being torn loose. Even though decay of proper pruning wounds is seldom a threat to trees, Shigo suggests that autumn, when most decay fungi are sporulating, is not a good time for pruning. This, along with Neely's (1970) findings, may make it wise not to prune in the autumn without good reason.

However, oozing sap, "bleeding," can be minimized if predisposed species are pruned in autumn and early winter instead of late winter and early spring. Wounds on mature trees, particularly deciduous trees such as birch, elm, and maple, can bleed heavily (Brown 1972). On susceptible trees, this can be minimized if only small cuts (less than 75 mm or 3 in.) are made. Bleeding is usually not harmful to plants, but if it is heavy and persistent it can cause bark injury below the pruning cut and can retard callusing in the lower portion of the wound. Brown (1972) describes a procedure for tightly binding small wounds to stop bleeding.

Cold injury may be increased if substantial pruning takes place before growth begins in the spring. Some plants, such as roses and subtropicals, can be stimulated into new growth if pruned in the fall and early winter, only to be injured when the weather turns cold. When winter temperatures go below −18° C (0° F), the hardiness of tissue near pruning cuts may be reduced even though growth is not stimulated by pruning. This is particularly true of some conifers (Brown 1972). If cold injury is a danger, it is best to delay pruning until just before growth begins in the spring.

The incidence of some diseases can be affected by the season of pruning. Ceanothus that is native to California is much more susceptible to fungus dieback (*Sclerotinia fructicola*) if pruning in the winter or early spring is followed by rain, which spreads the fungus; light annual pruning in the summer is best. Fire-blighted shoots and branches (those affected by *Erwinia amylovora*) should be pruned back heavily, at least 300 mm (12 in.) into healthy wood, as soon as possible to minimize the spread of infection down branches. Himelick and Ceplecha (1976) were able to stop or greatly retard the development of Dutch elm disease (*Ceratocystis ulmi*) by pruning out limbs that showed symptoms (yellowing and wilting of leaves) before 5 percent of the tree was affected. Pruning during or just before beetle flight, however, greatly increases the chance of infection, because fresh pruning wounds attract beetles. Check the influence of timing on infection and development of diseases before you prune species that are susceptible to serious vascular and foliage diseases and to boring insects that are attracted to pruned plants.

Solar heating capacities will be enhanced if you prune in the early fall to open up trees and allow more sunlight to reach exposed windows, solar collectors, and patios (see Chapter 5). The bare branches of deciduous trees can block up to 70 percent of the incoming radiation.

Types of Pruning Cuts.   The type of pruning cut not only affects the initial appearance of a branch or plant but also to a large measure determines growth. *Heading* and *thinning,* **two types of pruning cuts, produce quite different plant responses.**

Heading or heading-back **is cutting a currently growing or one-year-old shoot back to a bud, or cutting an older branch or stem back to a stub or a tiny twig** (Fig. 15–3). There are several variations of heading to a bud: *Pinching* is the removal of the terminal 20 to 50 mm (1–2 in.) of a growing succulent shoot; *tip pruning* is the selective heading of shoot terminals that may or may not still be growing; *shearing* is tip pruning without selecting individual laterals or buds, as when a hedge or topiary form is maintained. Heading a large branch or trunk is often called *stubbing.*

In response to heading young branches and leaders, new growth develops from one or more buds just below the cut; lower buds usually do not grow (Fig. 15–4). Depending on pruning severity, the new growth is usually vigorous, upright, and

Figure 15–3 Heading back is pruning to a stub (lower branch), a small lateral (trunk), or a bud (terminal on small lateral).

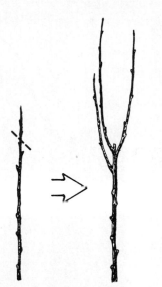

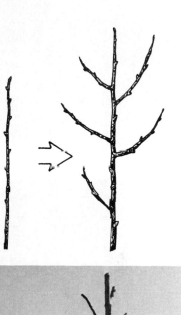

Figure 15–4 Heading a one-year-old shoot (left pair above) will force two to four buds just below the cut into vigorous upright growth. Growth from a similar but unpruned shoot will be more uniformly distributed along the shoot and will be less vigorous (right pair). The London plane tree in the foreground of the photograph (right) was headed back (stubbed) shortly before the photograph was taken; the tree immediately to its left was headed back the winter before. Note the vigorous upright shoots stimulated below the pruning cuts.

dense. New foliage and branches may be so thick that lower leaves, and nearby plants, are severely shaded. Shoots that grow from the trunk or large branches (epicormic shoots) after stubbing come from latent buds and are attached by only the thin layer of current season's wood (Fig. 15–5). Branches from these shoots, especially when young, are weakly attached and can break off easily.

Thinning or thinning-out **is the removal of a lateral branch at its point of origin or the shortening of a branch by cutting to a lateral large enough to assume the terminal role** (Fig. 15–6). Reducing the height of a tree or branch by thinning the terminal to a large lateral is called *drop-crotching*. The lateral to which a branch or trunk is cut should have at least one-third to one-half (N.A.A. 1987) or at least

General Principles of Pruning

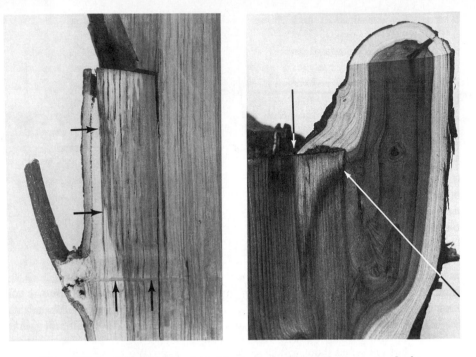

Figure 15-5 A latent red oak bud (note the bud trace [lower arrows]; the bud was formed the first year the branch grew as a shoot) was stimulated to grow by the wound inflicted above it the year before (left). The epicormic branch formed is attached only by the current-year's xylem. The barrier zone (upper arrows) formed after wounding further physically weakening the branch attachment. Vigorous sprouts forced to grow after heading cuts are similarly attached (right). Note the lack of union (between arrows) of the lateral branch that grew after this coast redwood was headed back (right); even though at least ten years old, the branch is weakly attached. (Left photo courtesy Alex Shigo and Kenneth Dudzik, U.S. For. Serv.)

one-half (Perry 1988) the diameter of the cut being made. I prefer at least one-half because such a lateral would be about one-fourth the size of the trunk, while one-third would be only one-ninth the size. "Heading" has been used to denote cutting back to a large lateral. **"Thinning," however, seems to be the more appropriate term, because little or no stub is left and the plant responds as it would to other thinning cuts.** The response to thinning is distributed more evenly throughout a plant than is the response to heading. A plant becomes more open but retains its natural form. More light will penetrate a plant that has been thinned, and foliage will grow more deeply inside it. For a given severity of pruning, growth of individual shoots will be less vigorous after thinning than after heading.

Shigo equates "cut at nodes" as a thinning cut as defined above (1986b, 1990). This might be true for whorl-branched conifers, which form only one whorl of branches a year, but not for other trees. It is true that branches arise at nodes, but there can be many nodes between branches. The term "node" refers to the location on a stem where leaves and buds arise and branches originate (Weier and others

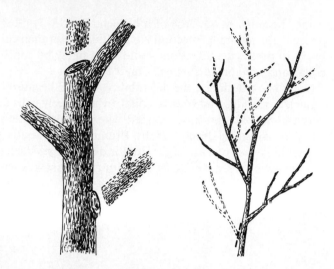

Figure 15-6 Thinning-out is removing a branch at its point of origin or shortening a branch or trunk by cutting to a lateral large enough to assume the terminal role, commonly called "drop-crotching." This applies to mature (left) as well as young (right) trees.

1982, Horticultural Research Institute 1971). Nodes are not easily seen on stems more than a few years old. A node would be a valid location for a thinning cut only if there were a large enough branch at the node; all other "nodal cuts" would be heading cuts. The term "node" is useful primarily for young stems where leaves or buds are visible.

Location of Pruning Cut. The type of pruning cut not only influences subsequent growth, but its location in relation to the branch attachment determines the size of wound and also affects callusing and woundwood formation, exposure to decay, and the possibility of ring shakes (circumferential separation of xylem along an annual ring) (Shigo and others 1979). The closeness of a cut also affects the amount of regrowth from the base of a cut and the strength of attachment when the end of a branch is thinned to a lateral.

Natural-target pruning (NTP) (Shigo 1983) and *conventional pruning cuts* (CP) (Neely 1988b) differ from "flush" cuts. Before 1979, several authors recommended the final cut be flush with or as close as possible to the trunk or mother branch. **Flush cuts are no longer recommended;** they are unnecessarily large and expose trunk tissue to the possibility of decay. Shigo and coworkers (1979) observed that flush cuts on black walnuts often resulted in multiple ring shakes (xylem separation circumferentially along annual rings). A flush cut removes trunk cambium and severs phloem and xylem pathways for food and water in trunk tissue that was near the branch removed; trunk tissue above and below a flush cut facing the afternoon sun is exposed to more intense radiation with little or no sap moving upward to keep the tissue moist and cool. Canker and/or death of the tissue is likely.

Since 1983, *natural target pruning* (NTP) has become the goal of most informed arborists. Pruning cuts are to be made close to but beyond the *branch bark ridge* and the *collar* at the base of a branch (Fig. 15–7) (Shigo 1983). Such pruning cuts retain "the natural protection zone within the branch collar"; should a pruning cut begin to decay, the infection would most likely be confined to branch tissue in the collar and not spread further (Shigo 1989).

Neely has questioned the desirability of NTP because in order to be sure that the branch bark ridge and branch collar (which Neely calls and is hereinafter called the *branch shoulder*) are not violated when pruning with a chain saw, there is a tendency to leave stubs (Neely 1988b). Pruning stubs are unattractive, slow to callus, and some think they offer a larger site for decay to get started. Compared to a NTP cut, a *conventional pruning* (CP) cut with a chain saw is easier and faster to make

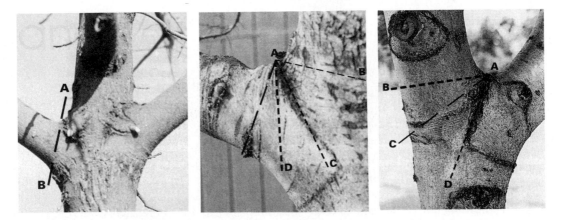

Figure 15–7 When the branch bark ridge (A) and the branch collar (B) can be located, remove the limb by making the final cut (A-B) just to the outside of the branch bark ridge and the collar (left).

When the branch collar is not visible or you wish to check its location (center), two estimates will help locate the proper final pruning cut: (1) the angle between an imaginary line parallel to the branch to be cut (A-B) and the branch bark ridge (A-C) should equal the angle between the branch bark ridge (A-C) and the line of the pruning cut (A-E), or (2) the angle between the branch bark ridge (A-C) and an imaginary vertical line downward from the branch bark ridge (A-D) should equal the angle between the imaginary line A-D and the line of the pruning cut (A-E). (Line A-D is often parallel to the upper portion of the branch being cut to.) Both methods indicate practically the same pruning cut (A-E) and the possible location of the branch collar (note the line of rough bark below D is not the branch collar).

The pruning cuts (left and right photos) are made just outside the branch bark ridge and the branch collar (Shigo 1986a). If a left stub is longer than wanted and/or the final pruning cut with a chain saw must be upward because of the angle of the cut, the pruning cut can be made just outside the branch bark ridge through the branch collar (Neely 1988c).

When thinning to a lateral branch (drop-crotching) (right), the final cut should be made on a line (A-C) that bisects the angle formed by an imaginary line perpendicular to the leader or branch being removed (A-B) and the branch bark ridge (A-D). A cut closer to A-D might weaken the attachment of the limb remaining; it could split out more easily.

with less chance of tree and worker injury (Neely 1988b); because of the angle of the final cut, most CP chain saw cuts can be downward while many proper NTP cuts must be upward.

Neely (1988b) compared the closure rates of 20 pruning cuts through branch shoulders (CP) with those made outside the branch shoulder (NTP) on three tree species for each of four years, a total of 240 comparisons. The branch diameters ranged from 25 to 34 mm (1.0–1.3 in.). The CP cuts were similar to cutting through *shoulder rings* as described by Davey (1967). Even though CP wounds were 27 to 42 percent wider relative to branch diameters than NTP wounds, by the end of the first growing season the area of exposed wounds was approximately equal between the two pruning methods. After the second growing season, more CP cuts than NTP cuts were fully closed.

Shigo acknowledges faster closure of cuts through branch shoulders but considers speed of wound closure less important than not violating branch shoulders because of the possibility of decay. Neely, however, thinks the possibility of infection because of cutting through a branch shoulder is slight. Both agree that the branch bark ridge should not be cut, and if there is an obvious collar around the base of a branch (because it is probably declining or dead), the collar should not be violated (Fig. 15–16).

Natural target pruning is illustrated in Figure 15–7; CP cuts are described in the caption. **Which of these two pruning methods is used is not as important as which branches are cut and their size.** The mark of a satisfactory pruning cut is the uniformity of its closure.

Although the foregoing remarks pertain primarily to large pruning cuts, Shigo and coworkers (1979) indicate that branch bark ridges should not be violated even on young trees. An exception might be made with vigorous, young trunks on which a close flush cut would reduce protruding buds and thereby reduce the number of shoots that might be stimulated to grow. Shigo (1989) disagrees, even though such wounds would be small and close quickly. Some species in cold winter areas might be subject to radial shakes (xylem separation along a radial plane) from flush cuts that could lead to frost cracks (see Chapter 4).

Leave a short stub and cut upward when you thin the terminal of a young plant to a lateral (Fig. 15–12). The new terminal is less likely to split out. Preserving the branch bark ridges may decrease the likelihood of decay.

Making the Pruning Cut

Pruning shears come in a variety of sizes and shapes but are essentially of two types. One has a curved blade that cuts by passing close to a curved or hooked anvil (Fig. 15–8). The other has a straight blade that cuts against a flat anvil. The straight shears are commonly available only as hand shears (secateurs). The curved shears make closer cuts and are less likely to crush stem tissue, particularly if the shears are dull.

Hand and power saws are used to cut branches that are usually too large for pruning shears (Fig. 15–9). Chain saws are used to remove large branches (Fig. 15–10). They have greatly eased the effort needed in pruning, but often the chain saw

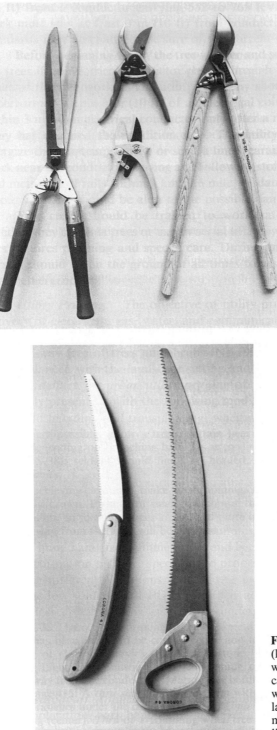

Figure 15–8 One-hand shears (secateurs) may be of the hook and blade (top center) or the anvil (bottom center) type. Long-handled shears also come in varying lengths and are of both types, though the hook and blade is more common (right). Hedge shears have two blades of similar design (left).

Figure 15–9 A hand saw that folds (left) is convenient when working with young trees because the saw can be folded and put in a pocket when not in use. A larger saw with larger teeth is more effective for more frequent sawing of larger limbs (right).

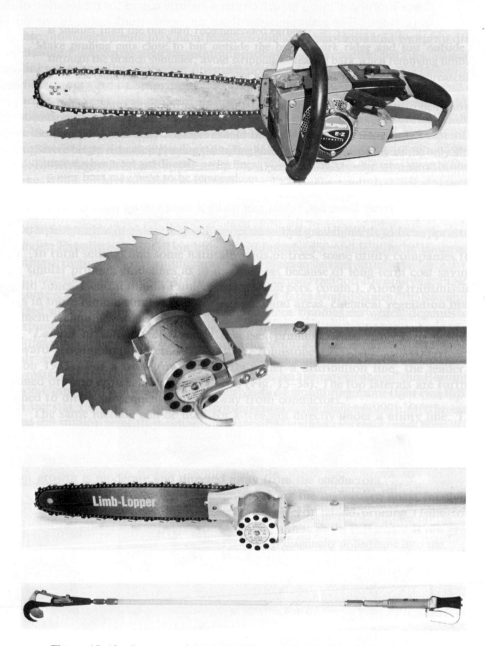

Figure 15-10 Power tools greatly add to the versatility and speed with which pruning cuts can be made. Gasoline-engine powered chain saw (top) and hydraulically operated circular and chain saws and lopping shears (bottom); photos are not shown at the same scale. (Photo of three lower tools courtesy of Fairmont Hydraulics, Fairmont, Minnesota)

blade is dropped into a branch crotch with little regard for the location of the branch bark ridges. The resulting large flush cut opens trunk tissue to the possibility of decay and usually takes longer to close than a smaller cut.

Other mechanically powered pruning tools, particularly those on poles (Fig. 15–10), greatly increase the versatility of the arborist. Even though most pruning cuts will not be accurately placed, the cuts cause few or no problems because they are usually made in the smaller branches. These tools can be powered by pneumatic, hydraulic, electric, and gasoline-driven equipment adapted to be used with aerial-lift equipment. A boom-mounted power saw with a limb clamp now available can cut limbs up to 170 mm (6.75 in.) in diameter, hold the cut portion, and remove it from the tree. This equipment is particularly adapted for utility-line clearance and other situations in which limbs would be difficult to remove and lower safely.

Small Branches. You can make a close cut by placing the blade against the branch bark ridge of the limb to be removed; an even closer cut can be made by placing the blade against the trunk or branch that is to remain (Fig. 15–11). A close cut reduces the number of shoots that might grow near the pruning cut. Conversely, if a short stub is desired, place the anvil against the trunk or branch to remain. However, unless the shears are sharp, the anvil will bruise the stub. Leaving a short stub and cutting upward when thinning a terminal to a lateral (Fig. 15–12) will lessen the chance of the new terminal splitting out.

Less effort will be required to remove a branch and less tissue will be crushed

Figure 15–11 Make a close cut by placing the blade of hook-and-blade shears just to the outside of the branch bark ridge of the branch to be removed (left). Less effort will be required if the blade cuts up (left) or diagonally (center) instead of down. Similarly, less effort is needed to make a diagonal heading cut (right); the branch is less likely to be crushed with a diagonal cut with a sharp blade nearer the branch base than the anvil. The top of the diagonal cut should be about 5 mm (0.25 in.) above the topmost bud left.

Figure 15-12 When cutting back to a lateral, place the anvil of the shears in the crotch (left) and cut up parallel to the direction of the lateral; the shears should be sharp. Leave a short stub so the branch bark ridge of the lateral is not cut (center). If you place the blade in the crotch and cut down, however, you will usually cut into the branch bark ridge and often split out the selected lateral (right).

if the blade cuts up or sideways (Fig. 15-11). A heading cut can be made with less effort when the cut is made diagonally instead of at right angles to the branch cut. It is also easier to cut a limb when you take a deep bite and place the fulcrum of the shears near the limb to be cut. If you cut through a limb too large for the shears or twist them while cutting, you can strain the shears and permanently damage them.

Where appearance is important, you can hide pruning cuts on shrubs and low-branched trees somewhat by angling cuts away from the direction of most frequent viewing (Fig. 15-13). The plants will appear more natural if they are cut to a lateral arising on the top of the branch; if the pruning cuts are horizontal, or parallel to the ground, they will be hidden from view. If you prune just before growth begins, the cuts will be covered by new growth most quickly.

Large Branches. Limbs larger than 25 mm (1 in.) in diameter will usually need to be cut with a saw. Branches much larger than 50 mm (2 in.) should be cut in three steps to avoid splitting back the branch and tearing the bark (Fig. 15-14). Make the first cut on the underside of the branch about 300 mm (12 in.) from the crotch. Cut the branch about one-fourth of the way through. Begin the second cut on top of the limb within 25 mm (1 in.) of the first cut and saw until the limb breaks off. The position of the second cut is usually recommended to be beyond the first (farther from the crotch). However, the cut branch should "jump" away from the crotch equally well on whichever side the second cut is made. The split is less likely to tear if the top cut is beyond the first (Fred Roth, California State Polytechnic University, Pomona, 1990, pers. comm.). On the other hand, when using a chain saw, the top cut is sometimes made closer to the crotch for safety reasons (Capel 1987).

Place the third cut at the crotch, making as small a wound as possible, but do

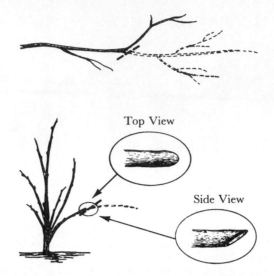

Top View

Side View

Figure 15-13 On low-growing shrubs, you can often hide pruning cuts by cutting back to a horizontal lateral growing from the top of the branch (top) or cutting to a bud so the cut surface is toward the ground and away from a viewer's angle of vision (bottom).

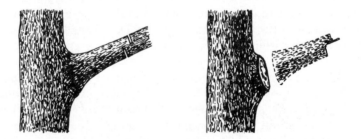

Figure 15-14 Remove a large limb by making three cuts. Make the first cut on the bottom of the branch about 300 mm (12 in.) from the branch attachment (left). Make the second cut on the top of the branch within 25 mm (1 in.) of the under cut. Make the final cut just beyond the outer portion of the shoulder and the branch bark ridge (right).

not leave a stub that will be slow to close. As described earlier, this is best done by cutting outside the branch bark ridge and in or just beyond the branch shoulder. Observations of old pruning cuts usually reveal that the most uniform closure is on those that were cut in or just beyond the branch shoulder area.

If the branch stub is heavy, it should be removed with two cuts: the first from the bottom and the second from the top. Until cut, a heavy stub should be supported by a rope sling which can later be used to lower the stub safely (Bridgeman 1976).

Remove a branch with a sharp "V" crotch in a similar three-step process (Fig. 15-15), being aware that the actual union of the two branches is often much lower

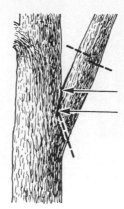

Figure 15–15 When removing a large branch with a sharp branch attachment, angle the third cut upward toward the top of the actual union of the branch with the trunk. Although they touch, the branch and trunk are not united between the two arrows.

than the apparent junction. The cut should slope upward to the point of attachment at a 40° to 50° angle from the horizontal.

Many books and articles on pruning give directions for paring the edges of large pruning wounds with a sharp knife or chisel. Paring means cutting the thick bark around a wound to give a smoother, more even surface. The process is thought to speed wound closure. However, although the wound will look neater, paring does not seem to be worth the effort. Cambium may be damaged by a mallet and chisel and will be subject to greater desiccation if much bark is cut. If you do pare bark around a wound, pare only thick bark and cut no more than one-half its thickness at the wound edge. Feather out the pared bark to the bark surface around the wound (Keith Davey, Belmont, CA, 1981 pers. comm.).

Trunk xylem may not completely enclose a branch on the underside so that the branch collar is incomplete; the trunk below such a branch will be indented or sunken. If the branch is removed, the tissue in the sunken area usually dies due to inadequate sap flow. Such results from removing a branch probably led to the recommendation by Thompson (1961), Bernatzky (1978), Pirone (1978a), and Fred Roth, California State Polytechnic University, Pomona (1989 pers. comm.) to trace (remove bark) around pruning wounds. This practice, however, is not recommended by Davey (1967), Brown (1972), Bridgeman (1976), or Shigo (1989). If the cambium dies below a pruning wound, then the dead bark can be removed to live tissue. No live tissue should be cut.

Dead Branch Stubs. Advice conflicts somewhat on how to remove a dead branch stub on which a collar has formed. Bartlett (1958) states the wound will close more quickly if the cut is made into the collar of woundwood. Shigo and coworkers (1979) dissected the trunks of seven 36-year-old black walnut trees from which dead branch stubs had been pruned 11 years earlier. Ring shakes (tangential and longitudinal separation in the wood) were associated with 14 of 17 stubs that had been flush-cut; they were not associated with four stubs whose branch collars had not been removed from the tree. In addition, more discolored wood was associated with the flush cuts and was more subject to decay. The authors recommend that when pruning is done late in the life of a tree, care must be taken to preserve branch

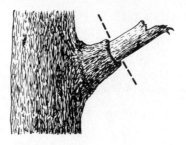

Figure 15-16 A dead branch stub that has a collar of live wood should be cut just at the outer edge of the collar.

collars that form around the bases of dying and dead branches (Fig. 15–16). When this zone is removed, the stub is more vulnerable to infection.

Protecting Pruning Cuts

It is doubtful that pruning cuts can be protected from decay by an asphalt emulsion or other materials (Neely 1970, Shigo and Wilson 1977, Harris and others 1969). The purpose of these coverings is to protect the cut surface from wood-rotting organisms and to reduce surface checking. Upon exposure to the sun, however, all but the thinnest coverings may crack. Moisture from rain, sprinklers, or dew can then enter the cracks and accumulate in pockets between the wood and the wound covering. These circumstances are even more favorable for wood-rotting organisms than an uncovered wound. If you wish to paint pruning or bark wounds for the sake of appearance, bonding of the paint will be strongest if you allow the wound to dry before applying a thin coating. Examine the wound several times in the first year and retreat it if the coating cracks.

A growth retardant, NAA (naphthaleneacetic acid), has been added to some asphalt emulsion and aerosol paints for application to pruning cuts. This will reduce the number of water sprouts, as well as the vigor of those that grow, by about 50 percent (Ashbaugh 1968). For NAA to be effective, it must be applied to the bark around the pruning wound. It is ineffective if it is placed only on the cut xylem surface (see Chapter 16).

Fresh pruning wounds attract female boring insects ready to oviposit and are vulnerable to infection when canker-forming fungi are sporulating. Asphalt paints applied to pruning cuts reduce borer attacks, and fungicidal paint or sprays can protect pruning wounds from certain canker fungi. Chapter 17 discusses these treatments as well as the treatment of wounds and cavities.

PRUNING TREES

Most trees grow quite well with little or no pruning; they have done so for centuries. But if trees are taken from their natural settings or if their natural settings are changed, a number of new requirements for growth and form are imposed. Many trees are no longer protected by other trees in a grove or forest but are exposed to the elements. Low branches may hamper activities that take place under and around trees. Trees may grow into utility lines, block views, obstruct the sun's rays, grow too close to buildings, become deformed or destroyed by strong winds, grow too large, or assume an unattractive shape. Growth patterns started in the nursery may be undesirable; structure may be weak. All of these developments are reasons for pruning.

Many tree species naturally develop a crown with desirable branch spacing and characteristic structure. Even though many more branches initially grow than will survive, competition and shading allow some branches to develop more rapidly than others. Smaller, weaker branches die and drop, in a natural pruning process. Trees allowed to develop with minimal pruning will often require only the correction of obvious structural faults, such as poorly positioned or strongly competing limbs, weak branch attachments, or limbs that are damaged or dead.

Structural Strength

Certain features contribute to the structural strength of the trunk and main branches of a tree.

Branch Attachments. For a strong attachment, a branch must be smaller than the trunk or limb from which it arises (see Fig. 2–12). Relative branch size is more important than the angle of attachment (see Chapter 2). Unless bark is included in the crotch, the growth of large, potential scaffold branches should be slowed to ensure strong branch attachment and lessen competition with the leader. Wherever a leader or branch forks, one of the branches should be substantially larger than the other. If there is a choice among permanent branches, usually those with wide angles of attachment and more horizontal growth should be selected; their less vigorous growth and smaller ultimate size will ensure a strong attachment.

The diameter of lateral branches should be less than three-fourths that of the parent branch or trunk. If the branch is close to the size of the parent branch, thin the branch's foliage by 15 to 25 percent, particularly near the terminal (Perry 1988). Laterals on the branch can be thinned; if the branch has few laterals or none at all, it should be headed. In many decurrent tree species, the lowest branch outstrips the growth of the trunk and the upper branches. When developing a trunk, therefore, examine lower laterals regularly during the growing season to ensure that they do not outgrow the leader. Branches from latent buds, even though they are much smaller than the trunk, are usually weakly attached (Fig. 15–5). If such branches are to be kept, prune them to decrease growth and permit the attachment to strengthen.

Branch Spacing. When the branches of broadleaved trees are well spaced on the trunk, both vertically and around the circumference, they are more likely to have strong attachments than when several branches arise at about the same level (Fig. 15–17). Vertical branch spacing is more critical in large trees than in smaller ones. If the main leader is headed in the nursery or at planting, close branch spacing will develop in most trees; in other trees even if not headed, the leader will tend to form several branches close together at the base or near the tip of one season's growth.

Figure 15–17 This Modesto ash was headed in the nursery or at planting; the many shoots that grew from below the heading cut were not thinned. Most of these branches are weakly attached and are in danger of splitting out as the tree grows larger and continues to spread.

Attachments with Included Bark. Branch attachments with included bark (see Fig. 2–13) are inherently weak and should be removed in young trees. In older trees, pruning to reduce the weight and spread of the branch, and rodding and cabling may be a solution to make the tree safe (see Chapter 18).

Included bark often occurs in sharp-angled branch attachments and between double leaders (codominant stems). The trunk is not able to grow around the branch or the other stem. Limbs or stems with included bark can grow to large size before they begin to spread and increase the stress on the weak attachment. It is usually only a matter of time before failure occurs.

Attachments with included bark often occur at the height at which permanent branches are wanted. On young trees, if only weakly attached branches are at the desired height for a scaffold, prune them off. A second shoot will usually grow free of included bark and smaller in size than the trunk (Fig. 15–18) (see Chapter 17).

Some tree species have branches with extremely narrow angles of attachment, acute enough to form indentations in the trunk. Certain trees, such as Lombardy poplar, have been selected for their erect branching (fastigiate) habit. These branches remain relatively small in relation to the trunk, and there is little or no need to remove them.

Tapered Trunks. Trees with tapered trunks can withstand greater stress from wind and vandals (Leiser and Kemper 1973). A tapered trunk decreases in diameter with height, and when it bends, the curvature is fairly even throughout its length, allowing for a uniform distribution of stress. The tops of well-tapered trunks bend farther under the wind than those with less taper, reducing the danger of broken

Figure 15–18 On young trees, shoots with sharp angles of attachment or included bark should be pruned off when they are about 150 to 200 mm (6–8 in.) long. A second bud will usually form a shoot with less vigor than the more upright shoots and will grow at a wider angle free of included bark; in the photograph, a short stub indicates the position of the removed shoot (left). A seven-year-old Modesto ash whose first laterals were removed during its first year in the landscape and later laterals thinned for vertical spacing is pictured on the right. (Compare with Fig. 15–17.)

trunks and other deformation. During the growing season, the tip of a leader may bend far enough to be nearly parallel to the wind load, relieving almost all stress on the immature wood of the tip.

Temporary branches on the trunk will strengthen and protect it, and the trunk will increase in base diameter more rapidly if laterals grow along it. The leaves and growing points provide food and auxins for rapid trunk growth. Branches along the trunk also shade it and reduce the likelihood of sunburn injury to the bark and cambium. Leiser and Kemper (1973) found that **stress is most effectively reduced if at least one-half of the foliage grows on branches originating on the lower two-thirds of the trunk.** Branches along the trunk will increase total tree growth, even though height will not increase quite as rapidly as it would if no branches grew along the trunk (Chandler and Cornell 1952, Larson 1965, Harris and Hamilton 1969). This slight reduction in total height is a definite advantage in developing a structurally strong tree.

Training Young Trees

Training young trees provides the greatest benefit of any cultural practice in influencing the future performance, safety, and maintenance costs of landscape trees. In the last five years, I have travelled on four continents; I am appalled at the pitiful condition of most newly planted trees, even in expensively developed landscapes.

Many trees have been headed, up to three times, to force low, multiple branching so that they appear dense and well proportioned in the nursery (Fig. 15-19). Even field-grown trees are often headed so that they are easier and cheaper to transport. Few or no laterals (feathers) remain along the lower trunk. The large branches are too close together, usually too low, and have outgrown the leader. Most of these trees cannot stand upright without support. Even trees brought from the wild are often straggly, have poor branch attachments and distribution, and have multiple leaders (codominant stems).

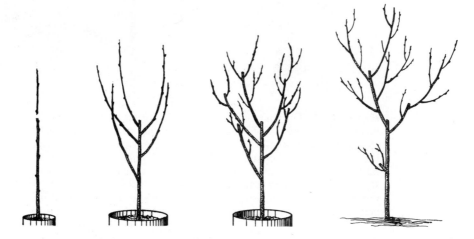

Figure 15–19   Container-grown trees are often headed in the nursery (extreme left) to encourage branching and to develop proportions that will be attractive at the time of sale (two center sketches). The laterals are usually headed a second time to ensure another set of branching (right center). Even for small-growing trees, most of the branches are too low; for large trees, they will be too close together also. Once the tree is in the landscape, select the top branch as the leader, prune another high branch so that it will become the lowest permanent scaffold, and remove or severely cut back the lower laterals on the trunk (extreme right).

Most of these trees should never have been planted (see Chapter 3), but all too often such trees are all that are available. Proper early training is essential!

Even though a newly planted tree may have an acceptable structure, you may want to prune it to achieve a desired form or improved structure. **A tree owner or landscape architect should inform the arborist if a tree is to serve a special purpose so the tree can be appropriately trained.** Usually little or no pruning is needed on young trees that have lateral branches on current growth of the leader (weak apical dominance). Such trees usually form a strong central leader and a conical shape, as do liquidambar, pin oak, and most conifers. On the other hand, trees with no laterals on current growth (strong apical dominance) may have quite irregular growth habits, as does Chinese pistache, or may form vigorous laterals in the second year, as do many flowering fruit trees. Poor branch attachments may also be common, as in Modesto ash. These trees have decurrent growth and usually need considerable pruning while young in order to develop a central trunk and a strong branch structure. *Scaffold* or *permanent* main branches are those that make up the framework of a decurrent tree. In contrast, most excurrent trees naturally have strong central trunks with many laterals that are much smaller in diameter.

The irregular, zigzag growth of some treetrunks can be minimized by generous application of nitrogen, prudent watering, and attention to soil aeration, which will increase tree vigor. You may also head the trunk back to its vertical portion so that new growth will be straighter. In Oregon nurseries, a number of deciduous tree species are headed close to the ground after the first season to develop straight, vigorous trunks.

Pruning at Planting

Although widely recommended in the past to compensate for root loss, pruning at planting has been brought into question. Maggs (1960), Evans and Klett (1985), and Whitcomb (1987) found that pruning at planting either had no effect or decreased subsequent growth. On the other hand, five bare root, deciduous tree species in England each made more growth the more severely they were pruned after planting (presumably they were unirrigated) (Gilbertson, Kendle, and Bradshaw 1987).

Unless a drought is severe, pruning may not increase survival or growth due to restricted rooting. Richardson (1958) found that removing the terminal bud of a one-year-old sugar maple seedling whip (single stem) just as it was beginning to grow delayed the initiation of root growth until another bud began to develop. When pruning removes most terminal buds, the reduction in potential leaf area may be accompanied initially by a smaller root system.

If sufficient moisture will be available, the pruning of bare root and B & B plants should be restricted to the removal of damaged branches and the correction of structural weaknesses. Thinning-type pruning can reduce leaf area or potential leaf area of most container-grown plants 25 percent without an apparent reduction in plant size; transpiration may be reduced somewhat. Earlier growth of the unpruned terminals and developing leaves should favor more rapid root development.

Training in the Landscape

The first three to five years in the landscape are critical in the training of most trees. An arborist should have a clear understanding of the landscape function of each tree in order to train it to the desired form. Pruning should be only extensive enough to direct a tree's growth and correct structural weakness. Trees that are pruned most lightly will make the most total growth. A young tree should be left with more branches than will ultimately be wanted; some of these will grow more and dominate the others. Dominant branches are usually fairly well spaced. Severe pruning of young trees may remove potentially dominant limbs, leaving some that will not develop as readily. For this reason, it is best to remove or prune back only those branches that are clearly unwanted and those that compete with the leader or other desirable limbs. Alternately, the growth of these branches can be retarded by pruning.

Branches to be removed or retarded on young trees during the first two to four years in the landscape are those that are too low to be permanent, those that grow from the same node as or within 100 to 150 mm (4–6 in.) of a potential scaffold branch, those that outgrow the leader or other potential scaffolds, and those that have sharp angles of attachment with included bark in the crotch (see Fig. 2–13). Occasionally, a tree will be so vigorous that the pruning needed to retard growth of less desirable branches would excessively stimulate potential scaffold branches. Extremely vigorous young branches are susceptible to breakage, particularly in windy sites. In such cases, about one-quarter of the current growth of temporary branches should be headed back. In order to provide some support for the future main branches, leave most of the temporary branches on the tree unless they severely compete with potential scaffolds. Do not fertilize the trees; in extreme situations, withhold irrigation to curb excessive growth.

Temporary Branches. Branches below the lowest permanent branch will strengthen and protect the trunk (Fig. 15–20). When young trees have not yet reached the desired height of the lowest scaffold, treat laterals as temporary branches. At planting and at each dormant pruning, select laterals of weak to moderate vigor to remain as temporary branches. Remove vigorous low-growing laterals if less vigorous ones can be selected. Short, horizontal laterals can be left unpruned.

Figure 15–20 Temporary branches, particularly on the side exposed to the afternoon sun, can shade and nourish a trunk (left); unprotected bark, particularly on young trees of low vigor, is often killed by the sun (right).

More vigorous laterals, if not removed, should be headed back to two- or three-bud spurs during dormant pruning.

Although their angle of attachment and spacing along the trunk are not particularly important, temporary branches should be 100 to 300 mm (4–12 in.) apart because closer spacing may unduly retard overall increases in height. Temporary branches on the western side of the trunk, which is exposed to the afternoon sun, will reduce the chance of sunburn. During the growing season, pinch the tips of vigorously growing temporary branches to keep them in bounds and to reduce competition with the leader and potential permanent branches. Examine vigorous trees at least two to four times during the growing season. Pinch back new shoots about 50 mm (2 in.) when they become 150 to 200 mm (6–8 in.) long. Vigorous shoots may have to be pinched more severely to keep growth within bounds. Pinching back temporary growth requires little time and provides an opportunity to observe other problems that may develop.

As a young tree develops a sturdy trunk and a top that shades the trunk, temporary branches can be reduced in number and eventually eliminated. Three or four years after planting, when the trunks of small trees (such as crape myrtle and Japanese maple) are 50 to 75 mm (2–3 in.) in caliper, and those of large trees (such as elm and sycamore) are 100 to 150 mm (4–6 in.) in caliper, the number of temporary branches can be reduced over a two- to three-year period. If, at each pruning, you remove the largest temporary branches, you will minimize the total surface of the pruning wounds.

Unfortunately, many young trees are grown with no laterals left along the lower trunk. Thus, any shoots that do begin growth in this region should be encouraged. If a trunk is spindly and could benefit from the nourishment and protection of temporary branches, it may be possible to stimulate their growth by notching the trunk. The removal of a small slit of bark above or below a bud will influence its differentiation into a leaf or flower bud (accomplished soon after bud formation) or will stimulate a dormant bud into growth. Removing a slit of bark 1 to 2 mm (0.04–0.08 in.) wide around 15 to 25 percent of the trunk circumference and 10 to 15 mm (0.4–0.6 in.) above a bud a month before expected growth will usually assure the growth of that bud. In Europe and the Orient, the technique has been used on young fruit trees, particularly apples and pears, that sometimes branch irregularly. The notch reduces the flow of bud-inhibiting substances from the terminal bud and developing leaves higher on the stem.

A less precise method of encouraging shoot growth is to bend the trunk so the side where the shoots are wanted is uppermost. The terminal portion of the leader is held nearly parallel to the ground. This should take place about two weeks before buds are expected to grow and should continue until buds break, possibly within three to six weeks. The trunk must be protected from sunburn with white latex paint or wrapping. After the buds break, the tree can be returned upright. Container-grown trees may be laid on their side before planting to encourage new shoots along the upper side of the trunk; again, the plant must be protected from the sun. Inclining the trunk from the vertical decreases bud-inhibiting substances on the upper portion, thereby releasing the buds.

Notching and bending are not so much recommended here as used to exemplify the manner in which a basic knowledge of plant responses can be used to develop growing techniques. Temporary branches have often been pruned off by uninformed people for a variety of misguided reasons. More education is needed.

Height of the Lowest Permanent Branch.   Height of the lowest branch is usually prescribed by the function of the tree in the landscape. The lowest permanent branch can be several centimeters (or inches) or more than 4 meters (12 ft) from the ground, depending on how the tree is to be used. A certain clearance is needed over streets or patios, but limbs may be lower if they will not interfere with traffic or the use of ground underneath the tree. The position of a limb on a trunk remains essentially the same throughout the life of the tree. In fact, as a branch increases in diameter, the distance between it and the ground actually decreases (Fig. 15–21).

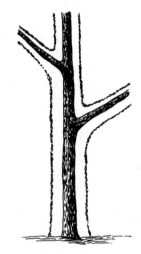

Figure 15-21 As a tree grows, branches retain their position on the trunk and at the same time increase in diameter, becoming closer to the ground.

Vertical Branch Spacing.   In many decurrent species, branch spacing is important for future leader dominance, structural strength, and appearance of the tree. Two or more vigorous branches arising at or near the same level on the trunk are apt to "choke out" the leader and limbs above. This is especially true in "fast-growing" trees whose laterals grow from buds formed during the previous season, such as flowering fruit trees, mulberry, and callery pears. On mature trees, closely spaced scaffolds break out more easily than those with greater spacing (Fig. 15-17). Closely spaced scaffolds usually have fewer laterals and thus develop as long, thin branches with little or no taper and little structural strength (Fig. 15-22). These branches are particularly susceptible to snow and ice breakage; if one is lost, the others are more likely to break out.

Vertical spacing between permanent branches should be greater on a large-growing tree with large-diameter branches than on a tree of smaller mature size. Major scaffold branches on large trees should be spaced at least 450 mm (18 in.) vertically, and preferably 600 mm (24 in.) or more. Many attractive mature trees have branches 1 to 4 meters (3–12 ft) apart. On excurrent trees, the vertical spacing of branches is less critical because the more vigorous branches are naturally well-

Figure 15-22 Closely spaced scaffolds on large trees are usually long, thin, and unbranched. If one should break out, the remaining branches are particularly vulnerable.

Figure 15-23 Occasionally a young vigorous branch (water sprout) will grow more upright than the others and will compete with the leader. Unless the sector in which it grows is devoid of limbs, the upright branch should be removed (broken line).

spaced, while others become relatively weak. Little or no pruning is needed except when an occasional limb becomes overly vigorous and competes with the leader (Fig. 15-23).

Radial Branch Distribution. Select five to seven scaffolds to fill the circle around a trunk without undue crowding (Fig. 15-24). This can be done in one or two rotations around the circumference. Although an ascending spiral may appear more symmetrical and pleasing to the eye, branches seem to grow equally well when they do not arise from the trunk in a spiral. Avoid allowing one limb to grow directly over another, which prevents both from developing properly (Fig. 15-25): The lower branch will be shaded and develop few or no ascending branches; the upper one will be less vigorous in the presence of the lower, which competes with it for water and nutrients.

Selecting Scaffold Branches. Examine the vertical and especially the radial distribution of potential scaffold branches. One sector of the tree may have few branches from which to choose. If so, choose a scaffold in that sector first and

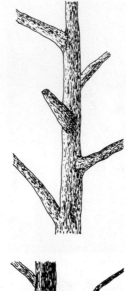

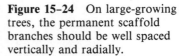

Figure 15-24 On large-growing trees, the permanent scaffold branches should be well spaced vertically and radially.

Figure 15-25 Proper radial distribution will allow limbs adequate space for development.

choose others in relation to it. This ensures the best radial symmetry and selection. Depending on the tree's growth rate, scaffold branch selection may take two to three years to complete.

Pruning During the Growing Season.   Growth should be directed when the tree is active as well as when it is dormant. Pinching the growing point (heading) or complete removal of a shoot (thinning) will direct growth into the leader and remaining shoots. Pruning during the growing season is usually confined to temporary shoots and branches. On a young tree, the leader or a scaffold will only occasionally need substantial pruning. Shoots that are too low, too close, or too vigorous in relation to the leader and selected scaffolds should be pinched or removed.

In many species, shoots do not branch during the season in which they form. Even in the second year, some trees may develop few or no laterals, except near the previous season's terminal. On vigorous leaders, it is possible to obtain branches during the growing season by pinching the growing point when it has grown a little above the desired height of a lateral (Fig. 15-26). Remove about 50 mm (2 in.) of the tip, and buds below the pinch will begin to develop. One will usually grow more

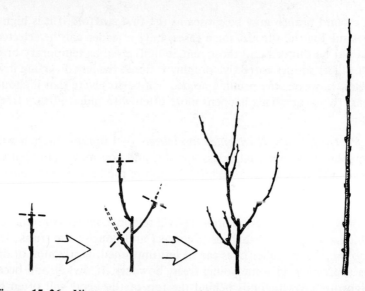

Figure 15-26 Vigorous young trees with strong apical dominance may grow 2 m (6–7 ft) with no lateral branches (extreme right). On the other hand, if the leader is pinched during the growing season when the terminal is 50 to 100 mm (2–4 in.) above the height desired for the first scaffold (extreme left), two or more shoots can be encouraged to grow. Select the most vigorous and upright shoot as the leader; select a second one as a lateral and head it lightly if necessary; head the others more severely (left center). Repeat this process as long as a vigorously growing leader can be selected (right center).

vigorously than the other shoots, and this can become the leader, although it may need encouragement. Choose a second developing shoot as a lateral by pinching the tips of the other shoots that were forced. The new leader should arise on the pruned trunk above the selected lateral; if it does not, the selected lateral may not develop into a vigorous scaffold. On a vigorous tree, the new leader may in turn be pinched when it reaches a height that is suitable for another lateral branch. It may be possible to force as many as three well-spaced laterals in one season. Without such pinching, the leader would require severe heading during the dormant season, usually to the height at which the lowest lateral is desired.

You can keep the development of scaffold branches in balance with the rest of the tree either by thinning laterals or by pinching the tips of the most vigorous laterals during the growing season. Growth can be channeled where it is most effective. For a tall, upright trunk, keep the leader in control by preventing laterals from outgrowing it. Uncontrolled laterals often outstrip the leader in decurrent species and occasionally in excurrent species.

Low Branches. Nursery trees with low, large laterals close together can be a problem in many landscapes. Such branching may be satisfactory for small trees in areas with little activity, but not for large trees or in active areas. At planting, it may be possible to select the most upright and vigorous branch to become the leader.

A second branch may be chosen as the first scaffold if it is high enough above the ground for the site. In some cases, only a leader can be selected. Other branches should be thinned and those remaining treated as temporary branches.

The sooner corrective pruning is done, the less dwarfing it will cause. In some cases, however, the pruning needed will be so severe that it should be done over at least two years. This happens more often with older nursery trees than with young ones.

Upright vs. Horizontal Branches. An upright branch will usually be more vigorous than one that is less upright, and you may want to use it as a permanent branch if its position is desirable. Because the branch may, however, compete with the leader, a more horizontal branch should usually be selected.

In contrast to upright branches, those growing more horizontally are usually of low vigor. Horizontal branches will seldom compete with the leader and are desirable as temporary branches to protect and nourish the trunk. Unless they become too long, the smaller ones can be left unpruned. Horizontal or drooping limbs may be a problem in some young trees, however. If they droop because of excessively vigorous growth, buds behind the top of the bend will often grow, and the new shoots will usually be more upright. Select the well-placed shoots from among these by thinning the lateral back to the selected shoot (Fig. 15-27). Thin out other new shoots that might compete or interfere with the one selected. If the horizontal or drooping limb has no well-placed upright laterals, head the branch to an upward-pointing bud slightly behind the top of the bend or to a point where you wish a lateral to form. Certain trees (such as weeping willow and Chile mayten) are chosen for their drooping branches, and the characteristic is exploited.

Windy Situations. Prevailing winds can deform trees so that most of the growth is on the downwind side. (Planting in windy locations is discussed in Chapter 5.) Depending on wind conditions and the type of tree, the leader may or may

Figure 15-27 To cause a horizontal branch to grow more upright, prune it back to an upright lateral (bottom), cut it back to an upward facing bud (center), or cut it back near the top of an arch (top).

not be bent by the wind. Many trees, such as conifers, liquidambar, and plane tree, resist deformation by moderate prevailing winds.

In windy sites, open up the top of the tree by removing moderate-size branches; this cuts the tree's wind resistance. Laterals develop more on the downwind side. In certain situations, such a condition may be picturesque and desirable. If it isn't, thin out branches on the downwind side to laterals to keep the tree more symmetrical (Fig. 15–28). Head curving branches on the windward side near the point at which they begin to bend with the wind; prune them to a lateral or a bud pointing into the wind. Repeat this each time the endmost new shoot starts to bend from the wind. Branches so pruned will be stockier and more resistant to bending. In some locations and for some species, windbreak protection may be needed.

Figure 15-28 A tree deformed by the wind can be made more symmetrical if branches on the windward side are pruned back to a lateral or a bud pointing into the wind, if the leader is thinned to a more upright lateral, and if some of the downwind branches are shortened.

Maintaining a Leader. Sometimes a leader may lose its control and become outgrown by a vigorous lateral, which may make a better leader than the original one. If so, remove the original leader by pruning back to the new leader (Fig. 15-29), which should be the topmost lateral on the trunk. Do not leave part of the original leader above the new one. It will seldom amount to much because new growth will go into the new leader. If caught young enough, laterals can be headed to keep the leader dominant (Fig. 15–30).

In species with moderate to strong apical dominance, several buds may begin to grow near the tip of an otherwise branchless leader or scaffold branch late in the season. They may grow 50 to 150 mm (2–6 in.) in length and as thick as the terminal. Unless these branchlets are thinned out on young trees, they and the terminal will grow weakly the next season. It is best to head to a bud below this tuft of branchlets or to thin the tuft, leaving one branch and the terminal.

Other Tree Forms. Although the previous discussion emphasizes the development of a well-tapered single trunk with well-spaced scaffolds, this does not mean

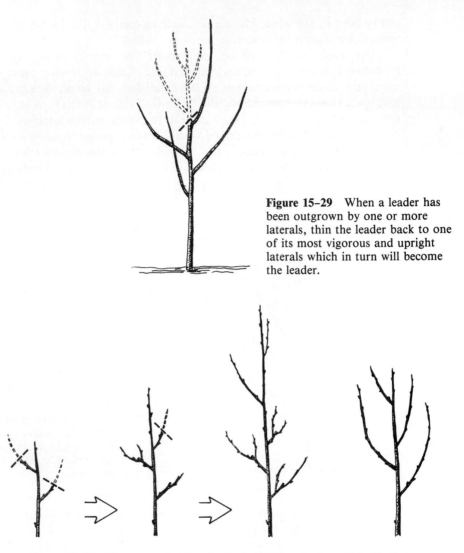

Figure 15–29 When a leader has been outgrown by one or more laterals, thin the leader back to one of its most vigorous and upright laterals which in turn will become the leader.

Figure 15–30 You can maintain a leader by heading back any laterals that may compete with it (extreme left); prune these laterals fairly severely if they are temporary (left center). The tree will grow taller (right center) than if it had not been pruned (extreme right).

that other tree forms are not recommended. The comments focus on developing a structurally strong tree and can be applied to whatever structure is desired. Trees pruned according to these guidelines should perform well and long. Plants can become fine sculptures in the landscape either naturally or through the skill of a horticulturist (Fig. 15–31).

Training Is More Than Pruning. Although pruning is the primary method of training young plants, other procedures may be used. Staking may be used to

Figure 15-31 A tree can be a sculpture in a landscape: The leader of this 40-year-old California sycamore was staked to the ground for the first three years after planting.

encourage a straight trunk or a more upright growth habit, although support staking has limits for most species if a sturdy trunk and well-proportioned top are to develop (see Chapter 9). Branches may be tied to supports to create special forms, cover a wall, or form a screen; espalier is a prime example.

The vigorous branches of young flowering fruit trees, particularly pear and apple, are sometimes bent and tied so their tips are at or below the horizontal to encourage flowering and fruiting at a younger age. Such a practice also increases the spread of trees that might otherwise be quite narrow. An upright limb can be made to spread if a length of branch pruning is wedged between the trunk and the limb (Fig. 15-32). First ascertain that the branch attachment is strong and the limb smaller than the trunk. The spreader can be left in place for two or three years and will be more effective than pruning to an outside bud, which in turn will most likely grow upright. Spreading branches can be held more upright (see Chapter 18).

Pruning Mature Trees

Scaffold limbs and the main structure of a decurrent tree can usually be selected by the third or fourth year, depending on the type of tree and its growing conditions. If scaffolds are well placed, the tree may need little or no pruning for several years. Temporary branches along the trunk should have been thinned out or reduced in number by this time. Staking to support a tree should not be necessary, but short stakes can be used to protect the trunk from mower, auto, or other damage.

Inspect young but mature trees annually to ensure that the main scaffolds are growing well and in balance with each other and the trunk. Remove low, broken, interfering, and diseased branches. A tree with dense foliage and the plants growing under it may benefit if you open up the tree to allow for the passage of light. Depending on tree species and growing conditions, inspections can be less frequent as the tree matures.

Pre-Pruning Procedures. In order to prune safely and efficiently, follow these guidelines before starting work:

Pruning Trees

Figure 15-32 A lateral branch that it too upright (arrow, left) can be inclined more if you insert a *spreader,* cut from a pruned branch of similar caliper, between the branch and the leader. The branch should be smaller than the leader and have a strong attachment. This is a flowering fruit tree, so the closely-spaced scaffolds will not be a problem.

All equipment and tools should be checked, serviced, and in top operating condition, preferably the night before use.

Public arborists (Hudson 1981) and nurseries (McCarthy 1991) have found that 10 to 30 minutes of stretching and calisthenics improve work performance, reduce accidents and injuries, and reduce insurance costs.

Each tree site should be checked for possible electrical hazards and appropriate action taken.

Each tree, particularly the root collar area (see Chapter 17), should be examined for possible structural hazards.

Pruning Guidelines. The National Arborist Association (N.A.A. 1987) in the United States and the British Standards Institution (BSI 1966, 1989b) in Great Britain have published standards for various pruning operations. The N.A.A. standards describe hazard pruning, crown reduction, and two severities of general pruning. The BSI standards in 1966 describe more categories according to pruning purpose than either the N.A.A. or the 1989 BSI standards. The International Society of Arboriculture (ISA) (Britton 1991) are developing guidelines incorporating categories of the N.A.A. and the 1966 BSI standards. In addition, ISA guidelines also will cover training young trees and utility pruning and view restoration of mature trees.

The important features of the N.A.A. and BSI guidelines that are discussed below are in the ISA Western Chapter standards (Perry 1988) and Appendix 6.

Crown Cleaning (BSI 1966).   Cleaning out or "dead wooding" consists primarily of removing broken, diseased, dying, and dead limbs and those that cross, are weakly attached, or are of low vigor. Water sprouts (epicormic shoots) and suckers can be quite a problem on some species and individuals and on trees that have been severely pruned; they should be cut as close to their base as possible in order to reduce the possibility of renewed growth. Treating these cut surfaces with a growth retardant also minimizes new growth (see Chapter 16). Climbing vines should be removed. Wire, rope, nails, and other foreign material should be eliminated. A tree can also be examined for defects that might require additional attention.

Pruning can quite effectively reduce insect infestation and infection and stop or slow the spread of infection within a plant. Brown (1972) describes pruning treatments to guard against 20 diseases and pests. Take care in pruning, because a few diseases can be spread by contaminated pruning tools. When you prune plants that are infected with a canker-forming disorder, prune uninfected parts of the tree first, then remove infected branches by cutting into healthy wood; if necessary, prune infected limbs, then disinfect tools before pruning the next tree to avoid spreading the disease to healthy trees.

Crown Raising.   Lifting the crown (BSI 1966, 1989b, N.A.A. 1987), raising the head, raising the canopy, and lift-pruning: All these terms refer to the removal of lower branches from the trunk or lower limits of a tree (Bridgeman 1976). Lower branches may block pedestrian and vehicular traffic, may obstruct a view, may grow too close to buildings, or may eliminate sunlight or breezes. As most trees increase in size, their branches bend lower under increased weight. Lower branches may also tend to grow downward because light intensities are higher below the tree canopy than within it. Lower branches should be removed only for valid reasons, however, because they have certain advantages if allowed to grow near the ground.

Plan the height of the lowest permanent branch when training a young tree so that you do not need to remove large branches later on, when pruning will leave large wounds. Some arborists prefer to prune for clearance in late summer, when branches are heaviest with foliage and will therefore reveal the extent of the problem.

When raising the crown, thin back to a more upright large lateral, or remove the branch entirely. All too often, low branches are cut back to small, more upright laterals (Fig. 15–33). These low branches are usually heavily shaded from above, weakening a low branch so that it becomes unattractive and needs further pruning. If there is a limb above, the low one should usually be removed to provide long-term clearance and to improve appearance.

The removal of low branches may increase stress along the lower trunk, immediately below the new lowest branch, and on the root system (Leiser and Kemper 1973). Particularly when you raise the head of a young mature tree, it may be wise to remove low branches over several seasons (Brown 1972); keep in mind that at

Figure 15–33 If there is a suitable branch above, completely remove a low, spreading branch (arrow) instead of pruning it back to less drooping laterals.

least one-half of the foliage should be on branches originating on the lower two-thirds of the trunk (Leiser and Kemper 1973). In order to leave a tree well balanced, both in appearance and weight, some arborists remove branches opposite those removed in raising the head. This is primarily a matter of aesthetic preference. Most trees can withstand considerable asymmetry without hazard, but removing lower branches, even to balance the tree, may increase trunk stress.

Crown Thinnning (BSI 1966, 1989b, N.A.A. 1987). Opening up the top of a tree permits deeper light penetration, which benefits inner leaves and branches. Moderate to high light intensity is necessary for active and productive leaves. If branches arise close together along the trunk, they will grow long and slender with few laterals (Fig. 15–34). Thinning out some of these branches will give the remaining ones more room, help initiate new laterals, and increase the taper of remaining branches. Wind resistance and weight at the top are reduced by crown thinning, which can be especially important for trees with weak branch structure or insecure roots.

The structural features of a tree can be emphasized by moderate thinning that opens the trunk and branches to view. Pittosporum, dogwood, olive, and ginkgo are particularly suited to this treatment.

The first branches to remove in crown thinning are those that would normally be removed in cleaning out the crown. After broken, weak, crossing, diseased, and dead limbs have been removed, it is easier to see what further thinning is required to open up the crown. On a medium to large tree (12–18 m or 40–60 ft tall), moderate (25–50 mm or 1–2 in. in diameter) thinning cuts of limbs are effective. Somewhat smaller cuts are appropriate on smaller trees. Make these in the top and around the

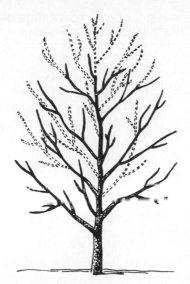

Figure 15-34 You can reduce the height and spread of a tree and yet maintain its natural shape. Branches that have been thinned are outlined by broken lines.

periphery. Remove branches that are close to others. In some large trees, you may remove limbs of up to 150 mm (6 in.) in diameter. Such large cuts, however, should not be needed unless the tree has been neglected or improperly pruned or its use in the landscape has changed.

Brown (1972) points out that crown thinning, which may involve the removal of one-third or more of the branch system, is generally confined to deciduous trees. Smith (1962) reports that several investigators found that 25 to 30 percent of the live crowns of a variety of conifers can be removed without reduction to height growth or serious decline in diameter growth.

Most arborists start thinning in the top and work their way down the tree. Prunings can be cleared as the work progresses downward; if falling branches damage those below, it is not too late to modify the selection of branches.

Crown Reduction (BSI 1966, 1989b, N.A.A. 1987). Many trees become larger than is desirable or safe: They may grow into overhead wires, block views, grow into buildings or other trees, shade solar collectors or other areas where sunlight is wanted, or become hazardous because of size or condition. You may control tree size by pruning, but it will be a continuing task. You may also slow growth without greatly affecting tree appearance by stopping or reducing nitrogen fertilization and irrigation. Size can be maintained most effectively if you prune as a plant begins to reach the maximum acceptable size. If you delay pruning until the tree is much larger than wanted, pruning will be more difficult, wounds will be harder to hide and slower to close, and renewed growth will be encouraged. Species vary in the severity of pruning mature trees can withstand if they are to remain healthy (Table 15-1).

Thinning, as a means of crown reduction, can reduce the height and spread of a tree while retaining its natural shape (Fig. 15-34). Prune branches back to lower laterals that are at least one-half the size (diameter basis) of the portion removed

TABLE 15-1 RELATIVE PHYSIOLOGICAL TOLERANCE OF SPECIES TO SEVERE PRUNING OF MATURE TREES (Arboricultural Association 1976)[a]

High	Intermediate	Low
Horse Chestnut	Acacia	Beech
Elm[b]	Maple	Birch
Lime tree	Ailanthus	Hornbeam
Oak	Alder	Eucalyptus
Mulberry[b]	Ash	Walnut
Poplar	Cherry	Most conifers
Willow	Catalpa	
	Sycamore[b]	

[a] Virtually all broadleaved trees withstand heavy pruning during their formative years.

[b] Severe pruning stimulates water sprouts (Fred Roth, California State Polytechnic University, Pomona, 1989 pers. comm.).

(Britton 1991). Pruning cuts will be less obvious and the tree less subject to water sprouts than if the branches are headed (lopped). Observation indicates that a thinned tree will usually take longer to grow back to the critical height than a headed tree. The finest compliment an arborist can receive after materially reducing the size or density of a tree is when passers-by fail to notice it has been pruned. Thinning requires greater skill and time than heading, but in most situations it is worth the effort: It will retain a tree's characteristic form, will minimize the problems of decay and regrowth, and will reduce future pruning (see Utility Pruning).

Even when severe pruning is needed, thinning can usually accomplish the task more satisfactorily than heading. Thinning to a large lateral is often called "drop-crotching." Tall, decurrent trees, such as elm, hackberry, and eucalyptus, occasionally branch in a pattern that provides a natural lower top to which the tree can be pruned (Fig. 15–35) (Keith Davey, Belmont, CA, 1980 pers. comm.). This second, lower top usually occurs between two-thirds and three-fourths up the tree. Except for the large thinning cuts to the branches that form the lower head, the tree reduced to such a lower head will appear unpruned. When the height of a tree must be reduced even though a natural lower top is not readily apparent, you can usually find appropriate lower branches to accomplish the same results. Treating the large cuts with NAA should keep regrowth to a minimum (see Chapter 16).

Although most excurrent trees can be reduced in width and still retain an unpruned natural form, it is difficult to reduce tree height without hastening development of a round head. Bridgeman (1976) cautions that mature beeches and birches do not respond to crown reduction and may die back from cuts (Table 15–1).

Crowded trees can become misshapen and their branches weakened by shading and rubbing. In many situations, the smaller, more deformed, or less desirably located trees should be removed. The remaining trees will grow into the new space with improved health and appearance. Trees are often planted closer together than will be desirable when they reach mature size. As these trees begin to crowd, the less desirable ones should be pruned back more severely each year until they are removed (Fig. 15–36). This will allow proper development of the more permanent trees while retaining the value of the temporary trees. Tree removal decisions may

Figure 15–35 This plane tree, in the Royal Botanic Garden at Kew, England, has a natural lower crown (broken line) to which the tree could be pruned should one want to lower the height of the tree.

Figure 15–36 The center tree has been pruned back every few years to allow more room for adjacent trees and can soon be removed with little or no loss.

Pruning Trees

be difficult, and many people oppose the practice, but its omission will be to the detriment of a planting.

Old trees that are of low vigor and have failing branches can often be kept healthy and attractive for several additional years by severe pruning. Wise fertilization, irrigation, and pest control may also be beneficial.

Heading (also called stubbing, dehorning, or lopping on older trees), unfortunately, is often used by well-intentioned but ill-informed people to reduce tree size. In such pruning, main branches are cut to stubs with little regard for their location. Regrowth from below the cuts is dense, vigorous, and upright. New shoots form a compact head, cast dense shade, and are weakly attached to the older branches (Fig. 15-5). They are held only by the surface layers (rings) of wood formed after the shoots begin to grow; these layers are similar in thickness to the outer layer of a sheet of plywood, but are not attached as securely. Shoots usually develop more rapidly than the strength of their attachment to the older wood. Branches from such regrowth are weakly attached, particularly if the heading cuts are large, and can be hazardous throughout their life. Regrowth near a heading cut on a vigorous small branch (less than 50 mm or 2 in.) may, however, cover over the cut surface and unite around the wound to form an intact cylinder of new wood in direct contact with the older wood. In such cases, there is little or no weakness and the actual lopping line may disappear (Brown 1972); however, Shigo (1989) thinks such wounds remain a point of weakness.

The unprotected surfaces of large heading cuts are also vulnerable to decay. Exposed branches may sunburn and be vulnerable to decay. If a branch has already begun to decay, cutting into it will speed the process. Large stub cuts seldom cover. Some arborists recommend that a large heading cut should be made at a slant (about 30° from the horizontal), with the upper edge above a small branch or in the direction in which regrowth is most desired. Field observations, though not verified by research, indicate that a slanted surface will remain drier, will decay less easily, and will be covered more quickly.

Pollarding is a training system used on some large-growing trees that are severely headed annually or every few years to hold them to modest size or to give them and the landscape a formal appearance. **Pollarding is not synonymous with topping, lopping, or stubbing** as stated by the BSI (1989b). Pollarding is severely heading some and removing the other vigorous water sprouts "back to a definite head" (Brown 1972) or knob of latent buds at the branch ends (Fig. 15-37)—normally the knobs are not removed.

Species that grow quickly and in difficult conditions are usually used for this purpose. The species most commonly pollarded are London plane, linden, black locust (Free 1961), and fruitless mulberry. A pollarded tree can be trained to an excurrent form, with short laterals distributed around the upper third of a central leader, or to a more decurrent form, with fewer laterals arising closer together on the upper trunk (Fig. 15-37).

Pollarding can be started soon after the lowest permanent branches are selected or at a subsequent dormant period during the first few years. For a decurrent shape, the laterals are vertically spaced within one meter (3 ft) or so at the top of the trunk. Every one (preferred) to three years, one to three shoots that grow from

Figure 15–37 A pollarded London plane tree at the University of California, Berkeley. Enlarged knobs develop at the branch ends to which the shoots are cut back each year (inset).

each of the headed laterals are in turn headed back to within 10 to 15 mm (about 0.5 in.) of the earlier pruning cut; the other shoots are removed close to their origin. After a tree has been pruned this way for several years, a knob of cut stubs and callus forms at the end of each branch. As would be expected with such severe heading, the regrowth is vigorous, upright, and dense. Anthracnose (*Apognomonic platani*) on American sycamore is sometimes minimized by pollarding because it allows less disease inoculum to be carried over to the next spring, but the dense growth often causes early fall of interior leaves because of heavy shading.

Crown Restoration, Crown Renewal, and Corrective Pruning (BSI 1966, 1989b). These terms refer to the practice of reshaping a tree that has been severely headed into one that will have a more natural form, improved health, and greater structural strength (Bridgeman 1976). A tree is probably worth saving if the main scaffolds and the trunk are sound or can be cut back to sound wood. Branches that have grown from headed scaffolds should be thinned until only one to three remain on each scaffold.

Selecting less vigorous sprouts will slow growth somewhat and favor more secondary laterals along the sprouts next season. Even though thinning out branches

Pruning Trees

411

opens the top so that the tree has less wind resistance, the remaining individual limbs are more exposed to wind damage. Therefore, the remaining branches should be thinned back to large, low laterals. The reduction in number and size of the branches helps develop their attachment to the main scaffolds, particularly in relation to their size.

Such severe pruning might best be done over two to four years to minimize its side effects, particularly the vigorous regrowth. Pruning during the growing season should reduce excessively long growth, strengthen branch attachment to the scaffolds, and dwarf total growth somewhat. In areas subject to fall frosts or winter cold, pruning should not be done so as to prolong growth and the beginning of cold hardening. Fertilization, irrigation, and other practices should be adjusted to minimize excessively vigorous growth on healthy trees. Pruned trees must be examined annually for structural development, presence of decay in framework branches, and general health. The safety of pedestrians and property is paramount.

Pruning Trees Suffering Root Loss or Damage. Opinions differ as to how much, if at all, to prune large trees that are being transplanted or have had severe root loss or damage. The following discussion assumes that the trees are in good condition and will be supplied with adequate water.

Common practice has been to prune the top of root-damaged trees rather severely. One California company specializing in moving large trees thins about 30 percent of the leaf area before boxing the roots (John Mote, Sylmar, CA, 1990 pers. comm.). The rationale for pruning is to reduce the leaf area to help compensate for root loss, to favor branches that are wanted to remain, and to minimize subsequent leaf drop and limb dieback. As much as 50 to 75 percent of the leaf area has been removed from some broadleaved trees transplanted in the heat of summer.

Others hold the view that when mature trees have suffered severe root loss, little or no pruning should be done. This opinion is reasoned from research findings on the effects of different pruning severities on newly planted young trees. Retaining as many leaves as possible is thought to supply maximum carbohydrates for root growth and compartmentalization. Also, since failure-prone limbs can seldom be identified at the time root damage occurs, it is thought best to wait and let the tree indicate which limbs to remove.

How a mature tree responds to a given severity of root loss depends on its health and its energy reserves. Trees have great redundancy in their tops and roots. Even though leaves, current shoots, and absorbing roots are lost, there are many others to carry on with little effect on the tree. Sinclair, Lyon, and Johnson (1987) state that a tree can lose half of its roots and still survive.

When a tree is properly and moderately pruned, leaves within its canopy are exposed to more sunlight and to slightly more carbon dioxide than before. Photosynthesis of the remaining leaves increases and offsets the carbohydrate loss that would have been supplied by the leaves removed. Just as transpiration is little affected by pruning until considerable foliage is removed, photosynthesis also is little affected unless pruning is severe. For a given plant, the rates of transpiration and photosynthesis are closely related (Farquhar 1979).

As with most things, moderation would appear to be wise in caring for root-damaged trees. Thinning weak and crowded branches that contribute little to a tree's energy reserves exposes leaves within the canopy so they are more productive. The well-being of limbs important for the structure and beauty of a tree can be better ensured. A pruned tree should be more stable. Another advantage is that the owner of the tree will interpret the arborist's efforts as being in the best interests of the tree.

View Enhancement and Restoration. Residential, vacation home, restaurant, and roadside sites are often selected and developed to exploit beautiful and interesting vistas. Skill is needed to create, enhance, and maintain such views. Mature trees are most commonly involved, but new tree plantings also need to take future view encroachment and enhancement into consideration.

A number of cities have ordinances which allow a person to prune offending plants on another person's property in order to restore a previous view. Some arborists predict that in time, there will be few or no tall trees in these "vista" areas as the old trees die and the young ones are kept cut back. Guidelines may help delay or prevent this from happening.

Properly pruned, well-placed trees can frame and enhance scenic views through strategically selected branches and foliage. Pruning established trees calls for variations of several of the pruning categories already discussed, particularly *crown thinning.* Large limbs may need to be removed to open the canopy, followed by selective thinning of some of the laterals along remaining branches which intrude into the vista. A person at the viewing spot can best direct the operation. A friendly relationship with and a concern for the landscape of the trees' owner will make the task more pleasant and effective.

When a site is initially developed, some trees may need to be removed. Young trees need to grow up through the view, partially blocking some of the view for a time, so that branches can develop above the view. Some selective branch removal can then begin without opening obvious viewing holes. Views can be quite intriguing when seen through sparsely leaved branches.

Pruning Near Energized Lines. Energized lines can be hazardous to arborists who work near them. **Any power line can be lethal,** depending on the contact that completes the circuit between an energized conductor and the ground. Because it is not possible to anticipate when a power line may energize other lines because of a storm or accident, the American National Standards Institute (ANSI) recommends that "All overhead and underground electrical conductors and all communications wires and cables shall be considered to be energized with potentially fatal voltages and shall never be touched either directly or indirectly" (cited in Haupt 1980). This includes fire alarm, cablevision, telephone, and house drops, as well as utility transmission and distribution lines.

The American National Standards Institute (1988) has established minimum distances between electrical conductors and working arborists (ANSI Z133). Workers whose primary responsibility is pruning trees for line clearance must have special training in working near conductors, and they are permitted to work closer to power lines than are other arborists. Line clearance workers must stay at least 0.6 m (2 ft) from a primary conductor carrying up to 15 KV (15,000 volts) and at least 4.5 m

(15 ft) from a conductor carrying 552 to 765 KV. Arborists who do general tree work must stay at least 3 m (10 ft) from conductors carrying up to 50 KV. Since standards may be changed, be sure of the current ones.

Before beginning work, the tree worker and supervisor should inspect individual trees if an electrical conductor passes through the trees or within reaching distance of the tree worker. The utility company should be notified before any work is performed within 3 m (10 ft) of an electrical conductor. Work should be started within 3 m of an electrical conductor only after a representative of the utility company has declared the condition safe. The utility may move the conductor, de-energize the line temporarily, or send a line-clearance tree-trimming crew to do the work near the conductor. During and following storms, telephone lines, wire fences, and metal guard rails within 1 km (0.6 mile) of damaged lines may be lethally energized. Workers should be alert to the possible dangers.

Tree crews should be trained to work safely near energized conductors, whether they climb in trees or use an aerial lift. Lowering cut branches near conductors requires planning and special care. During any work in large trees, a trained person should be on the ground at all times to assist the person in the tree and to act in emergencies.

Utility Pruning. The objective of utility pruning is to provide safe, reliable delivery of electricity, gas, water, and communications. The electric utilities in the United States spent an estimated $1 billion dollars in 1990 to keep power lines and rights-of-way free of trees and shrubs (Ng 1990). The quality of utility pruning is being balanced with the landscape setting and the costs involved by many utilities.

Guidelines for urban utility pruning of trees should follow those guidelines previously presented, with the following modifications:

> Where possible, allow a tree to attain near normal height and locate pruning cuts depending on branching habit so as to direct growth away from utility lines (Fig. 15–38); seldom should limbs be headed, especially near pre-established clearing limits.
>
> In achieving clearance, make most thinning cuts back to medium-size branches within the crown. Pruning of small branches in the outer crown should be kept to a minimum. In so doing, trees are pruned more quickly, and more terminal buds growing away from the lines will be retained.

Figure 15–38 Directional pruning of mature trees under utility lines has been found to be cost effective and to increase reliability of electrical service (R. A. Johnstone, Delmarva Power & Light Co., Salisbury, MD). Thinning cuts direct growth away from the conductors. Mature tree before pruning (right). Page 415: A *thru-side trim* or *side trim* with *"under-build"* (branches under the conductor) (left). *Undertrim* style of pruning (center). *Thru* or *V* trimming when tree is directly under the conductors (right).

To the extent possible, on decurrent trees, make cuts back to the trunk or a branch which is at least one-third the diameter of the portion being removed (note that this is smaller than the one-half recommended earlier).

Make pruning cuts close to but outside the branch bark ridge and just outside or through the branch shoulder; avoid stripping or tearing bark when removing limbs.

Do not use climbing irons or hooks except when removing a tree or in a life-threatening emergency.

Remove the entire branch if more than 50 percent of live foliage needs to be removed from it.

Severed limbs and dead branches should be removed from the tree.

Severe height reduction pruning (including heading) may occasionally be the only alternative when trees are directly under lines; vigorous, tall-growing trees directly under power lines may need to be removed.

Utilities should sponsor educational programs for the general public about planting compatible tree species close to utility lines.

In rural settings and some natural stands of trees, some utility companies follow similar pruning guidelines as in urban areas because of long-term cost savings (Keith Jones, Central Illinois Public Service, 1990 pers. comm.). Along transmission lines in more remote rural, forested, and wildland areas, chemical vegetation management techniques are used.

Lateral or directional pruning involves pruning a leader back to a lower outward-growing limb (lateral) (Blair 1940, Johnstone 1988, Holewinski, Orr, and Gillon 1983). For a tree growing directly under a distribution line, the leader is thinned or *drop crotched* to lower laterals (Fig. 15–38). The top laterals are further pruned to direct subsequent growth away from conductors.

The same method can be used for a tree not directly under a utility line. The tree is pruned by making lateral cuts on those branches which threaten conductors (Fig. 15–38). Branches growing above the conductors are directed up and away from the wire while those below are directed down and back. In both situations, subsequent growth is moderate and directed away from the conductors.

A five-year study by Delmarva Power along the mideastern coast involving thousands of mature trees demonstrates the value of lateral pruning (Johnstone 1988). Initially, lateral pruning may take longer, although the difference is not great

when arborists gain experience. The real benefits are realized in the following pruning cycles: The basic structure is established, problem trees and branches are removed, regrowth is less vigorous, the interval between pruning cycles increases, and many trees do not need to be pruned every cycle. In 1987, only 15 crews were used to maintain the same circuits that required 20 crews three years previously. Electrical service reliability was improved with 56 percent fewer tree-related interruptions in 1986 than in 1980. Considering inflation, productivity was increased 15 percent (Johnstone 1988).

The use of plant growth regulators coupled with directional pruning are further increasing the efficiency and effectiveness of utility line clearance (see Chapter 16).

CONIFEROUS TREES

Most conifers have an excurrent growth habit, particularly while young, and their training and pruning needs are similar in many respects to those of excurrent broadleaved trees. Since most conifers have strong central leaders, they need little or no training unless you desire some atypical effect. Conifers are pruned primarily to control the density of branching, the shape of young trees, and the size of older ones. Double leaders should be thinned to one; dead, diseased, crowded, and structurally unsound branches should be removed. See that branches are also smaller in diameter than the trunk.

Conifers have several characteristic, though not unique, growth patterns, which should be kept in mind during pruning (Table 15–2). The main leader of a conifer is seldom subdued by the several branches that arise at or near one level on the trunk. Therefore, branches arising in whorls or close together along the trunk can be left. Branches may be thinned, however, to reduce wind resistance or enhance appearance.

Laterals on conifers radiate from the trunk either randomly or in whorls (Table 15–2 and Fig. 15–39). In species that branch in whorls, shoot elongation is usually determined by the number of preformed initials in the terminal buds; those that branch randomly continue growth as long as conditions are favorable. To a large measure, the distribution of latent buds or growing points limits the severity with which conifers can be pruned (Table 15–2). **Few conifers have latent buds below the foliage area in older wood; if branches are headed back to older wood with no foliage, the branch stub usually dies.** Pruning to control size or direct growth is easiest when you can cut within foliage alone. More severe pruning can be done by cutting back to active laterals while retaining a semblance of natural form. On the other hand, when conifers are able to form new shoots from older wood, pruning is much simpler and trees are better able to recover from injury by storm or fire.

Whorl-Branching Species

These should be selectively pruned back to a bud or the branch from which the pruned portion originated. In vigorous trees, branch whorls are sometimes farther

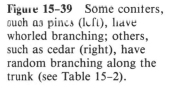

Figure 15-39 Some conifers, such as pines (left), have whorled branching; others, such as cedar (right), have random branching along the trunk (see Table 15-2).

apart on the trunk than will be desired. You can create a denser tree by heading current growth back to a bud. This can easily be done with pine trees, whose buds elongate into candles before starting their period of greatest elongation (Fig. 15-40). As candles approach full length, needles begin to elongate; this is the time to pinch them (USDA 1969). Growth will be inhibited according to the severity of the pinch. Break or pinch candles, if feasible, because cutting the needles with shears causes them to develop brown tips. If you must shear, shorten candles after they have hardened.

Figure 15-40 You can easily reduce the growth of pine shoots by pinching (breaking) their candles back proportionately to the growth reduction desired.

TABLE 15-2 PRUNING GUIDELINES FOR CONIFERS (Leiser, Harris, and Matheny 1981, unpublished)

Genus and branching pattern	Distribution of latent buds or dormant growing points	Type of growth	Method of pruning for a given response[a]		Comments
			Reduce size, direct growth	Increase density	
Usually whorled					
Pine	Almost none. See comments for exceptions.	From preformed initials in most species. Some have more than one flush of growth.	Tip or cut to lateral shoots only. Do not cut below terminal buds.	Pinch candle when it expands in the spring. This may be done on each flush of growth.	Canary Island pine and a few others have latent buds and may be pruned more severely if necessary.
Spruce Fir Douglas fir	Some	Growth from preformed initials.	Tip or cut to lateral shoots or visible dormant buds.	Pinch lateral shoots in spring when they expand. The terminal (leader) may or may not form multiple leaders if pinched.	These usually require little pruning. Remove bottom limbs for clearance over several years.
Random-branching					
True cedar Larch	Some latent buds on short spur-like shoots.	Long shoots have new growth from preformed initials but continue growth with favorable conditions.	Cut to lateral on long shoots. Short shoots may break into long shoot growth.	Pinch expanding shoots.	These usually require little pruning. Remove bottom limbs for clearance over several years.

Podocarpus	None or few.	More or less continuous under favorable conditions.	Thin to laterals or visible foliage.	Pinch expanding shoots.	Removal of leader may produce a multistem crown.
Bald cypress Dawn redwood Cryptomeria Giant sequoia	May have some growing points capable of new growth.	More or less continuous under favorable conditions.	Thin to laterals or visible foliage tufts.	Pinch expanding shoots.	These species have both persistent and annual (deciduous) shoots. Prune to persistent shoots.
Arborvitae False cypress Incense cedar Cypress	No true buds. Numerous dormant growing points where foliage persists.	More or less continuous under favorable conditions.	Thin to laterals or within visible foliage.	Tip prune.	Do not prune into bare wood. Cypress are very slow to make new growth after pruning.
Juniper Yew	Numerous dormant growing points in foliage areas, some on older wood.	More or less continuous under favorable conditions.	Thin to laterals, tip prune.	Tip prune.	Those withneedle-like foliage may not respond to pruning as well as those with awl-shaped or scale-like foliage.
Coast redwood	Numerous latent buds.	More or less continuous under favorable conditions.	Thin to laterals, tip prune.	Tip prune.	Remove multiple leaders.

[a] For maximum dwarfing effect, prune in any of the suggested ways in late summer.

419

You can shape a pine by removing most of the candles on shoots you wish to discourage and leaving most or all on those you wish to encourage. Reduce the size of whorl-branching conifers by thinning a branch tip back to a laterally growing shoot. Spruce, fir, and other whorl-branching trees whose buds elongate into shoots can be kept dense and small if you pinch the expanding shoots in the spring. Trees that have short internodes, as do firs, may need little or no pruning. In fact, Scarlett and Wagener (1973) warn that the topmost whorl of firs should not be pinched because regrowth may be poor.

Pinching shortens the distance between the topmost branch whorl and the next one to be formed, but excessive distance between existing whorls cannot be decreased. Bare trunks between widely spaced whorls can be unattractive on vigorous pines. Heading to shorten a long distance between whorls usually results in a stub that does not resprout. Some arborists and Christmas tree growers induce branching between whorls by girdling, or removing needles between the whorls. To girdle a trunk, they remove a narrow (3 mm or $\frac{1}{8}$ in.) band of bark at the height at which branching is desired. One or several branches may grow below the girdle and help fill the void. Alternately, one can remove the needles 25 to 50 mm (1–2 in.) above the level at which branching is wanted; this may stimulate shoot growth from the band of needles just below the exposed area (Francis, Breece, and Baldwin 1974). Branches developed in this way have a relatively late start and will not grow as large as older limbs. The younger the tree and the closer to the terminal girdling or needle removal is done, the more likely success will be.

Random-Branching Species

These can usually be sheared or pinched to control size, branching, and form. When latent buds occur along older branches as well as on younger shoots, pruning cuts are usually made near a latent bud, which will then become active and develop a new growing point.

To control the spread of most random-branching conifers, prune back new growth of side branches to their halfway point in late spring. The branches can be taken back more severely and still appear natural if each lateral is pruned back to a shorter shoot growing on the top of it. The small shoot will hide the pruning cut and give a tip-end appearance to the shortened lateral. When conifer branches reach 250 to 300 mm (10–12 in.) in length, new growth can be cut back to 20 to 30 mm (1 in.). Small side shoots will develop, making the tree more dense. You can then maintain the tree at about that size by cutting back new growth. Williamson (1972) suggests that even tall-growing deodar cedar can be kept as low as 2 m (6 ft) if you cut back the leader each year and tape a side branch near the cut into a vertical position to form the new tip.

Pruning Severity

As mentioned, response to pruning severity is largely determined by the presence or lack of latent buds on older wood. Pruning severity can also be based on the duration of growth during the season. If conditions are favorable, some species (including some pines) with preformed shoot initials in their buds may experience more than one growth flush during a season. Young, expanding shoots can be pruned during any or all of these flushes. If there are no visible latent buds, however, pruning into old wood will usually produce a stub from which no new growth will arise.

At the other extreme are species with buds or dormant growing points (with no bud scales) whose shoots continue to elongate even after their preformed initials are fully expanded. These species grow almost continuously or in successive flushes during the growing season. Their usually abundant latent buds produce new growth even when pruning is severe and extends into old wood. Branching is usually spiral or random, and growth may be either excurrent or decurrent, at least when the trees mature. Even though plants in this group can be pruned severely, thinning will produce the most attractive results.

As with all natural things, there are intermediate forms. Many conifers continue growth in a series of flushes under favorable growing conditions or the stimulation produced by pruning or the removal of adjacent trees. These conifers usually have latent buds randomly spaced along stems; they may retain active laterals or short shoots for many years on older wood. If pruning is needed on such conifers, it should be moderate. These species may either maintain or develop a decurrent growth habit with age.

Columnar Conifers

A number of upright conifers, such as arborvitae, cypress, false cypress, and yew, whose branches arise from the ground to the top, are often used in formal settings. Many are clipped into "unnatural, spectacular ugliness" (Chandler and Cornell 1952), while most can be pruned to have a natural, textured appearance. The branches of columnar conifers that start near the ground may become quite tall, growing up on the outside of the branches above; in such circumstances, the inside foliage is shaded and may be lost. Outer branches that bend away from the column are unattractive. Most of these conifers can be trained or retrained to have short branches if upright branches are cut back to short, spreading laterals. These spreading laterals can be spaced to provide pleasant shadows and glimpses of the trunk. Weak and dead foliage can be washed out with a hose.

Outside branches are often held close by wires wrapped around the tree at several elevations to keep them from bending out of the columnar form. Even though some of the outside shoots will be sheared, the plants accumulate many dead twigs and leaves between the inner stems. As such trees mature or lose vigor, the multitude of bare, slender branches becomes quite unattractive.

Yew and podocarpus may be too slender or sparse if branches are cut back to spreading laterals. In these cases, the branches should be headed at varying lengths above their laterals; the branches of yew might be left 300 mm (1 ft) or more long, and those of podocarpus cut to stubs with six to 10 leaves and as many buds, several of which will grow and increase the number of branches (Chandler and Cornell 1952). Overgrown yews can be cut back to main branches to force new shoots (Brown 1972).

Nursery-Grown Trees

Young nursery-grown conifers may develop asymmetrical or decurrent growth in a normally excurrent species, as well as particularly compact branching. Unless you want these features, do not select such trees for planting. Such growth is the result of vegetative propagation from lateral shoots (buds or cuttings) of species that have strong radial symmetry. A compact conifer, trained by severe pruning or shearing in the nursery, may look well-proportioned when young, but unless similar pruning is continued when the tree grows in the landscape, the new branches will be more widely spaced, giving the appearance of a tree growing on top of a shrub.

To correct asymmetry in a small tree, choose the most upright and vigorous shoot to produce a new leader, so that the growth habit will be changed from decurrent to excurrent. This shoot should have strong two- or three-year-old wood below it and a cluster of buds at the tip. If the chosen branch is not the longest, remove competing branches or reduce them in size to direct major growth into the new leader. If you must plant an asymmetrical conifer, partially tip (rotate) the root ball in the planting hole to keep the selected leader vertical. You may also have to stake the selected leader into a vertical position. Continue follow-up pruning and staking until radial symmetry is obtained.

You can correct excessive and closely spaced laterals by selectively thinning to match the more normal later growth. Thin crowded laterals lightly until tree growth in the landscape indicates what the new spacing between whorls or branches will be. Even then, extend the thinning over two to three years. The tree will usually be most attractive if this thinning is done early in its life, so that the selected branches develop without crowding. If remedial pruning is delayed, much of the interior foliage may die from shading.

Young Conifers in the Landscape

Conifers are most attractive when the tips of low branches almost touch the ground, but if clearance is required under the canopy, remove lower limbs more gradually than you would with broadleaved trees, about 250 to 350 mm (10–14 in.) at a time on young trees. Some pruning back of laterals that are scheduled for removal will keep them relatively small so that later pruning wounds will also be smaller than they would be otherwise.

Conifers that normally have an excurrent growth habit often form multiple leaders; all but one of them should be removed. Some conifers, however, including a number of pine, some cypress, cedar, juniper, false cypress, and arborvitae species, form decurrent crowns. In these conifers, you can allow multiple stems to develop into their characteristic form, unless the branches are too low or their attachments weak. These main stems should be differentially pruned, however, so they develop unequally to increase structural strength and present an interesting shape.

Conifers occasionally lose their leaders. When this occurs, a new leader may develop from one of the uppermost branches or from a latent bud near the top. In many conifers, when the top dies, is cut, or breaks out, branches in the topmost whorl will bend upward as reaction wood forms. One branch usually dominates to become the leader, and the others return to approximately their original orientation. Alternately, a latent bud may grow into a new leader. When a leader is lost, latent buds are more likely to grow if a 50- to 70-mm (2- to 3-in.) stub is left above the top whorl (Scarlett and Wagener 1973). A leader from a latent bud should usually be selected over one developed from a lateral; the former will usually be more vigorous and more symmetrical. Shorten laterals in the topmost whorl to reduce competition with the developing leader. If no leader develops naturally, tie one of the topmost branches upright to induce it to become the new leader. If no leader is selected or encouraged, numerous leader-like candles may develop randomly from upper branches producing a particularly unattractive top (Fred Roth, California State Polytechnic University, Pomona, 1989 pers. comm.)

PALM TREES

Even though palms are pruned primarily to remove old and unsightly fronds and fruiting clusters, several of the most commonly planted species must be pruned frequently and are consequently some of the most expensive trees to maintain. **The terminal growth is never removed because single-trunked palms have only one growing point.** Entire stems are sometimes removed in clustering species that produce more than one stem.

Except to minimize disease or fire damage, you need not prune established palms for their own well-being. In public landscapes, however, many palms must be pruned frequently to keep them safe and attractive (Fig. 15–41). A palm normally maintains only a certain number of live fronds, depending on the species and growing conditions. As new leaves develop at the terminal, the lower, older fronds die. Some palms shed their old leaves; others retain dead fronds until they eventually rot away, are blown off by the wind, or are removed. Fronds can sometimes cover the trunk to the ground (Ledin 1961). In some palms the leaf blade will break off, leaving the petiole base or boot remaining on the trunk for many years. Dead fronds or their bases harbor insects and rodents and can be a fire hazard. Falling fruit and fronds can be a serious nuisance. Low-growing palms may have spines on their trunks or spiny leaflets that can be dangerous.

Multiple-stemmed palms form new shoots below ground. Though there is little need to prune cluster palms, an entire stem is sometimes removed. You may have to remove individual stems if they grow close to a building, if the cluster occupies too much space, or if you want a specimen plant. In some monocarpic clustering palms, the old stems should be pruned out after the final fruiting, because they will die near their base shortly afterwards (Ledin 1961). The new shoots that surround the old will maintain the clump. Stems should be cut off as close to the ground as possible.

Palms that produce hazardously large fruit or frequently drop old fronds may have to be pruned every three to six months. Though it is one of the most commonly planted palms in tropical landscapes, the coconut palm must be pruned every six months to keep it safe. Both fruiting stalks and fronds are removed at each pruning.

Figure 15–41 Mexican fan palms are pruned frequently in intensively maintained landscapes to minimize litter from the flowering stalks. The green petioles are removed in two cuts of a sharp carpet knife (with a hooked point) leaving V-shaped tips (inset). The petiole bases will be removed a year or so later.

The fruiting stalks are pulled down and cut as close to the base as possible. On regularly pruned coconut palms, only the bottom one or two rings of fronds are removed. A thin, fibrous sheath surrounds the frond base and makes cutting difficult unless the sheath is first pulled down and cut off. The exposed fronds are green and easily cut with a pruning knife or machete. An experienced pruner can climb and prune a mature coconut palm in less than 10 minutes. Unfortunately, most pruners use spurs, which create holes in the palm trunks that do not close (Fig. 15–42); shorter spurs cause less injury.

Some palms, including royal palms, readily shed their old leaves. When the fronds begin to turn brown, they can easily be pulled from the trunk. This should be done fairly regularly on tall trees. The fronds of other palms are usually removed less frequently, depending on the species and the amount of effort required. Some of the date palms are pruned with a chain saw every two years. The cut frond bases can be quite acceptable if pruning is done carefully. The chain housing may have to be enlarged to minimize clogging with palm fibers. However, do not use a chain saw on *Phoenix* spp. due to the possibility of transmitting the deadly fungus disease Fusarium wilt (*F. oxysporum*). Use a reciprocal saw whose blade can be sterilized after pruning each tree (Ohr 1989).

In landscapes that receive more intense care, palms are pruned at least annually and with greater attention to detail (McClements 1988). When yellow fronds

Figure 15-42 Holes punched in palm trunks by using climbing spurs remain for the life of the tree. In humid regions, these holes could become sites for decay.

are cut, 200 to 300 mm (8–12 in.) of the base is left. The ends of the frond stubs may be shaped to a point by two cuts with a sharp knife (Fig. 15–41). The frond stubs left by the previous pruning are cut off close to the trunk.

Take care when cutting fronds so as not to cut into live growth, which could cause unsightly scars. Most fronds should be cut from below with a chain saw to minimize ripping fibers down the trunk.

Palms injured by cold will usually suffer most damage in the lower leaves. Ledin (1961) recommends removing all injured fronds, except the inner two or three, at the terminal. Even when all the fronds are injured, do not remove the tree for at least six months to determine if it will survive. If the terminal bud is alive, it will send out new leaves. If the top leaf bases have hardened, the new growth may become stunted as it tries to push up. To prevent this, loosen the dried leaf bases as the new growth begins to show (Ledin 1961).

SHRUBS

Shrubs are woody plants with several stems originating at or near the ground. They are usually smaller than trees, but their distinctive features are form and branching pattern. Most shrubs have abundant latent buds at their base and along their branches; new shoots from the base replace weak and dying branches and, in effect, keep a shrub young. A mature shrub that normally forms few or no new shoots from its base will do so readily if its top is pruned or its base opened to light.

Shrubs range in height, spread, vigor, and flowering characteristics, each of which may influence pruning techniques. Slow-growing evergreens need little or no pruning, even in confined spaces. Many vigorous evergreen shrubs may do well with no pruning if they grow in ample space and will be viewed from a distance or at high speed. Most fast-growing evergreen and deciduous shrubs require moderate or extensive pruning to contain them, maintain their attractiveness, and balance vegetative growth and flowering. Pruning frequency for mature shrubs depends on their vigor, flowering habit, and desired size.

Shrubs are pruned for essentially the same reasons as trees are pruned. Pruning for structural strength, however, may not be as critical except for large shrubs and those exposed to wind or wet snow. Unfortunately, people too often prune shrubs primarily to control their size; in many situations, this is a sign that the wrong species was planted. Nurseries like to market fast-growing shrubs because they are easy to propagate and grow, are attractive in the nursery, and are popular. Many landscape architects and horticulturists similarly favor fast-growing plants because they generally do well and look attractive early. It is not until several years after planting that the shrub becomes overgrown and difficult to contain. For small landscapes and moderate-sized beds, shrubs should be chosen according to their mature size.

Pruning should begin at planting and be part of the regular maintenance program for shrubs. Plants need not be pruned at each inspection or even annually, but their growth and form should be assessed. All too often, shrubs receive little or no attention after initial care at planting until they are too large for their allotted space. By then, inside and lower foliage has probably been weakened and shaded out. If the shrub is pruned to a smaller size at that point, large cuts will result in a stiff, unattractive plant with little or no low foliage; regrowth will be vigorous and upright, further detracting from the natural appearance of the shrub. Pinching the tips of vigorous, nonbranching shoots will make the shrub more compact and symmetrical. **Shrubs whose growth must be contained should be on a regular pruning program before they reach the size desired.**

As many shrubs mature, shaded lower branches weaken, lose leaves, and become unattractive. The solution is not to cut the low branches partially or completely off but to thin the shrub so light can enter it. If plants are thinned in time, existing branches and new ones will usually be stimulated enough to keep the shrub full to the ground. Blue blossom ceanothus and Victoria tea are exceptions, because they do not sprout readily from old wood (Chandler and Cornell 1952). If you head branches instead of thinning them, more dense, upright growth in the top will further shade lower growth.

As stated, pruning severity and frequency depend in large measure on shrub vigor and flowering habit. More specific comments can be made about pruning slow- and fast-growing shrubs.

Slow-Growing Species

Slow-growing shrubs, usually broadleaved evergreens, make most of their growth from terminal buds or buds near shoot ends. With many shoots forming, the growth of each is reduced, producing a slowly expanding, fairly symmetrical, compact plant. Most plants of this sort develop a dense outer shell of foliage with few leaves in the center. They require little pruning, except when an occasional branch outgrows the general contour of the shrub. Such a branch should be cut clear to its point of attachment or to a lateral arising deep within the plant.

Slow-growing shrubs develop more of a main-branch framework than do many of the fast-growing shrubs that sprout primarily from the base. Slow-growing shrubs seldom require severe pruning; usually only a few shoots are cut back to their attachment or to another lateral. A light thinning will open up a compact shrub so more light and air can reach the center. Such pruning can create pleasing branch and leaf patterns and stimulates inside leaves. Slow-growing plants should seldom be headed back and should never be sheared.

Fast-Growing Shrubs

Both deciduous and broadleaved evergreen shrubs occur in fast-growing species. Some species form a main framework from which most new shoots grow. Other species, particularly many deciduous shrubs, sprout vigorously from the base, which in a few years can become a tangle of branches unless thinned. On some species, shoots are so vigorous that they cannot remain upright and bend over instead. New laterals on these vigorous shoots are usually stimulated at, and just behind, the top of such a bend. The original shoot bends even more with the added weight of laterals. With proper pruning, the shrub can have a graceful arching form; if left unattended, it will continue to enlarge and become a confusion of branches. Fast-growing shrubs need regular, fairly severe pruning to keep them attractive and contained. You can maintain the natural shape by thinning out branches and cutting others back to a bud or lateral to control the direction of future growth. This pruning should also enhance renewal of the plant.

When pruning a fast-growing shrub, remove unwanted branches first, branches that lie on the ground, have little foliage, are weak, broken, crossing, dead, diseased, or infested with insects. Extremely vigorous unbranching shoots that originate from the base and outgrow the general outline of the shrub should be removed close to the base. Next, cut back some of the largest and oldest branches to their origin. **Before each cut, visualize what the plant will look like with the branch in question removed.** Pruning should not leave a void in the shrub form. Distribute succeeding cuts to provide a uniform thinning (Fig. 15–43). On vigorous, mature shrubs, remove at least 25 to 30 percent of the older branches annually. Finally, you may need to shorten some of the younger, smaller branches to maintain shrub size or shape. Take care to vary the depth of these cuts in order to leave the shrub with a textured, natural look. A shrub so pruned will produce and maintain vigorous shoots with low foliage and healthy appearance. A well-pruned shrub will add more character and interest to a landscape than one that has become a dense tangle of branches. If a mature, overgrown shrub is to be considerably reduced in size, reduce it over a two- to four-year period.

Severe pruning is recommended only for old, overgrown shrubs in certain situations, but it will usually not endanger the plants. In fact, most fast-growing shrubs can be cut close to the ground (50–150 mm or 2–6 in.) (Brown 1972). New shoots

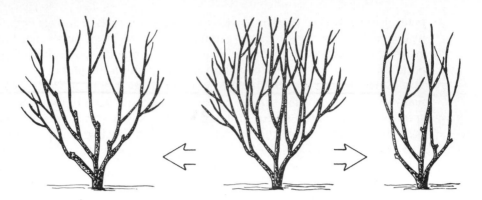

Figure 15-43 A shrub (center) can be pruned so that it becomes more spreading (left) or more upright (right) in habit. Short stubs have been left to indicate where the cuts have been made.

soon establish a shrub with renewed vigor and foliage clear to the ground (Fig. 15-44). Plants pruned so severely would have less vigorous regrowth if some of their roots were cut, but roots are seldom pruned. Vigorous regrowth will develop more girth if it is headed a few times during the growing season. Some of the new basal shoots should be thinned to allow the greater development of others and to give the

Figure 15-44 Many overgrown shrubs can be pruned close to the ground when there seems to be no other way to reduce their size without exposing bare branches. The escallonia on the left was pruned within 150 mm (6 in.) of the ground two years before the photograph and the one on the right one year. Compact symmetrical spheres result; frequent thinning will be necessary to develop any character in the plants.

shrub a more natural look. Without thinning, the shrub might consist of many spindly upright shoots growing close together.

Shrubs that must be kept from spreading or overgrowing must be pruned with care if they are to maintain an informal, natural look. If shrubs are sheared back, the resulting surface will reveal many dead twigs. The offending shoots should be cut back varying distances to make the surface irregular and the plant more natural looking. Prostrate shrubs can be cut back considerably without appearing to have been pruned. Cuts should be buried back in the shrub center and pruning should be to a lateral growing from the top of its parent branch.

Shrubs into Small Trees

A shrub that has grown too large to be easily reduced can often be trained to become an attractive small tree (Fig. 15–45). Low-growing branches that do not carry into the top of the plant, and usually some of the main branches as well, should be removed; the remaining branches will then become more prominent visually. You will seldom need to create a new branch structure but should thin the existing one. Severe pruning and exposure of lower trunks to more light may stimulate water

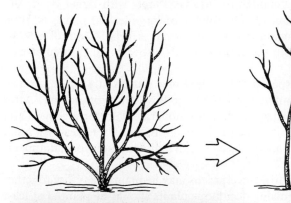

Figure 15–45 Occasionally, an overgrown shrub can be pruned into an attractive small tree (top). A more attractive plant will usually result if fewer branches are left than on the callistemon (lower left), though they can be more than the single-trunked pyracantha (lower right).

Shrubs

sprouts. If you make close cuts and treat the wounds with naphthaleneacetic acid, water sprouts will be minimized (see Chapter 16). If several shrubs in the same planting are to be trained into small trees, some of them should probably be removed to provide room for the others to develop individually.

Hedges

A hedge is a row of closely spaced shrubs or trees, usually of a single species, grown either informally with little or no pruning, or formally, trained and sheared. Hedges serve as visual and physical screens and dividers. They are more common now in public and large private landscapes than in small residential plantings. Informal hedges are allowed to grow naturally and usually require considerable space; Wyman (1971) suggests that hedges, particularly conifers, be allowed to grow nearly as wide as they are high. Formal hedges are sheared fairly regularly to retain their form, size, and attractiveness.

Plants—particularly for formal hedges—should have small leaves, short internodes, and dense branches, and should be able to sprout from old wood. Evergreens are preferred for effective year-round screening. In colder climates, conifers may be the only evergreens hardy enough to survive. Plants with larger leaves are often used for informal hedges, but the cut foliage of large-leaved plants is often unattractive when it has been sheared for more formal effects. Grounds (1973), Wyman (1971), and Dunmire (1988) recommend specific species for particular uses and geographical areas.

Informal hedge plants are spaced 500 to 900 mm (20–36 in.) apart, and all growth is headed back at least halfway to encourage low, dense branching. After this, only occasional pruning is needed to maintain moderately uniform growth and symmetry and to keep the hedge in bounds. Pruning individual branches with hand shears instead of shearing many branches at once will maintain an informal appearance.

For a formal hedge, small-growing evergreens may be planted as close together as 450 mm (18 in.), while large-growing deciduous plants should be planted at least 900 mm (36 in.) apart. Depending on the size of the shrubs or trees at planting, small evergreens should be cut back to within 75 or 100 mm (3 or 4 in.) of the ground and large deciduous plants to within 150 or 300 mm (6 or 12 in.) of the ground. No further pruning takes place in the first growing season, so plants can become well-established.

Before growth begins in the second year, the shoots of broadleaved plants should be headed to within 100 or 150 mm (4 or 6 in.) of the pruning height at planting. Conifers should be tip-pruned. Until the desired height is reached, head back new shoots one-half to two-thirds of their length each time they grow 150 to 300 mm (6–12 in.). Severe heading of new growth encourages plants to spread and develop a dense, low-branched structure. When the developing hedge has filled in fairly well at the base, it should be allowed to grow more in height than width. Hedges, particularly conifer hedges, should be wider at the base than at the top to ensure that the lower foliage will receive enough light (Fig. 15–46). In areas subject to early frosts or cold winters, pruning should not take place after midsummer.

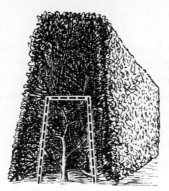

Figure 15–46 To maintain low foliage, shear a hedge so that it is wider at the base than at the top, particularly if oriented in an east-west direction. When a hedge has become too tall and wide, prune it back to about the size indicated by the dark broken line; after two or three shearings, the hedge will be well clothed in leaves.

After a hedge has nearly reached the desired height, time shearing according to the amount and cycle of growth. The effects of shearing will last longest if it is done after growth has ended for the season. Growth may be so vigorous, however, that the hedge becomes unattractive and must be sheared during the growing season. With experience, you will be able to compromise between minimizing the number of shearings and maintaining an attractive hedge. Most small formal hedges begin to look unkempt when new growth is 50 to 100 mm (2–4 in.) long.

Shear new growth on a mature hedge to within 10 mm (0.4 in.) of the previous shearing, leaving only one to three new leaves (and buds). This will clothe the hedge in new foliage and slow its increase in height and spread. After each shearing, the hedge will grow a little larger than it was after the previous shearing. Eventually most hedges become too large for their sites or lose their lower foliage. Light pruning removes most foliage without greatly reducing the size of the hedge. In contrast to many conifers, most broadleaved hedges can be pruned back severely. It is best to cut broadleaved hedges back to about half the desired mature height and width (Fig. 15–46). The pruned plants will have little or no foliage but should soon develop new shoots. After one or two shearings, the hedge will return to its original beauty. **Such heavy pruning should be done just before growth begins in spring,** so the hedge is bare for the shortest time. Conifers must be pruned with greater care; be sure new shoots will grow from older wood (Table 15–2).

Roses

Roses are usually pruned more severely than other shrubs, not only to keep them in bounds but to balance vegetative growth and flowering. Balance vegetative growth and flowering by varying the number and length of canes selected at pruning. In deciding how severely to prune, evaluate the plant's vigor and the size of the previous season's roses. If you want more vigorous growth or larger blossoms, leave fewer or shorter canes than previously. On the other hand, if growth has been vigorous with large blooms, you may leave more wood to take advantage of the plant's flower-producing capacity.

You can find details for pruning roses in many books specializing in roses and pruning, particularly those by Brown (1972) and Dunmire (1988).

VINES

Vines, also called climbers or wall shrubs, have vigorous, slender, flexible stems that require support in order to grow upright. Vines are versatile and can grow quickly as a ground cover, a screen on a trellis, a cascade of foliage and flowers against a wall, a leafy cover on an arbor, or a climber on a pole or tree. Since they are vigorous growers, most vines must be pruned annually to keep them attractive, healthy, and in bounds. To be effective in the landscape, most climbing vines require as much attention as roses do. Vines can greatly enhance even a small landscape, but the species must be selected with regard to function and space.

Not all vines can climb. While most have some means of holding onto a support, some need help. Vines climb upright in four basic ways (Fig. 15–47): aerial roots, holdfasts, or sucker discs (ivy and Virginia creeper); tendrils (clematis and grape); twining stems (star jasmine and wisteria); or artificial means of support (bougainvillea and rose). Thorns and hook-like projections on stems and leaves help other plants attach themselves to supports. Vines with aerial roots or holdfasts can climb almost any surface, while other vines need a trellis, fence, or wire for support.

Managing vine growth consists primarily of directing the stems on a support (training), pruning out old growth, and containing plants in their allotted space. Particularly when plants are young, you may wish to arrange stems on their support to form a pleasing pattern. If the stems do not branch readily, pinch their tips where branching is wanted. Pinched stems cover a trellis more densely, especially stems that radiate from the base. Many vines overgrow their space quickly and become unattractive tangles unless they are pruned regularly. New shoots that will not be useful replacements must be pruned back or out to keep a vine from becoming too dense and heavy. Older branching systems that fail to produce vigorous flowering shoots must be cut back to main branches or all the way to the base. Vigorous stems that grow beyond their allotted space or begin to climb where not wanted should be cut back or trained back onto the vine's support.

The growth and flowering habits of vines provide clues for training and pruning the plants to carry out their landscape function. Vigorous vines that cling to a wall or support, such as ivy, can be clipped close to the wall at least once a year. Prune them before growth begins and again as needed during the growing season. This keeps the foliage close to the wall and, in ivy, prevents mature branches from forming. Some vines, such as wisteria, form a heavy structure on which vigorous shoots grow; buds near the bases of shoots flower during spring following shoot growth. The first shoots of a young vine should be trained to become the permanent framework. Later shoots from the framework branches should be pinched or cut back to two or three leaves during the growing season. Succeeding growth flushes, in turn, must be headed (Fig. 15–48). Before spring growth, cut back each headed shoot system to the lowest one or two buds. Short shoots will grow from these basal spur buds and form terminal racemes (short shoots) of flowers.

On vines that flower along vigorous one-year-old canes, such as honeysuckle and rambler rose, the canes should be pruned out or back to new low-growing shoots after they have flowered in the spring. The pruned stems must be removed carefully so that the new shoots are not damaged, particularly if the shoots twine

Figure 15-47 Vines attach themselves for support in several ways: English ivy by aerial roots (top left); Boston ivy by holdfasts (top right); grape by tendrils (bottom left); and wisteria by twining stems (bottom right).

Figure 15-48 A 30-year-old wisteria in bloom. A wisteria shoot system that was headed to two to four buds during the growing season (lower right); the spur system could be cut back to the second spur before spring growth. Short flowering shoots grow from the basal spur buds (lower left).

around each other. Other vines, such as weeping forsythia, develop main branches from which pendulous shoots grow readily for several years. Flowers are borne on these one-year-old shoots. The main branches should be trained to provide height from which the long shoots can hang. Although annual pruning is recommended, it need not be as severe as most vines require; remove about one-third of the older wood each year after flowering. Likewise, thin about one-third of the drooping shoots on remaining branches. Vines that flower on vigorous current-season's growth, as some clematis do, should be pruned severely each year, often close to the ground, before growth begins in the spring. This annual pruning keeps the plant vigorous and in bounds.

A number of vines, such as ivy and star jasmine, are also grown as ground covers. These spread quickly, and some, such as ivy, root where shoots contact moist soil. These vines must be trimmed two or three times a season to keep them within their borders. Within a few years, the mass of canes builds up so high that it must be cut back. If ivy canes are not too large, a rotary power mower or hedge shears can be used to remove almost all the canes back to where they have rooted. Do this just before spring growth begins, so that new growth quickly covers the bare canes. Many vigorous vines must be planted with care near buildings. Small shoots can grow between cracks in walls, roofs, fences, and other structures and can cause considerable damage as they increase in diameter. They can clog downspouts and rain gutters. If not regularly pruned, some vines become so heavy they collapse their support. Vines can engulf shrubs and small trees and sometimes kill them by blocking sunlight.

SPECIALIZED PRUNING

Espalier

Some trees and shrubs can be trained in a vertical plane on a wire or wooden trellis, either free-standing or against a wall (Fig. 15–49). An espalier is a trellis or a plant grown on a trellis; in the United States, espalier is also used as a noun or verb to describe a method of training. Fruit trees have been espaliered along walls and stone fences that face the equator in European gardens for centuries. The heat capacity of walls and stone fences usually hastens the bloom and fruit maturity of plants growing against them, and it protects the plants from frosts and winter cold. Espaliered plants, sometimes referred to as wall shrubs, also require less space, an important factor in small gardens.

Species that flower on spurs, such as apple and pyracantha, are best suited for the formal espaliers, especially if branches are to be horizontal. Species that flower on vigorous one-year-old shoots, such as peach, or on current growth, such as citrus, are better suited to more informal or upright patterns.

Espaliers are usually trained on tightly stretched wires or wood supports. Horizontal and vertical supports are usually 400 to 500 mm (16–20 in.) apart. A trellis 100 to 150 mm (4–6 in.) from a wall allows the sun to create interesting shadows on the wall, particularly behind deciduous plants in winter. Fruit injury due to high temperatures can be reduced by encouraging more foliage to shade the wall or espaliers of fruit can be grown in the open. Species that must be sprayed for pests should not be espaliered on wooden walls, painted or unpainted, because pesticides can discolor or stain the surface.

Palmette and *cordon* are forms of espalier whose distinctions have become unclear, at least in America. A palmette plant has one vertical trunk with parallel pairs of laterals arising at about equal intervals along it (Fig. 15–49). A proper

Figure 15-49 A pear tree at Wisley Gardens, England, trained as a palmette espalier (left). (Photo: H.T. Hartmann and others. *Plant Science: Growth, Development, and Utilization of Cultivated Plants,* 2nd ed. © 1988, p. 352. Reprinted by permission of Prentice-Hall, Inc., Englewood Cliffs, NJ) A candelabra or double-U espalier of pear at Cornell University, New York (right). (Photo courtesy Clarence Lewis, East Lansing, MI)

cordon is a plant grown to a single trunk or to two main branches trained in opposite directions (Bailey 1916).

Palmette Training. Palmette training is the most common form of espalier (Fig. 15-49). Laterals should arise from the leader just below the horizontal support on which they are to be tied. Where laterals are available, tie one in each direction from the leader and remove those not tied to horizontal supports. At planting, if no laterals can be satisfactorily spaced, head the leader at or just below the lowest horizontal support.

During the growing season, when the leader has grown about 100 mm (4 in.) above a horizontal support, head it back to the support or just below it. Develop the new topmost shoot as the leader and train two laterals onto the horizontal support in opposite directions. Pinch off other shoots that grow. Tie each shoot down about every 150 mm (6 in.) as it grows, but do not tie down the growing tip; allow it to grow freely and maintain vigor. Repeat this process until all horizontal supports have laterals; then keep the leader headed at the top laterals.

While laterals are kept horizontal, vertical branches may be allowed to develop along them to carry the main foliage, as well as flower and fruiting spurs. Frequent pinching of a vertical branch checks its growth, allowing the horizontal branch beyond the vertical to maintain vigor. The shoots pinched to hardened wood will not grow back as quickly as they would from a softer pinch.

Cordon Training. Cordon training is best for fruit trees that form or are pruned to spurs. You can develop a cordon by bending an unbranched trunk (whip) about two meters (6 ft) tall onto a single-wire trellis about 500 mm (20 in.) above the ground. The trunk should usually be twisted at the bend to slow growth and hasten flower-bud formation.

Some cordons are trained to one straight trunk, either vertical or inclined at about a 30° angle from the vertical. Many cordons have one vertical trunk terminating in two lateral branches that grow horizontally in opposite directions.

Fan Training. To form a fan, head the trunk 200 to 300 mm (8–12 in.) above the ground; select about ten shoots and tie them to a trellis in a narrow fan (Bailey 1916). Before the next growing season, cut back the top two branches about halfway to reduce their total growth and to encourage branching. Head back the other branches to mature wood. Tie remaining branches to the trellis in a broad fan (Fig 15–50). Species that flower on vigorous one-year-old wood or on current growth are more productive when fan-trained than when trained as palmette or cordon espaliers.

Figure 15–50 This pear tree at Wisley Gardens, England, has been trained as a fan espalier. (Photo: H. T. Hartmann, and others. *Plant Science: Growth, Development, and Utilization of Cultivated Plants,* 2nd ed. © 1988, p. 352. Reprinted by permission of Prentice-Hall, Inc., Englewood Cliff, NJ)

Informal Training. Many plants are better adapted to informal or more upright espalier training (Fig. 15–51). Informal espaliers are trained similarly to fan espaliers except that shoots of informal espaliers are tied in a desired pattern. Shoots of species that do not readily form spurs should grow at an angle of at least 30° above the horizontal or should have their terminal growth so inclined. This will help maintain branch vigor and flower production.

All espaliers require persistent attention so growth will be directed where it is

Figure 15-51 A young ginkgo informally espaliered against a wall at the Morton Arboretum, Lisle, Illinois. (Photo courtesy of Clarence Lewis, East Lansing, MI)

wanted and unwanted growth removed before it weakens the rest of the plant and mars the beauty of the espalier.

Pleach

To *pleach* is to weave or intertwine. Occasionally, suitable trees can be grown in one or more rows to a height of 3 to 5 meters (10–15 ft), where they are headed and the more horizontal vigorous shoots are tied to horizontal wires or another support. Eventually, as the shoots from adjacent trees intermingle, they are pleached (woven together) (Fig. 15-52). Upright branches and those from along the trunks are cut off. Species suitable for pleaching have supple branches that can be woven without breaking. Beech, hornbeam, buttonwood, apple, peach, and pear have been successfully pleached (Free 1961).

A double row of closely planted trees can be arched and pleached upon a strong wood or metal frame to form a covered walkway, or allée. Upright branches are arched over, tied to the frame, and eventually tied and pleached with branches

Figure 15-52 This pleached allée of goldenchain trees at the Royal Botanic Garden, Kew, England, has been trained on a pipe and cable trellis (left). The branches are intertwined and could eventually support the allée without the trellis. Three London plane trees in a Whittier, CA, park have been pleached to form a pergola (right).

from trees in the other row. When these branches graft together, the arch can stand without support. The upright shoots should be cut back so leaves on the inside of the allée will receive enough light.

Pollard

Pollarding is heading back one- or two-year-old branches to approximately the same place every year or two, usually at the terminals of framework branches. Trees can be kept small in this manner and will have a lush, formal appearance. Pollarding was described earlier in this chapter (Fig. 15-37).

Topiary

Topiary is the practice of training and shearing plants into various formal shapes, geometric or mimetic. Topiary was practiced by the Romans and is widespread even today (Fig. 15-53). Small-leaved evergreen plants that grow readily from latent buds are best adapted to topiary. Creating the more intricate designs requires skill and patience. Wire may be used to hold developing branches in desired positions. Trees may also be sculpted to simple geometric shapes. Plants must be clipped often during the growing season to keep them attractive.

Small-leaved evergreen vines are sometimes trained over wire-mesh frames to

Figure 15-53 English holly topiary (mimetic) in a home garden near Stratford-upon-Avon, England (left); waxleaf privet topiary (geometric) at Oki Nursery, Sacramento, CA (right).

Specialized Pruning

439

create topiary-like forms. This is a quicker way to develop garden sculptures, and few people will be aware that the plants are not actual topiary forms.

Bonsai

This ancient Chinese (Penjing) and Japanese (bonsai) art produces miniature replicas of mature trees and landscapes (Fig. 15–54). The essential procedures are careful pruning of the top, the occasional use of wire to bend a trunk or branch, maintenance of a small but healthy root system in a shallow container, and sensitive placement in a complementary microlandscape, which includes rocks, moss, and lichen. Species selected for bonsai should have small leaves, flowers, or fruit that will be in scale on a plant less than 600 mm (24 in.) tall. Frequent pruning of shoots not only keeps the plant small but increases the taper of the trunk and branches and makes branch growth more tortuous, to simulate old trees. Roots should be pruned back and repotted with new soil every two to four years. Plants are kept in moderate vigor, but growth is pinched back to control size and form. Bonsai plants must be protected from extreme heat and cold and must be watered frequently because of their limited roots. A number of good books are available on the art of bonsai (Wyman 1971).

Figure 15–54 Japanese maple (left) and ginkgo (right) grown as bonsai by N. William Stice, Davis, CA.

FURTHER READING

BRIDGEMAN, P. H. 1976. *Tree Surgery*. London: David & Charles.
BROOKLYN BOTANIC GARDEN. 1981. Handbook on Pruning. *Plants & Gardens* 37(2).

BROWN, G. E. 1972. *The Pruning of Trees, Shrubs, and Conifers.* London: Faber and Faber.

HALLIWELL, B., J. TURPIN, and J. WRIGHT. 1979. *The Complete Handbook of Pruning,* 2nd ed. London: Ward Lock.

MCCLEMENTS, J. K. 1988. *Sunset Pruning Handbook.* Menlo Park, CA: Lane.

ROTH, S. A. 1989. *All About Pruning.* San Ramon, CA: Ortho Books.

SHIGO, A. L. 1989. *Tree Pruning: A Worldwide Photo Guide.* Durham, NH: Shigo and Trees, Associates.

STEBBINS, R. L., and M. MACCASKEY. 1983. *Pruning: How-To Guide for Gardeners.* Tucson, AZ: HPBooks.

CHAPTER 16

Chemical
Control
of Plants

A number of chemicals are used to stimulate, reduce, or stop the growth of shoots and to initiate, delay, or prevent flowering and fruit set. Those who use chemicals for these purposes want to interfere as little as possible with the health and appearance of plants. Some of the same chemicals, as well as others, are used to kill unwanted plants. Certain chemicals have been employed for years, some are recent, and others are still experimental. An arborist should understand the effects of growth-regulating chemicals and the conditions under which they can be used safely and beneficially.

STIMULATION OF GROWTH

When water or a nutrient is limiting, addition of the limiting item usually increases plant growth. Irrigation may be necessary for normal, vigorous growth of young plants and to maintain plants when soil moisture is limiting (see Chapter 13). Fertilization has been perfected to a science for many food, fiber, and greenhouse crops and is particularly important for stimulating the growth of young landscape plants (see Chapter 12). There are, however, chemicals other than nutrients that stimulate plant growth.

Growth Stimulants

A number of chemicals can increase the growth of shoots or the dry weight of leaves and stems. Some substances, normally used as growth retardants or herbicides, can stimulate growth if diluted.

442

Gibberellins, a group of more than 50 closely related natural compounds, stimulate cell division and/or elongation (Weaver 1972). They can stimulate stem elongation in many plants that are genetically dwarf or rosette.

Used experimentally, sprays of gibberellic acid stimulated a second flush of growth of sour cherry trees (Hull and Lewis 1959). In southern California, gibberellin sprays were able to overcome the growth-suppressing effects of long nights in autumn and winter (Goodin and Stoutemyer 1962).

Gibberellin is used commercially on a few crop plants (to encourage melon fruit set and grape enlargement, for example), but a use has not been found for it in the landscape.

OONTROL OF PLANT HEIGHT

Chemicals can control shoot growth of many herbaceous and woody plants. Choose and use growth-regulating chemicals with care. Each has specific properties; different species may react differently to the same concentration of the same chemical. Their effectiveness and possible toxicity may vary depending on the weather conditions when they are applied and last for up to four years. Long-term environmental effects must be considered. Even though a chemical is mentioned or even recommended, **check locally to be sure the chemical is currently registered for that use on the subject species.**

Growth Control Mechanisms and Chemicals

Chemicals can restrict shoot elongation in at least two ways (Sachs and Hackett 1972).

> *Terminal-bud inhibitors* kill terminal buds or severely inhibit apical meristematic activity, or reduce apical dominance—allowing lateral buds to grow, which in turn reduces terminal growth
> Some terminal bud inhibitors are
>> Naphthaleneacetic acid (Tree-Hold®)
>> Chlorflurenols (Maintain CF 125® [injection]; Maintain A® [paint])
>> Dikegulac (Atrinal®)
>> Maleic hydrazide (Royal Slo-Gro®)
> *Subapical inhibitors* retard internode elongation without disrupting apical meristematic functions
> Some subapical inhibitors are
>> Daminozide (Alar®), primarily for flowers and nursery plants
>> Flurprimidol (Cutless®)
>> Paclobutrazol (Clipper®, Bonsai®)
>> Uniconazole (Prunit®, Sumagic®)

In addition to retarding stem elongation, the newer subapical inhibitors have some of the effects of the earlier ones described by Cathey (1975). These effects include

Increase in the green color of leaves
Enhancement and acceleration of flowering
Greater formation of yellow pigments in flowers
Reduction of visible injury from ozone and sulfur dioxide
Increased resistance to cold and transplanting shock

As would be expected, the effectiveness of growth-retarding chemicals will vary, depending on

Plant species and clone
Formulation and concentration of the chemical
Method of application
Time of application in relation to growth cycle
Frequency of application
Weather during and after application

Until the early 1980s, many of the plant growth regulators (PGRs) that were available for reducing the growth of woody plants were not widely effective; their methods of application were not always acceptable (e.g., spraying); their application-timing window was narrow, especially if many trees had to be treated; the extent of their phytotoxicity was not always acceptable; and some were too expensive for landscape use. Some (such as chlorflurenol) are still used in the landscape for reducing the growth of shrubs and turf and for reducing fruiting. A few are still being used to suppress tree growth, primarily because the newer generation of PGRs (the anti-gibberellins) are either awaiting registration or were just becoming available in the late 1980s.

The Anti-Gibberellins

The biosynthesis of gibberellins in plants can be disrupted by several compounds; the most obvious consequence is reduction of internode elongation (Fig. 16–1). Node or leaf number per shoot and leaf size are usually reduced though not to the extent that internode elongation is. As would be expected, the amount of reduction varies with species, chemical, and chemical concentration.

A darker-green, more compact plant results. The time of leaf emergence and leaf fall can be affected but is quite variable depending on species. In some species, flowering and fruiting are enhanced and/or epicormic shoots increased. **These PGRs produce more consistent growth suppression with fewer undesirable effects than the terminal-bud inhibitors.**

The Anti-Gibberellin PGRs. Flurprimidol, paclobutrazol, and uniconazole are compounds that have been shown to effectively regulate tree growth, primarily by reducing internode elongation (Szeto 1989, Kimball 1990). Even though the three compounds have similar growth-regulating effects on plants, they differ in the concentration of the active ingredient (a.i.) necessary to obtain equal growth inhibition (Table 16–1). Not only do species vary in the degree of inhibition due to a PGR,

Figure 16-1 Young mature trees of two species in southern California were pruned and the left tree in each of the top photographs was sprayed with paclobutrazol 18 months before being photographed—*Eucalyptus grandis* (left) and *Ulmus parvifolia* (top right). The sprayed elm trees had only a fraction of the prunings (inset) that the unsprayed trees did (lower right). (Photos courtesy Roy Sachs, University of California, Davis)

but the relative effectiveness of the PGRs may vary with species. Comparisons of more species treated by the different application methods are needed. A more meaningful comparison would be relative effectiveness on a cost per tree basis as long as secondary effects were comparable. At present, selection is based primarily on which PGR is registered for which method of application on which species.

These PGRs are only slightly soluble in water (Table 16-1). This complicates application but probably is responsible for their long period of effectiveness, low phytotoxicity, and minimum mobility in soil. The PGRs are dissolved in an organic solvent, such as methanol. It is thought that when a PGR is diluted in the water in the xylem, the active chemical precipitates, remaining close to where it was injected. PGRs move almost exclusively in the xylem. The low solubility of a PGR provides a continuous, low concentration in the transpiration stream; in field use on more than two thousand trees in the eastern United States, paclobutrazol remained active for at least four years (Environmental Consultants 1990).

TABLE 16-1 COMPARISONS OF ANTI-GIBBERELLINS
AVAILABLE FOR REDUCING THE GROWTH OF TREES

Compound Trade name Company	Effectiveness[a] Percent inhibition by trunk injection	Solubility[b] in water ppm
Flurprimidol Cutless® Elanco	51	130
Paclobutrazol Clipper® Monsanto	42	35
Uniconazole Prunit® Valent[c]	32	14

[a] Trunk injections at recommended rates. (Environmental Consultants 1990).

[b] Sterrett, Tworkoski, and Kujawski (1989).

[c] Chevron/Sumitomo.

PGRs appear to be relatively safe if used as directed. In Pennsylvania, to test the toxicity of paclobutrazol, Richard Wells (Philadelphia Electric Company, 1990 pers. comm.) applied it at 20 times the recommended rate to a mature ash tree; new growth was severely stunted but the tree was still alive five years later. One ground-cover plant grower routinely sprayed plants with one of the early subapical inhibitors shortly before delivery. This slowed growth and enabled the plants to better withstand handling and transplanting, when water is most likely to be limiting.

Weak or stressed trees should not be treated with PGRs. Even on healthy appearing trees (particularly in autumn) if PGR uptake is slow, treatment should be stopped (Michael Watson, Potomac Edison Company, MD, 1990 pers. comm.). Apparently the PGR concentration remains so high near the application site that lethal amounts of the alcohol carrier diffuses to the cambium. For example, it has been found unwise in the midAtlantic states to inject maples if it takes more than 1 to 2 minutes or black locust 30 to 45 minutes. Response to PGRs may be different between humid and arid-summer areas and between irrigated and nonirrigated trees. More information on more species as to proper concentrations and application timing under different conditions is needed.

The low solubility of PGRs and their adsorption on soil and organic matter greatly decrease their being leached into underground water strata. No flurprimidol was detected in the leachate from oleander growing in 4-l (1-gal) containers with soil treated with up to twice the amount of PGR that gave commercially acceptable growth reduction (Sachs and others 1990). This offers the possibility of safely applying PGRs by soil injection.

Methods of Application. PGRs can enter a tree through wound dressings, foliar application, bark banding, trunk injection, or soil application (Chaney 1986). *Wound dressings* cannot provide adequate distribution of a PGR in a large tree (see

Control of Trunk Sprouts later in this chapter). *Foliar application* is not a viable alternative for large landscape trees: Public opposition to foliar spraying can be intense; nontarget plants may be affected; soil may be contaminated by spray drift and falling foliage; repeated applications probably will be needed because little active ingredient can be applied at one time.

Bark banding by spraying or painting trunks or branches of young trees, or branches with green bark, reduces tree growth (Kimball 1990). Applications to older trees with thick bark, however, is not effective. Although not reported for PGRs, bark banding may be more successful on conifers than on hardwoods because of a bark-suberin layer in some hardwoods (Backhaus, Sachs, and Paul 1980).

To be successful, bark applications must penetrate to the xylem without injuring the cambium. In the past, the toluene/diesel oil carrier could be toxic; it can be replaced by a nonphytotoxic spray oil mixed with isopropyl alcohol, kerosene, and an emulsifier (Sachs and others 1986). The main drawback of this method is lack of bark penetration in older trees—the ones on which growth needs to be suppressed the most. Exposure to a PGR on a tree would not be acceptable in most landscape plantings. Even so, bark banding may find a place in certain situations.

Trunk injection is presently the most widely used method for applying PGRs; for some species, it is the only method registered in the United States. PGRs can be injected under pressure with special equipment or by infusion which depends on transpirational suction from the leaves (Fig. 16–2). Infusion is slower and can only be done when leaves are present. Pressure injection is more rapid and more commonly used.

Much of the following discussion is based on the experience of the Potomac Edison Company (Watson 1987) and four years of monitoring more than two thousand trees at Potomac Edison by Environmental Consultants (1990). Of the estimated 2.3 million trees to maintain in parts of Maryland, Virginia, and West Virginia, Watson reported that 1.5 million trees (large and fast-growing) were candidates for injection. Most of the trees were injected with paclobutrazol.

Figure 16-2 An injector port is being placed in a shallow, diagonally drilled hole into the xylem of the tree. After the injector ports are in place, a pre-measured amount of growth regulator (PGR) in each of the six chambers in the box on the cart can be forced through the small hoses and injector ports into the tree. Pressurized carbon dioxide propels the PGR. The Arborchem® equipment is a closed system; once the PGR is put in the reservoir, the PGR is measured and injected by turning valves. (Photo courtesy ICI Americas, Inc., Willington, DE)

Control of Plant Height

Of the five species monitored (black locust, black walnut, hackberry, Norway maple, and silver maple), the four-year growth of treated trees averaged 40 percent less than control trees. Some species exhibited as much as an 80 percent reduction in growth. Evidence suggests growth reduction will continue beyond the fourth year. Injection with a PGR can extend the period between trimming trees two- to fourfold without impacting safety or reliability of utility service. Cost savings from using a PGR depend on the average tree/line clearance that is maintained. For 2-m (6-ft) clearance that would normally require a one-year trim cycle, a four-year trim-and-inject cycle could annually save 60 percent ($42 vs. $16/tree). For 3.5-m (12-ft) clearance, normally requiring a four-year trim cycle, an eight-year trim-and-inject cycle could save 25 percent ($12 vs. $9/tree) annually.

Tree species greatly influence growth response to a PGR; the large, fast-growing trees are usually the most responsive. The concentration of PGR that should be used varies with species (follow manufacturer's directions). The amount of time to inject a tree ranged from 4.5 minutes for sycamore to 28.2 minutes for white ash (Watson 1987). Generally, diffuse-porous species are the fastest to inject and ring-porous the slowest. Conifers, which have nonporous xylem, are difficult, if not impossible to effectively trunk inject with PGRs (Sachs and others 1986).

A battery-powered drill was used to bore holes 5 mm ($\frac{3}{16}$ in.) in diameter and 50 to 60 mm (2–2.5 in.) deep just above the root flare. To minimize PGR leakage, bark splitting, and injured cambium, drill the holes at a 30° to 45° angle tangent to the bark—the thicker the bark, the greater the angle.

PGR injection is 2.5 times faster into a hole drilled with a high-quality brad-point bit than with a standard high-speed twist bit—70 vs. 190 seconds for 75 ml (2.25 oz). Injection into a hole drilled with a sharp bit is twice as fast as with a dull bit. A larger volume of PGR at a lower concentration has a much greater influence on distribution within the tree than does the pressure at which it is injected. An injection pressure of 5 kg/cm^2 (70 psi) was used. Each hole was sealed with a vinyl plug to prevent any material from backflushing.

Originally, the same crew pruned and injected a tree in one operation. Almost twice as many deciduous trees that were injected when dormant had cambial damage as those injected during the growing season. Now pruning crews work year-round, and special crews inject during the growing season, usually before the trees are pruned (Watson 1990 pers. comm.).

Three brands of trunk-injection equipment have been compared by Fuchs (1988). The location and method of application use would most likely determine which equipment would be the best in a particular environment.

Soil application of a PGR can be done by injecting it into the soil or applying it to the soil surface as granules or a liquid drench on or near the trunk. Soil injection is preferred to minimize human exposure to the material.

Sterrett, Tworkoski, and Kujawski (1989) found that a root collar drench of flurprimidol to trees of four 11-year-old species in Maryland reduced shoot growth for two years with only a slight reduction in photosynthesis and none in transpiration. Subsoil injections of flurprimidol in Alabama markedly reduced shoot growth of newly planted bare root seedlings of *Acer rubrum* for two years (Gilliam, Fare, and Eason 1988).

Soil-injected PGRs obtain uniform and consistent response in a tree (Curry and Williams 1984). Soil injection is simple and minimizes exposure to humans. Tree trunks are not injured by injection holes or the toxicity of the PGR solvent.

Growth retardation has been delayed up to 18 months when a PGR was soil injected (Webster and Quinlan 1984). The soil may dilute the effectiveness of the chemical, though this does not appear to be a problem. Contamination of ground water is a concern, but the PGRs have a relatively low solubility (Table 16-1); and if they are properly applied, they should not be a pollution problem.

General Comments. The PGRs should not be applied to weak or stressed trees. Sugar maple, fruit, and nut trees should not be treated with PGRs until resi due information is available. Until there are data available about retreating plants with PGRs, they should not be retreated while under the influence of a previous application.

The anti-gibberellins offer the potential of being a great aid in more effectively and economically managing trees under utility lines and in other landscape situations. Little has been done on the possible uses of the new PGRs on shrubs, vines, and ground covers. Much work still needs to be done to tailor PGR application to the species, the environment, the climate, and goals of the management organization. The most promising aspect of the use of these chemicals is that they may be able to be soil-injected with minimum likelihood of contamination.

There remains the possibility that the manufacturers of the PGRs may deem it unwise to continue producing them due to environmental concerns of the public and financial concerns of the manufacturers.

CONTROL OF TRUNK SPROUTS

Sprouts on the trunks of trees and suckers from roots are a nuisance to nursery persons, fruit growers, and arborists. The only long-term solution to trunk sprouting is to plant trees known to sprout little or not at all. Species which root easily from cuttings appear also to produce the most sprouts (Patch, Coutts, and Evans 1984). Young trees, however, benefit when shoots of low vigor grow along their trunks to provide shade and speed trunk development. When a trunk is 75 to 125 mm (3-5 in.) in diameter, you can begin to remove the shoots along the trunk below the first scaffold branch. On grafted trees, keep the rootstock free of sprouts.

In many species, however, pruning vigorous sprouts from the trunk and trunk roots does not eliminate the problem and may even encourage the development of more sprouts, particularly if pruning has not been close to the trunk. Pruning water sprouts from English oak trees increased the number of sprouts the next year by 40 to 55 percent (Patch, Coutts, and Evans 1984). Even though some trees continue resprouting for years or throughout their lives, for many trees the periodic removal of sprouts as trees mature will overcome this problem.

The chemicals most commonly used for this purpose are naphthaleneacetic acid (NAA) (Fuller, Bell, and Kazmaier 1965) and chlorflurenol (Maintain A) (Domir 1978), both carried in an asphalt paint. A one percent solution of an ethyl ester of NAA reduces the number of sprouts growing near pruning cuts by about 50 percent. The sprouts that do develop usually achieve about 50 percent of their normal size (Fuller, Bell, and Kazmaier 1965). A one percent NAA solution sprayed on the trunks of northern California black walnut, olive, and crape myrtle sharply reduced the number of new sprouts for four to five months after treatment (Harris, Sachs, and Fissell 1971). This solution has produced successful results on nearly all species tested: The length as well as the number of sprouts were reduced; sprouts present at the time of spraying were killed; and there appeared to be little translocation, even though NAA was applied at high concentrations (Fig. 16–3). Even when treated areas were only 25 to 50 mm (1–2 in.) away from untreated areas, the former exhibited almost complete suppression of sprouts, while adjacent bark areas whose sprouts had been removed but not treated resprouted. In most cases, annual treatment is necessary.

Figure 16–3 The bases of these coast redwoods were excavated about 150 mm (6 in.) to expose basal sprouts. The sprouts to the left of the pencil in each photograph were cut close; those to the right were headed near the original ground level. The left tree was not sprayed; the base of the right one was sprayed with NAA in midspring. The photographs were taken eight weeks later. The close-cut bases produced only 20 percent as many sprouts as the stubbed bases produced; NAA reduced the sprouting in both areas to 5 percent. (Harris, Sachs, and Fissell 1971)

Paul Bairley of Ann Arbor, MI, and Dennis Ceplecha of Evanston, IL (1990 pers. comm.) report better control of water sprouts and suckers on flowering crabapple trees with chlorflurenol (as Maintain A®) than with NAA. Ceplecha reported that the chemical controlled water sprouts better than root suckers. Unfortunately, chlorflurenol may not be reregistered for use in the United States. In England, NAA (Tipoff®), maleic hydrazide (MH, Burtolin®), and glyphosate (Roundup®) were sprayed on young sprouts on English oak trees in early summer. Sprout regrowth the next year was 52, 7, and 67 percent respectively: "Current shoot growth was halted by the application (of MH), and there was a high degree of epicormic suppression in the following year" (Patch, Coutts, and Evans 1984).

Trials that would compare the effectiveness of the materials now used would provide better information on which growth regulator to choose and when and how much to apply. Until then, it would be wise to compare them yourself.

KILLING WOODY PLANTS

Techniques for clearing land of unwanted brush and trees for agriculture and forestry can be used to kill woody plants in the landscape and utility transmission rights-of-way. Resprouting of stumps or large roots left in the ground after tree removal is a frequent problem.

Various forms of 2,4-D (2,4-dichlorophenoxyacetic acid) and ammonium sulfamate (Ammate®) are commonly recommended for killing woody plants. The low-volatile ester forms of 2,4-D are applied as basal trunk sprays to shrubs. The water-soluble amine forms are poured into cuts in the bark or on the stumps of trees (Leonard and Harvey 1965). Glyphosate (Accord®, Roundup®) is also applied as a spray to some trees or as a cut-surface treatment on stumps after cutting. Vegetation management of utility transmission rights-of-way "feature primarily the phenoxy and selective plant growth regulator compounds, i.e. Tordon® (picloram), Garlon® (triclopyr), 2,4-D, dicamba. Next to water we use glyphosate in the form of Accord®" (Keith Jones, Central Illinois Public Service, 1990 pers. comm.). Chemical vegetation management cost only 25 percent of the cost of manual clearing and/or mechanical mowing practices.

Nonchemical Control of Sprouting. In the midwestern and eastern United States, trees are sometimes girdled a year or two before being cut down to prevent stump sprouting. Some arborists remove a ring of bark at least 75 mm (3 in.) wide; others strip the bark from the lower trunk. After a growing season, the stump will be less likely to sprout than if it had not been girdled. This process is slow and usually not acceptable once a decision has been made to remove a tree from a landscape. Grinding a stump and basal roots with a stump cutter reduces sprouting from the remaining roots except on species that sucker readily from roots.

CONTROL OF ROOTS IN SEWERS

Tree roots constantly break and clog sewer lines, a perennial problem that wastes millions of dollars each year. Roots cannot enter intact sewer lines, but as they enlarge they may break lines and later enter the cracks. A sewer district in the state of Washington reports that two distinct types of roots create problems. The roots of trees such as poplars and willows enter sewers through the lower quadrant of joints and grow longitudinally along the bottom of the sewer. The roots of trees like honey locust, on the other hand, enter any part of a joint and form a ring in the pipe but seldom extend far from their point of entry (EPA 1977). Once a root enters a sewer, the conditions of aeration, moisture, and nutrients are so favorable that it inevitably grows until it clogs the sewer (see Fig. 1–5).

The standard procedure for unplugging a sewer line is to use a mechanical router (a powered rotary blade attached to a flexible steel cable) to physically cut and remove the roots. A crack in a sewer line is usually not repaired unless the same line must be routed out more than once a year; roots are cut back only to their place of entry. Unless some other action is taken, the sewer clogs again in one to three years.

Metham and Dichlobenil Treatment

Since the early 1970s, a mixture of two herbicides has been used to kill roots in sewer lines and to delay root regrowth. A solution of 1000 ppm metham (Vapam®) and 100 ppm dichlobenil (Carsoron®) was first used as a root soak (Leonard and Townley 1971). Later an air-aqueous (19–1) foam of the chemicals was found to be at least 20 times more effective than the root soak (Leonard, Bayer, and Glenn 1974). In hilly terrain and for large drains, sprays required less chemical, though they were not as effective as soaking or foaming. The chemical mixture is marketed as *Vaporooter Plus*® (Airrigation Plus® Engineering Co., Pleasanton, California 94566).

In the root-soak method, a 75 to 125 m (250–400 ft) section of sewer is blocked with an air plug and filled with the solution for an hour. By starting at the upper end of a sewer line, the same solution can be used several times. In problem lines, mechanical routing usually take place annually, whereas chemical-soak treatments last for two years and sometimes longer.

Foam uses less of the chemicals and is easier to apply, particularly in uneven terrain, but special equipment is required to create the foam. A hose is placed in the sewer between two access holes. The herbicide foam is pumped through the hose as it is retracted from the sewer, completely filling the sewer with foam. Placing the foam this way is least likely to cause back-up into adjacent buildings. The metham kills roots a short distance beyond the area treated; injury to other roots or to the tops of plants has not been a problem. Dichlobenil also inhibits regrowth. The foam mixture is available as Sanafoam Vaporooter® and can be applied with foaming equipment such as the Foamaker®.

As mentioned, sprays of Vaporooter Plus® are used in hilly terrain and in large drains where soaking or foaming would be more difficult or use excessive amounts of chemical. The spray kills roots in the drains but not in the cracks or joints. It also requires special equipment. Spraying must be done more often than soaking or foaming.

The dead roots become brittle and slough off into the liquid flow. Two applications may be necessary for a complete root kill if the root masses are dense (Water Pollution Control Federation 1980). Large root masses should be removed after they have been killed to prevent stoppages.

One herbicide-foam treatment may be 25 to 40 percent more expensive than the mechanical method. However, the mechanical method has to be repeated almost every year while the foam is effective for three to five years (Monck 1980). The

Water Pollution Control Federation (1980) indicates that with proper treatment, regrowth and new root growth may be inhibited for two to seven years.

The results of metham and dichlobenil treatment vary with tree species, soil conditions, and thoroughness of application. The chemicals apparently create no problems in sewage treatment plants.

Copper Sulfate Treatment

Copper sulfate inhibits root regrowth in sewers if the solution is brought into contact with the roots for a sufficient time. Copper sulfate should be used sparingly because it is toxic and can interfere with sewage treatment (Water Pollution Control Federation 1980). However, the Environmental Protection Agency (1977) states that copper sulfate can impede root growth for three years with no detrimental effects reported at sewage plants. If an entire community were to use as much copper sulfate as some cities have, sewage sludge contamination might result.

CONTROL OF FLOWERING AND FRUIT SET

Inhibiting the flowering and/or fruit set of certain trees has been a goal of many people in order to minimize allergies, safety hazards, or unsightliness and odor caused by falling flowers and fruit. Naphthaleneacetic acid (NAA) has been the standard spray for preventing or reducing fruit set on olive. It must be sprayed more than once in cool areas where bloom is prolonged (Table 16–2). If bloom is short, one spray at full bloom can eliminate most of the olive set (Opitz 1970).

Several chemicals have been shown to prevent flowering or fruit set on a wider range of species. NAA is reported to have fruiting reduction effects on more species than other chemicals (Table 16–2). This is probably due to its having been available since the 1950s. On species where it is effective, NAA reduces fruit set but does not prevent bloom. This is of little help to pollen-allergy sufferers, since the trees return with extremely heavy bloom the next year. Dikegulac, mefluidide (Embark®) and chloroflurenol applied before bloom are reported to inhibit flowering, particularly of olive.

Maleic hydrazide (MH; Royal Slo-Gro®) and mefluidide, not listed in Table 16–2 but registered as growth regulators, are reported to be effective in inhibiting fruiting of ginkgo and flowering and fruiting of olive respectively. MH was applied to ginkgo at 750 ppm in the eastern United States (Hartman 1980) and 5000 ppm in California (Hamilton 1977). The best control of fruiting was to spray both male and female ginkgo trees. Sevin®, an insecticide used to blossom thin commercial apples, has been suggested for fruit control on similar species (Hartman 1980).

Tender plants growing under or near those to be treated should be covered with a plastic tarp. If nontargeted plants are sprayed with growth regulator, spray them immediately and thoroughly with water. To reduce spray drift, use moderate

TABLE 16-2 CHEMICALS TO PREVENT OR REDUCE FLOWERING AND/OR FRUIT SET

	Naphthalene-acetic acid App-L-Set®, Sta-Fast®, Stix®	Dikegulac Atrinal® Atrimec®	Ethephon Florel®	Chloro-flurenol Maintain CF 125®
Apple Crabapple Pear-fruiting	40–60 ppm FB + 10 day[a] (H)[b]		750–1000 ppm 1000 ppm B–FB[a] (L)[b]	
Pear-Callery Aristocrat			1000 ppm FB[a] (P)[b]	
Carob			500 ppm B–FB (L)	
Catalpa	60 ppm–FB (H)			
Elm	40 ppm–FB (TS)[b]			
Holly Japanese		1000–2400 ppm B–FB (L)		
Honeylocust	60–100 ppm Frt 2–5 cm (H)			
Horsechestnut	40–60 ppm FB (H)			
Liquidambar				150 ppm FB (P)
Olive	150 ppm −3 days to FB + 10 days (H)	4000–8000 ppm B–EB (L)	1000 ppm B–FB (L)	300 ppm −4 weeks before Fl buds form (L)
Maple	40 ppm FB (TS)			
Plum	150 ppm FB (S)[b]			150 ppm FB (S)
Privet, Glossy and Rose, Multiflora		1000–2400 ppm B–FB (L)		

[a] FB = Full bloom; B–EB = Flower bud to early bloom; B–FB = Flower bud to full bloom

[b] (L) = concentration and timing are on product label

(H) = Hartman 1980; Ohio

(P) = Perry 1990, unpublished; California

(S) = Svihra 1990; California

(TS) = Smiley 1990; unpublished; North Carolina

Be sure chemical is registered for use on species in region. Follow label directions for use of chemicals.

pressure; apply a coarse spray (avoid fogging); spray only on cool, calm days; and direct the spray into the tree or shrub.

As with all growth regulators, these chemicals should not be applied to weak-growing plants or those under stress. Check with authorities to be sure a chemical is registered for use on the species in your region. Trial applications may be wise to ascertain the response of particular varieties under local conditions.

CHAPTER 17

Tree-Hazard Management _____

Millions of trees in public and private landscapes are maturing; many are declining. Some people insist that all trees, on private as well as public land, be preserved with little concern for the cost or tree safety. Although highly prized by many, trees can lose limbs, break apart, or fall. Property can be damaged and people injured or killed when trees fail or even obstruct a critical view, such as a driveway. Trees, however, cannot be neatly separated into hazardous and nonhazardous groups. Nearly every tree has some potential to cause injury. Complete tree safety could not be attained without removing most trees. Arboricultural managers must be able to not only evaluate tree-hazard potential but also to establish an acceptable level of safety at a reasonable cost with minimum loss of trees or damage to the environment.

A tree is considered to be hazardous if it is *structurally unsound* and there is a possible *target* (Fig. 17–1). An unsound tree in an area with no target is not a hazard.

Trees become structurally unsound due to weak structures, decay of trunk and branches, cankers and canker-rot, and root loss or root decay. Low-hanging branches and tall shrubs can dangerously obstruct road signs and intersections. Fruit that drops and roots that crack pavement can cause personal injury.

Structures, vehicles, and people are possible targets. The extent of possible property damage depends on the likelihood of a tree or a part of a tree striking the property and how serious the damage might be. Injury to people depends on the likelihood of a tree striking a specific area when people are present. Views of road signs and intersections obstructed by plants can result in serious accidents.

Although most tree failures occur during storms, a well-structured tree free of

Figure 17–1 A 23-m (75-ft) tall smooth-leafed elm tree leaning about 25° from the vertical (left). The 70-year-old tree had an 0.8 m (33 in.) dbh (diameter at breast height) trunk. Almost all of the crown of the tree was to the leaning side of the trunk base. The trunk base had five openings into a 0.75-m (28-in.) diameter cavity (note the handle of 0.75-m [28-in.] long probe protruding from an opening at the base of the trunk [upper right]). A few of the large roots were dead and one broken (lower right). A bus stop was located at the street light and the two walks are heavily used at this college campus (left). The weight of the vigorous growth increased the likelihood of tree failure. It is amazing the tree was able to withstand strong winds even though its loss of trunk strength was estimated to be greater than 50 percent. This tree was a definite hazard.

significant weakening defects is best able to withstand strong winds and the weight of snow and ice or rain.

TREE-HAZARD MANAGEMENT PROGRAMS

Some forest and park agencies have developed tree-hazard management programs to reduce accidents caused by tree failures (Paine 1971, Mills and Russell 1981, Wallis, Morrison, and Ross 1980, and Johnson 1981). These programs focus primarily on native trees in forested recreation areas. Even though few of these trees receive regular maintenance, the programs provide a framework for developing urban programs. An effective tree-hazard management program should

- Prepare and maintain a written management program that is approved by the appropriate administrative body
- Establish the level of hazard control (acceptable risk) which depends on budget and management philosophy
- Have a systematic inspection procedure
- Train inspection and maintenance personnel
- Record inspection and control recommendations
- Control hazardous situations in a timely manner
- Maintain records of inspections and control measures
- Review the program and records and make recommendations for improvement

Influenced by tree-failure information from federal and some state recreation areas throughout the United States, Paine (1971) based a tree-hazard rating system on (1) probability of tree failure, (2) probability of a target being hit, (3) expected damage to the target, and (4) monetary value of the target. The first three, expressed as percentages, and the target value are multiplied together. If the product, monetarily expressed (dollars in the United States), exceeds the established *hazard control level,* corrective measures are to be taken. The tree-hazard management programs of the California Department of Parks and Recreation (Bakken 1986) and the San Francisco Recreation and Park Department are based on Paine's system.

Rating systems developed in British Columbia (Wallis, Morrison, and Ross 1980) and in the state of Washington (Mills and Russell 1981) are based on numerical ratings for (1) probability of failure (ratings 1 to 3 or 4) and (2) expected damage to the target (1 to 3). The sum of the two ratings for a tree determines the remedial action to be taken.

The system developed by Johnson (1981) in Colorado is based on numerical ratings (1 to 3) for (1) tree species, (2) probability of failure, and (3) expected damage to the target. The product from multiplying the three ratings together determines the remedial action to be taken.

In each of the systems described, dead trees are not rated—they are to be removed if there is a possibility of their striking a target. Only live trees which might pose an appreciable hazard to people and property are examined; in some heavily used sites, all trees are examined. The location of trees given a hazard rating should be described or shown on a map. An identification tag should be placed inconspicuously on each hazard-rated tree.

Annual inspections are required by the U.S. Forest Service, although two per year are recommended—one during the summer to include the leaves and one during the dormant season (Robbins 1986). Trees in California state parks are to be inspected every two years (Bakken 1986). Annual inspections are generally recommended for forest recreation areas; these areas are seldom used in the winter. Wallis, Morrison, and Roth (1980) recommend inspecting evergreens in the spring after they have been exposed to winter storms. Inspect deciduous trees during leafing-out, when dead tops and branches are more easily seen.

Mature trees in urban areas may need to be inspected more frequently. Occurrence of tree failures between inspections will indicate the adequacy of the interval between inspections. Some species may need to be inspected more frequently than

TABLE 17-1 COMPARISON OF THE LOCATIONS OF TREE FAILURES OF HARDWOOD AND SOFTWOOD TREES IN URBAN AREAS AND IN FORESTED RECREATION AREAS

	Hardwood		Softwood	
	Urban[a]	Forest[b]	Urban[a]	Forest[b]
Tree failures				
Number	363	153	137	1153
Percent	73	12	27	88
Location of failures (%)				
Branches	42	10	40	5
Trunk	31	30	29	13
Butt + root	27	60	31	82
Total	100	100	100	100

[a] Reported failures of public and private urban trees primarily in northern California 1988–1990 (Costello and Berry 1991).

[b] Reported failures in Rocky Mountain forest recreation areas, USDA Forest Service 1965–1980 (Johnson 1981). Accidents resulted from 3.4 percent of the tree failures.

others. Additional inspections should be made following heavy rain, snow, ice, or wind storms in target areas.

For each hazard-rated tree, any recommended corrective measures, the dates the tree was inspected, and the corrective measures performed should be entered on a hazard-rating sheet. **These rating sheets, maps, and other information should be safely filed in the appropriate office.** They should be periodically reviewed to be sure that the indicated corrective action has been taken. These records are also valuable (1) to note changes in tree condition between inspections, (2) as a record of which species fail and why—to help predict likely failures and which species to avoid in future plantings, (3) as evidence of inspection frequency and thoroughness, in case of litigation involving damage or injury, and (4) as a record of costs involved in hazard-tree inspections and treatments (Robbins 1986).

Trees should be inspected as or after corrective action is taken to evaluate the accuracy of the hazard-ratings given. It would be wise to keep records of tree failures and the possible reasons for them. This information can be used to assess and improve the inspection program.

Public and private horticulturists in California are cooperating with the University of California in documenting tree failures and the probable cause of each failure. Even though only 500 tree failures were reported by the end of 1990, interesting relationships are emerging between the various types of information recorded (Table 17–1).

FACTORS INFLUENCING TREE FAILURE*

The extent of hazard depends on the type, size, and location of the defect; tree species, size, and age; site characteristics; wind patterns; and the target. Familiarity

*Adapted from Mills and Russell (1981).

Factors Influencing Tree Failure **459**

with the site, species at the site, and previous problems will indicate what types of defects to look for. Some species are more prone to certain diseases and defects than are others. Maintained, solitary urban trees are more susceptible to certain disorders than are uncared-for forest trees, and vice versa (Table 17–1).

Defects

The degree of hazard depends largely on type, size, and location of a defect. Root rot, loosened or cut roots, or severe decay in the trunk (bole, stem) present a greater and more immediate hazard than a dead conifer top or branch canker. Should a failure occur, the entire tree would probably fall in the first case, but only a portion of the tree in the second.

Species

Some species are weak wooded or more subject to decay than others. Root and butt (trunk base) rots result in the highest percentage of tree failures in conifers. In general, hardwoods (broadleaved trees) undergo more stresses than conifers because of differences in growth habit. Most hardwoods have large spreading branches; branches of most conifers are relatively small and arranged around a central trunk. Most hardwoods are phototropic, growing toward openings in the canopy; such trees become lopsided. Branch attachments of hardwoods are usually not as strong as they might be because many branches are or are almost as large as the trunk (see Chapter 15). The relatively high mechanical strength of most hardwoods compensates in part for the great stresses they must withstand (Wagener 1963).

Heart rots in hardwood trunks commonly extend into the main branches, whereas in conifers such extension is less frequent (Wagener 1963). These characteristics of hardwoods result in most failures occurring in branches or in crotches where they join the trunk (if butt failures are not considered). Hardwoods are more prone than conifers to crown damage from ice and snow. Basically, branch weight, taper, leverage, and size relative to the trunk are critical to the safety of hardwoods. Trunk butts and roots are the most common failure site for conifers.

Size and Age

Large, old trees are more likely to fail than are smaller or younger trees of the same species. Old trees are less able to adapt to unfavorable conditions and are more subject to decay and other disorders. Fast-growing trees are usually weak wooded and may be hazardous even when young. Vigorous trees, especially trees with poor branch attachments, may be more subject to limb breakage than less vigorous ones of the same species, due to the greater weight of their limbs. On the other hand, root rot can be a problem at any age.

Site

An urban setting is much different from a wooded or forested area. Trees in a forest protect one another from wind. When a stand of trees is opened up for a road, ski trail, or home site, trees adjacent to the cleared areas are more subject to windthrow. Urban trees with roots confined by narrow planting strips, containers or vaults, or root barriers are more subject to windthrow during high winds than are trees without confined roots (Fig. 17-2).

Trees on heavy, shallow, and poorly drained soils have shallow roots which are subject to windthrow, particularly when the soil is saturated.

Figure 17-2 Wet soil and a narrow planting strip contributed to the blowdown of this deciduous hardwood during a severe winter wind storm in the San Francisco Bay area in 1988. (Photo courtesy Pavel Svihra, University of California Cooperative Extension Service)

Weather

Wind accompanied by heavy rain can be particularly destructive (Fig. 17-2). A storm in England in October 1987 blew down an estimated 55,000 street trees in 24 London boroughs and an additional 35,000 trees were damaged in 11 of those boroughs (Department of the Environment 1988). In 1989, Hurricane Hugo devastated the tree population in the southeastern United States. Less severe storms can also cause extensive damage, particularly for trees with defects.

Prevailing winds and those from the opposite direction are important factors. A defective tree is more hazardous if it is upwind from a target. Often a tree that has withstood strong prevailing winds will fail when buffeted by a wind of less velocity from the opposite direction.

As already mentioned, ice and snow can cause serious limb breakage and tree failure.

Maintenance Practices

How trees are grown in the nursery and planted in the landscape can have a great effect on the extent of girdling roots. The effect of circling roots may not be evident for many years—until a tree blows down in a storm or begins to show poor growth.

The terminals of many trees, particularly those grown in containers, are headed in the nursery at 1 to 2 m (3–6 ft) above the soil. Several shoots grow close together from below the pruning cut. These are weakly attached and many will split out unless the tree is properly trained (see Chapter 15, Fig. 15–17).

Target

The safety of a target is the major reason for a tree-hazard management program. Chances of an accident increase as occupancy of a site increases. Therefore, heavily and frequently used areas should be inspected more often than those of little activity.

Moving a target, such as a picnic table, may be the best answer for minimizing hazard from a defective tree.

NONINFECTIOUS CAUSES OF TREE FAILURE

Tree Structure

Multiple leaders (codominant stems), several large branches arising near the same level on a trunk, and weakly attached branches are becoming more frequent causes of branches and tops breaking out of trees. Two or more leaders about the same size or branches near the same size as the trunk are more likely to fail than if one leader or the branches were only half the size (75 percent of the diameter) of the main trunk (Fig. 17–3). A trunk is not able to grow around a branch when both are near the same size. Several relatively large branches arising near the same level on the trunk are even more vulnerable to failure. Also the weight and leverage of such limbs are great in relation to the strength of their attachments. As trees age, their branches usually continue to spread, further increasing the stress on their attachments.

Many of these structural problems in planted trees begin in the nursery when the leader (terminal) of most decurrent trees is headed to force branching. This results in what appears to be a well-proportioned tree when purchased, but most branches are too low, too close together, and about the same size—and the tree seldom has a leader. These problems are seldom corrected in the landscape, even by "professional" gardeners. **The practice in the nursery of heading trees to force compact branching should be stopped.**

The angle of branch attachment is not a problem unless there is included bark

Figure 17-3 Double leaders (codominant stems) of large trees can be hazardous, even in upright conifers; gusty winds sometimes create torque forces that can split out one of the leaders. Included bark further weakens the attachment of the double leaders.

in the attachment crotch (see Fig. 15–15) or the branch is about the same size as the trunk. Most branches with included bark are fairly upright while young, so there is little stress on their attachments. As the trees mature, however, the branches become more spreading and greatly increase the likelihood of crotch failure.

Occasionally during a storm, a branch attachment will crack near the top of the attachment, but the branch does not fall. Callus forms at the edges of the crack. New xylem may begin to rejoin the branch and trunk in the crack area. Adventitious roots may form in the crack. Decay may get a start. Indications of a split attachment are cracks, indentations in the trunk bark below a crotch, and swelling on either side of a crotch. If any of these are seen, the branch and its attachment should be examined carefully to assess the hazard. Necessary steps should be taken to eliminate or reduce the danger.

A moderate wind occasionally blows down a tree that has withstood gale-force winds—but from the opposite direction. The strength of a trunk or branch depends on its ability to resist cracking under a force applied at right angles to the grain or crushing under a load applied parallel to the grain. Wind exerts a compressive force on the wood on the lee side of the trunk and a tension pull on the opposite side. In a strong wind, xylem on the lee side of the trunk can be damaged by compressive forces without the trunk failing. Sometime later, however, the trunk may fail from a lighter wind from the opposite direction because the compression-damaged xylem cannot withstand the tension pull (Wagener 1963).

Crowded main branches (see Fig. 15–17) and those that have had most of their lower laterals removed may have less taper than necessary to distribute branch weight and stress evenly along the lower portion of the branches. These branches are more likely to break a short distance out from their attachment in a storm or as the branch grows longer and heavier. The possibility of such a failure is increased if a branch has an upward-growing lateral removed, a crook, or other weakness in this vulnerable area.

Most of these structural hazards can be prevented by proper training of the trees while they are young (see Chapter 15). Wise pruning and possibly cabling and bracing can minimize these hazards in most mature trees (see Chapter 18). Sometimes, however, it is best to remove the hazardous branch or the entire tree.

Summer Branch Drop

Apparently sound limbs up to 1 m (3 ft) and trunks up to 1.3 m (4 ft) in diameter have broken and fallen during or following hot, calm summer afternoons (Harris 1972, 1983). Falling branches have caused fatalities, and many have caused serious injury and property damage. This phenomenon, known as *summer branch drop,* has been reported in Australia, England, South Africa, and California. Since 1983, I have received reports of such incidents from British Columbia and from New York to Texas in the United States. In England, branch drop is associated with calm weather following a heavy rain after a period of increasing soil dryness (Rushforth 1979). Kellogg (1882) first reported summer branch drop in California.

Branches that drop are usually more horizontal than vertical and extend to or beyond the edge of the tree crown. The break occurs most often out on a limb some distance from its attachment. The wood at the break may appear sound or the branch center may be brash (weak, blocky structure) or decayed. Most failed branches had previously withstood severe storms only to fail during or following a calm summer day. Such branches seldom exhibit any outward appearance of a hazardous condition.

In England, summer branch drop has been reported in oak, sweet chestnut, beech, ash, poplar, willow, and horse chestnut (Rushworth 1979). Harris (1983) reported similar limb breakage in several species of eucalyptus, oak, elm, ash, plane, pine, cedar, and in *Ailanthus altissima. Erythrina caffra, Ficus microcarpa, Olea europaea, Grevillea robusta, Sequoiadendron giganteum,* and *Sophora japonica.* Since 1983, failures have been reported in silver maple, sweetgum, and several poplars. Young and mature trees of good vigor of susceptible species are less prone to branch failure than are overmature, senescent trees (Rushworth 1979). A tree that has lost one limb is more likely to lose another than a similar nearby tree of the same species is to lose its first branch.

In California, where summers are dry, limbs have fallen from native and planted trees in both irrigated and nonirrigated landscapes (Harris 1983). J. William Roach of Roseville, CA (1988 pers. comm.) observed 80 native oak trees, *Quercus lobata,* 250 to 600 mm (10–24 in.) in trunk diameter, growing in the Sierra foothills. In eight years, wind blew down one tree and broke limbs out of two trees; however, during the summers, under calm conditions, nine trees lost limbs up to 500 mm (20 in.) in diameter.

Although no explanation is readily apparent, the phenomenon appears to be caused by some type of water stress. Limb failures on hot afternoons are somewhat unexpected since branches become lighter due to transpiring more water than they receive. Branches rise and trunks and branches shrink during summer afternoons

(Kozlowski and Winget 1964). Drought conditions have caused conifer trunks, with spirally oriented xylem, to split (Rushworth 1979) as a wisteria seed pod does. These observations are borne out since most of the breaks are relatively dry; but in oak and eucalyptus, liquid has been observed to "flow" from both sides of some breaks. Most likely, the liquid is from wetwood in the branch center even though the fluid in the outer conducting xylem is under tension.

Shigo (1989) suggests that internal cracks in large branches, caused by wounds or flush pruning cuts, may spread outward and break the bark. As the cracks spread outward, "the branch becomes similar to two cantilevered beams; one on top and the other on the bottom. So long as the wetwood along the secondary cracks keeps the wood moist, the branch will bend. When cracks begin to dry, the branch may fracture." (Shigo 1989). A similar condition might develop when transpiration greatly exceeds moisture uptake but only in the outer conducting xylem. This explanation would not account for limbs failing that have wetwood under pressure.

Another possibility is that ethylene, produced by plants (increasingly so when under stress), could begin to dissolve the cementation of cell walls, weakening a limb to the point of failure. If wood actually weakens under hot, calm conditions, the process must be reversible or few mature trees would have many large branches.

Summer branch drop is serious enough for the Royal Botanic Gardens at Kew to post a large sign at each entrance warning visitors that "the older trees, particularly beech and elm, are liable to shed large branches without warning." Many Australian parks have similar warnings.

To minimize the hazard of summer branch drop on mature trees of susceptible species

- Remove or shorten and lighten long horizontal branches
- Prune trees to encourage tapered branches
- Maintain tree health and moderate vigor
- Inspect for and correct structural defects, removing low-vigor limbs and those with decay or cavities
- Remove overmature trees of low vigor or restrict activity under them
- Place heavily used facilities in "hazard-tree–free" areas
- Do not plant species known to be susceptible to this problem in areas that may be frequently used

Increased Exposure

Trees that have been protected by structures or adjacent trees are particularly vulnerable to windthrow if they become exposed to the elements. Trees on the edge of recent clearings for roads, houses, and open space are easily blown down. Such trees lack large, spreading, deep root systems, have little trunk taper, are tall, and have most of their foliage (wind sail) at the top. Seldom are they attractive landscape plants; pruning to try to improve their stability does little for their appearance.

Even if exposed trees do not fail initially, the crowns of hardwood trees on the edge of a clearing often grow into the opening. The lopsided growth may cause the tree to fall. The possible consequences of exposing previously protected trees must be carefully considered.

Root Loss

Soil excavations which sever large roots near trunks of mature trees often create hazards. It is best not to cut large roots, but the potential for a tree uprooting depends on the specific situation. Factors to consider include potential targets; height of the tree; density of the foliage; exposure to wind, including speed and direction; soil depth, texture, and moisture; and depth and spread of large roots. The spread of moderately deep, large roots provides greater stability than deep roots with little spread do.

A tree is more likely to be blown down if roots are cut on the upwind side. Thinning out the top and reducing the height of a tree reduces the likelihood of windthrow (see Chapter 18).

Unstable Rooting

Closely akin to the problems of *increased exposure* and root loss is *unstable rooting*. In many situations, the roots are not able to adequately anchor a tree against wind and weight. Trees on shallow soil or those from which considerable soil has eroded from the roots are particularly susceptible to windthrow during or following a heavy rain or irrigation. In shallow soil, it would be wise to plant small- to moderate-size trees and to minimize erosion before it becomes serious. If large trees are present in such situations, however, consider thinning their tops and reducing their height.

Girdling Roots

Roots that begin to circle a tree trunk at or before planting in the landscape are more likely to jeopardize tree stability than are circling adventitious roots which form 15 to 20 years after planting. Not as many nursery-grown trees with girdling roots become hazardous as one might expect because the most seriously affected trees fail before they become very large. Watson, Clark, and Johnson (1990) suggest inspecting Norway maple trees for girdling roots four to six years after transplanting in the landscape. Problems can be minimized by requiring girdling-root-free nursery stock and should removing girdling roots occur on trees in the landscape handle as suggested in Chapter 20 (Fig. 3-6 and 20-4).

Leaning Trees

A leaning tree may or may not be likely to fall. A tree that has grown most of its life in a leaning position develops trunk xylem and a root system to compensate for the lean. Trees that originally grew upright but now lean due to wind, root damage, or being struck by a falling tree may have inadequate root anchorage. A wind-, snow-, or icestorm or the weight of continued growth can cause root or trunk failure. Even a tree that has always leaned can become so heavy with foliage as to be hazardous (Fig. 17-1). Carefully check leaning trees for decay and root stability.

Unfavorable Soil Conditions

Construction and other activities can seriously affect trees. Soil is compacted by any activity that puts pressure on the soil. Paving and raised soil levels around trees occur all too often. Land grading can block surface and subsurface drainage of water. All of these reduce water penetration and gaseous exchange. Root growth, function, and well-being are reduced depending on the extent of soil degradation. A tree so stressed may be slow to show symptoms of an unfavorable root environment, but the roots may be so severely affected that some die and begin to decay. If large lateral roots are dead, even though the top of the tree appears alive, the tree should be removed.

Frost Cracks and Lightning Scars

Lightning can kill one or more trees with little outward indication of injury or it can rip a tree asunder. Trunk cracks and scars from cold or lightning, however, are seldom a problem in themselves but may be entries for decay organisms. Such trees near targets should be regularly inspected.

Dead Trees and Branches

Species and climate help determine how long dead trees can safely be left standing. Most recently killed conifers without root rot can remain structurally sound for up to three years (Mills and Russell 1981). Resin-impregnated species are more resistant to failure than those not so impregnated. Dead hardwoods begin to lose branches earlier than conifers and are a greater hazard because of branch size.

Dead branches in live trees are relatively safe as long as they do not become decayed. Dead branches are lighter than live ones because they are dry and without leaves.

In public areas and near structures and parking lots, dead trees, dead branches, and broken branches should be removed soon after discovery. A dead tree or branch may not be hazardous, but it is not attractive and may give the impression of poor maintenance. The failure of a dead tree or branch might raise suspicion of negligence. A broken branch more than 50 mm (2 in.) in diameter that might hit a person should it fall should be removed.

In areas free of targets, dead trees and branches that will not be a source of infection or pests may be left for wildlife. Visitors should be informed where areas of potential hazards are and why dead trees remain there.

INFECTIOUS CAUSES OF TREE FAILURE

Decay is an essential process that converts complex organic matter to simpler forms and releases nutrients. These protect and enrich the soil and provide energy for a host of beneficial organisms.

Some of these microorganisms, however, can destroy the structural integrity of trees. They can cause trunk and branch decay as well as cankers that kill trunk cambium, rot cankers and adjacent trunk tissue, and kill and rot roots. Trees so affected can become hazardous. Fungi are predominantly responsible for these problems.

Decay

Foresters, fruit growers, arborists, and builders—practically all people who grow trees and use tree products—are concerned about tree wounds and decay. Bark wounds not only interfere with movement of organic compounds between the top and roots but also open wood to microorganisms that can cause decay. Bark can be wounded by animals, wind, lightning, sun, ice, cold, fire, cars and other machinery, vandals, and even arborists who expose internal tissues when they prune, cable, and inject trees. Unless a wound practically encircles a trunk or branch, the main concern is decay.

Decay in trees is the breakdown of wood cell walls. Although many pathologists have studied this process and ways to prevent it, Robert Hartig's 1878 publication on the subject was supplemented by only a few significant discoveries in the following 50 years. Hartig's concept was that wood deterioration followed fungus infection of a fresh wound into heartwood (Merrill, Lambert, and Liese 1975). Welch (1949) sums up what was learned about decay during the 50 years after Hartig.

> Some tree species decay faster than others, and individuals within a species differ in their susceptibility to decay.
>
> Fungi vary greatly in their ability to attack and disintegrate wood.
>
> Decay is influenced by the amount of available water and air. Moisture in the wood of living plants is probably always present in quantities large enough to support the growth of fungi. If water can enter a plant through wounds, fungus spores and other microorganisms can also enter. The amount of air in the wood is probably more significant than the amount of water that enters from outside. Superficial wound rot and sapwood rot require a great deal of air, whereas deep rot is able to get along with very little; some rot does quite well when completely sealed up inside a tree. It appears, therefore, that sealing a cavity may stop some types of decay but not others.
>
> The activity of secondary agents—insects, rodents, and birds—may introduce air and water so as to influence the development of decay.

A big advance in understanding decay was the concept introduced by Hepting (1935) and refined by Shigo (1979b) that a tree can confine infection by compartmentalization (see Chapter 2). This concept can help in assessing the potential hazard of decay, its possible progress, and what should be done.

Infection does not necessarily lead to decomposition, even in wood that is near a wound. How far the succession of organisms progresses depends on the tree's genetic makeup and possibly its vigor; the type, size, and position of the wound; the virulence of microorganisms and their antagonists; and environmental conditions. Decay need not be completed. Antagonism among organisms, environments unfavorable to decay organisms, the tree's defenses, and other forces may stop the process in any stage.

Discoloration Preceding Decay

After wounding, wood will discolor before decay begins (Shortle 1979). Discoloration involves the alteration of living cell contents in response to metabolic changes regardless of the presence of microorganisms. Discoloration itself does not weaken wood but is a warning that decay may not be far behind.

Not all discolored wood is predisposed to decay, however; heartwood darkens in the normal process of aging. Some species (oak, pine, and black walnut) have heartwood, and some (birch and maple) do not. Although heartwood no longer transports water and nutrients, it retains strength and its ability to resist infection by an accumulation of chemicals previously secreted by parenchyma cells as the wood aged (Shigo 1986a, Sinclair, Lyon, and Johnson 1987) and the reduction in microelements and moisture (Shigo 1986b). Normal heartwood and wood that has discolored and become predisposed to decay are quite different. Heartwood, however, can discolor and decay if exposed to infection; active sapwood (formed before the wounding) adjacent to discolored heartwood will usually discolor also.

Fungi decay in two ways. *White rots,* although their color ranges from whitish to reddish or brown, involve all wood tissues, resulting in soft and flaky or stringy decay. *Brown rots,* on the other hand, are caused by fungi that remove the cellulose of the wood but leave the lignin. Rotted wood ranges from light to dark brown; it is friable in texture, breaks apart into rectangular blocks when dry, and reduces to a powder when rolled between the fingers.

Aerobic fungi grow and function in dead branches or exposed sapwood—locations with ample oxygen. They are more common in hardwood trees than in conifers. Most of them can extend from a decaying branch into a larger live branch or trunk to which the decaying branch is attached. Due to compartmentalization and possibly a lack of oxygen, few progress far in the live member. Aerobic fungi seldom cause extensive decay within a tree unless the trunk has numerous openings (Wagener 1963). However, they can seriously weaken branches and their attachments.

Anaerobic fungi are adapted to grow and function with a limited oxygen supply if necessary. Unless confined by compartmentalization, they can invade heartwood for considerable vertical distances, regardless of openings. Extensive heart rot in trees usually develops from anaerobic fungi.

Heart rots usually occur either in the trunk or the butt (trunk/root collar), depending on the place of entry of the fungi. Fungi that normally enter through branch stubs on the trunk develop as trunk rots; those that enter through basal wounds develop as butt rots or rots of the lower trunk (Wagener 1963). The location of decay within a tree is an important factor in identifying decay that may influence tree safety.

Detecting Decay

Decay may be indicated by open or callused wounds, cavities, frost cracks and other trunk splits, broken or dead branches, or fruiting bodies of decay fungi (conks) on exposed wood. Even when the surface of a wound appears sound, it may be decayed underneath. Check the trunk base closely for signs of root injury. It is possible for a dangerously decayed tree to appear healthy: Its bark may be intact and top growth fairly vigorous. Striking the bark of a hollow trunk with a mallet will often reveal the defect with a hollow, drum-like sound.

A more accurate field method to detect the presence of decay and its extent is sorely needed. In addition to using a mallet, drills, increment borers (see Fig. 19–5), X-rays, ultra-sound, and electrical probes have been tested and some used (Costello and Peterson 1989). At best, a mallet can only detect loose bark and cavities. Increment borers are difficult to use in hardwood species and usually require a microscope and skill to examine borings. X-ray and ultra-sound equipment is not yet practical for field use.

Costello and Peterson (1989) compared the drilling and electrical probe (Shigometer—see Chapter 19) methods with that of a *pick test* (wood splinters "picked out" with a sharp tool for examination) on mature eucalyptus trees that were to be removed. Both the drilling method and the Shigometer predicted more decay than that found by the pick test. The two methods correctly predicted trunk condition, decayed or sound, in only 55 and 65 percent of the comparisons at 176 sites in 9 *Eucalyptus globulus* and 100 sites in 5 *E. viminalis* respectively. With these eucalypts, the Shigometer was no more accurate than examining the borings from drilling.

Wilkes (1983) reports that the Shigometer correctly predicted decay in 70 to 80 percent of radii probed in several eucalyptus species but concluded that a 20 to 30 percent error was too large for reliable decay detection. Wilson (1983) had similar results with red beech. On the other hand, Gallagher and Sydnor (1983) found the Shigometer useful in detecting discolored and decayed wood with silver maple as did colleagues working with Shigo (Costello and Peterson 1989) on several other species.

There is concern that the drilling and the Shigometer methods, as well as the increment borer, create holes in wood, some of which may be sound and some decayed. The drill hole for the electrical probe is only 2.4 mm (0.09 in.) in diameter compared to 11.5 mm (0.45 in.) for drilling and 12.5 mm (0.5 in.) for an increment-borer. Even though the probe hole is smaller and less wood is injured than for drilling or an increment borer, each method opens up a tree to the possibility for outside infection as well as for decay to spread internally if it is already present. **The difference in hole size as far as the possibility of infection is concerned is minor.** The rate of wound closure has little effect on whether infection will take place (Shigo 1986a).

Whatever method is used to detect decay, you should use extreme caution in evaluating the results, even if the method has been previously validated for the tree species and you have considerable experience with the method.

Cavities

In the advanced stages of decay, wood is consumed by fungi and insects or falls from the wound. Cavities develop from bark wounds, breakage, and pruning wounds. Older trees occasionally have large cavities in their trunks and main branches.

A vigorous tree will usually more than compensate for wood lost to decay. Due to compartmentalization, a cavity seldom becomes larger than the diameter of the tree at the time of the injury, unless the tree is further injured (Shortle 1979). New xylem forms as fast or faster than decay weakens old internal xylem. Sapwood and bark active in transporting water, nutrients, and organic material are unaffected except for the cavity opening. As a treetrunk grows in girth, most of the new wood is continuous with the spreading surface roots that provide tree stability. Decay and cavities are slow to develop except in softwood trees of low vigor. Wood-decaying fungi are not the pioneer invaders of wounds and often do not infect a wound for several years (Shigo 1979b). A tree of low vigor, however, may be doomed unless decay is arrested and tree vigor improved.

Experience indicates that on the average a conifer can lose up to 70 percent of its total wood diameter—equivalent to about a one-third loss in trunk strength (Fig. 17–4)—without materially affecting the safety of a tree, if the weakening defect is heart rot or a cavity uncomplicated by other defects (Wagener 1963). A stricter standard may be advisable in highly exposed situations or in the vicinity of buildings occupied much of the year.

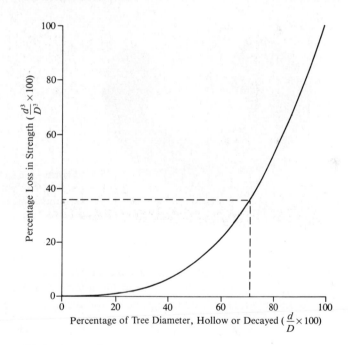

Figure 17–4 Approximate percentage loss in the strength of a tree trunk with a cavity or advanced center heart rot compared with the strength of a sound tree. The loss in strength is equal to the cube of the diameter of the cavity (d^3) divided by the cube of the diameter of the trunk inside the bark (D^3) multiplied by 100. Experience indicates that a conifer can suffer up to one-third loss in strength—equivalent to approximately a 70 percent loss in total wood diameter inside bark—without materially affecting its safety if not complicated by other defects (Wagener 1963).

Infectious Causes of Tree Failure

The specific standard for trunk strength in conifers is less applicable to hardwoods. Trunk decay and cavities often extend into the main branches, weakening them and their attachments to the trunk. In hardwoods the condition of main branches and the trunk/root collar are usually more critical from a safety standpoint than that of the main trunk.

Cavities usually follow the development pattern of wood decay in that they are long and narrow. Cavities almost always open to the outside: Some openings will callus over and others may be concealed in branch crotches, but the cavities to which they connect may still be enlarging. Most cavities extend farther above and below their openings through the bark; water, debris, and insects collect in the lower portions. The cavity opening often restricts access so that it is difficult to assess the extent of decay or take remedial steps without enlarging the opening and damaging sound wood.

The tissue around a cavity opening will callus and form woundwood, particularly along the vertical edges. Woundwood extends in a smooth, thin layer over a firm wound or fill surface, but it rolls inward on itself in a cavity and can become quite thick (Fig. 17-5). If a cavity is large, a "callus roll" may provide needed mechanical strength to the trunk (Armstrong 1925, Bernatzky 1978, Pirone 1978a).

Figure 17-5 At the edge of a cavity in this eucalyptus cross-section, the woundwood rolls inward on itself.

You can usually determine the extent of a cavity by sight if the cavity is open or by sound if you tap on the bark with a mallet—a hollow sound indicates a possible cavity. The amount of decay and discolored wood associated with a cavity and sometimes the size of the cavity itself can be determined only by probing through openings with a slender rod or using one of the techniques discussed earlier.

Root and Butt Diseases

Root and butt diseases and rots are caused by a number of fungi and cause some of the most serious tree hazards. Root decline and decay are often advanced before symptoms become apparent. Above-ground symptoms are similar to a host of other soil and root problems: reduced growth, thinning of foliage, pale foliage, highly colored leaves and fruit in the autumn that drop early, and a heavy crop of fruit or cones.

Some trees fail even though they appear to be in fair health. Britton (1990) reports that mature valley and black oak trees can fail even though they exhibit few or no above-ground symptoms. Some trees produce adventitious roots while experiencing extensive root loss from disease (Fig. 17-6). The new roots are able to grow fast enough to supply the tree with adequate water as the main supporting roots continue to fail and supply less water. Eventually, the main roots are no longer able to support the tree and neither are the adventitious roots. This usually happens where the base of the trunk is kept moist by rain, irrigation, a deep mulch, or dense ground-cover plantings.

Figure 17-6 An adventitious root has been stimulated to grow from live trunk tissue after bark and cambium died at the root collar of this mature valley oak tree. Adventitious roots often supply enough water and nutrients to keep the crown appearing reasonably healthy while the dead main structural roots deteriorate. (Photo courtesy John Britton, John Britton Tree Service, St. Helena, CA)

An increasing number of arborists do a *root-collar inspection* of mature trees before climbing them or when inspecting for hazards in maintained landscapes and frequently used areas (Fig. 17-7).

Most root-rot fungi spread from one tree to another where the tree roots intermingle (Wallis, Morrison, and Ross 1980). Consequently, in wooded areas, diseased trees commonly occur in groups, resulting in openings in the tree canopy. Old openings frequently contain fallen trees, with standing dead and dying trees around the margins.

Trunk and Branch Decay

Decay is the main cause of trunk failure in most species. Conifers are more often affected with trunk failure than hardwood trees. Trunk decay in hardwoods is most often associated with branches, resulting primarily in branch failure at the branch-trunk attachment.

Although differences occur among species of fungi, the number, location, and size of *conks* (fruiting bodies) on the trunk of a tree are usually indicative of the extent of decay in the trunk (Wallis, Morrison, and Ross 1980). Heart-rot fungi,

Figure 17-7 The soil has been removed to expose the root collar of a mature valley oak tree. John Britton checks the health of the structural roots. Root-collar inspection has proven to be important in diagnosing tree health and safety. (Photo courtesy John Britton, John Britton Tree Service, St. Helena, CA)

infecting broken tops 75 mm (3 in.) and larger in diameter, frequently advance rapidly down the stem, particularly in deciduous species. Large branches adjacent to broken tops often break out when rot reduces the stem to a thin shell.

Cankers

A number of fungi invade the trunk and branches of trees killing localized areas of bark, cambium, and wood; these are called *cankers*. Most cankers enlarge as the fungi infect adjacent healthy tissue. Since the cambium in a canker is dead, no new xylem or phloem is formed. The growth of healthy wood and bark around a canker creates a sunken area at the canker site. If a canker involves enough of the circumference of a trunk, a serious weakness is created. The trunk or branch is smaller in diameter at the canker and has lost flexibility due to loss of live wood. A trunk or branch with a large canker often breaks at the canker.

Should cankers become infected by decay fungi, they are likely to fail sooner than if not infected. Decay infection on the canker face is easy to detect; some decay fungi are internal, however, and not externally visible (Tattar 1989).

FURTHER READING

Diseases of Trees and Shrubs (Sinclair, Lyon, and Johnson 1987) is an excellent diagnostic reference with extensive descriptions and color photographs of decay- and canker-causing microorganisms.

HICKMAN, G. W., J. CAPRILE, and E. PERRY. 1989. Oak Tree Hazard Evaluation. *J. Arboriculture* 15:177–84.

JOHNSON, D. W. 1981. *Tree Hazards: Recognition and Reduction in Recreation Sites.* U.S. For. Serv. Tech. For. Pest Mgmt. Rpt. R 2–1. (Lakewood, CO).

MATHENY, N. P., and J. R. CLARK. 1992. *Photographic Guide to the Evaluation of Hazard Trees in Urban Areas.* Urbana, IL: Intl. Soc. Arboriculture.

MILLS, L. J., and K. RUSSELL. 1981. *Detection and Correction of Hazard Trees in Washington's Recreation Areas.* Dept. Natl. Res. Rpt. No. 42.

PAINE, L. A. 1971. *Accident Hazard Evaluation and Control Decisions on Forested Recreation Sites.* U.S. For. Serv. Res. Ppr. PSW–68 (Berkeley, CA).

SHARON, M. E. 1987. Tree Health Management: Evaluating Trees for Hazard. *J. Arboriculture* 13:285–93.

WAGENER, W. W. 1963. *Judging Hazard from Native Trees in California Recreational Areas: A Guide for Professional Foresters.* U.S. For. Serv. PSW Res. Ppr. PSW-P1 (Berkeley, CA).

WALLIS, G. W., D. J. MORRISON, and D. W. ROSS. 1980. *Tree Hazards in Recreation Sites in British Columbia.* B.C. Min. Lds., Parks and Housing, and Canadian For. Serv. Jt. Rpt. No. 13.

CHAPTER 18

Preventive Maintenance and Repair

No matter how well plants are maintained, some will be injured, develop weak structure, cause damage to buildings, or require removal. These problems usually become serious only as plants grow and age; most can be avoided or at least minimized or postponed by proper selection, planting, and care. Problems must be anticipated in order to safeguard the plants, people, and property. In this chapter, some of the more important plant maintenance problems are discussed, especially how they might be avoided or minimized or treated should they be suspected.

CABLING AND BRACING

All too often trees develop weak or poor structure and require special care to preserve them or to prevent further injury. Even trees that are routinely maintained may develop weaknesses that affect their own safety and that of people and property. The most common problem is the weak attachment that may occur when two or more branches of about equal size arise at approximately the same level on the trunk. Horizontal branches can become heavy and dangerous, particularly those of conifers in snowy regions. Branches can also be weakened by decay and storms. Proper pruning can shorten, lighten, and thin hazardous branches, but it may be insufficient to keep certain limbs or trees safe. In such cases, cabling and bracing may be required to reduce stress.

Cabling involves the attachment of a cable between branches to limit excessive limb motion and to reduce stress on a crotch or branches (N.A.A. 1987). Sometimes trees are cabled together for support. Bracing uses bolts or threaded rods to rigidly secure weak or split crotches, unite split trunks or branches, and hold rubbing limbs together or apart (Thompson 1959). Before you undertake cabling or bracing, however, assess the internal condition of the trunk and main branches and the value of the tree as compared with the cost of labor, supplies, and continued maintenance. Is the tree worth the effort?

Almost all cabling and bracing recommendations are based on those given by Thompson (1959), including the standards of the National Arborist Association. Alex Shigo (Durham, MA, 1981 pers. comm.) recommends that cabling, bracing, and other practices that wound a tree should not be done during spring growth flush or leaf fall. Use sharp tools to make a clean-edged wound, which will close faster than a ragged wound.

Cabling

One of three cabling systems is usually used to safeguard main branches and stems. If a single cable (*simple direct system*) is used, two limbs of about the same size arising from a weak or split crotch can directly support each other, and a large limb or the trunk can support a weak or heavy horizontal branch (Fig. 18-1). Three cables attached to three limbs in a *triangular* arrangement provide direct support to weak crotches and lateral support to minimize branch twisting, which can put considerable strain on branch attachments. A *box* or *rotary* system of cables connects the main limbs of a multibranched tree and permits movement of individual branches within safe limits. A box cable system can be made more rigid if interior cables are placed to create a series of triangles. The system chosen will depend on the tree and the branches that need support.

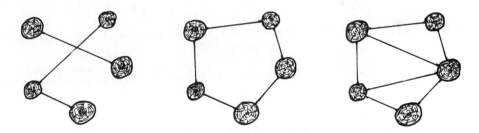

Figure 18-1 Cabling systems: Simple direct cabling involves a single cable between two branches of approximately equal size (left); box cabling connects the large branches of a tree and allows movement within limits (center); triangular cabling gives direct support and minimizes twisting (right). (Adapted from Thompson 1959)

It is important to position cable attachments so as to combine adequate support and flexibility. Cables should be attached to limbs at about two-thirds of their length from the crotch, but each cable attachment should be approximately the same distance from a crotch, so some compromise will be required when limbs vary substantially in length. To support a horizontal branch, however, Bridgeman (1976)

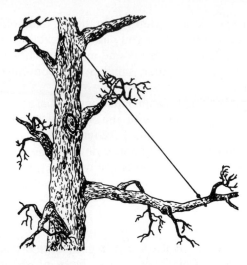

Figure 18–2 When supporting a weak horizontal branch, attach the cable as far out on the branch and as high on the trunk or support branch as possible without unduly distorting the branch; the cable and branch should form an angle of at least 45°. (Adapted from artwork by M. J. Whitehead in Bridgeman, P. H. 1976. *Tree Surgery.* London, England: David & Charles)

recommends that the cable be attached as high as possible on the supporting branch, so that its angle from the weak branch is not less than 45° (Fig. 18–2). If the supporting branch leans toward the one supported, it may be wise to install a second cable from the support branch to a third branch for counterbalance. Place cables so that they do not rub against limbs or one another.

Thompson (1959) found that lag screws (Fig. 18–3) are as satisfactory as bolts for cabling sound hardwood branches, but bolts are safer for softwoods or decayed hardwoods. Shigo and Felix (1980) confirm that lag screws should not be put into

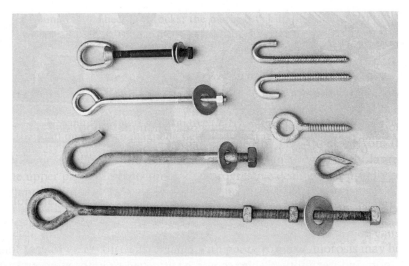

Figure 18–3 Cabling hardware includes (clockwise from bottom): forged eyebolt, hook bolt, bent eyebolt, bolt with amon nut (left end), right-hand threaded lag hook, left-hand threaded lag hook, forged lag eye, and thimble; the bolts have washers and nuts.

decayed wood: When a lag or rod touches internal decayed wood, decay can spread to surrounding tissues. Where there is decayed wood, insert a bolt through a branch and secure it with double washers and nuts. Trace the bark so one round or oval washer is set against exposed wood; tighten a nut against the washer. Then place a second washer and nut on the bolt (see Bolting or Rod Bracing and Fig. 18-8).

Lag screws are cheaper and easier to install than bolts, but they are not as secure in softwood and do not allow for the adjustment of cable tension unless right- and left-hand lags are used (Bridgeman 1976). Lag screw hooks or bolt hooks are commonly used in the United States even though they do not have the tensile strength of drop-forged eye bolts or ring nuts (Amon nuts) (Table 18-1). Lag screw hooks ("J" lags) should have at least a 12-mm ($\frac{1}{2}$ in.) unthreaded shank above the threads and should be inserted the full length of the unthreaded shank, because the threaded portion is more vulnerable to sway damage and crystalization (Mayne 1975). Bernatzky (1978) advises against the use of lag screw hooks and hook bolts for attaching cables because they will not hold a slackened cable as securely as screw eyes or eye bolts will. Cables seldom come loose, however, if they are properly tensioned, if the hooks are tightened until they almost touch the bark, and if the wood is sound. Mayne (1975) claims that a cable without a thimble will secure as much weight as a cable with a thimble and that it takes less time to install and splice a cable without thimbles, but the cable without a thimble may stretch more and could wear thin if there is much branch movement.

Lag screws should be screwed into holes drilled 1 to 2 mm ($\frac{1}{16}$ in.) smaller in diameter than the lag and slightly deeper than the length of the threaded shank. Mayne (1975) recommends that the hole in softwood should be 3 mm ($\frac{3}{32}$ in.) smaller than the lag. Drill the hole so that the cable and lag will form a straight line when attached (Fig. 18-4). The closer to 90° you can make the angle between each lag and its branch, the more mechanical advantage the cable will have. Lag screws should be screwed in far enough to just allow the cable to be slipped into place. A lag screw should penetrate at least two-thirds the diameter of the smaller branch. Tighten each screw the full distance without stopping, or the wood may expand and hold the screw so tightly that it cannot be moved. Spraying rods and screws with silicone will make turning easier. Align each hook or eye with the grain of the wood for smoother

TABLE 18-1 RELATIVE STRENGTHS OF CABLE ANCHOR HARDWARE 15 MM ($\frac{5}{8}$ IN.)[a] IN DIAMETER (Thompson 1959)

Cable anchor hardware	Safe loads[b]	
	Kilograms	Pounds
Bent hook eye bolts	540	1200
Drop-forged eye bolts	1350	3000
Ring nuts (Amon nuts)	1485	3300

[a] This represents the smallest size available in all three types of hardware.

[b] The weights listed are 25 percent of the maximum static load each type was able to support.

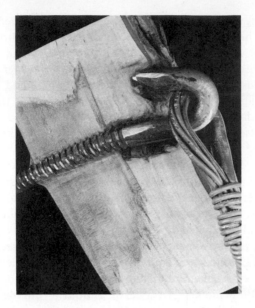

Figure 18–4 Install the lag screws and cable in a straight line; otherwise, the lag will be pulled laterally and/or vertically, further injuring the wood and reducing the holding strength of the lag. (Photo courtesy Alex Shigo, U.S. Forest Service)

coverage as the branch grows. Do not align a lag screw or bolt with the crotch junction or place it within 300 mm (12 in.) of a weak crotch. Attach only one cable to each lag screw or bolt. Place lag screws or bolts at least 250 mm (10 in.) apart on the same branch to avoid weakening the branch.

If hook or eye bolts are to anchor cables, install them as you would install lag screws, but drill holes that are the same diameter as the bolts, and countersink the nuts and washers to the cambium. If a bolt is 50 mm (2 in.) longer than the diameter of the branch in which it is placed, the extra threads can be used to adjust cable tension. Place a bolt through a weak or decayed branch for greater support; use a lag screw in a support branch or trunk so large that it would require a bolt much longer than needed to support the weak branch.

The required strength of cable and other hardware will depend on the species, size, and general condition of the tree; the size and weight of branches to be supported; the presence of decay and cavities; and exposure to wind and snow. These factors have not been integrated into a formula for determining cable size. Thompson (1959), however, examined several hundred cabled trees and concluded that a 7-wire galvanized cable 6 mm ($\frac{1}{4}$ in.) in diameter is safe for limbs up to 150 mm (6 in.) in diameter at the point of cable attachment. A 7.5-mm ($\frac{5}{16}$-in.) cable is satisfactory for limbs up to 250 mm (10 in.) in diameter at the point of attachment. Although common grade galvanized cable is not as strong as steel cable of higher tensile strength (Table 18–2) and is more subject to stretching, the common grade is regularly used in the United States because it is easy to splice. Improved methods of joining cables, however, should increase the use of extra high strength (EHS) cable.

Before cabling, prune the tree to correct any hazardous structure, reduce weight at the ends of long branches, and balance the tree. The effectiveness of a given cable tension is determined by the size of the limbs, the weight of the limbs

TABLE 18-2 SAMPLE SUBSTITUTIONS OF EXTRA HIGH STRENGTH (EHS) CABLES FOR COMMON GRADE CABLES WHEN PREFORMED TREE-GRIP DEAD ENDS ARE USED (Jeffers and Abbott 1979)[a]

7-Wire cable diameter sizes (mm)	Minimum cable strength (kilograms)		Weight of cable (kg/100 meters)	
	EHS	Common	EHS	Common
48 (3/16)[b]	1795 (3990)[c] ─┐	┌─ 515 (1150)[c]	11 (73)[d] ─┐	┌─11 (73)[d]
64 (1/4)	2945 (6550) ─┐│	│┌─ 855 (1900)	18 (121) ─┐│	│├─18 (121)
80 (5/16)	││	│└─ 1440 (3200)	││	└─30 (205)
96 (3/8)	│└────────	┌─ 1910 (4250)	│└────────	┌─40 (273)
112 (7/16)	└─────────	└─ 2565 (5700)	└─────────	└─59 (399)

[a] Figures in parentheses are English equivalents.

[b] Inches.

[c] Pounds.

[d] Pounds per 1000 feet.

and foliage, and distance to the crotch, but guidelines to proper cable tension are not very specific. Thompson (1959) advises that a slack cable may have to bear sudden, heavy loads during gusty winds but that tree shape may be distorted if the upper foliage is drawn too tightly together. A cable should be snug but not tight when the branches are least strained (during winter for a deciduous tree). Bernatzky (1978), on the other hand, recommends that a cable be slackened so that in an unloaded condition it sags 2 to 3 mm for each meter (0.1 in./3 ft) of cable length. During a gusty wind, such a practice might create too much slack, followed by extreme stress as the cable is pulled taut. Since very few cables or cabled limbs have failed regardless of which of these approaches was used, most cables and their attachments and the cable tensions must be adequate, if maintained and adjusted for tree growth.

A tree that is rigidly cabled may lever the trunk and roots unduly, possibly causing windthrow. Compression spring inserts (see Fig. 10-10) in cables would reduce this stress on the tree, the tension necessary, and the maximum load under wind. Cables with compression springs would allow more movement of the branches, would reduce stress on trunk base, and would foster strong branch growth. In a triangular or a box cable system, a compression spring in one cable would provide flexibility to all cabled limbs (Mayne 1975).

After the lag screws or bolts are in place, pull the limbs together to install the cable on the hooks or in the eyes. You can pull limbs together by tying a rope loop around them and twisting it tight or by using right- and left-hand lag screws, bolts with extra long threads, a block and tackle, a lineman's come-along, or a running bowline (Thompson 1959). Cable tension can be adjusted more precisely if at least one of the anchors is a bolt or if right- and left-hand lag screws are used.

Common grade cable can be attached to an anchor lag screw or bolt by means of an eye splice, cable clamps (bulldog or U; N.A.A. does not include clamps in its 1987 Standards), or Tree-Grip® dead ends (Fig. 18-5). Extra high strength steel cable (EHS) can be held with two or more cable clamps or Tree-Grip® dead ends.

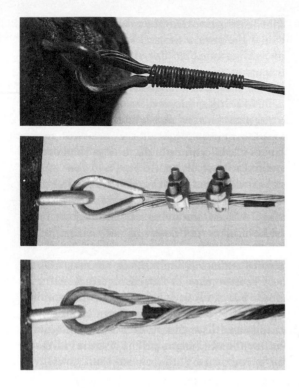

Figure 18-5 Cables are joined together by (top to bottom) splicing, two clamps with the tightening plates against the incoming cable in order to provide maximum cable strength, and a twisted cable loop (Tree-Grip®) whose ends are wrapped around the incoming cable. A thimble protects each cable from wear by an eye-bolt or J-hook.

Thompson (1959) reports that a properly constructed eye splice is as strong as common grade cable, whereas a single cable clamp will slip when the strain exceeds one-seventh of cable strength. That is the reason for using two or three cable clamps. Tree-Grips® will hold EHS cable securely well beyond the rated breaking strength of the cable (Jeffers and Abbott 1979). Cable clamps or Tree-Grips® allow for the use of EHS cable and a consequent reduction of cable weight for a given cable strength. With these fasteners, fewer cable sizes are needed.

An eye splice in a seven-wire strand is made as follows (Fig. 18-6) (Thompson 1959): Make a loop by bending the cable about 250 to 300 mm (10–12 in.) from the end, then insert a thimble in the loop, pass the cable through the eye, or loop it on the hook. Unwrap the wires of the 250-mm section and lay them parallel to the main piece of cable. Using pliers, wrap one strand tightly two times around both

Figure 18-6 (See opposite page) Steps in making a seven-wire splice that will provide a strong and attractive union for a looped cable end (top). (Thompson 1959) Vertical photos, upper left to lower right: Taped end of the cable is laid in the Tree-Grip® dead end's short leg (painted black for visibility), slightly above the cross-over paint mark; the short leg of the dead end is wrapped around the cable; a thimble is inserted and the longer leg of the dead end (painted white) is wrapped around the cable. Horizontal photos: A Tree-Crotch-Grip® ready for installing between branches too close for the regular Tree-Grip®; a Tree-Crotch-Grip® as it would appear when installed; in order to obtain the desired tension, hook or eye bolts or left- and right-hand threaded lag hooks need to be used.

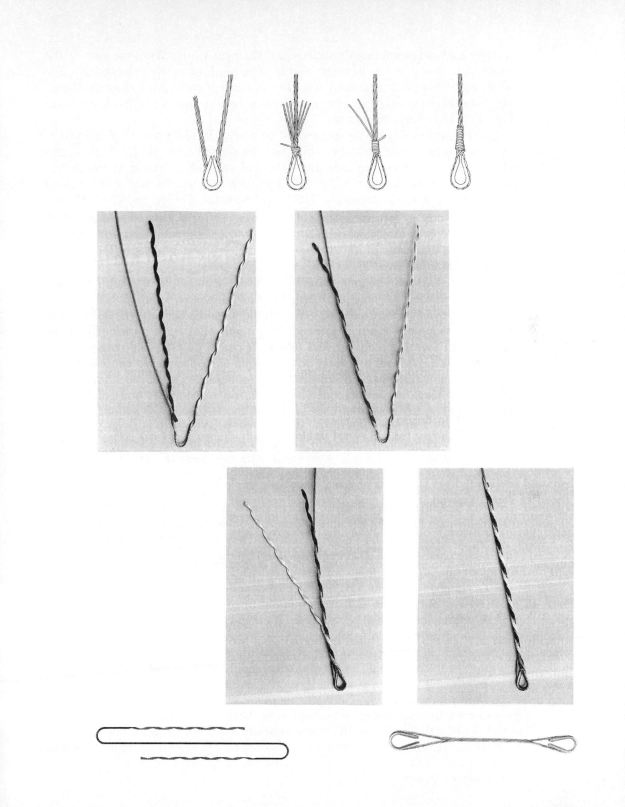

the cable and the remaining six strands and then cut it off. Wrap the rest of the strands, one at a time, in the same fashion. The resulting tapered eye splice is not only effective but inconspicuous. Although one wrap of each strand has been shown to be adequate, two wraps are practically as easy to make and will provide a margin of safety and improve appearance.

Cable clamps (bulldog grips) should be placed 50 to 75 mm (2–3 in.) apart. The clamp yoke should hold the cable end (the pigtail) (Fig. 18–5). If the clamp yoke is against the incoming cable from the other branch (as shown in the *British Standard Recommendations for Tree Work* BS 3998:1989, p. 7), the incoming cable will be weakened by the crimp put in it when the clamp is tightened (Erik Haupt, West Hyannisport, MA, 1990 pers. comm.). The cable end is under less than half the pulling stress as the incoming portion of the cable. Tighten the clamps securely but not so tightly as to flatten the cable.

Tree-Grips®, introduced in 1977, speed installation, particularly for less experienced workers, by simplifying cable attachment (Jeffers and Abbott 1979). They come in four sizes for EHS cable from 4.5 to 9 mm ($\frac{3}{16}$–$\frac{3}{8}$ in.).

Since the Tree-Grip® dead ends are left-hand lay, the cable must be the same for proper holding. Tape EHS cable before cutting it to prevent unraveling. Lay one end of the cable in the short leg of the dead end and wrap it with the short leg (Fig. 18–6). Insert a thimble in the Tree-Grip® loop; then wrap the longer leg around the cable between the wraps of the short leg. Additional instructions are provided with the Tree-Grips. A shorter version of Tree-Grips® is available for use with cables too short for the standard length Tree-Grips; for the dead ends to function properly, their length must not be reduced.

Inspect cabling at least annually to check cable tension and the stability of the anchor lags and bolts. If a tree is vigorous, you may need to raise the cabling in eight to 10 years to provide adequate support.

Wiring Small Trees.   Occasionally small trees have long slender branches that are or will begin spreading more than desired. Wiring branches to a center ring is a training technique or can be a permanent support (Fig. 18–7). The height of the wires is similar to that for large trees—about two-thirds the distance from the crotch of a branch to its tip. A branch should neither sag below nor bend out above the wire attachment. Branches at the height of support would probably be 20 to 60 mm (0.75–2.5 in.) in diameter. Galvanized solid wire, 1.5- to 2.5-mm diameter (16–12 gauge), should be strong enough; a washer can be used for a center ring. It takes little strength to hold the branches.

Drill a hole in the branch twice or slightly larger than twice the wire diameter. The hole should be aligned toward the center ring location and at a height to make the wire horizontal when the branch is in the desired position. From the tree center, thread about 50 to 75 mm (2–3 in.) of the wire through the hole in a branch. Double the end back on itself and insert it back into the hole 15 to 25 mm (0.5–1 in.) on top of the wire there. Pull on the long end of the wire until the wire loop is tight in the branch. Attach the long wire end to the center ring. Repeat for each branch to be supported. With all the wires in place, pull each branch to its desired position, and secure each wire to the ring.

Figure 18–7 Small trees are often misshapen because slender branches bend outward due to increasing weight of continuing height growth, wind, rain, or snow. A 60-year-old Irish yew six months after being remedially pruned to reduce tree spread and to renew more compact lower branching; new shoots are beginning growth along the branches (top left). Branches of the yew have been pulled more upright (top right) by a small-diameter wire from each branch to a center ring (lower right). A wire is secured to a branch by drilling a hole in line with the center ring and slightly larger than twice the diameter of the wire. One end of a wire is fed through the hole, doubled back on itself (left center), and pulled snuggly into the hole (lower left). (The wire is white for visibility in the photograph.) Branch position is adjusted by the tension on the wire at the center ring (lower right).

The branches will grow and engulf the wire loops. Xylem growth will strengthen the branches in their new positions. The wires can be left on for several years or indefinitely.

Bolting or Rod Bracing

Rod bracing provides rigid support for weak branch attachments and split branches and trunks. It also provides cavity bracing and holds limbs together or apart. Machine threaded steel rods of various diameters can be used alone or with nuts and round or oval washers. For maximum holding power, a rod is screwed through a hole that is 1 to 2 mm ($\frac{1}{16}$ in.) smaller in diameter; a washer and nut are then tightened at each end. When two branches are to be brought closer together, the rod must be able to move in one of the holes, and washers and nuts are used.

Weak Crotches. Split crotches are usually rigidly braced with a screw rod to supplement cable bracing higher in the tree. Weak but unsplit crotches of large trees are often braced in a similar manner. A single rod is usually sufficient to hold the two sections of an unsplit crotch together securely (Fig. 18–8). Large split or weak crotches usually require two or more rods, instead of only one, to hold the two sections together and to minimize twisting. Holes should be centered on the trunk. When two rods are used to brace a single crotch, they should be placed just below the crotch, parallel to each other in the same horizontal plane and at a distance from one another approximately equal to a radius of the trunk or branch (Fig. 18–9). Thompson (1959), however, cautions that parallel rods should rarely be placed closer than 125 mm (5 in.) or farther apart than 450 mm (18 in.). If the rods are more than 500 mm apart, use a third rod in the same plane.

If at least 150 mm (6 in.) of sound wood is available at each end, self-threaded holes and rods should provide sufficient holding power (Thompson 1959), but if any of the wood in the crotch region is decayed, use round or oval washers and

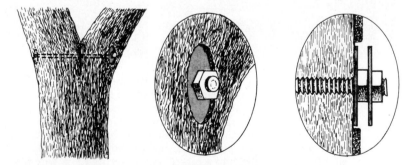

Figure 18–8 If only one rod is to be inserted, it should be just below the crotch (left). If nuts are used, round or oval washers should be countersunk flush with the cambium (center and right). When rodding decayed wood, secure a second washer separated by a nut or spacer from the first washer (right); woundwood will grow between the washers strengthening their holding power.

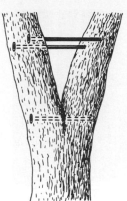

Figure 18–9 For large, weak, or split crotches, insert a rod through the crotch and two more above it, separated by a distance equal to twice the diameter of the limbs. The two rods give direct support and reduce twisting strain on the crotch.

nuts on the rods (Shigo and Felix 1980). Fred Roth, California State Polytechnic University, Pomona (1990 pers. comm.) suggests that for rodding decayed wood, secure a second washer separated by a nut or spacer from the first washer at each end of the rod (Fig. 18–8). Woundwood will grow between the two washers strengthening their holding power, particularly if decay continues. In a large tree where the branches to be braced are close together for a considerable distance above the crotch, use an additional safety bolt 0.3 to 1 m (1–3 ft) above the crotch to provide extra rigidity (Fig. 18–9).

When a screw rod is to be inserted at only one level, Thompson (1959) recommends inserting it directly through the crotch or just above it. Bridgeman (1976) and Pirone and others (1988), on the other hand, recommend that one or more rods be inserted at or just below the crotch. From an engineering standpoint, insertion at or just below a crotch would provide the greatest stability in gusty winds. Cables will not prevent the branches from being blown together, and self-threaded rods located above a crotch can act as a fulcrum; in the unlikely event that the tops of the bolted limbs are blown together, the trunk or branch below the crotch might split. Rods above the crotch will be effective if they are situated at more than one level or are combined with rods inserted through the crotch.

Split Trunk or Branch Bolting. A long split in a branch or trunk can be held together by one or more bolts or rods (lip bolts) (Fig. 18–10). You may also be able to close or reduce the split in width before bolting. You may partially close a split in a stressed branch by moving the branch into a favorable position. Frost cracks in the trunk (see Fig. 4–1) usually close when temperatures rise above freezing.

Rods are usually installed with washers and nuts and are not self-threaded through the wood. "Lip bolts" should be placed about 0.3 to 0.4 m (12–16 in.) apart; take care so that consecutive bolts are not located in the same wood grain but are staggered to minimize the chance of weakening the wood. A bolted or rodded limb or trunk is less flexible so will be subject to more stress at its base in a storm.

Bracing Rubbing Limbs. Occasionally, two limbs grow so that they rub together, to the detriment of both. Usually one can be removed to the advantage of

Figure 18-10 Bolt a frost crack closed during warm weather when the crack is closed. The washer and nuts have been painted with asphalt on this London plane tree. (Photo courtesy E. B. Himelick, Illinois Natural History Survey)

both the tree and the surviving limb. If both crossing limbs must be preserved, however, you can protect them by bolting them tightly together or bracing them a short distance apart.

When branches are to be held together to foster grafting, remove the bark so that the two cambiums will be in contact. Drill a hole through the two limbs for a bolt. After the two branches have been bolted together, seal the slit around their union and around the washers and nuts with asphalt emulsion or grafting wax. The dressing will protect the cambium from drying out. This approach-grafting should be done in the spring just as growth begins.

When branches are to be held apart, block them at the desired distance while you drill a hole in a direct line through both. Washers and two nuts or a short length of pipe with washers at either end can constitute the permanent spreader on the rod. Washers and nuts tightened on the outside of either branch will hold them as desired.

Rodding Standards. Thompson (1959) gives extensive instructions for the selection and use of rods. The most important are the following:

Drill holes 1 to 2 mm ($\frac{1}{16}$ in.) smaller in diameter than the rods, except when limbs or split trunks must be drawn together; in this case, drill at least one of the holes the same diameter as the rods.

Countersink round or oval washers to the cambium (Shigo and Felix 1980); cut the bottom of the holes so that the washers will lie flat and at right angles to the rod.

Except for countersinks, ream all rod holes below the cambium so as to avoid loosening the bark when the rod is inserted.

Cover exposed rods with tightly fitting pipe that extends the full length of the exposed area to provide extra strength and protection.

When rods and nuts are used to draw split crotches together, use two or more washers under each nut for additional strength. Prevent the washer against the wood from twisting by separating it from the nut with a second washer. Shigo and Felix (1980) recommend round or oval washers instead of diamond-shaped washers, whose sharp points can lead to cambium dieback and trunk splitting.

When a rod is to be broken off below the cambium, cut it about halfway through before you insert it into the hole. Make this cut at least 40 mm (1.5 in.) from the rod end so there will be space to clamp a pipe wrench. Take care when inserting a rod so that the wrench bites do not damage the threads that have to engage wood or nuts. When it is screwed in far enough, break the rod by bending the exposed portion to and fro.

When you drill for screw rods, line up the holes in the opposite limbs exactly. A long extension bit is useful for this purpose.

Take care when you tighten nuts on screw rods so that you do not injure the bark and cambium. A socket wrench is good for this purpose. Expose some of the threads after you tighten each nut so that the rod end can be spread with a ball peen hammer; this will prevent the nut from being worked loose or easily taken off.

Thompson (1959) recommends that all rods be covered with a mastic or tree wound dressing as they are being inserted and that all wounds be dressed after exposed metal has been covered with a metal preservative paint. He suggests that exposed rods, pipe, nuts, and washers also be painted. Except in especially wet regions or near the seacoast where rust and corrosion are prevalent, painting the metal is of doubtful value. Wound dressings are probably not worth the effort either, except for cosmetic purposes.

Propping

If pruning alone does not make low branches and leaning trees stable and safe, these trees may have to be propped after pruning (Fig. 18-11). The main reasons for propping are

The upper branches and trunk are missing, too weak, or wrongly positioned to support a cable.
The entire tree leans or exerts leverage against the root system.
Supporting cables would be unsightly (Bridgeman 1976).

Metal props are usually of pipe, angle iron, or I, U, or box beams. Wooden props, particularly straight branches with open forks selected from felled trees, are also common. If wood is used, it should have a straight grain. Metal props are more permanent and usually less obvious than those made of wood. The top end of a prop can injure or girdle the bark at the point of contact. A bolt inserted horizontally through a branch and supported by a U end on the prop will carry a load without pressure on the bark (Bernatzky 1978).

Another solution is to use a single metal prop embedded into the trunk or branch to be supported (Fig. 18-12) (Kenneth Meyer, San Mateo, CA, 1990 pers.

Figure 18–11 A Japanese pagoda tree supported with three single and two double pipe supports at the Royal Botanic Garden, Kew, England. Some of the props have "V" crotches to more securely cradle the limb or trunk they are supporting.

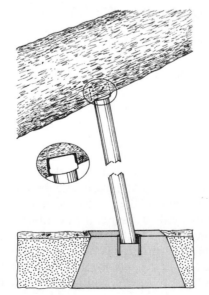

Figure 18–12 A single pipe prop for a large low horizontal limb on an otherwise stable tree. The capped pipe is inserted into a hole bored into the underside of the branch. The bottom pipe end fits into a reinforced hole in a concrete support base. (Adapted from information supplied by Kenneth Meyer, Mayne Tree Expert Company, San Mateo, CA)

comm.). On the top end of a galvanized metal pipe (40 to 100 mm [1.5–4 in.] in diameter) screw on a cap. About two-thirds of the way out on the branch, bore a hole in the under side, 40 to 50 mm (1.5–2 in.) deep, to take the pipe cap; the hole should be in line with the final setting of the prop on its support base. A precast concrete support base with a reinforced center hole (about 50 mm [2 in.] deep) to hold the pipe should be firmly placed in the ground under the branch. When the pipe is in place, it should be long enough to raise the branch 50 to 100 mm (1–2 in.) so the prop will not dislodge in a wind. Raise the branch with a hydraulic jack (lift) to install the prop. Asphalt around the prop in the support hole will allow some movement but still keep water out.

Place props so that they provide the maximum support and are least noticeable. If the load is heavy, place the prop on a firm base of concrete, stone, or other durable material. A prop should be almost vertical to support a low branch and angled slightly to support a leaning tree.

Occasionally, a tree may be partially uprooted from wet soil during strong winds or a flood. Often, a small or medium-size tree can be rehabilitated if it is straightened soon after the mishap and if the root system was fairly well formed beforehand. If most of the roots on the upwind (upstream) side are broken, dig out soil from that side to make room for the root portions still attached to the tree. Take care not to injure unbroken roots. Make sure the soil is very muddy; then attach a strong cable to a strong branch or the upper trunk at a good angle for leverage and use a winch or block and tackle to pull the tree upright while you work the exposed roots into the muddy soil. When the tree is in a reasonable position, guy it and possibly prop it upright. Roughly level the soil surface and prune the tree to a more upright structure. Irrigate carefully during the following growing season. At the end of that season, determine the stability of the tree without its guys and props; it will probably need some support for two to three years.

Improper braces are all too common. Many have been used as temporary expedients and subsequently forgotten; some are used with no reference to the way trees grow. Whatever its cause, improper bracing can jeopardize the tree it is supposed to help. The most common mistake is to wrap supported branches or encircle them with cable. The weight of the branches or the constriction of the cable begins to cause girdling as branches and trunk increase in girth. Growth beyond the attachments may die back or break off.

Take care when removing cable or bands that constrict branches; not only for the tree's well-being but for your own safety. The constricting collar may be under such pressure that fragments of bark, wood, and metal will be hurled at dangerous speeds when it is cut. If the collar is deeply embedded, it will be difficult to remove without injuring the embedding tissue. Occasionally, woundwood will overgrow and enclose the collar. Even though new continuous wood is laid down outside the collar, that area will always be a point of weakness. In such cases, the portion beyond the girdling collar should be thinned and shortened to reduce the stress at the collar.

After removing a constricting collar, you may hasten recovery by cutting longitudinally through the bark in several places on either side of the girdle. Extend cuts to the cambium at the bottom of the girdle (Bernatzky 1978). It may also be wise to thin the branch to reduce its weight.

WOUNDS

Plant wounds do not "heal" in the sense that animal wounds heal. An animal wound usually heals from the inside out, replacing dead and damaged tissue; a surface scar may be the only sign of an earlier wound. In woody plants, however, wounding breaks or destroys the cambium so that the original wound remains, and though it may eventually close, wood may decay in what appears to be a sound trunk. It is more appropriate to speak of wound closure than wound healing.

Most wounds are naturally or accidentally caused, although some are deliberately made in the course of normal tree care, particularly when pruning (Fig. 18-13), cabling, injecting, or diagnosing is improperly performed. Some of these practices can be performed at a certain time or in certain ways so as to minimize the amount or seriousness of wounds. Early pruning, for example, will minimize the size of pruning wounds and hasten their closure.

Figure 18-13 Pockets of decay have begun in the discolored sapwood in a black walnut stub (large arrows). The barrier zone will confine decay to the inner side of the barrier (small arrows). (U.S. Forest Service photo, Shigo and others 1979)

Characteristics of the Tree. The rate of wound closure is correlated with tree vigor as measured by radial stem growth at the wound site (Neely 1979). Vigorous trees callus more quickly than those of low vigor. Practices that encourage growth not only speed wound closure but also reduce the possibility and the progress of decay. Decay spreads more easily in some species, such as mulberry and willow, than in others. Individuals within a species also differ: The clones of poplar hybrids

(Shigo, Shortle, and Garrett 1977) and red and silver maple (Santamour 1979a) exhibit quite varied resistance to the spread of decay. You may wish to consider this when you select trees for landscape planting.

Wound Characteristics. Wound closure depends on the wound's location, the season of injury, and which tissues are exposed. Most commonly, callus develops primarily from the sides of a wound, much less from the top, and least from the bottom (Marshall 1931, McQuilkin 1950, Neely 1970). When Alex Shigo (Durham, MA, 1981 pers. comm.) examined elliptical wounds that were parallel to the wood grain, he found three times as much cambial dieback (and hence less callusing) above and below those with pointed apexes as above and below those with rounded apexes. Neely (1970), however, made square wounds and reported that callus formed first at the corners and less along the sides. He also found that **the shape of a wound has little influence on rate of closure:** After the first year, elliptical, circular, and square bark wounds of the same width closed at essentially the same rate.

If, however, the injury occurs early during the growing season, the vascular cambium remains attached to the xylem, and the exposed surface is kept from drying, new bark will usually form over the wound the same season (Zimmermann and Brown 1971). Even if little cambium remains on the exposed xylem, enough proliferating ray-parenchyma cells may remain in the xylem to form woundwood over the wound surface (Neely 1988b). Fred Roth (California State Polytechnic University, Pomona 1990 pers. comm.) points out that this type of wound closure can accurately be called "healing" in contrast to that described in the previous paragraph.

Since dry loose and torn bark can harbor insects and disease organisms, cut it back cleanly to the edge of a wound, taking care not to increase unnecessarily the width of the wound. Peninsulas of live bark in a wound speed closure; therefore, leave as much live bark as possible that is firmly attached (Fig. 18–14).

Pruning wounds close at about the same rate whether the pruning cuts are made outside the branch collar or are cut through the collar (though not a flush cut) (Neely 1988a). This was true even though the pruning wounds through the collars on pin oak, sugar maple, and sycamore were 25 to 38 percent larger than those pruned outside the collars. However, avoid pruning too close (flush), which might open trunk tissue to infection.

Callus and woundwood growth is most rapid when xylem growth is most rapid. Trunk wounds directly above large, vigorous roots close more rapidly than those located in depressions between roots. On vigorous trees, wounds near the base usually close more rapidly; on trees of low vigor, wounds higher on the trunk probably close more quickly than lower wounds. Neely (1970) found that trunk wounds between 0.6 and 1.8 m (2–6 ft) above the ground closed at about the same rate.

Season. Bark is easily damaged or torn in the spring and will slip when the cambium is actively dividing. Bark can easily be knocked loose from the trunk or torn when holes are drilled or pruning is done in the spring (Felix and Shigo 1977). Torn bark, however, can often be replaced or the exposed wound covered before the cambium dries.

Figure 18–14 Even though the peninsulas of bark do not callus as rapidly as the edges of the wound, they protect wood and speed the closure of this 200-mm (8-in.) wide wound on the trunk of a mature lime tree in England.

Neely (1970) examined ash, honey locust, and pin oak during the growing season after wounding: Wounds made in the spring, summer, and winter suffered little dieback, but those inflicted in the fall closed about 20 percent more slowly. Three to six times more area was covered during the first growing season on wounds inflicted in spring than those made in summer. The latter had less time to close. The bark died back around wounds made in summer and fall, especially on honey locust. It would appear that most of the wounds were not exposed to direct sunlight, because closure did not vary with the side of the trunk on which a wound was made. Wounds made late in the growing season, when many species of wood-invading fungi are sporulating, may be infected more rapidly than are wounds made at other times (Felix and Shigo 1977).

Other Factors. Microorganisms antagonistic to wood-decaying fungi may prevent or at least delay decay. In red maples, heavy inoculation of the fungus *Trichoderma harzianum* may delay the invasion of wood-decaying Hymenomycetes for up to two years (Shortle 1979). Attempts to sterilize a wound may do as much harm as good because antagonistic organisms will also be affected. Decay usually develops more rapidly when a wound is alternately exposed to wet and dry conditions.

Injection Wounds

Trees may be injected with insecticides, fungicides, fertilizers, and growth regulators. In urban areas and in plantings of mixed species, the injection of chemicals that would otherwise be sprayed or dusted minimizes toxic hazards and usually improves effectiveness. Injection holes, particularly those that are large and deep, can lead to decay and weakening of the tree, especially if annual treatments are necessary. The possible damage from injection wounds must be weighed against the possible benefits of the chemical treatment.

Injection wounds alone seldom create problems in vigorous trees, but the chemicals injected are not without their hazards. When Shigo, Money, and Dodds (1977) examined injection holes in red maple, white oak, and shagbark hickory in West Virginia, they found little discolored wood and cambial dieback associated with one-year-old injection holes (no chemicals added) 5 mm ($\frac{3}{16}$ in.) in diameter and 18 mm ($\frac{3}{4}$ in.) deep. Columns of discolored wood and some cambial dieback, however, were associated with injection sites of Bidrin® or Meta-Systox-R®. Less injury resulted when Fungisol®, Stemix®, or micronutrients were injected singly.

Shigo and Campana (1977) found discolored wood associated with every injection wound made for the control of Dutch elm disease in 80 large American elm trees in the northeastern United States. Injection wounds made in several successive years caused severe internal injuries. More than half of the trees exhibited decay near the injection sites. In one case, obvious decay began within one year of injection. Their study indicates that

> After an injection hole is made, discolored wood develops, microorganisms infect the wood, and decay may set in
>
> If holes are made in the same trees in successive years, the individual columns of discolored wood associated with each wound begin to coalesce to form large columns of dead discolored wood; these may contain some early decay
>
> The rate and extent of infection, decay, and coalescence of injured tissue vary greatly among trees, even those of the same species

Shigo and Campana recommend that, until better methods of injection are developed, holes be made shallow, clean-edged, small in diameter, and few in number. Injection holes should be drilled infrequently (once every three to five years) on a given tree. Make the first injections on the ridges of roots, not in the depressions between the ridges, and as low as possible on the trunk. Establish subsequent holes at least 500 mm (20 in.) above and at a diagonal to the grain from old holes. Shigo and Campana (1977) conclude that the injection of chemicals might be a reasonable treatment for a tree infected with Dutch elm disease, but that the use of injections for preventing the disease may be a poor trade-off.

Treating Wounds

Decay problems can be largely prevented if you plant species or clones known to compartmentalize effectively, to close wounds rapidly, and to resist decay. Wounds are inevitable, but their number, size, and severity can often be minimized. Felix and Shigo (1977) recommend that needed pruning, cabling, scribing, and injection be done at times other than spring growth flush and leaf fall, when plants seem to be most vulnerable. If you choose safe locations for planting and use protective stakes or guard rails, you can prevent or reduce plant injury from cars, machinery, and people. Early training of trees can prevent sunburn and reduce the number and size of later pruning cuts; well-trained trees should not have to be cabled or braced. When possible, use methods other than injection for applying chemicals to trees.

Maintaining Vigor. **Tree vigor is the best treatment for wounds and decay.** Vigor, particularly in conifers, greatly reduces the possibility of borer infestation. Vigor can be maintained and enhanced by wise fertilization, irrigation, and pest management.

Cleaning and Shaping Wounds. Remove loose, dry bark from in and around a wound; the wound will look better and insects will not have protective cover. Arborists commonly recommend that the bark around wounds and even around pruning cuts be shaped, but experiments have brought the value of this practice into question, particularly in cases where it widens the wound. Shaping a wound is of little or no value in hastening wound closure (Neely 1970).

If you do shape a wound, take care to increase its width as little as possible and leave no sharp apexes. Cambium at the apexes of pointed ellipses may die back, and the new xylem often develops cracks (Felix and Shigo 1977, Mercer 1979). Arborists usually shape bark wounds so that the long axis is parallel to the grain of the wood immediately under the cambium; this may or may not be parallel to the length of the trunk or branch. Shaping a wound otherwise increases its width and the time of closure. An elliptical wound supposedly favors the flow of organic compounds around and along its edge so as to enhance closure. A circular wound, however, will callus as quickly as an elliptical one (Neely 1970). Neely found that the rates of callus growth from a straight edge (parallel to the grain) and a convex edge were about 50 and 25 percent respectively of the growth rate from a concave edge (the shape usually recommended) (Fig. 18–15). In no case did it prove worthwhile to widen a wound.

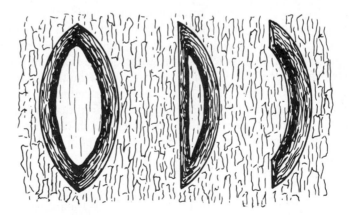

Figure 18–15 The value of tracing wounds in the shape of an ellipse has been refuted by experiments performed by Neely (1970). Neely cut a series of wounds in the bark of several tree species: The right edge of each wound was half an ellipse, the left side was half an ellipse or concave (left), a straight line (center), or a convex curve that reduced the width of the wound to one-half that formed by half an ellipse (right). Even though the convex edge callused (dark-lined areas) only 25 percent as quickly as the concave edge (standard ellipse), the wound it surrounded (right) closed in one growing season while the others did not close until the next. (Adapted from Neely 1970)

Leave peninsulas of live bark in a wound to speed callusing. Even though callus formation at the edge of a peninsula of bark is only 10 percent of that on the inside of an ellipse, the peninsula callus has to fill less area than if the entire wound had been "properly" shaped. When cleaning a wound, use a sharp wood chisel, gouge, or pruning knife to cut the loose bark at right angles to the wound surface, usually the cambium layer. Cut to firm bark. If the wood is damaged, smooth it so it will not trap water and debris, hamper closure, or present an unpleasant appearance. **Do not unnecessarily enlarge a wound.**

Dressing Wounds. Ideally, a wound dressing should prevent decay, stimulate closure, and improve tree appearance. To date, however, no dressing commercially available in the United States has been shown to prevent decay. Shigo and Wilson (1977) applied asphalt, polyurethane, and shellac to separate wounds on red maple and American elm and found that none affected vertical extensions of discolored and decayed wood or the presence of decay fungi. In fact, the electrical resistance of discolored wood indicated that wounds dressed with asphalt were most serious and untreated wounds the least serious. Of more than 40 wound paints tested in the laboratory by Mercer (1979), all those containing a fungicide, except one, were phytotoxic. The one exception was Santar®; because it contains mercury, it is not registered for use in the United States and probably never will be. Neely (1970) and Shigo and Wilson (1977) concluded after testing six wound dressings available in the United States that the materials had no effect on the rate of wound closure. Lac Balsam®, a latex emulsion available in Europe, was the only material tested by Mercer (1979) that stimulated callus formation. Alex Shigo (Durham, NH, 1980 pers. comm.) maintains, however, that unless a wound closes within three years, the rate of closure has little or no effect on decay. Once decay begins, closure will not reduce its development.

The main function of wound dressings, therefore, is cosmetic (Shigo 1981). Many wounds are less noticeable when painted with a dark-colored covering. Many people regard wound dressings as the sign of professionalism. The most common dressings are asphalt-based; they should be applied in thin coats to minimize cracking. If wound coverings crack, moisture can accumulate and foster decay.

Applying an asphalt paint to fresh pruning and bark wounds has been successful in protecting those wounds on oak, pecan, fruit, and other trees from borers and beetles that cause damage directly or transmit certain vascular and canker diseases (Wiener 1982, Sinclair, Lyon, and Johnson 1987, Johnson and Lyon 1988). The addition of a fungicide is recommended for some species. In a few cases, fungicides may be applied directly for disease control (Riley and Okie 1983). Some of these problems can be minimized by not pruning when the insects are active and/or the fungus sporulating.

Since wound paint contaminated with canker-stain fungus can infect London plane, it is wise not to paint fresh pruning wounds where this fungus is common (Pirone and others 1988).

Replacing Loose Bark. Many bark wounds occur in the spring and early summer, when cambium is active and bark easily separates from wood. Loosened bark can often be replaced with little or no damage to the injured area, but it must be replaced soon after it has been loosened. Keep the moist cambial regions of both the wood and the bark from drying out. Carefully remove all shredded bark and debris from the wound so that the bark will fit tightly against the film of cambium on the xylem. Press the patch of loose bark into place and hold it with a few small lacquered nails or duct tape. Cover the wounded area with a moist pad of cloth, paper towel, or peat; wrap a sheet of polyethylene film around the pad and the trunk or branch. Hold the film in place with a tight tie of tape or line above and below the wound; you may have to smooth rough bark to ensure a snug fit. Shade the wound from the sun.

If you perform this operation carefully and soon after an injury occurs, much or all of the loosened bark may reunite with the wood. Reunion will be most likely during the period of rapid cambial growth. Within a week or two, inspect the wound to see whether the bark patch is still alive and has begun to unite with the wood. You can remove the plastic film and the moist pad if callus has formed around the cracks, but shade the wound. If the bark patch does not reunite with the trunk or if the wound is an old one, free the wound area of dead bark and leave it exposed to the air.

On the other hand, if the torn bark is mutilated and cannot be used, carefully clean and cover the wound with a plastic film, keeping the film from touching the exposed surface. Secure the film around the wound with duct tape to keep the wound from drying. Follow the steps above for covering replaced bark.

Grafting Bark. Girdling wounds can be bridged with bark implants or bridge grafts. Bridge grafts have been used by fruit tree growers to reunite the top and roots of trees whose trunks have been damaged by rabbits or other rodents (Hartmann, Kester, and Davies 1990), but the technique is seldom used in landscapes. Bark implants, however, have been used to bridge girdling bark wounds, sometimes caused by winch cables or vandals. If the cambium is still active, you can use bark from other parts of the tree to bridge the break (Fig. 18–16).

Figure 18–16 A girdle injury can often be repaired by bark implants. A girdle through the bark and into the wood is shown at the left. When the bark will slip easily (during the early growing season), cut bands of bark about 20-mm (0.75-in.) wide from both sides of the girdle (center). Cut a bark patch the width of the enlarged girdle from another area of the trunk or a branch and tack in into place, covering the girdle and newly exposed cambium (right). Keep the exposed cambium, new bark patch, and the new patch wound from drying out.

Begin by cleaning the girdled wound of shredded bark and debris. Remove a 60 mm (2.5 in.) length of bark about 20-mm (0.75-in.) wide from above and below the girdle. Cut a bark patch the same height as the newly enlarged girdle and about 50-mm (2-in.) long from another place on the trunk or branches. Place the patch on the newly exposed wood and tack it in place with small lacquered nails. Repeat these steps until the entire girdle has been covered with bark patches. While doing the patch transfers, keep the girdle and patch area moist, and cover the new wounds on the trunk and branches with duct tape. Place a moist paper towel or cloth over the patched girdle extending over about 15 mm (0.5 in.) of undisturbed bark above and below the patches. Hold the moist covering snugly in place with duct tape with about 25 mm (1 in.) of tape on bare bark above and below the covering. Shade all the wounded areas.

Within a week or two, the bark implants should be callused in placed. The cambium on the patches will unite with that of the trunk and form woundwood over the girdled area. The patch wounds should also form woundwood from the film of cambium that remained on the xylem. Keep the patches shaded from the sun for most of the summer.

By using bark implants, Herbert Warren, former park director of Victoria, British Columbia, saved two 0.4-m (15-in.) diameter trees that had been girdled with a chain saw.

CAVITIES

The origin and development of cavities in trees was described in Chapter 17. When deciding whether and how to treat a cavity, consider the importance of the tree; its species, age, and condition; and the size and location of the cavity. A cavity may be so large or close to large branches that the tree is irremediably unsafe. On the other hand, a seriously cavitated tree may be of such historic or landscape value that considerable work is justified. Short-lived softwood trees are usually of questionable value compared with longer-lived hardwoods and conifers.

Decaying wood in cavities is sometimes infested with termites and ants. The most destructive species in the United States are the subterranean termites (*Reticulitermes* spp.) and carpenter ants (*Camponotus* spp.) (Pirone 1978a). Termites feed primarily in decaying wood, seldom in sound wood.

Ants and termites, however, are usually contained within the chemical and physical barrier zone formed by the cambium after wounding. Carpenter ant larvae can create large tunnels in decayed and sound heartwood. Ants can be distinguished from termites by the pronounced constriction of the body between the thorax (where legs and wings are attached) and the abdomen; termites have no such constriction. Ants can be controlled by dusts or sprays of Diazinon® in the infested cavity (Pirone 1978a); termites are more difficult to control. The solvents in some materials used to protect buildings against termites are toxic to woody plants and should not be used on them. Consult a knowledgeable specialist (see Chapter 19) about the chemicals registered to safely control ants and termites in your particular area.

Treating Cavities

Arborists have recommended that cavities be cleaned, sterilized, braced, sealed, and filled or covered so that decay would be stopped, the tree strengthened, and a hard surface provided for wound closure (Collins 1920, Bernatzky 1978, Pirone 1978a). In the early 1900s, sterilization of a cavity after cleaning was added as a treatment, although the value of filling cavities has been questioned. These procedures do little to help the tree and can in fact be harmful if they cause the protective barrier to be broken (Shortle 1979). Although some may be of questionable value, the various procedures will be discussed.

Cleaning. The essence of almost all recommendations is to clean all decayed wood out of a cavity. Many people also advise removing discolored and water-soaked wood. However, decay fungi have been found 0.3 to 1.4 m (1–4.5 ft) in advance of decayed wood (Welch 1949), and removing infected wood from a cavity is impractical if not impossible. Any attempt to remove even all of the rotten wood will probably breech the compartmentalized barriers to decay and open sound wood to infection (Shortle 1979).

Most cavities collect water, but providing drainage would probably cause more harm than the water, particularly if sound wood is exposed (Shigo and Felix 1980). Alternating wet and dry conditions usually promote wood rotting more than continual saturation; air is more often a limiting factor than water (Welch 1949).

If cavities are to be cleaned, take care not to cut into sound wood even if it is discolored (Wilson and Shigo 1973). Even more importantly, do not cut into healthy wood formed since the wound originally occurred. A pocket fill would probably be most satisfactory to minimize cavity water (Fig. 18–17).

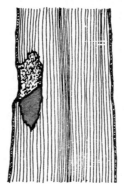

Figure 18-17 Small cavities can be partially filled to keep water from accumulating. (Adapted from artwork by M. J. Whitehead in Bridgeman, P. H. 1976. *Tree Surgery*. London, England: David & Charles)

Sterilizing. None of the recommended sterilizing chemicals, such as creosote or copper sulfate solution, penetrates wood more than a few millimeters (a fraction of an inch). As already noted, decay infection can extend more than a meter (3 ft) into sound wood. In addition, most sterilizing agents are poisonous or corrosive and must be handled with care.

Bracing. A cavity opening may need mechanical strengthening. Bracing will partially replace the lost strength of the decayed or removed wood and will hold the cavity walls in position. Use screw rods in thin-walled wood, with washers and nuts on each side of the wood walls to provide rigidity and keep the walls from being pulled together (Thompson 1959). Cross-bracing is used to keep the cavity walls from spreading and to minimize twisting strains that can damage fill material. Screw rods are placed in alternating diagonals across a trunk cavity: If seen from above, the rods form an *X*. Cross-bracing has limited benefits, however, because the stresses set up in a large cavity and the leverage exerted by a tree in strong winds are so great (Thompson 1959). In many cases, it will be better to replace the tree.

Sealing. No wound dressing available in the United States has been shown to inhibit decay; in some trials, asphalt dressings actually increased the extent of infection (Shigo and Wilson 1977). The main reason for painting the inside of a cavity is to improve appearance if the cavity is to be left open.

Filling. Most arborists agree that filling a cavity is of little or no value in promoting the health and longevity of a tree.

Fill material seldom strengthens tree structure as much as would the "callus roll" (also called "ram's horn") that develops around an open cavity. Both the movement of a tree and differing expansion characteristics loosen a fill in its cavity and decrease its efficacy as a support; and air, moisture, and spores can diffuse or seep into the openings. Further decay is favored. Tree appearance remains the main reason for filling cavities.

Cavity fill material should be durable, nontoxic to trees, flexible, and waterproof (Pirone 1978a). Many substances, including patented fill materials are used to fill cavities. *Concrete* has been the standard fill material, primarily because of its availability, low cost, and durability (Fig. 18-18). However, it is not very flexible or waterproof; it is heavy, requires skill to install, and is difficult to remove. Concrete is most satisfactory in small cavities and trunk cavities subject to little or no movement in storms. Place in a clean cavity (decayed wood removed), a relatively dry concrete mix in layers 75 to 125 mm (3-5 in.) thick separated by heavy tar paper to allow for expansion and some movement. Pack the concrete firmly and form a smooth surface even with the cambium.

A mixture of *ashpalt and sand* is sometimes used, particularly for basal trunk and pocket cavities (Bridgeman 1976). Asphalt mixtures are better fillers than concrete but are difficult to prepare and apply (Collins 1920). The mixture must be stiff so that it will not slump at the base when it is tamped into place. Asphalt mixtures are most satisfactory for pocket cavities (Fig. 18-17) or trunk cavities that will not be warmed excessively by the summer sun, which could cause the mixture to soften and slump. Since asphalt mixtures are malleable, no expansion joints are necessary.

Urethane foam is credited with several advantages over other fill materials: It is easy to use, is relatively inexpensive, and requires little time to install (King, Beatly, and McKenzie 1970). When installed, the foam is lightweight, nontoxic, and somewhat flexible. Hamilton and Marling (1981) and others, however, report that foam fills have not held up. Bridgeman (1976) reports that polyurethane foams tend to shrink and draw away from the cavity wall, but that urethane foams do not. Gordon King, University of Massachusetts (1978 pers. comm.) found that cavities filled with polyurethane and covered with auto body putty were in good shape after eight years.

However, the number of foam fills that have not held up raises questions as to their value. Improper finishing with autobody filler and hardener, particularly near the edges of the cavity, may be a problem. It may be wise to try one or two cavities before doing more. Follow the manufacturer's directions for installation of

Figure 18–18 Although seldom done any more, trunk cavities were occasionally filled with cement to improve the appearance of trees, such as these two elms: The basal cavity was filled and the main branches rodded on the elm (left) on the University of Illinois campus shortly before being photographed. (Photo courtesy E. B. Himelick, Illinois Natural History Survey). The elm on the State Capitol grounds in Sacramento in the right photograph is actively covering the filling installed 35 years earlier.

the foams; and King, Beatly, and McKenzie (1970) give additional pointers for filling cavities in trees with foam.

Covering. A cavity can be covered with sheet metal when you do not want to leave the cavity open or fill it in (Collins 1920). However, there is little to recommend this practice. Unless thick enough, sheet metal covering can be collapsed inward by the pressure of the growing woundwood (Keith Davey, Belmont, CA, 1975 pers. comm.). Such a covering may not serve its purpose long.

Recommendations

A tree with a cavity will be best served if you do everything possible to improve tree vigor so that new wood and bark grow faster than decay advances within the compartmentalized barrier. A cavitated tree may have to be thinned to reduce the weight and wind resistance of the top. If a cavity is not conspicuous or endangering the tree, you need only remove branch stubs or decayed wood that might interfere with woundwood growth. If a cavity is so extensive that it weakens tree structure, you may also have to brace the tree or remove it. Decayed wood that can be easily removed should be extracted from the cavity before bracing rods are installed.

If cavity appearance is important, remove enough decayed wood to give a smooth, firm surface, taking care not to cut into sound wood. Clean the portion of the cavity that is below the opening, dry it (with a blow torch if necessary), and fill it with asphalt (if the cavity is small), an asphalt-sand mixture, or possibly urethane foam so that water will be shed. Paint the entire cavity and any fill installed.

Trees with large cavities may need to be protected from people who climb on bracing rods, build fires in the cavities, or otherwise deface them. Protect the entire trunk with a free-standing screen fence or cover the cavity with a screen, always making allowance for trunk expansion.

MANAGING ROOT GROWTH

Root growth varies with tree species, soil conditions, surface covering, rainfall, and irrigation practices. Root growth and function are essential for healthy, sturdy plants, but roots can also cause problems. Roots crack and plug sewer lines; they lift and break curbs, sidewalks, pavement, and building foundations (Fig. 18–19). When roots dry the soil under pavement and buildings, certain soils shrink and settle so that structural damage occurs. Shallow roots in lawns can be unsightly and difficult to mow around. Root suckers in lawns and flower beds are a nuisance.

Figure 18–19 Even though the curb and sidewalk have been replaced at least once, this eucalyptus tree again has broken and raised the curb, sidewalk, and driveway.

Roots and Sewers (Drains)

People rarely give thought to possible root problems when they plant a young tree or shrub; not until a sewer is blocked do most people become aware of the problem. Roots can crack a sewer line and invade it if there is an opening. When a sewer is plugged, roots need to be removed or chemically treated (see Chapter 16).

The problem is often difficult to address until failure occurs, but there are measures that will prevent or at least delay the invasion of sewer lines. One of the most important measures is to avoid planting any fast-growing species that are known to have invasive roots, such as willow, poplar, and silver maple, near a sewer or a septic field. Unfortunately, home owners often find utility trenches the easiest places to plant. Less than four years after a fruitless mulberry was planted in a newly filled sewer trench in Sacramento, California, two of its roots crushed an orangeburg sewer line 100 mm (4 in.) in diameter (see Fig. 1–5). Only small trees and modest-size shrubs should be planted near sewer lines, particularly if the soil is shallow or poorly drained and the sewer trench is deeper than the surrounding soil.

Typar Biobarrier ⓦ (Reemay, Inc., Old Hickory, Tennessee) holds promise of being an effective protector of sewer lines and septic fields. This root barrier combines a geotextile fabric with a time-released herbicide (trifluralin). For maximum protection, a sewer pipe would be wrapped wherever roots might come in contact with it. This would protect the joints and any cracks from invasion and greatly reduce the possibility of nearby expanding roots cracking or collapsing the pipe. Less expensive protection would be to wrap just the pipe joints with the Biobarrierⓦ. Although it is unlikely, an enlarging root could crack an unprotected pipe section, creating an entry for roots.

Water-tight flexible telescopic joints for sewer lines in new building sites can minimize root problems (Morling 1963). Some contractors wrap sewer joints with copper wire screen, which is toxic to small roots. Copper may delay sewer invasion, but it will do little to prevent cracks caused by enlarging roots. Wrapping a sewer line with a root-resistant geotextile will not prevent joint or pipe cracking by roots, but it should keep invading roots from entering cracks. Moisture from a sewer crack could result in excessive root growth adjacent to the crack causing further displacement of the sewer.

Roots and Pavement

One of the costliest operations associated with municipal trees is repairing curbs and sidewalks damaged by tree roots (Wagar 1985b). The cost of repairing sidewalk damage by one tree is often $500 or more per incident. Once started, damage often recurs at about five-year intervals.

Large buttress-rooted trees in small planting sites are almost certain to cause pavement and sidewalk damage (Hamilton 1976, Wagar and Barker 1983). Compact, poorly aerated soil and excess water can also exacerbate the situation. In a survey of 2232 street trees in Manchester, England (13 percent of the total), Wong, Good, and Denne (1988) found 30 percent were causing sidewalk damage and 13 percent damaged curbs. More damage occurred to asphalt walks than to 300 ×

600 mm (1 × 2 ft) slabs (pavers) set on soil or sand. Trees in sidewalks caused more damage than those in planting strips. Trees in planting strips less than 3 m (10 ft) wide caused more damage than those in wider planting strips.

In the San Francisco Bay area, sidewalk damage begins when trees are seven to 20 years old (Hamilton 1976). Damage to curbs takes about twice as long (Gordon Mann, Redwood City, CA, 1988 pers. comm.). In Manchester, England, most trees started to cause damage when they were 100 to 200 mm (4–8 in.) trunk diameter, but not oaks and horse chestnut until they were larger than 200 mm (8 in.). The distance between trees had no significant effect on sidewalk or curb damage.

New Plantings. Tree-rooting characteristics, size and configuration of planting site, soil condition, and root-directional devices should be taken into consideration in the planning and installation of a landscape. Some of the pavement planters described in Chapter 11 should create conditions favorable for deeper root growth. Wong, Good, and Denne (1988) recommend that trees in England be planted in planting strips that are wider than 3 m (10 ft.). Those planted in pavement should have an opening, ideally at least 2 × 2 m (6.5 × 6.5 ft). Curb problems can be reduced if the trunk is at least 1 m (3 ft) from the curb when the tree is mature.

Soils are usually severely compacted before sidewalks and pavement are installed. When roots encounter a paved area, the only entry is often a crack between the soil and pavement caused by differential temperature expansion and contraction. Once a root gets started, it is only a matter of time before something gives, particularly if there is water and air beyond the paving.

Where planting space is limited, several techniques are used to minimize root damage to pavement. Since the early 1970s, the city of Santa Maria, California, has planted many of its street trees in wells with the soil level 0.5 m (20 in.) below the sidewalk. Wagar (1985b) found that fruitless mulberry and zelkova trees planted in 450-mm (18-in.) deep wells had fewer and smaller roots in the surface 200 mm (8 in.) of soil 0.6 to 0.9 m (2–3 ft) from their trunks than did trees planted on the surface. Such planting depths could be achieved by lowering the soil surfaces of planters as shown in Figure 11–1.

Several devices are available that direct roots down in the soil. A square polyethylene planter, wider at the base than at the top, is marketed by the Deep Root Corporation in Beverly Hills, California. It is 450 mm (18 in.) deep and open top and bottom. When a root strikes a wall of the planter, it is directed downward. Century Products of Anaheim, California, manufactures a cylindrical planter with watering vents to apply water 450 to 600 mm (18–24 in.) below the soil surface as a tree becomes established. Trees have been planted in large plastic garbage cans with the bottoms cut out and turned upside down. In Vallejo, California, about 100 *Photinia* × *fraseri* and *Pyrus kawakamii* were so planted on business streets between 1973 and 1976, and they are doing well. The roots of a 185-mm (7.4-in.) diameter trunk pear tree (it had been damaged) had grown down the sloping sides of the garbage can. One 75-mm (3-in.) root grew out from the bottom of the root barrier under the sidewalk but did not come near the surface.

Wagar (1985b) concluded from experiments with young mulberry and zelkova trees that planters and wells substantially reduced surface rooting. Individual trees differed considerably, however, in the extent to which their roots returned to surface layers after having been forced deeper. Root-control devices appear to be least effective where most needed—that is, where poor soil aeration or compaction encourages shallow rooting.

Trees in root-control devices do not grow quite as fast as those not confined (Wagar 1985b). In compacted soils, roots may be confined to the device and the trees become unstable as they increase in size. As roots enlarge, they occasionally lift the device in the ground. If the devices are set low, roots often grow over the top.

For trees in pavement or between curb and sidewalk, the hardscape can be protected by a sheet barrier placed against it (Fig. 18–20). Interlocking, semirigid sheets of plastic are marketed for this purpose; their width is usually 0.45 or 0.6 m (18 or 24 in.). Barriers of other materials can also be used. They should be sloped to direct roots downward with the upper edge above the soil level or integral with the pavement or curb.

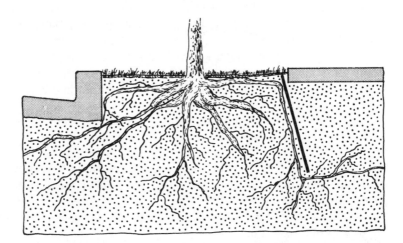

Figure 18–20 Tree roots can be deflected downward if a thin impervious sheet of wood, metal, plastic, fiberglass, or Biobarrier® is installed at a slight angle from the vertical near paving or foundations to be protected.

The Biobarrier® should provide more complete exclusion of roots than a barrier without an herbicide, because few roots would grow under a Biobarrier®. The Biobarrier® is more expensive. Even though it is considered safe, some people may not want to put such a persistent chemical in soil. The Biobarrier® should not be used to confine large-tree roots to a limited volume of soil because of instability when the tree becomes large.

Mature Trees. Arborists are now using a variety of techniques to manage roots of mature trees in confined areas, including the following:

Remove damaged walk and
>	Bend narrower walk around tree
>	Bridge over offending roots
>	Remove roots under old walk, repave with more flexible material which is easier and less expensive to maintain and replace
>	Root-prune (usually before walk is removed or as part of a pavement replacement operation), place a sloping-vertical barrier, and repave

Root-prune close to walk and install a sloping vertical barrier

Remove offending tree and
>	Replant with a smaller-rooted, smaller-growing tree

Remove offending tree and damaged walk, and
>	Curve walk to the curb and plant replacement tree behind the monolithic walk/curb.

Root-cutting machines are used on a regular schedule by some cities. Do not root-prune weak or stressed trees. Be sure to check for underground utility and irrigation lines. Surface roots are cut when a trench is dug between the trunk and the walk or curb (Fig. 18–21). The trench should be close to the pavement. The trench is usually 50 to 100 mm (2–4 in.) wide and 200 to 300 mm (8–12 in.) deep.

Figure 18–21 A Vermeer root-cutting machine speeds the cutting of large surface roots. (Photo courtesy Vermeer Manufacturing Company, Pella, IA)

Most of the large roots will be in the upper 300 mm (1 ft); cutting roots deeper may make a tree more subject to windthrow with little additional protection for the pavement. Cut roots on only one side of a tree; wait at least three to four years before pruning the other side (Daniel Condon, Santa Barbara, CA, 1989 pers. comm.). To minimize risk of windthrow, the top of the tree should be thinned before roots are cut—a year before root pruning is best. Crown reduction pruning may be advisable.

Few arborists are satisfied with present root-pruning practices (Hamilton 1976); regrowth is extremely rapid from most cut roots (Hamilton 1988). Sidewalks have been lifted within two to three years after they were repaired and the trees root-pruned. Installing barriers as described above should greatly reduce further sidewalk damage. The top of the root barrier must protrude above the soil surface to prevent roots from growing over the barrier top.

John Britton (St. Helena, CA, 1990 pers. comm.) recommends that when roots are hand trenched or cut, they should be thinned back to where they arise or to a large root, so that growth of the remaining roots is directed parallel to or away from the pavement. He believes this will minimize future problems.

The long-term solution to root damage should combine selection of better species, more thorough preparation of compacted soils, root barriers or deflectors, wise irrigation, and root-pruning when necessary.

Roots and Buildings

Clay soils, in particular, shrink and swell due to moisture extraction by roots and subsequent rewetting. Alternate shrinkage and expansion of certain clay soils have caused serious cracking and subsidence of pavement, foundations, and walls (Morling 1963, Cutler and Richardson 1989). Doors, windows, and drawers commonly stick due to such building movement. Surface cracking of a soil indicates its susceptibility to shrinking and swelling.

Soil Subsidence. Soil shrinkage is particularly severe during droughts. Following the drought of 1975–1976, English insurance companies paid about $70 million to cover building damage caused by the shrinkage of clay soils (Reece 1979). Even though soil subsidence rarely makes a building unsafe, it may seriously affect its appearance, value, and marketability (Pryke 1979). Uneven subsidence is the most disruptive effect. After a dry summer, a clay soil may subside 50 to 75 mm (2–3 in.) (Biddle 1979, Pryke 1979). Soils under buildings may also dry and shrink after repair of leaking drains or installation of underground central heating pipes.

In a root survey of about 11,000 trees and 1300 shrubs, Cutler and Richardson (1989) have compiled interesting and useful information about roots of the more common trees in England and their damage to buildings and drains. The two genera most frequently reported to cause building subsidence were oak (primarily *Quercus robur* and *Q. petraea*) and poplar. Oaks were estimated to be 2.1 percent of the street and garden tree population but accounted for 11.5 percent of the trees causing damage. In 10 percent of the cases, oak trees were farther than 18 m (60 ft) from the buildings damaged. Poplars estimated to be 3 percent of the tree population accounted for 8.7 percent of the damage. In 10 percent of the cases, poplar trees were farther than 20 m (65 ft) from the buildings damaged.

Birch and the flowering fruit trees caused the least damage; in only 10 percent of their cases were the trees farther than 8 m (25 ft) from the buildings damaged.

Morling (1963) estimated that fast-growing, large trees caused problems in clay soils over a horizontal radius equal to 1.5 times their mature height. Cutler and Richardson (1989), however, found poor correlation between tree height and the distance that roots were a problem. Seldom was the maximum distance that a species' roots caused damage as great as the height of the tree. Young trees, with rapidly enlarging roots, usually cause more damage than mature trees. Shrinking and swelling are greatest near the surface, where most roots are located, and decrease with depth to about 1.5 or 2 m (5 or 6 ft) below the surface, where soil movement is slight. During a dry year, shrinkage near large trees may extend to 3 m (10 ft) below the surface (Morling 1963, Pryke 1979).

Soil Expansion. Most cracks in buildings caused by soil subsidence will nearly close during the next rainy season or irrigation period. However, the swelling of a dry clay soil after a building has been constructed may cause even more serious problems. Morling (1963) reports that the surface of a clay soil increased 100 mm (4 in.) in elevation in the six years after two large elm trees were removed and still rose 10 mm (0.4 in.) in the seventh year. Soil moisture in the root zone must have been near the wilting point when the trees were cut down. Even with ample rain or irrigation, many clay soils under buildings are slow to rewet and will continue expansion over long periods of time.

Minimizing Shrinking and Swelling Problems. A number of things can be done to reduce the problems caused by trees growing in clay soils.

> Provide buildings on expansive clay soils with particularly deep and strong foundation footings. Short-bored pile foundations (augered holes filled with cement so they are integral with the foundation) 3 m (10 ft) long can provide needed stability at reasonable cost (Pryke 1979).
>
> Near buildings and pavement, plant moderately vigorous trees that will grow to a small or medium size. Avoid trees that become large when mature, especially oak, poplar, willow, elm, and horse chestnut, particularly if building foundations are inadequate.
>
> If shrubs, vines, or young mature trees begin to cause damage around existing buildings, consider removing them. Do not repair settling cracks in buildings or pavement until the soil has had a chance to rewet and the cracks close, at least partially.
>
> While a tree is young, or before planting, improve the tilth, fertility, and moisture conditions of the soil under what will be the mature canopy. Between the building and the treetrunk, treat only that half of the soil area that is nearer the trunk. This should increase root branching in the improved soil and minimize root growth toward the building.
>
> Biddle (1979) suggests that heavy pruning (reducing trunk growth 75 percent) will reduce water use (transpiration), root growth, and soil subsidence. Transpiration would not be reduced proportionately to the severity of the pruning. If heavy pruning were done every year or two, subsidence might be reduced. More information on this strategy is needed.
>
> Install a barrier around the house, similar to that described for protecting pavement.

Where subsidence causes only minor cracking, slowly soak the soil along the foundation or pavement. This would be especially important during a drought. I am reminded to irrigate an ivy ground-cover planting next to my house when the kitchen cupboard doors begin to stick. Unless it is near the foundation, drip irrigation may not be adequate.

Cutting down a large tree that has been creating subsidence problems may cause more damage due to heaving as the soil rewets.

Keep trees growing in shrinkable clay soil at a construction site adequately watered so that the soil does not subside or expand unduly.

If a shrinkable clay soil has been severely dried by plants, irrigate the soil thoroughly and allow it to drain before grading for construction.

TREE REMOVAL

A tree may be removed for a number of reasons; it may be

Unsafe because of old age, storm damage, poor structure, or death
Diseased or a host to a disease that can spread to more valuable trees
Located in a future building site or roadway or obstructing a view
Crowding other trees
Wanted for timber or firewood

If a tree is in the open, it can be pushed down with a large track-laying bulldozer, pulled down with a power winch, or cut down with a chain saw. In more restricted areas, a tree may have to be felled in a certain direction or removed in sections. Bridgeman (1976) gives directions for felling trees.

Sometimes large trees are cut close to the ground and the stumps allowed to

Figure 18–22 A stump cutter can chip a stump away to a depth of about 250 mm (10 in.) below the soil surface. (Photo courtesy Vermeer Manufacturing Company, Pella, IA)

remain. The stumps of trees and large shrubs susceptible to *Armillaria mellea,* however, should be removed even though they are not infected (see Chapter 21). As Rishbeth (1970) has shown, airborne spores of the fungus can infect stumps. You can control or minimize stump sprouting and root suckers by treating the stump chemically or girdling the trunk a year before the tree is cut down (see Chapter 16).

Stump removal is not easy. The stumps of small trees are usually dug out by hand with a shovel and ax; a winch makes the job easier. When the equipment is available and the stump accessible, a stump-cutting or grinding machine is the quickest and most satisfactory means of removal (Fig. 18–22). It can chip out a stump to 200 or 300 mm (8 or 12 in.) below ground level in minutes. A stump-cutter can remove stumps in confined spaces, like those between curb and sidewalk. You may lift the stump out with a hydraulic lift—a "Stumpmaster" can lift stumps up to 375 mm (15 in.) in diameter (Bridgeman 1976). Blasting the stump out of the ground with explosives should be done by a specialist and only when the stump is in open country. A dynamite blast can kill nearby trees. Finally, you can burn the stump in place by pouring saltpeter (KNO_3) and paraffin in holes in the stump and igniting them—this method is slow and can be a fire hazard in organic soils.

CHAPTER 19

Diagnosing
Plant Problems _____

Most landscape plantings get off to a good start, particularly in the first season, and begin growth under nearly ideal conditions of spring-like weather and moist soil. Both new and existing plants may also retain vigor and food reserves from earlier care and conditions. Their new surroundings, however, may be less favorable than their previous environment. This is particularly the case in inner cities and other intensively developed sites. Plants can be harmed as their roots grow out into surrounding soil and as new shoots encounter the increasingly hostile environment of summer and following seasons. It is remarkable that many plants grow as well as they do.

Problems encountered in new or redeveloped sites include

Stratified and compacted soil with construction debris
Infestation by pests and disease-producing organisms
Reflection and reradiation that increase air temperature
Light intensity that varies from deep shade to full sun
The extension of daylength by night lighting
The reduction of humidity by arid expanses of pavement
A sparse or excessive water supply
Insufficient nutrients
Air pollution
Accidents, vandalism, and neglect

You can minimize or avoid most such hazards by choosing a favorable site, ensuring that construction does not damage soil or plants, preparing the site for planting, selecting plants that are tolerant to the site and relatively pest-resistant, and caring for plants to ensure vigorous growth. Be alert to environmental changes that may jeopardize existing plantings.

Even in what appear to be the best growing situations, plants may not perform as desired: They may become unattractive, grow slowly, or become so seriously diseased that they must be removed. Plant problems are diagnosed so that they may be corrected or minimized. Diagnosis may also establish responsibility in the event of litigation or for insurance or tax purposes.

THE DIAGNOSIS

Efforts at treatment may be ineffective or may even compound a problem if plants are treated for the wrong malady. Accurate diagnosis of a plant problem depends on

> The observation of subtle differences from the normal in plant appearance or the surrounding environment
>
> A knowledge of plants, soil, climate, cultural practices, pests, diseases, and their interactions
>
> Accurate information about the recent history of affected plants, the site, the climate, and cultural practices
>
> The availability of a few simple tools
>
> An analytical approach to difficult problems

Careful observation is essential and can be developed with practice and interest. Familiarity with normal plant appearance and growth and the effects of environment and treatment will increase with experience. *Diseases of Trees and Shrubs* (Sinclair, Lyon, and Johnson 1987) and *Insects That Feed on Trees and Shrubs,* 2nd ed. (Johnson and Lyon 1988) are excellent diagnostic references with descriptions and color photographs.

Normal Plants

Inexperienced observers often think they have encountered a serious problem when, in fact, the "symptom" that alarms them is only a normal feature of a plant, a seasonal change in its appearance, or the sign of a nonparasitic agent. Thus, recognizing the appearance of healthy plants at different stages of growth and seasons can be as important as recognizing the symptoms of particular disorders.

Leaf Color. The leaves of a number of plants may vary from the green, gray-green, or reddish color that is normal for most of the growing season. The early spring growth of some plants, such as apricot and Chinese photinia, may be copper or reddish in color, giving way to green as the leaves and shoots mature. Other plants may have yellowish foliage early in the season, particularly in springs that are cool and wet.

In some plants, the older leaves lose their green earlier than younger leaves;

the oldest year's leaves of some evergreens will lose their color just before they drop. Half of the needles of Eastern white pine, which holds its needles for only two years, will turn yellow in the fall before they turn brown and drop (Tattar 1989). Trees may look quite unhealthy during this normal transition. Many species develop yellow, orange, or red foliage in the fall as chlorophyll fades. The onset and intensity of leaf color change will vary for a given plant or clone from year to year, as summer growing conditions and fall weather vary.

Yellow foliage is usually thought to indicate a nutrient deficiency, some other soil problem, or root abnormality. Frisia locust, sunburst honey locust, and several junipers, however, are often selected for landscape use because of the normal yellow in some of their foliage. A number of shrubs, trees, and vines are vegetatively propagated for their characteristic variegated patterns of yellow or white on otherwise green leaves. Gold spot euonymus has dark-green leaves with golden yellow blotches and edges; tricolor dogwood has variegated leaves of white, pink, rose, and green in spring and summer.

On a single plant, some branches will have green foliage, others yellow, and still others variegated yellow and green (Fig. 19–1). In such cases, the variegated combination of green and yellow is the result of a *chimera,* a genetic vegetative mutation that involves only a sector of the tissue. Chimeras, when propagated vegetatively to preserve the variegated leaf characteristic, may not be stable. That is, leaves may form from tissue that has all (yellow), none (green), or only some (varie-

Figure 19–1 Most of the foliage on this Gold Spot euonymus is variegated yellow and green, but one portion has reverted to leaves that are entirely green and another portion has all yellow foliage.

gated) of the mutation. This mixture of differently colored branches may be more likely after a vigorous variegated plant has been severely pruned. If you wish to preserve the variegated characteristic in a specimen, remove shoots with pure yellow or green leaves as soon as you notice them; cut as close to the parent branch as possible. If you wait until severe pruning is necessary, many more or all of the new shoots will have leaves that are all green.

Occasionally, the leaves on one or more branches of a tree or shrub will differ in color, size, or some other characteristic from leaves on other branches. This could be the result of a mutation. More likely, one set of the branches derives from the rootstock and the other from the top grafted onto the rootstock. You can remove the branches with the less desirable characteristics or can allow both kinds of branches to remain, depending on your preference.

Juvenile and Mature Leaves. Many plants, such as ivy, acacia, and eucalyptus, have distinct juvenile and mature phases, which are reflected in the characteristics of their leaves (see Fig. 2–18). These differences are normal. If desired, juvenile types of foliage can sometimes be maintained by certain practices.

Leaf Abscission. Some deciduous plants drop their leaves at the first indication of water stress. California buckeye often drops its leaves in early summer unless well supplied with water. These plants essentially go dormant until spring, with no apparent ill effects. Broadleaved evergreens vary in the season of year when the oldest leaves are shed. Southern magnolia, depending on plant and climate, may drop its oldest leaves at almost any season over a period of only a few to many weeks. Cork oak usually drops its oldest leaves just before or as new growth begins in the spring. Some evergreens may drop a few leaves all year. If a plant is stressed, leaf drop will be earlier and heavier than normal, and leaf color will be more intense.

Most deciduous plants shed their leaves in autumn. Leaf fall may be hastened by an early frost or if the plant is stressed. Some plants, however, such as beech, chestnut, and certain species of oak, retain many of their dead leaves through the winter (Tattar 1989). A few leaves drop with each wind and storm until spring, when the rest are shed and new growth begins. Some plants, however, both deciduous and evergreen, retain dead brown or black leaves into the winter because branches and leaves were killed quickly by a disease such as fire blight (*Erwinia amylovora*). Inspect buds and bark on such plants to ascertain whether they too are dead.

Bark Characteristics. Most young plants have fairly smooth bark, but as trunks and branches grow, most bark eventually peels or cracks. The bark of canoe birch, plane tree, and Russian olive flakes or peels. The bark of black walnut, black locust, and mulberry cracks and furrows as trees grow. Although it is entirely normal for such bark to crack or peel as a tree increases in size (Fig. 19–2), inexperienced observers may become concerned when they first notice this phenomenon.

Flowers, Fruit, and Seeds. Most woody plants produce flowers that set fruit with viable seeds; this is the primary way by which plants have reproduced themselves through the ages. Many landscape plants, particularly shrubs, are selected and grown for their attractive flowers and fruit. On the other hand, plants have

Figure 19-2 The peeling bark of red gum (top left), the dark rough areas of birch bark (top right), and the rough cracked bark of cork oak (bottom left) are normal as these trees mature. The bark of young liquidambar branches can be deeply furrowed (bottom right) and is particularly prized by arborists.

been selected and vegetatively propagated to produce showy blooms but set very little or no fruit. Some of the flowering fruit trees, such as the double and semidouble flowering cherry, peach, and plum, are examples. Some trees that have objectionable (messy or smelly) fruit are *dioecious*—that is, staminate and pistillate flowers occur on separate plants. The male trees of dioecious species are fruitless. When such trees are wind-pollinated, as is the Chinese pistache, they can aggravate allergies. Fruitless mulberry, ginkgo, and honey locust are dioecious trees often selected for their fruitlessness.

On the other hand, the absence of fruit may be a concern when plants are prized for their berries, as are holly plants. This absence may occur when male plants are not close enough to fertilize the female flowers. Even when perfect flowers or both male and female flowers occur on a single plant, the flowers may be incapable of setting fruit without cross-pollination. The female flower or flower part may not be receptive when the pollen is shed. These forms of unfruitfulness are normal for some plants.

Young plants may not flower for several years. This is most common on plants whose flowers grow on spurs. When young plants with a spur flowering habit respond to severe pruning and heavy nitrogen fertilization with extreme vigor, spur formation and flowering can be delayed for several years. Plants under low light conditions can also be slow to set fruit. Trees and shrubs may also develop an alternate bearing cycle, so that a year of heavy fruiting is followed by a year in which few flowers and fruit are produced. Pyracantha, crabapple, and apricot may develop alternate bearing cycles. The poor crop in the "off year" can worry an inexperienced grower. Of course, flowering and fruiting can be reduced or eliminated by unfavorable weather, nonparasitic disorders, diseases, insects, birds, rodents, and raccoons. It is important to distinguish between inherent and invasive causes.

A number of plants normally set fruit with no seeds or no viable seeds. Most edible figs, some grapes, and navel oranges may mature without seeds. Perfectly good seeds may fail to germinate because they have impervious seed coats or contain inhibitors that must be changed chemically or leached out.

Pollen shed by conifers, particularly pines and junipers, may be so abundant that it coats foliage with what may appear to be spores of disease organisms. Many people are bothered by the pollen, but it is harmless to plants and will soon disappear with wind or rain.

Galls. Insects, nematodes, and disease organisms can cause swellings or galls on the leaves, fruit, twigs, branches, trunks, or roots of plants. These may or may not impair the growth and appearance of a plant (see Chapters 21 and 22). On the other hand, swellings may be normal for certain plants or caused by beneficial fungi and bacteria.

Certain mycorrhizal fungi cause small roots to enlarge until they are lobed, beaded, or brain-like in shape. Mycorrhizae protect plants from root infection and help plants to absorb nutrients, particularly in infertile soil (see Chapter 6). The small roots of legumes, such as black locust and silk tree, may have nodules containing bacteria that fix nitrogen from the air into nutrient forms. This ability can be especially beneficial in infertile soils. A number of species, such as redwood, manzanita, and eucalyptus, may develop *burls* or *lignotubers* on their branches or trunks (see Fig. 3–10). These swollen masses of woody tissue are reservoirs of latent buds. A plant with well-developed lignotubers is able to sprout quickly if its top is injured. In contrast, no sprouts have been reported to grow from crown galls (caused by *Agrobacterium tumefaciens*) or the knots caused by nematodes.

Lichens, Mosses, Algae. Certain lower order plants grow harmlessly on the bark of woody species. Algae and mosses, both green plants, are often found in moist, shady areas on the bark of trees. Some mosses, such as Spanish moss, hang from branches in wooded areas with mild winters.

Lichens are composed of algae and fungi growing in symbiotic relation for their mutual benefit. They grow in a variety of colors on soil, rocks, bark, and other moist surfaces. Lichens are common on woody plants throughout the southern and southwestern United States but are less numerous in the northern states (Pirone

1978a). Since lichens are sensitive to air pollution, they are seldom found in or downwind from large cities or industrial complexes. The presence of lichens may indicate clean air (Tattar 1989) or poor growth of bark-shedding species (Pirone and others 1988).

An objection to their appearance is the only reason to remove algae, moss, or lichens from plant surfaces. For this purpose, a copper fungicide should be effective (Pirone and others 1988).

APPROACH TO DIAGNOSIS

Diagnosing the problems of landscape plants can be complex: The characteristics and requirements of these plants are more diverse than those associated with agricultural crops. Even so, landscape gardeners often attempt to grow plants at the limit of their tolerance. In a given area, however, you can become familiar with the plants commonly grown, their problems, and their responses to environmental and cultural practices. Talking with experienced growers in the area can be invaluable.

Not all disorders require an analytical approach. Many are easily identified and remedied by routine corrective measures. Plants that wilt during a prolonged drought likely need watering. Shoot dieback of pyracantha during a warm, wet spring is probably fire blight, caused by *Erwinia amylovora,* and infected shoots should be promptly removed. A preliminary diagnosis is often possible even without inspection of the plants involved, on the basis of plant species, season, and recent weather alone. It may be more difficult to diagnose other problems or to predict how a malady will affect a particular plant. In such cases, you will need a systematic approach for quick and accurate identification.

Plant problems can be classed or grouped according to several criteria: symptoms, causes of the problem, kinds of plants affected, or plant parts involved. Neely (1972) proposes that plant disorders be placed in one of three categories: (1) injury, (2) noninfectious (physiogenic) disorder, or (3) infectious (pathogenic) disease. Diseases and noninfectious disorders result from continued irritation or association with a causal agent and involve abnormal plant responses and observable symptoms (Table 19–1). The causal agents of infectious diseases are living organisms that become intimately associated with tissues of the host plant. Noninfectious disorders are caused by nonliving agents. An injury is the result of a single or discontinuous event that is detrimental to plant health. There are, of course, exceptions to this scheme. If, for example, the amount of a chemical involved or the speed of plant response is great enough, a problem may be considered an injury instead of a noninfectious disorder, analogous to acute or chronic toxicity in animals. A girdling wire might be classed as an injury but a girdling root as a noninfectious disorder, since the former is associated with a specific act while the latter may occur naturally. Both girdling problems might exist for a long period before symptoms appear. Similarly, moderate air pollution may cause a general weakening of plants, whereas a single exposure to higher levels of the same pollutants can cause serious leaf symptoms and dieback. Despite ambiguities, these three categories should assist in identifying specific problems.

TABLE 19-1 STRESS FACTORS THAT MAY CAUSE PLANT INJURY OR DISEASE[a]
(Adapted from Smith, W. H. 1970. *Tree Pathology: A Short Introduction.* San Diego: Academic Press)

Causes of injury		Causes of disease	
Abiotic	Biotic	Abiotic	Biotic
Temperature extremes	Insects		Bacteria
Wind	Mammals		Fungi
Snow	Birds		Mycoplasmas
Ice			Viruses
Lightning		Mineral deficiencies and excesses	Higher plants
Salt		Salt	Nematodes
Air pollutants		Air pollutants	
Moisture extremes		Moisture extremes	
Pesticides		Pesticides	
Radiation		Radiation	

[a] Certain abiotic factors may cause either injury or disease (noninfectious disorder), depending on the severity of the stress factor.

Plant Injury

Mechanical, chemical, or thermal injury may result from specific actions, chemical excesses, or temperature extremes.

Mechanical Injury. Mechanical injury is usually the easiest to diagnose. Leaves are tattered by wind or chewed by insects or animals; limbs are broken by wind, ice, snow, or high temperatures; bark is lost, pulverized, cut, or separated from the wood by rodents, insects, accidents, or vandals. Insects can injure plants by feeding or ovipositing on leaves, stems, trunks, or roots. They can chew, suck, mine, bore, or cause galls and dehydration of tissue. Live insects or their remains may often be present on the affected parts of the plant. Snails and slugs feed on leaves and immature shoots. They hide during the day but may betray their presence by a trail of dried slime.

Chemical Injury. A wide range of chemicals, including certain insecticides and fungicides, can injure or kill leaves, particularly if temperatures are above 25° C (77° F). Emulsifiable liquids cause more injury than wettable powders because the carrier solvents may be phytotoxic (Fred Roth, California State Polytechnic University, Pomona, 1989 pers. comm.).

The most obvious and frequently observed chemical injury of landscape plants is caused by careless or inadvertent application of herbicides (see Fig. 20–8). Leaf and shoot distortion, discoloration, or death may result from spray drift, spray vaporization, or root uptake of chemicals.

Ocean spray and salt used to reduce ice on pavement can cause leaves to develop necrotic spots and margins. Since brown leaf margins may develop quite gradually from increasing soil salinity, the problem would properly be classed as a noninfectious disorder.

Thermal Injury. Low temperatures injure plants more often than high temperatures. The season of the year and the stage of growth greatly influence a plant's ability to withstand cold. Low temperatures can produce a wide range of symptoms.

Some plants are injured when air temperatures are high. If the soil is dry, the root system inadequate, or the trunk restrictive to water movement, transpiration demand may exceed absorption and translocation of water from the soil to the leaves. In this situation, the plant becomes desiccated and leaves may wilt, become dry and brittle, or fall off. (Similar symptoms on grape may be produced by infection with the bacterium *Xylemella fastidiosum,* Pierce's disease [Fred Roth, Pomona, CA, 1989 pers. comm.].)

Fire is a frequent cause of high temperature injury. Plant parts exposed to fire may shrivel, blacken, and become misshapen.

Noninfectious Disorders

Most noninfectious disorders involve the entire plant and result from air pollution and soil and root problems. Soil and root problems usually uniformly affect leaf color and size and rate of growth. Proper assessment of shoot and trunk growth is discussed in Chapter 2. Nutrient deficiencies often result in reduced growth and leaf chlorosis.

Mineral excesses and chemical pollutants in the soil or air, like other noninfectious problems, generally have uniform effects on a plant, but iron deficiency symptoms may be confined to particular limbs (see Chapter 12).

Excess water, natural gas, compacted soil, and fill soil reduce the level of oxygen and in some cases increase carbon dioxide in the root zone. This reduces root activity and, if conditions are severe, can injure or kill the roots. Lack of water can result in leaf fall or necrosis of leaf tips or margins. Circling roots or wire can girdle the trunk of a plant, causing swelling above the constriction, limiting growth (particularly of roots), and causing a physical weakness that can eventually result in death. Plants grown in relatively small soil volumes may become "pot bound." The limited capacity of the rooting medium may combine with infrequent irrigation to deny the plant moisture and nutrients sufficient for normal growth.

If temperatures in winter are below 7°C (45°F) for an insufficient period, the growth and flowering of some temperate plants may be delayed the following spring (see Chapter 4).

Infectious Diseases

Bacteria, fungi, viruses, mycoplasmas (more correctly called *mycoplasma-like* organisms), nematodes, and parasitic flowering plants can result in unsightly appearance; poor growth; necrotic tissue of leaves, branches, and fruit; abnormal growth; and even death of an entire plant. Any part of the causal organism that can be seen on the outside of the infected plant is called a *sign:* Rust, powdery mildew, and mistletoe are examples. Other disease organisms may cause distinctive *symptoms* or abnormalities of the host plant: necrotic leaf spots, shoot dieback, cankers on branches, mottling or mosaic patterns on leaves, and wilting (Fig. 19–3). With study

Figure 19-3 Leaves are susceptible to many diseases: powdery mildew (*Oidium euonymijaponici*) on euonymus (upper left), anthracnose (*Gnomonia plantani*) on sycamore (upper right), and peach leaf curl (*Taphrina deformans*) (lower left). Insects, however, such as aphids (*Periphyllus*), can similarly deform leaves like these of green ash (lower right). (Sycamore photo: McCain 1979a)

and experience, you can readily identify distinctive symptoms of the more common diseases on plants in your area. When infectious diseases attack a mixed planting of woody plants: (1) Only one genus or species is likely to be affected; (2) only one plant in a group may be affected; (3) symptoms should be of the same type on corresponding plant parts; and (4) diseased plants will not succumb overnight (Neely 1972).

Several causal factors may be involved in a plant problem. Each may mask the typical symptoms of the others, and some may interact. There is mounting evidence that nonparasitic stress and injury can increase the disease susceptibility of plants, especially to canker diseases (Fred Roth, California State Polytechnic University, Pomona, 1989 pers. comm.). Diseased plants are certainly more vulnerable to environmental stress (Schoeneweiss 1973). Trees and shrubs subjected to drought or freezing are often prone to stem diseases, which are usually not a problem in similar unstressed plants. Plants weakened from any cause may be more vulnerable to insects, environmental stresses, and disease-producing organisms of low virulence.

A Diagnostic Approach

Most arborists use similar approaches for diagnosing the more difficult problems (Neely 1972, Pirone and others 1988). The following might be used as a diagnostic checklist:

Depending on the situation, obtain information about
  The history of the plant(s) and the area
  Cultural practices used on affected and nearby plants
  Any unusual weather conditions
Look at the plant(s) and surroundings for yourself
An owner or caretaker may withhold certain facts or give incorrect information—careful questioning and observation should reduce the likelihood of your being misled
If you are familiar with the area, you should be aware of
  Problem soil areas
  Previous weather conditions
  Recent air quality levels
  The more serious pest and disease problems
  The characteristics of the host species
Survey the general site and its plants
  How many plants are affected? If more than one, are they of the same genus and species?
  Are plants affected uniformly or on only one side or branch?
  Do they form any pattern? Are they clustered together or in a line? Are they grouped in relation to the topography or wind?
  Does the site appear to be unfavorable for the plants? Are there low areas where frost or drainage could be problems? Has there been recent construction, paving, excavation, or soil filling?
  Where are the gas, water, and sewer lines and the septic field? You may have to inquire about these features.
Examine the leaves and shoots or the dormant twigs
  Are the leaves and twigs normal in appearance and size?
  Has annual growth been normal for the last few years?
  Are dead leaves on some branches, live foliage or none on others? Dead leaves may indicate that the condition set in too quickly for leaf abscission to occur. Healthy plants of some deciduous species, however, may retain leaves into the winter even though they are brown and dry.
Check the trunk and main branches
  Does the bark appear normal? (Is the trunk sound?)
  Are there abnormal swellings, sunken areas, or signs of girdling?
  Does the trunk flair uniformly around the base as it enters the soil? If not, a girdling root may be a problem or the grade may be raised.
  Do vigorous suckers emerge from the trunk base? Do water sprouts emerge below the affected portion of the plant? The portion below such vigorous growth is probably healthy, and the cause is probably above ground.
  Is top growth generally weak, characterized by many short twigs, 250 to 300 mm (10–12 in.) long and small leaves growing along the upper surface of one or more main branches (Fig. 19–4)? Such symptoms indicate a severely weakened

Figure 19–4 A blue oak tree surrounded by asphalt and severely weakened from lack of water and oxygen has many weak spindly shoots on large branches and the trunk. Low auxin levels, caused by weak growth and increased light on the bark, release latent buds.

plant that has been in poor health for some time. The cause is probably a noninfectious agent, a slow-acting disease, or old age.

Determine root stability and health

    If the plant is not too large, try to move the trunk to and fro. If the trunk is not securely anchored, suspect a defective root system or a poorly prepared planting hole.

    If necessary, expose and inspect the main roots. Are roots girdling the trunk or each other? Are the roots solid? Is the bark firmly attached?

    Expose a small area of the trunk cambium below the soil level. Do the xylem and cambium appear normal? A fan-like white fungal growth between the bark and wood indicates shoestring root rot (*Armillaria mellea*).

Probe the soil under the plant with a soil tube or shovel

    Is the soil soggy, moist, or dry?

    Is moisture uniform with depth?

    Does the soil have a normal color and odor, or is it dark-gray with a smell like rotten eggs? The latter symptoms indicate extremely wet soil, a natural-gas leak, or landfill gas.

Take a closer look at nearby plants, particularly those of the same species

    Are similar symptoms on those plants? Could the condition be normal?

    Could these plants be in the beginning stages of the same difficulty?

    Look at other species of plants. Similar symptoms could indicate a noninfectious disorder or injury; such symptoms are less specific than those of infectious disorders.

You may want to check or obtain additional information from the owner or caretaker of the plants.

TABLE 19-2 TOOLS AND SUPPLIES RECOMMENDED FOR DIAGNOSING
AND EVALUATING PLANTS (Adapted from CTLA 1992)

To locate and describe plants		
Camera equipment[b]	Distance/height meters	Report forms[b]
Clipboard[b]	Engineer's scale[b]	Tape measure[b]
Compass[b]	Pocket calculator	Tape recorder

To collect and examine specimens		
Disinfectant	Labels[a]	Vials[b]
Dissecting microscope	Plastic bags[a]	

To diagnose plant problems		
Binoculars	Increment borer	Resistance or impedance meter
Chisel/gouge[a]	Mallet[a]	Shovel/spade[a]
Entrenching tool[a]	Pole pruner[a]	Small saw[a]
Hand lens[a]	Pruning shears[a]	Sturdy knife[a]
Ice pick[a]	Battery-operated drill	

To diagnose soil problems		
Gas detector	Soil auger/profile tube[a]	
pH meter[b]	Soil-moisture meter	

[a] Basic equipment

[b] Extremely helpful

Diagnostic Tools

Various tools will help in diagnosing the condition of a plant (Table 19–2). A sturdy pocketknife is probably the most useful and easily available piece of equipment. Even better is a folding linoleum knife, which is sturdy and versatile. With either, you can slice shoots and small branches to examine young wood and bark; you can cut into bark to check the condition of larger branches and the trunk; you can explore cankers, decayed areas, and insect holes to determine the extent of injury; when no other implement is available, you can even use a sturdy knife to remove soil next to the trunk for a look at the trunk root.

You can use a shovel or spade and an entrenching tool or sturdy trowel to uncover the trunk base and roots and to examine roots and the texture, compaction, moisture, color, and odor of soil. A soil tube is particularly useful for sampling soil below 250 mm (10 in.). You can use an auger instead, but soil changes with depth are easier to assess from a soil core than from a pulverized auger sample.

Use pruning shears to collect twig and branch samples and to cut them for internal inspection. A folding tree saw will conveniently remove limbs that are too thick for shears. You may need a pole pruner to remove high twigs and branches for closer observation. A pruner with a sectional pole can be easily carried and stored.

A folding 10-power hand lens may be helpful for detecting mites or inspecting plant tissue. With binoculars, you will be able to examine specimens that are out of reach.

You may need a wood chisel and hammer or mallet to cut the thick bark of large trees. You can use an ice pick to probe bark and wood for decay. Use any of these three tools to tap the bark if you wish to determine whether it is sound.

With a camera, you can record plant symptoms and surroundings for future reference or evidence. Plastic bags and vials will keep specimens separate and prevent them from drying out. Use labels, string tags, and rubber bands to identify specimens.

You may need measuring tapes to check shoot growth, trunk caliper, or other sizes and distances. Optic meters can measure heights and distances fairly accurately, even through or around obstructions and voids. A compass may be useful for determining direction and location when you prepare a plot plan of a planting.

In some cases, an increment borer is the only satisfactory way of sampling the xylem of a large tree or shrub to check its condition or to determine its growth rate (by width of annual rings) for a period of years (Fig. 19–5). It removes a small core

Figure 19–5   An increment borer is useful for measuring tree growth over a period of years. The width of annual rings of the upper core indicates it is from a more vigorous pine than the lower one (the beginning of each growth ring has been darkened; the diagonal marks are near the trunk centers).

that represents the cross-section of bark and wood. Take such a sample only when necessary, since the process is time-consuming and injures the trunk. You will usually make such a wound in a tree of low vigor, and it will consequently be slow to close.

Up to this point, the diagnosis has been made after a plant has been affected seriously enough to show injury, symptoms, or signs. Knowledge of the prior condition of a plant would be helpful in order to judge the severity of a disorder and its prognosis, or to estimate the ability of a plant to resist infection or tolerate stress. The passage of an electrical current through plant tissue is affected by the condition of the tissue. Many people have developed equipment and techniques to correlate the electrical conductance of a plant to its condition.

Electrical methods are suitable for determining the effects of stresses which cause physical damage, such as temperature, water status, infection, and wounding (Blaze 1990). Two factors related to plant vigor can be assessed by electrical methods: ionic concentration and membrane condition. Pulsed direct current (DC) or a single frequency alternating current (AC) can be used to measure ionic concentration of the tissue. However, unless extremely high or low, a reading should be interpreted relative to that of another plant of the same species growing under similar conditions. Water content, temperature, depth of insertion of the probe, and so forth complicate interpretation. Considerable care and experience are necessary in the interpretation of measurements taken with pulsed direct current or single frequency alternating current.

The magnitude of an alternating current passing through a plant tissue is called the *admittance* of the tissue. The admittance depends on the frequency of the alternating current and the status of the tissue (Blaze 1990). The admittance of cell membranes varies with the frequency—it is higher with high frequencies. However, the admittance of the cell walls of similar tissue of a given plant is the same at all frequencies. Thus, in healthy tissues (which have cell walls and cell membranes), the admittance at high frequencies is higher than that at low frequencies. If the high frequency reading is divided by the low frequency reading, the ratio (called the *impedance ratio*) will be greater than one because the admittance reading is greater at the higher than at the lower frequency.

If cell membranes are damaged (lose their semipermeability—become leaky), the ions contained within the cell membranes leak into the walls. Thus, the cell walls become more conductive, so that readings made at a single frequency will be increased. But, assuming there to be no intact membranes present, the readings at all frequencies would be the same. Thus, if the high frequency reading is divided by the low frequency reading, the impedance ratio will be one or nearly so.

Admittance readings (and pulsed DC readings) are affected by temperature, water content, depth of probe insertion in homogeneous tissue, and so forth; impedance ratios are not. Therefore, **an impedance ratio provides an estimate of the condition of the cell membranes of a tissue.**

The *Shigometer* (available at most arborist suppliers) uses a pulsed direct current to measure the resistance of tissue to that current. The *PIRM* (Plant Impedance Ratio Meter, available from Biosensors, P.O. Box 331, Ringwood, Victoria 3134, Australia) utilizes alternating current to measure admittance at two frequencies (1 and 10 kHz). Both instruments can use similar twisted-wire probes for determining xylem tissue condition with depth and double-needle probes for measuring vigor of tissue near the surface (i.e., phloem and cambium).

Tippett and Barclay (1987) compared the PIRM with the Shigometer for detecting bark lesions on *Eucalyptus marginata* caused by inoculations of *Phytophthora cinnamomi* and reported the following:

> The electrical admittance measured at 1 and 10 kHz (PIRM) and the electrical resistance (Shigometer) of healthy tissue varied with both the depth of tissue probed and the water status of stems. However, the impedance ratios (calculated from the admittance values) remained relatively constant for healthy tissue and changes were independent of depth of probing. Hence, changes in ratios indicated a change in tissue condition or necrosis rather than changes in either tissue water content or depth of probing. The impedance ratios recorded for healthy bark tissues were consistently higher than those for the *P. cinnamomi* lesions in *E. marginata*. Trends in electrical resistance measured across the boundaries of the lesion with the Shigometer were variable depending on lesion age. The PIRM was used successfully to detect *P. cinnamomi* lesions in *E. marginata* and lesion fronts were predicted to an accuracy of +7.2 mm (0.3 in.) (n = 150), lesions being up to 1.0 m (40 in.) long at the time stems were harvested.

An accuracy of +7.2 mm indicates that most of the probings were accurate within 5 or 10 mm (0.2–0.4 in.), since probing was at 5-mm intervals.

Experience with these instruments and their newer versions, improved techniques in using them, and learning where they can be used most effectively should improve diagnosis.

Collecting Samples

You may have to take samples of affected and healthy plant parts when you cannot diagnose the problem in the field or must verify a field observation. **Choose leaf, shoot, branch, or root samples that include the transition from diseased to healthy tissue, and collect them as soon as you notice symptoms,** particularly when an insect or disease is involved. In any case, gather both diseased and healthy specimens.

Keep leaf samples fresh and do not allow them to dry or to become too wet; otherwise, they may decay or foster secondary organisms that mask causes or symptoms. Keep leaves cool and seal them in plastic bags from which excess air has been pressed out. A dry paper towel will absorb excess moisture and reduce mold growth. Immature fruit left on leafy twigs will provide water to the twig and leaves and will thereby delay wilting.

Specimens of branches, trunks, and roots should include both bark and wood. Plastic bags are convenient for transporting specimens. If samples are to be mailed for analysis, however, pack them so that they are not damaged in transit. Keep small bark and wood samples from drying by placing them in plastic bags before packing.

Take soil samples from representative sites of diseased and healthy plants, or from problem and favorable soils. You may need samples from different depths (see Chapter 12).

Label all specimens with the name of the plant species, location of the plant(s),

and the part of th   ι was collected. Also include the
date of the sample and the name of the collector.

Sources of Help

You can employ many sources of information to help solve difficult problems or to
confirm a diagnosis. University and governmental agriculture, forestry, and conservation extension services publish useful bulletins on the diagnosis and treatment of
unhealthy landscape plants. You can phone or visit most extension services to dis-
cuss specific problems. Local agricultural commissioners, who must make sure that
only healthy plants are sold and transported, are particularly knowledgeable about
insect and disease problems. Retail nursery personnel are usually familiar with problems common in the area. Some park and urban forestry departments provide assist-
ance. The staff members of botanical gardens and arboreta can be quite helpful,
particularly in identifying plants and many of their pests.

Sales representatives for distributors of landscape plant supplies will often
provide advice on plant problems, particularly if some of their products can alleviate the malady. You may have to employ a commercial arborist who can provide
treatment as well as diagnosis or an arboricultural consultant for difficult or extensive problems. Be certain, however, that the people you consult are qualified to
diagnose landscape problems. A good indication of competence is membership in a
recognized professional organization, such as the International Society of Arboriculture (primarily in the United States and Canada), the National Arborist Association (United States), the American Society of Consulting Arborists (United States
and Canada), the Arboricultural Association (Great Britain), or other recognized
professional organizations.

FURTHER READING

See Further Reading section at the end of Chapter 23.

CHAPTER 20

Noninfectious Disorders

Noninfectious disorders often lead to poor growth, damaged appearance, or even the death of plants. Such problems are more common in landscape situations than in commercial nurseries or fruit orchards, where plants are usually given better care and where sites are particularly suited to the species grown. Landscape plants may be subject to unfavorable soil, climate, cultivation practices, traffic, or a number of other hazards.

PROBLEMS RELATED TO SOIL OR ROOTS

Most noninfectious disorders are related to plant roots or the soil. You can usually determine the ability of a soil to support the adequate growth of woody plants by observing plants that grow at the site or nearby on the same type of soil. An absence of plants or signs of poor growth may indicate poor soil conditions. You may need to provide drainage, mound the soil to increase depth, amend the soil, or replace it with a more favorable rooting medium.

An established site may feature mature trees and shrubs that have begun to decline. Soil-borne diseases or pests may be responsible, or physical changes may have affected soil aeration and moisture. Changes in grade, new paving, construction, trenching, changes in drainage or moisture supply, soil compaction, and gas leaks can all affect the ability of a soil to support plant life. Other soil or below-ground problems include mineral deficiencies and excesses, toxic chemicals, girdling roots, and rodent damage.

Poor physical structure of the soil will usually restrict root systems and reduce their activity. As a consequence, plants may receive inadequate water, air, and nutrients and may generally decline, may show symptoms of nitrogen deficiency, and may temporarily wilt on hot days. Short twigs with small leaves may grow on the upper surfaces of branches (see Fig. 19–4). It should be emphasized that such symptoms are not necessarily caused by the physical condition of the soil. The same symptoms may be induced when roots encounter disease, nematodes, or nutrient deficiencies.

Adverse changes in the soil may decrease aeration, thus reducing oxygen and increasing carbon dioxide and other gases, some of which may be toxic. Poor aeration reduces the ability of roots to absorb water and nutrients. Certain practices may increase or decrease water in the root zone. Excess water impairs aeration and can encourage root-rotting organisms to infect susceptible plants. Lack of soil moisture reduces growth, wilts leaves and young shoots, and in arid areas can produce a more saline soil. In a seeming paradox, plants may also wilt because of excess water in the soil; damaged roots may be unable to supply the top of the plant with adequate water.

Impervious Soil Coverings

Aeration and moisture in the rooting zone can be seriously decreased by pavement; fill soil; buildings; soil compaction; and frequent, shallow waterings of heavy soil. Pavement and other construction near plants can accentuate the effects of drought and high temperatures. Depending on the species and health of a plant, the texture and depth of the soil, the thickness or permeability of soil covering, and subsequent care, some portion of the soil surface under a tree or large shrub can be covered or compacted without adversely affecting the tree.

Mature trees of many species survive fairly well even when asphalt or cement covers the soil around the trunk to more than 50 meters (150 ft) in all directions (Fig. 20–1). On the other hand, a mature oak growing with moderate vigor in a clay soil may be weakened if less than one-third of the soil surface within the dripline of the tree is paved or if the entire area is covered with a thin layer of fill soil. In general, healthy plants withstand adverse soil changes better than plants of low vigor, and young plants do better than old ones.

Symptoms. When you suspect poor physical condition of the soil, examine the tree or shrub trunk at ground level. The trunks of most woody plants flare out just above or below the original ground level, and the soil around many trunks will be 50 to 100 mm (2–4 in.) higher than the surrounding soil surface. If a trunk enters the ground with little or no flare and no raised soil surface, it is very likely that soil has been added to the planting area. If fill soil is coarser than the original soil, it should cause few or no problems unless it is compacted, unusually deep, or placed over an uneven or compacted surface. Fill soil of finer texture than the original will not only restrict air and water movement but will also form a perched water table above the soil interface, further impeding air movement into the root zone.

Impervious coverings (pavement, buildings, compacted surface soil) can

Figure 20-1 This mature New Zealand Christmas tree does well in Wellington, N.Z., city center even though the soil has been covered for at least 50 m (150 ft) in all directions.

greatly impede water and air movement into the soil below. Lateral movement through the soil is extremely slow without a positive flow of water from a higher elevation. Unless they are graded to drain in another direction, impervious surfaces can concentrate excess water around plants (Fig. 20-2).

Although compaction can seriously impair water infiltration and aeration in most soils, several studies have shown that compaction of some forest soils may be beneficial. The compaction of shallow, friable soil in forest campgrounds provides greater root firmness and improves the stability of trees and shrubs (Lull 1959, Magill 1970). It may also improve growth by allowing the soil to retain greater supplies of moisture and nutrients because microflora are reduced in compacted soil.

Recommendations. You can protect plants from impervious soil coverings by installing drainage (see Chapter 8). You can also take less expensive steps to increase aeration and water movement: You can bore holes into tight soil with an auger or a water jet. To remain effective, these holes must be open to the surface. Boring new holes every two or three years may be more effective and less troublesome than filling the original holes with gravel or other coarse material to slow the accumulation of soil and debris. Even when gravel is present, soil and roots can soon fill holes. You may help a plant adjust to improved soil conditions by lightly thinning the top, applying a moderate amount of nitrogen (1 kg/100 m² or 2 lbs/1000 ft²), and taking care to water in proper amounts. On the other hand, young or newly planted trees should adjust to confined and paved areas with little harm if paving and buildings do not funnel water to the planting basins or increase reflected heat unduly. Such trees may of course grow more slowly than they would in less confined circumstances.

If possible, keep soil away from the trunks of trees and shrubs. Soil covering the bark keeps it moist and makes the plant vulnerable to various collar-rot fungi. It may also seriously reduce aeration. Some tolerant species will put out new roots from portions of the trunk that are covered by soil (see Chapter 7). Place a tree well

Problems Related to Soil or Roots

Figure 20–2 The soil level has been maintained near these two mature live oak trees, but the soil has been mounded on either side (upper left). The turf was sprinkler irrigated. Within two years, the tree in the foreground was doing poorly (upper right) and soon died. Two years later, the remaining tree began to weaken (lower left); however, when irrigation was severely reduced during the droughts of 1975–1977, the tree made a spectacular recovery (lower right). (Photos courtesy Robert Raabe, University of California, Berkeley)

around the trunk to keep fill soil away; see that the soil surface in the well slopes away from the trunk, that drainage is adequate, and that the soil next to the trunk is not kept unduly moist (Harris and Davis 1976).

Impaired Drainage

Symptoms. Impaired soil drainage can produce symptoms similar to those caused by impervious soil coverings. As plants mature or lose vigor, they become more sensitive to excess soil moisture. Construction on slopes or in canyons can reduce the flow of water in the soil so that, at least at certain times, water accumulates closer to the surface on the uphill side of the construction site. Soil moisture may increase more quickly than it can drain away, especially if rainfall is higher than normal or unaccustomed sources (a leaking water or sewer pipe, a newly irrigated landscape, drainage from buildings and pavement) contribute. Without proper drainage and aeration, plants may suffer root suffocation or root rot (Fig. 20–2).

Recommendations. To improve plant health, identify the cause of the impaired drainage and correct it if economically feasible. Check soil moisture with an auger or probe. If the soil is too wet, block incoming water or improve surface drainage. It may be necessary to provide internal drainage (see Chapter 8).

Gas Injury

Plants can be injured and killed by gases in the soil; such gases may be directly toxic or may reduce oxygen to critically low levels. Manufactured gas derived primarily from coal and oil and gas produced by the decomposition of organic matter in landfills are directly toxic to plants at certain concentrations. Although natural gas may not be directly toxic to tree roots (Pirone and others 1988), a wide range of plant species have been killed or severely weakened by gas leaks which displace the oxygen in the soil.

Manufactured gas contains a number of toxic compounds, including carbon monoxide, hydrogen cyanide, and unsaturated hydrocarbons (such as ethylene and acetylene) (Hitchcock, Crocker, and Zimmerman 1932). Manufactured gas has been less commonly used in the United States since the 1940s, when natural gas was piped to many northern regions from the southern and southwestern states. As supplies of natural gas decrease, however, manufactured gas may become more common again.

The decomposition of organic matter, primarily wood and paper, in landfills produces methane and carbon dioxide (Leone and others 1977); the latter can reach levels that are toxic to roots. A leak of natural gas (saturated hydrocarbons, including methane) can displace much of the soil atmosphere, reducing the oxygen level. Bacteria in the soil consume methane and oxygen, producing carbon dioxide and water (Hocks 1972). This further reduces the oxygen and increases carbon dioxide levels. Garner (1974) has also found that under the anaerobic conditions caused by natural gas leaks, bacteria will transform sulfates in the soil to hydrogen sulfide. Nitrous oxides also form under these conditions. Both substances are toxic to plants.

Symptoms. Since most gas leaks or landfill gases build up slowly, symptoms usually develop over several months or years and may be similar to those caused by poor soil aeration, excess moisture in the soil, and certain soil-borne fungus diseases. Therefore, diagnosis may be difficult. The timing and severity of symptoms will depend on soil permeability, plant species, and the rate at which gas has evolved. If all plants in the area are affected, this is a good indication that an infectious disease is not involved. There have been cases, however, in which trees were killed at some distance from a gas leak while intervening plants were apparently unaffected (Davis 1971). In such a case, a compacted soil layer may prevent the gas from escaping near its source and cause it to diffuse beneath the impervious layer until it escapes to the surface or dissipates in the soil. Under pavement, frozen soil, or a compacted soil layer, gas can travel 50 m (150 ft) or more. Gas has been found to damage plants up to 30 m (90 ft) from the edge of landfills (Leone and others 1977).

Gas-damaged plants grow poorly, with sparse shoots and small leaves. Leaves and shoots wilt and turn brown, and the most seriously affected branches die. Roots develop a bluish, water-soaked appearance and are also killed. Many gaseous soils become blue-gray or black, similar in color to poorly drained swamp mud; these soils often have the "rotten egg" smell of hydrogen sulfide. The soil atmosphere can be tested for combustible gases by trained diagnosticians or employees of a gas utility. The preferred instrument for gas analysis uses the catalytic combustion principle (Leone and others 1977). Instruments used by many gas companies work on a thermal conductivity principle that is calibrated for methane alone. A landfill gas mixture contains primarily air, methane, and carbon dioxide. Meters that work on the thermal conductivity principle may give inaccurate results for methane from landfills because high concentrations of carbon dioxide also exist. Catalytic combustion gas meters are available from the Mine Safety Appliance Company and the Bacharach Instrument Company, both of Pittsburgh, Pennsylvania.

Other methods of detection might involve the measurement of the oxygen and carbon dioxide in a problem area (Leone and others 1977). Uneven settling and surface cracks in the soil may be clues that organic material is below the surface (Table 20–1). Gas injury is likely to become an increasingly common problem as gas lines age and corrode, manufactured gas is used in greater amounts, and improperly installed landfill barriers begin to leak.

Recommendations.   In a gaseous planting site, have the gas leak repaired, remove the weak and dead plants, and replace all discolored or seriously affected soil with uncontaminated soil. Then replant. On a landfill site, you may need an underground system to collect toxic gases (see Chapter 11). If large plants appear to have a chance of survival, remove the most heavily contaminated soil; the roots in that soil are probably dead anyway. Aerate the remaining root zone with an

TABLE 20-1 GUIDE TO EVALUATING GAS PROBLEMS IN LANDFILL SOIL
(Leone and others 1977)

Characteristic	Anaerobic soil	Aerobic (healthy) soil
Odor	Septic	Pleasant
Color	Dark	Light
Moisture content	High	Low
Friability	Poor	Good
Temperature	High	Low
Combustible gas levels	High	Very low or zero
Oxygen levels	Low	High
Carbon dioxide levels	High	Low

auger, water jet, or aerator and replace the contaminated soil with a good planting medium. Fertilize moderately with nitrogen (10 g/m^2 or 2 lbs/1000 ft^2). Lightly thin the tops of the plants and water wisely.

Loss of Roots

Roots may be lost directly when the soil surface near trees and large shrubs is lowered, when trenches are dug for utility pipes, or when storms cause breakage.

Symptoms. Plants so affected will be more subject to drought, will show symptoms of nitrogen deficiency (poor growth and light green foliage), or will be unable to stand in a strong wind. Unless the grade has been lowered too drastically, plants will adapt to new moisture and nutrition regimes after initial shock. A storm-damaged root system, however, may require considerable care while new roots become established.

In an area exposed to the sun and sky, removal of surface soil will force the tree to root into deeper soil, which is usually less fertile and not as well-aerated as the topsoil. If a mulch or low-growing plants are removed, the remaining unprotected soil will be subjected to drying and possibly extreme temperatures. The plant may be blown down in a wind if many large surface roots are cut. Lowering the soil level will have less severe effects in a deep, well-drained soil of a sandy texture than in a shallow, poorly drained clay. Vigorous, moderate-size young trees of species that normally root deeply should be able to withstand soil removal better than shallow-rooted large trees of low vigor. The less soil removed, the smaller will be the effect on a tree. Several cultural practices will materially enhance the survival and growth of plants from which soil has been removed.

Recommendations. Remove as little soil as possible and avoid injury to large roots. A raised planter on a level area or a terrace on a slope can retain the original soil surface around plants (see Figs. 7–10, 7–12, and 7–13). In some cases, large roots can be exposed for a distance of 1 to 3 m (3–10 ft) from the trunk provided that they then continue below the lowered ground level. Exposed roots can add visual interest to the landscape. Newly exposed roots must be protected from direct sunlight and freezing, however, until they have had time to acclimate. Protect the bark of exposed roots from foot and vehicular traffic.

You can usually remove one or more large surface roots without serious damage if vertical or other horizontal roots grow below them on the same side of the plant and if the tree is not generally shallow rooted. Each situation, however, must be appraised individually. Cut the roots smoothly and lightly thin the top of the plant to reduce the weight and wind resistance. No treatment of wounds has been shown to be beneficial in preventing decay. When a plant has had more than 20 percent of its root system disturbed, water it more frequently and apply nitrogen moderately (1 kg/100 m^2 or 2 lbs/1000 ft^2) on undisturbed or fill soil areas. To minimize injury to shallow feeder roots, apply only half as much nitrogen to excavated areas. After the soil level is lowered, protect plants further by placing a 50 to 100 mm (2–4 in.) organic mulch over the newly created surface. The temperatures of the surface and roots will thus be moderated, and moisture conserved.

Problems Related to Soil or Roots

Drought

Plants or portions of plants can become deficient in moisture for a variety of reasons: lack of soil moisture; poorly aerated or frozen soil; injured or diseased roots, which reduce water uptake; and injured or diseased trunk and branches, which restrict water movement. Sometimes, an excessively large leaf area will develop in cool spring weather; if the longer days of later spring are accompanied by high temperatures and wind, the root system may not be able to supply the top with enough water to prevent wilting or leaf scorch.

Symptoms. The first symptom of inadequate moisture in the top of a plant is a wilting of the leaves and the tips of young shoots. These symptoms will commonly appear on healthy plants during warm afternoons in late spring and early summer. If the leaves and young shoots are not turgid again by the following morning, however, lack of moisture could seriously affect the plant. Should wilting persist or recur frequently, the areas between veins and on the margins of leaves may fade and turn brown; entire leaves may turn brown and fall. Species most susceptible to leaf scorch are dogwood, many maples, horse chestnut, ash, elm, and beech (Pirone and others 1988). Lack of moisture usually affects all the leaves on one or more branches. The leaves affected first and most severely are those exposed to the afternoon sun and prevailing winds. Older leaves, leaves that are small, thick, and rigid, and most conifer leaves may not wilt visibly but may turn brown entirely or just at the tips or margins.

A number of broadleaved evergreen shrubs and conifers can withstand considerable periods of moisture stress without becoming unattractive (Sachs, Kretchun, and Mock 1975). In two trials in California coastal valleys, little or no leaf injury occurred even when transpiration was sufficiently reduced to produce leaf temperatures 6° to 15°C (11°–27°F) above ambient. The exposed leaves of plants with adequate soil moisture are usually at or slightly below ambient air temperature. In the California experiments, tree and shrub growth was markedly reduced; this can be an advantage when plants near their desired landscape size.

Relatively few trees and large shrubs were lost in California during the 1975–1977 drought, even though annual rainfall was below 250 mm (10 in.). Some surviving plants had been irrigated and some unirrigated before the drought; most could not be irrigated when water was scarce. Most trees that were lost had shallow root systems or were growing on shallow soil. Whether irrigated infrequently or not at all, trees and large shrubs growing in Mediterranean climates (with essentially dry growing seasons) can survive without rain or irrigation for relatively long periods, as contrasted to similar plants that grow in areas where summer rainfall occurs. The latter plants may have less extensive root systems than plants accustomed to arid seasons. It may be that some plants can grow in soils of limited moisture-holding capacity if they receive summer rainfall but are seriously affected when moderate drought occurs. Plants may appear to survive a drought in good shape only to show stress or succumb to a "secondary" malady two or three years later; this happened to many trees after the 1975–1977 California drought. Similar responses were reported by Holmes (1986) in Massachusetts.

Recommendations. If plants cannot survive in good health and appearance on their own during dry periods, one or two good waterings during the growing season will usually keep them attractive. The number of irrigations and the amount of water required will depend on the water-holding capacity of the soil and the rooting depth of the plants. One good irrigation that brings most of the rooting zone to field capacity would be better than the same volume of water applied more frequently in smaller amounts. Observations indicate that in areas where dry summers are normal, most woody plants with extensive root systems can survive on only half the water they would otherwise transpire (Harris and Coppock 1977) (see Drought in Chapter 13).

Girdling Roots

Trees and shrubs are weakened and can be killed by roots that girdle the trunk and main roots (Fig. 20–3). Girdling roots can start in the nursery (see Fig. 3–5), at planting in the landscape, or years later.

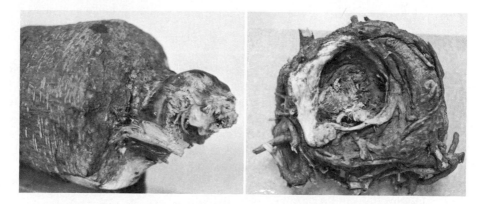

Figure 20–3 The Monterey pine trunk (left) and roots (right) exemplify the fate of more than 75 trees on a northern California golf course; the trunk diameter at the girdling root is less than half that immediately above. Circling roots were not corrected when the tree was planted.

Nursery practices, unless properly performed, can force roots to circle, leading to poor growth and inadequate root support in the landscape (see Chapter 3). Trees that are started in seed flats and transferred to liner pots are particularly vulnerable. By the time trees are ready to be planted in the landscape, little can be done to correct circling roots that began during these early moves. Tap-rooted and fast-growing trees are most often affected.

Girdling roots are easily seen on bare root plants but not on container or balled-and-burlapped (B & B) plants. The outer circling roots of container-grown plants can usually be corrected at planting, but seldom those inside the root ball of container and field-grown B & B and mechanically transplanted trees. The constrict-

Figure 20–4 Not all circling roots are present or appressed to the trunk when planted; they form or become appressed later. It is unlikely that the surface root on the mature oak tree was present when the tree was planted (left). The roots circling the mature Norway maple appear to have been laterals close to the trunk on primary roots that were stimulated when roots were cut when the trees were transplanted (right) (Watson, Clark, and Johnson 1990). (Photo courtesy Gary Watson, The Morton Arboretum)

ing effects of these girdling roots are seldom evident until trees have been in the landscape three to 15 years (Fig. 20–3).

Watson, Clark, and Johnson (1990) exposed the root crowns of Norway maple, sugar maple, red maple, green ash, honeylocust and littleleaf linden three to 10 years after planting. Girdling roots were common on young trees of the three maple species, but not on the other three. Formation of girdling roots was associated with transplanting and probably resulted from the stimulation of existing and new lateral roots when the main roots were cut as the root ball was dug. The lateral roots grew at about right angles to its primary root; some were close to the trunk. In mature trees, girdling roots were present in great numbers and caused damage on Norway maples but not on red nor sugar maples.

In the landscape trees can develop girdling roots. North of New York City, d'Ambrosio (1990) found that 93 percent of 832 30-year-old roadside Norway and sugar maple trees had girdling roots. He determined that improper planting and excess soil around their trunks caused the girdling in 52 percent of these trees. The trees with girdling roots for which there was no obvious cause had lower crowns, lower branches, more recent pruning scars on the trunks, and wider spreading branches than the trees without girdling roots. Each of these characteristics indicates a wetter and more shaded trunk at ground level in prior years, which would favor adventitious root formation in many species.

Most of the girdling roots on the 775 affected trees in d'Ambrosio's (1990) study were initiated sometime after planting and were tightly appressed to the trunk (Fig. 20–4). Roots usually grow radially away from a trunk, but not these. One explanation is that the soil around a trunk becomes compacted as the tree grows, but cracks between the soil and trunk occur due to soil expansion and contraction with changing soil moisture and temperature. An emerging adventitious root might easily be deflected to grow around the trunk in the space between the soil and trunk instead of growing radially outward through compacted soil. A well-shaded trunk and soil provide favorable conditions for root growth in the crack between trunk and soil. A tree with a higher crown and a trunk free of branches at a younger age creates less favorable conditions for adventitious root formation and growth (possibly due to drier soil and higher light at the trunk-soil interface).

Pirone and others (1988) identify Norway and sugar maples, oaks, elms, and pines as particularly subject to girdling roots. d'Ambrosio (1990) found girdling roots to be equally serious on Norway and sugar maples. In Illinois, on the other hand, even though young red and sugar maple trees had as many or more girdling roots as Norway maples, they had very few girdling roots when mature (Watson, Clark, and Johnson 1990).

Trees growing in confined spaces are thought to be more subject to girdling roots. Roots encountering a barrier might grow back toward the trunk or large basal roots and create a girdling problem. However, Tate (1981) in Michigan found no association between girdling roots of Norway maple and narrow tree lawns.

Symptoms. When a root circles a trunk or another root and they both continue to enlarge, each will be deformed at the point of contact, particularly when young (Fig. 20–5). Each will slow the caliper growth of the other, but the trunk tissue will be more compressed than root tissue (Hudler and Beale 1981).

Initially, the movement of food from the top to the roots is restricted and root growth is slowed, then the entire plant is weakened, and if severe enough the trunk will break at the girdle.

The symptoms of a weakened, ineffective root system have been described earlier. Except in newly planted plants, the symptoms of girdling roots may take

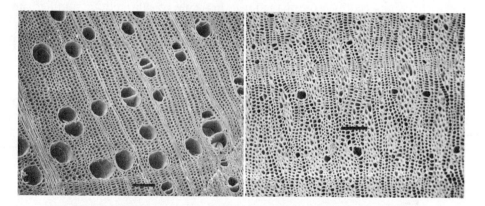

Figure 20–5 Transverse sections of Norway maple stem: normal xylem (left); malformed xylem from stem portion compressed by a girdling root (right). Scale of each bar is 100 μm. (Electron photomicrographs courtesy George W. Hudler and M. A. Beale, Cornell University, 1981)

years to develop. A gradual decline in growth and appearance occurs as the girdling becomes more intense. It is most apparent in autumn, when leaves color and fall earlier than is normal.

If you suspect girdling roots, inspect the trunk where it enters the ground. A normal flare at the ground will usually rule out girdling roots. If the trunk enters the soil vertically or you notice an indentation just below the ground around part or almost all of the trunk circumference, a girdling root or roots are likely. If the entire trunk circumference enters the soil vertically, the problem is more likely due to soil fill. If a trunk is completely encircled by a girdling root or roots, it will be larger in caliper just above the girdle than below. Do not be misled by a swelling of the trunk above a graft union, for that occurs often. Graft swelling may indicate a mild incompatibility between the rootstock and top but is seldom a problem.

Girdling roots are often visible at the soil surface. If they are not, remove soil carefully from around the trunk until you find girdling roots or encounter the main horizontal roots. If unencumbered roots radiate out from the trunk, girdling roots are not a problem.

Recommendations. Prevention is well worth the effort. Have good tree specifications. Inspect tree roots and reject any with serious kinking and circling (see Chapter 3). Planting must be done carefully to avoid circling roots. Avoid compacting and glazing the sides of the planting hole. The roots may need to be shortened and straightened to keep them from beginning to circle. Be sure the tree will not be too deep when the soil has finally settled after planting. Follow the planting procedures discussed in Chapter 9.

If poor nursery or planting practices have caused girdling of the trunk, there may be little that can be done to save a tree after it has been in the landscape a few years. It is usually best to start over.

For species likely to form girdling roots, initiation of adventitious roots may be minimized by pruning the lower branches so that midday summer sun strikes the trunk base and by keeping soil from being raised around the trunk base. However, the lower trunk of some trees may need to be protected from the sun, particularly if the tree is of low vigor.

On field-grown B & B and mechanically planted species that may develop serious root-girdling problems, it may be wise to examine the root collar of a few trees four to six years after transplanting (Watson, Clark, and Johnson 1990). With the species they examined, few girdling roots formed after the trees had been in the ground six years. At this age the roots can still be fairly easily removed. Careful removal of 150 to 200 mm (6–8 in.) of soil is usually enough to examine the roots if the tree was planted at the correct depth. If girdling roots are found to be a problem, other trees in the planting should be examined and root problems corrected.

Except for girdling roots in the trunk-surface zone and those that started circling in containers, girdling roots found on trees in the landscape can probably be removed to the tree's advantage. Take care not to remove so many roots that the tree will become stressed.

On larger trees, if the root is one-fourth the diameter of the trunk or larger, check to see if the root and trunk are grafted together. If they are grafted, it may be best to leave them; if not, d'Ambrosio (1990) suggests cutting a V notch halfway through the root midway along the girdling portion. The cut will weaken the root but still let it supply the top. Eventually, the expanding trunk will break the root and it can then be removed.

Smaller girdling roots can be removed most easily with a chisel and mallet. Cut each at its attachment to the trunk or to a large root and again beyond the girdled area. Do not remove the cut portion if grafted. Painting the cut root surface is of little or no value. When all of the girdling roots have been cut, use the soil removed to fill the excavation back to the original soil level. If the tree has been weakened by the girdling, apply about 10 g of nitrogen per m^2 (2 lbs/1000 ft^2), and water wisely. The girdling may be so severe that the girdled trunk cannot support the tree in a storm (Fig. 20-3). Besides thinning the top to reduce wind resistance, you may have to guy the tree (see Chapter 10). The tree may have to be removed if it is so severely weakened it will grow poorly or constitute a hazard.

Leaf Scorch

Symptoms. A number of conditions can cause the death of scattered areas between the veins or at the tip and along the margins of leaves (Fig. 20-6). Affected areas become light or dark brown and appear scorched. In severe situations, the entire leaf dries up. The primary causes are drought, a marked increase in light intensity, high temperatures, frozen soil, or drying winds, which either decrease the amount of water available or increase transpiration. Potassium deficiency or excesses of boron, chloride, or sodium can also result in leaf scorch.

Drought and increased transpiration were discussed earlier in this chapter. Light intensity or temperatures can increase markedly when a plant is moved from a cool, shady location to one of more intense sunlight, when the top of a plant is pruned, when a structure or another plant is removed, or when the sun nears the summer solstice.

Recommendations. Nothing can be done to correct the scorched leaves. You may keep the condition from becoming worse by ensuring that the plant has sufficient moisture and is not subjected to a hot, drying environment. To minimize future leaf scorch, encourage root growth by wise fertilizing and watering. If the plants are irrigated, be sure water reaches most of the roots.

Marginal or tip burn on leaves may be due to a mineral imbalance. Potassium deficiency can cause yellowing of leaves, followed by scorching of the margins or entire leaves. See recommendations for potassium application in Chapter 12. Particularly in areas of low rainfall, excesses of boron, sodium, and chloride can cause leaf scorch and reduced growth. Soil salinity and its correction are primarily irrigation problems (see Chapter 13). Excess boron is somewhat similar to excess salinity, in that the most feasible solution is leaching with good quality water. Successful leaching depends on adequate drainage.

Figure 20–6 Conditions that cause leaf scorch cannot be easily distinguished by symptoms alone: dessication (top) of Japanese maple (left) and of holly; salinity (center) on tulip tree (left) and hibiscus; and excess boron (bottom) on southern magnolia (left) and coast redwood. (Salinity photos courtesy Leland François, U.S. Salinity Laboratory, Riverside, CA)

Leaf Chlorosis

The general yellowing of a leaf, often referred to as *chlorosis,* can be caused by a variety of factors and is often associated with poor growth. Chlorosis may be caused by disease or insects, excess soil moisture, cold weather, air or soil pollutants, high levels of certain minerals in the soil, or nutrient deficiencies (Pirone and others 1988). Camphor trees, liquidambar, and pin oak are particularly sensitive to iron deficiency (see Chapters 6 and 12).

Animal Damage

In snowy areas, the roots of young plants and the trunks of large plants may be injured by pocket gophers, moles, and mice. Gophers can also feed on tree roots from their underground burrows. For most effective control, place Macabee traps in the main burrow runs. For large areas infested with gophers or for landscape plantings that border upon open fields and farmlands, underground tunnels similar to gopher burrows may be formed with special equipment and poison bait placed at intervals (Marsh and Cummings 1976). This effective control is commonly used in California in soil that is free of tree and shrub roots.

Moles are less serious threats than gophers but they disfigure lawns and ground surfaces with their burrows, heave up small seedlings, and eat or break small roots. Moles are also more difficult to control than gophers. Traps (choker, harpoon, or scissor-jaw) and poison baits are used in control attempts.

When a coarse organic mulch gives mice an opportunity to tunnel to the trunks of plants, they may girdle the bark around trees and shrubs, particularly when snow restricts their access to other sources of food. Therefore, mulch must be kept away from trunks of plants. You may place a ring of gravel about 200 mm (8 in.) wide and thick around each trunk, or use a sleeve of hardware cloth or fine wire mesh and bury the lower portion of the sleeve around the base of each plant. Either barrier should keep mice and other rodents from burrowing to trunks.

Squirrels, porcupine, beaver, cattle, deer, elk, horses, and other grazing animals can also severely injure bark of trees and shrubs. You can fence trees individually or in groups to keep larger animals away. A hardware cloth sleeve around the trunk will protect a plant from small and large animals. Dogs can injure the bark of young trees by urinating on it. A plastic sleeve around the lower trunk will protect a tree from such injury. Treeshelters will not only protect trunks of trees from dogs and other above-ground animals, but also have other benefits for young trees (see Chapter 9). If dog urination is excessive, tree roots may still be injured with either a sleeve or treeshelter. A repellent may be needed. The trunk should be protected from the sun when the sleeve is removed.

Sapsuckers and woodpeckers disfigure the trunks of many trees when they search for insects. Woodpeckers normally work only on decayed and dead wood; the only harm they cause is visual. Sapsuckers may damage live bark, even girdling small treetrunks. The most practical control appears to be the removal of nesting places in trees that have cavities or internal decay (Tattar 1989).

WEATHER-INDUCED PROBLEMS

Temperature and moisture are the environmental factors most influential in determining the distribution and well-being of plants. Irrigation can modify the influence of rainfall and evapotranspiration. Plants grown near their climatic limits are usually more subject to the vagaries of weather, particularly if temperatures fluctuate markedly near critical levels. Chapter 4 discusses the influence of climate on plants, the types of temperature-related plant injuries, and the ways many weather-induced problems can be avoided or minimized.

AIR QUALITY

Pollution

Symptoms. Leaves are the plant parts most likely to show symptoms of air pollution injury. On broadleaved plants, leaves may develop interveinal necrotic areas, marginal or tip necrosis, stippling of the upper surface, or silvering of the lower surface (Table 20–2). Conifer needles affected by pollution may exhibit patterns of necrosis and chlorosis. These are symptoms of acute toxicity, since they usually result from a short exposure to high concentrations of a gaseous pollutant.

Symptoms of acute injury can be highly variable, depending on the plant species and stage of growth, the type and concentration of pollutants, the length of exposure, the amount of moisture in the leaves, humidity, light, temperature, wind, and other factors (Scott 1973) (Fig. 20–7). Different leaf tissues respond with different sensitivities to the gaseous pollutants that enter the leaves through the stomates (Table 20–2).

Long periods of exposure to low levels of pollution result in chronic injury, which may be hard to distinguish from other forms of poor growth. Affected plants will be low in vigor and fruit yield, leaves will be pale green, and leaves and fruit

TABLE 20-2 MAJOR AIR POLLUTANTS: COMMON SOURCES, CONCENTRATIONS TOXIC TO SENSITIVE PLANTS, AND PLANT SYMPTOMS (After Scott 1973, USDA 1973a; Sinclair, Lyon, and Johnson 1987)

	Sulfur dioxide	Fluoride	Ozone
Source	*Point* (local)	*Point* (local)	
	Atmospheric	*Atmospheric*	*Atmospheric*
	Power plants and industries that burn coal and high sulfur oil or use compounds in processing	Industries that use catalysts and fluxes containing fluorides	Combustion of coal and oil; auto exhausts are the primary source
Toxic levels	0.7 ppm for 1 hour 0.18 ppm for 8 hours	1 ppb for 24 hours	0.05 ppm for 4 hours
Symptoms Broadleaved plants	Bleached or brownish interveinal necrotic areas on both leaf surfaces	Marginal or tip chlorosis or necrosis	Flecking or pigmented stippling on upper leaf surface
Conifers	Needle tip necrosis and chlorotic banding	Nonspecific needle tip necrosis and chlorosis	Needle tip necrosis, tip chlorosis, chlorotic banding and mottling

Figure 20-7 Right birch leaf exhibits typical interveinal loss of color followed by necrosis after exposure to sulfur dioxide (left). (Photo courtesy Edward Lee, USDA Beltsville Agriculture Research Center) Ozone causes mottle of ponderosa pine needles (right).

will color early and more brightly and will drop earlier. Affected plants are usually more easily subject to other disorders. Chronic injury from gaseous or particulate air pollutants occurs when pollutant concentrations are high enough to interfere with or change certain plant processes.

Airborne chemicals have been known for some 150 years to injure and kill plants. The earliest reports were of sulfur dioxide injury from coal-burning factories in Europe (Scott 1973). The major phytotoxic air pollutants are the gaseous compounds: sulfur dioxide, the fluorides (HF, SiF_4), and ozone. Ammonia, the chlorides (HCl, Cl_2), ethylene, hydrogen sulfide, nitrogen oxides, and peroxyacetyl nitrates (PAN) are also toxic to plants but are not as widespread or as frequently encountered at toxic levels. Carbon monoxide, a major product of gasoline combustion, does not cause plant injury in ambient concentrations (USDA 1973a).

Sulfur dioxide and the fluorides are generally considered point-source pollutants, since injury usually occurs within 15 km (10 mi) of the source (Scott 1973). Sulfur dioxide is produced by the burning of coal, refining of petroleum and natural gas, and smelting and refining of ores. Fluorides are released when fluoride-containing materials are heated or chemically treated in the production of steel, aluminum, ceramics, glass, and phosphate fertilizers. Concentrations of these pollutants have been reduced or at least contained by the use of specialized filters (which reduce toxic emissions) and tall chimneys (which allow greater dilution before susceptible plants are reached). Sulfur dioxide is now more an atmospheric pollutant. It and nitrous oxide are acid-forming pollutants from urban and industrial areas that can be carried great distances. They acidify rain and are blamed for the decline of conifers in the northeastern United States, Canada, and Europe. Although quite

Air Quality

toxic to fish, acid rain may be only one of many contributing agents to the decline of conifers.

Ozone and PAN are oxidants formed in the atmosphere by photochemical reactions between hydrocarbons and nitrogen oxides in the presence of sunlight. Photochemical oxidants can cause plant injury as far as 120 to 200 km (75–120 mi) from the origin of the primary pollutants. The hydrocarbons and nitrogen oxides come from the combustion of coal and petroleum; vehicular exhausts constitute the major sources. Although not as toxic as ozone, PAN can be highly toxic to plants but does not often accumulate to toxic levels (Sinclair, Lyon, and Johnson 1987).

Although the sources of the major air pollutants are the activities of people, natural phenomena release considerable quantities of these chemicals into the atmosphere. Volcanoes release sulfur dioxide. Ozone and other pollutants are generated photochemically by hydrocarbons volatilized from vegetation. Rasmussen estimated in 1972 that the world's forests emitted 192×10^6 metric tons (175×16^6 tons) of reactive hydrocarbons annually, six times the emission produced by human activities (Rich 1975). In rural areas of upstate New York, Stasiuk and Coffey (1974) found ozone concentrations as high as those in New York City. Ozone levels remained high at night in the rural areas but dropped to low concentrations after sunset in cities. Emergence tip burn of eastern white pine, now associated with ozone injury, was known before 1900 (Hepting 1968). Since the hydrocarbons produced by human activities at that time were only a fraction of what they are today, the toxic oxidants must have had natural origins. In some situations, natural sources may even account for toxic pollutant levels.

Particulates in the air may have chronic or acutely toxic effects on plants. Particulates are seldom acutely toxic when they are dry, but dew or fine mist may increase their toxicity. Rain may wash the particulates from leaf surfaces before injury occurs. Even from a single source, particulate matter is not homogeneous but is a mixture of different chemicals, so generalizations are difficult. Sulfuric acid mist may form in the air when sulfur dioxide oxidizes to sulfur trioxide, which quickly hydrates to form aerosols. Necrotic spots on the upper surfaces of leaves have been observed after periods of heavy air pollution accompanied by fog (Middleton, Darley, and Brewer 1958).

When particles settle on leaves, shoots, and flowers, surfaces become coated, reducing the amount of sunlight reaching the plant. In some cases stomates may be clogged with dust (Lerman and Darley 1975). Photosynthesis can be greatly reduced, leading to poor growth, leaf drop, and death of twigs or, in severe cases, the entire plant.

The source of particulate pollution is usually local industrial activity, refuse burning, or transportation. Specific particulates include kiln dust from cement plants; fluoride dusts; soot; sulfuric acid mist; lead, rubber, and oil particles from vehicles; and particulates from various types of metal processing (Lerman and Darley 1975). Cement-kiln dust primarily coats the upper surfaces of plant parts. If it remains dry, the dust will be removed by wind or hard rain. On the other hand, fog, dew, or light rain will form a crust that may continue to build up. Along heavily traveled streets and highways, plants can become coated with particles of oil, rubber, lead, soot, and soil. Plantings close to heavily-used roadways become grimy, unattractive, and sickly if they are not cleaned.

Recommendations. The most obvious solution is to reduce pollutants at their sources. Filters, modified processing, and strategic industrial location can accomplish a great deal. Much more is needed, however, to protect human health as well as the health of plants. In areas that are likely to be highly polluted, plant trees and shrubs that are most tolerant of the expected pollutants (Sinclair, Lyon, and Johnson, 1987). You can reduce particulates on shrubs and small trees by hosing them with water. If the leaf coating is oily or adheres to the leaves, use a detergent solution.

In regions with arid summers, plant sensitivity to gaseous air pollutants can be influenced by irrigation practices (Rich 1975). In almost all reported observations, moisture stress before and during exposure to toxic pollution levels has reduced injury. In a similar manner, low relative humidity has increased the resistance of plants to air pollutants. In both situations, stomata close more quickly and completely, thereby reducing pollutant uptake. Experiments have shown, however, that moisture stress following toxic exposures increases subsequent injury (Rich 1975). Rich also cites an observation that dew on tobacco leaves increases resistance to ozone by impeding the movement of gases into the leaves. You may, therefore, be able to reduce pollution injury to sensitive plants by spraying them with water during and after toxic exposures. Apply enough water to create some runoff but not enough to wet more than the surface soil, at least if the pollution level is likely to continue or recur soon.

Certain fungicides and growth retardants increase the resistance of plants to air pollutants. Several of the dithiocarbamate fungicides have been reported to protect plants against ozone injury (Rich 1975). These materials, however, gave protection only when the lower surface of the leaves (where stomata are located) was treated, and protection was restricted to the portion of the leaf covered. Benomyl and closely related compounds act systemically to protect plants against pollutants. Rich reports that naturally occurring oxidant injury on azaleas was suppressed by soil drenches and foliar sprays of benzimidazole and oxathiin compounds. To date, these fungicides are still used primarily for disease control, and the protection they afford against pollution damage is considered an added benefit.

The subapical growth inhibitors, such as daminozide (SADH, sold as Alar® or B-9®) and CBBP (Phosfon®), increase the hardiness of responsive plants in the face of a number of environmental hazards, including ozone and sulfur dioxide (Cathey 1975). Only the growth that develops after treatment has a special resistance to ozone. One large ground-cover nursery in California routinely sprays most of its plants with those chemicals as much to increase plant hardiness as to retard growth. These responses raise the possibility of a practical treatment to minimize pollutant injury to plants.

The relationship between nutrition and susceptibility to air pollution is unclear. Most observations have involved annual herbaceous plants, primarily tobacco and vegetables. Plants grown at moderate fertility levels are usually more tolerant than plants grown at either lower or higher fertility levels (Rich 1975).

CHEMICAL INJURIES

A number of chemicals used for various purposes in and around landscape plantings can cause poor growth, injury, and sometimes death. These may act by direct application, through toxic levels in the soil, or as air pollutants.

Herbicides

Herbicides are usually the most phytotoxic of the pesticides used around landscape plants. Herbicides are designed to inhibit or kill undesirable plants. Nontargeted plants, however, can be injured by too concentrated a mixture, too heavy an application, careless application, or subsequent movement in soil or water. In urban landscapes, treatments can cause damage through the misapplication of products that combine fertilizers and herbicides or from the drift of sprays. In rural areas, damage may be caused by farm or garden chemical applications. The chemical 2,4-D (2,4-dichlorophenoxyacetic acid) has injured more nontargeted plants than almost any other herbicide. There are three reasons: (1) 2,4-D is systemic, so that it may affect the whole plant even though it comes in contact with only a part of the plant; (2) early formulations (primarily esters) were quite volatile; and (3) since 2,4-D is used to eliminate broadleaved weeds in lawns, it can easily affect adjacent trees and shrubs.

Symptoms. Pirone (1978a) noted that pin oak produces distorted leaves the spring after a fall application of 2,4-D to nearby lawn. The leaves and shoots of affected broadleaved plants may exhibit epinasty, depending on concentrations of the herbicide, plant species, temperatures, and stage of growth at the time of herbicide contact (Fig. 20–8). The leaves become cup shaped, with margins curling up or down. Leaf petioles bend down, giving the plants a wilted appearance. Shoot tips become twisted. Cracked calloused areas may form on larger shoots. The leaves of seriously affected plants lose their bright green color and may die. Plants will recover, however, unless they have received an exceptionally heavy dose of the herbicide.

Although most of the preemergent herbicides can be sprayed on the foliage of landscape plants without injury, toxic amounts can accumulate in the root zone. That is particularly true in sandy soils when rainfall or irrigation is heavy or where the watering basin for a newly planted tree or shrub has been treated with a preemergent herbicide. Subsequent watering or rain can erode the surface soil on the berm into the basin, leaching toxic concentrations into the root ball. Plant symptoms vary with the herbicide, its concentration, and the plant species, but in most cases affected plants will exhibit veinal or interveinal and marginal chlorosis in the leaves, followed by necrosis and leaf fall.

Soil sterilants used to prevent weeds in and around pavement, around buildings, and along railroad rights-of-way, ditch banks, fences, and roads can seriously affect plants that grow nearby or are subsequently planted in treated soil. Water runoff and erosion can carry the herbicide a considerable distance, injuring plants at some distance from the treated area.

Figure 20–8 Herbicides can injure nontargeted plants if care is not taken. Carob shoot deformed by 2,4-D drift (left). Simazine injury to pyracantha (upper right). Green ash leaves injured by excessive application of Amitrol to the soil (lower right). (Carob and pyracantha photos courtesy Clyde Elmore, University of California, Davis)

Recommendations. You can minimize the chances of 2,4-D injury by using low-volatile acid or amine formulations and applying a coarse spray at low pressure on calm, cool days or during the cool part of the day. Spray equipment used for herbicide application should not be used to apply other materials to landscape plants. If leaf symptoms appear within a few weeks of preemergent herbicide application, reduce further injury by removing the top 50 mm (2 in.) of soil around the plants. Stir a 25-mm (1-in.) layer of activated charcoal or 50 mm (2 in.) of finely divided compost or fortified sawdust into the 25 mm (1 in.) of remaining surface soil so that it may adsorb some of the remaining herbicide. Sprinkle water lightly to moisten the mixture of soil and organic matter. Bring the soil up to its former level by adding clean soil. Do not irrigate for several days and take precautions to minimize surface erosion during rains or irrigations; erosion might accumulate the herbicide close to plants. Later, heavily irrigate sandy soils to leach unadsorbed herbicide from the root zone or to dilute it. Remove severely affected young plants and plant replacements after you have substituted clean soil for the contaminated soil.

Chapter 11 contains information about planting in soil with toxic levels of residual herbicide.

Salt

Symptoms. Sensitive plants exposed to saline soils or sprays will usually show marginal and tip leaf scorch somewhat similar to boron excess (Fig. 20–6). If the salt comes from the soil, injury tends to be more severe on the portion of the leaf that has the greatest concentration of stomata or vein endings. Spray-borne salt usually accumulates on the windward side of the plants. As severity increases, entire leaves are affected and may drop; buds fail to grow, and dieback occurs. In extreme cases the plant dies. The first and most common symptom of salt injury, however, is simply a slowing of plant growth that often goes unnoticed or is thought to be caused by something other than salinity. Reduced growth is usually accompanied by early fall color and defoliation. Lime street trees in Hamburg, Germany, had an annual tree-ring width of 2 mm (0.08 in.) in the mid-1950s; 710 tonnes (780 tons) of salt were applied to the streets in 1956–1957. By 1978–1979, 40,000 tonnes (44,000 tons) of salt were being applied; tree-ring width was 0.2 mm (0.01 in.) (Petersen and Eckstein 1988).

The extent of injury will depend on the species of plant, the concentration and kind of salt(s), how the salt reaches the plant, soil moisture and rainfall, and many other factors. Salts that affect plants may be present in the soil initially or may be added by irrigation water, fertilizers, salt water intrusion, or salts applied to icy or dusty roads, or ocean spray. Soil salinity can become a problem particularly in low rainfall areas subject to irrigation and fertilization (see Chapter 13).

Millions of metric tons of sodium chloride are used each winter in the United States to de-ice roadways and sidewalks. In 1989 in Milwaukee, Wisconsin, more than 38,000 metric tons (42,000 tons) of salt were applied to 2240 km (1400 miles) of streets; that was 17 kg per running meter (11 lb/running ft) of road (Robert Skiera, Milwaukee, WI, 1990 pers. comm.).

Salt is usually mixed with sand, volcanic cinders, or other gritty carriers to promote uniform distribution and improve traction. Even though the salt is applied on roadways, most of it eventually ends up in the soil or on plants. Melting snow and rain runoff carry the salt to low areas along the roadway. Salt is thrown from the road during snow removal and may drift considerable distances in the turbulent mists created along high-speed roads. As would be expected, the amount of salt in the soil and on plants and the severity of plant injury decrease with distance from the road, lower traffic speed, and lower traffic volume (Lumis, Hofstra, and Hall 1973). In New Hampshire, Rich (1971) found that most trees affected by salt were within 9 m (30 ft) of the road.

The effects of repeated annual applications of salt depend on the amount of salt applied, the amount blown on plants or concentrated by runoff, the soil texture and drainage, and the rainfall. In Maine, Hutchinson (1968) analyzed soil to a depth of 150 to 200 mm (6–8 in.) at different distances from the road. In the first five years of salting, most of the increased sodium and chloride was within 5 m (15 ft) of the road. After 18 years, however, the levels were much higher and extended at least 15 m (45 ft) from the road; salts were about half as concentrated at 15 m as at 2 m (6 ft).

Plants seem to be more seriously injured by de-icing salt that is thrown, splashed, or allowed to drift onto them than by salt that accumulates in the soil (Lumis, Hofstra, and Hall 1973). Injury will be more severe on the downwind side of a road and on the side of a plant that faces the road. Lumis, Hofstra, and Hall found that plants covered with snow or in sheltered locations were not injured. Damage became evident in late winter on coniferous species but not until spring growth began on deciduous species.

On conifers, the injury appears as tip necrosis of the needles and progresses toward the base as injury increases. Conifers with more cuticular wax on the needles are less injured by salt spray. Lumis, Hofstra, and Hall also found that deciduous plants with resinous buds, such as horse chestnut and cottonwood, were resistant to injury, as were those trees whose buds are submerged in the twig, such as black locust and honey locust. Within a given genus, plants with naked buds were injured more than plants with scaly buds.

Since the usual symptoms of salt injury are poor growth and scorched leaves, it is easy to confuse this problem with a host of other ailments. Careful observation and leaf analysis may be needed to ascertain the cause of poor growth and leaf scorch.

Not only are sodium and chloride ions toxic to plants but high concentrations of sodium disperse colloidal particles and produce tight, poorly drained silt and clay soils. Poor drainage aggravates salinity by reducing leaching. High sodium levels will have little effect on the physical properties of sandy soil and decomposed granite, however.

Calcium chloride is sometimes used on dirt and gravel roads to keep the dust down. Like de-icing salt, calcium chloride may be carried to nearby vegetation by rain runoff, splashing from puddles, or air turbulence caused by passing vehicles. Calcium chloride is not nearly as toxic as sodium chloride but can cause similar injury to plants.

Plants growing near the sea or other bodies of salty water can be injured by salt spray and salt blown by the wind. Again, plants nearest the water are affected most severely. Hurricanes and typhoons can carry salt considerable distances inland. Injury from salt spray will vary with the seasonal development of leaves, particularly on deciduous plants. Deciduous plants will be little affected by wind-borne salt just before leaf fall. On the other hand, evergreen plants may be more seriously injured during this period.

Recommendations. As always, prevention is usually the best course. Street and planter construction can minimize salt spray and infiltration. On Milwaukee's landscaped, divided roadways, the driving lanes are sloped to the median planter where the water drains to a common storm drain under the median (Robert Skiera 1990 pers. comm.). The median planter surface is slightly convex (mounded). A sprinkler irrigation system is installed in each median.

Salty water drains from the streets to the center storm sewer. The soil usually freezes in Milwaukee. In the spring, the accumulated snow with salt spray melts before the soil underneath. The sprinklers are turned on to wash the last of the salty snow and salt from the boulevard landscapes, including turf. Later irrigation leaches much of the remaining salt from the soil. Salt entry into sidewalk planters is minimized by having the edges of the planter slightly higher than the walk. The cost of the irrigation system is divided equally between the salt removal operation and irrigation during the usual midsummer drought.

Salt deposits on leaves and twigs can be washed off with a garden hose or high-pressure sprayer if heavy rain does not follow the salt build-up. Severely injured plant parts should be pruned off. In soil that is moderately or well drained, you can leach salt by applying large amounts of water (see Chapter 13). Frequent irrigation in summer will keep salt at lower concentrations in the soil.

Ben-Jaacov, Natanson, and Hagiladi (1980) were able to prevent leaf scorching of oleander by activating an overhead sprinkling system when onshore wind speed exceeded 22 km (13 mi) per hour. Even when plants were well watered, foliage that was not sprinkled was severely damaged by only one dry windstorm. The sprinklers applied 50 mm (2 in.) of water per hour during strong winds and for five minutes after the wind slowed below the critical speed. A lower application rate may be just as effective and will reduce water use and the possibility of waterlogged soil. Sprinklers should emit large droplets so that less water will become wind-borne.

Long-term measures to minimize salt injury may be more effective. Select species that are relatively salt tolerant, but be aware that plants tolerant of salt in soil may not tolerate salt spray, and vice versa. Grade salt-treated paved areas and adjacent planting areas and install barriers so that surface drainage water does not accumulate near plants. Apply de-icing salt sparingly and only when needed. To reduce slipping on pavement near plantings, use sand, decomposed granite, or sawdust whenever possible. Plant in the spring, so that plants may grow for a full season before they are subjected to winter de-icing salt. If the soil structure deteriorates because of high sodium, apply gypsum (calcium sulfate) to improve structure and drainage and to facilitate leaching.

Boron

In some areas, primarily those regions with low rainfall, excess boron in the soil can cause serious plant injury and even death (see Chapter 12).

Other Chemicals

Pesticides can sometimes injure plants. Since many species may be grown in a single landscape, take care that treating one species for pests does not injure sensitive plants nearby. Read pesticide labels carefully to learn which species are sensitive and what concentrations and timing are recommended. Do not assume that if a little is good, more will be better. Plants of low vigor are often especially susceptible to spray injury. Lime sulfur and sulfur sprays and dusts are more likely to cause injury if temperature and humidity are high. Toxicity of a pesticide may be due to the solvent or carrier, or the active ingredient.

Cinders of burnt coal should not be used as a soil covering near plants, since they contain several toxic chemicals, particularly if they have not been leached by rain or sprinklers. Rain or sprinkler irrigation could leach these chemicals into the root zone.

MECHANICAL DAMAGE

The bark of trunks is the part of trees most often subject to mechanical damage, especially in spring and early summer, when the cambium is active and the bark "slipping." A slight blow at that time can cause the bark to loosen or tear off. Autos and lawnmowers can debark a tree at any season of the year (Fig. 20–9). Particularly vulnerable are trees planted in parking lots and near street curbs and driveways. Trees in lawn areas are often partially girdled by lawnmowers. Such damage can severely weaken young trees. Even worse, string trimmers girdle young trees in turf without the operator being aware of damage because the grass returns upright after being cut and hides the damage to the bark. A guard on the trimmer or a protective sleeve around the trunk will prevent such injury. Bark wounds close to the ground are particularly susceptible to decay organisms; they may also force latent buds to grow.

Figure 20–9 The trunk of a two-year-old London plane tree is 75 percent girdled by continually being hit with a lawnmower (left). The bark of a mature London plane growing near a driveway has been repeatedly injured by vehicles (right).

Trees are often vandalized: Their tops may be damaged or destroyed; their bark may be cut, beaten, or torn from the trunk. Treetrunks have been girdled by saws and axes. Trees subject to bark injury should be protected by two or three guard stakes (see Chapter 9). Plant young trees so that they are least likely to be struck by vehicles. A turf-free area around the trunks of young trees will reduce the likelihood of lawnmower and string-trimmer injury.

You can often minimize bark injury if you take action before the cambium has had a chance to dry. Loosened bark can often be replaced, and even trees with girdled trunks can be saved (see Chapter 18). If a young tree or shrub is in good health when all or part of its top is broken off, a new top will usually grow. Prune with a slanting cut so that the cut surface faces away from the midafternoon sun. Keep the trunk and cut surfaces cool by shading them or painting them with white latex.

Mechanical Damage

Improper staking or guying can result in damaged bark and wood. Young treetrunks increase in girth quickly; and unless the tree ties are inspected regularly and adjusted when needed, they can girdle the trunks. This is a particular problem when low branches or leaves along the trunks hide the ties. A tie may occasionally break loose from the stake but continue to circle the trunk, becoming embedded and causing poor top growth and a ring of weak wood that is subject to breaking (see Fig. 9–8). When tree ties break, are fastened too low, or become loose, the trunk bark can be seriously damaged by rubbing against the stakes or tree guard (see Chapter 9).

Ice and wind storms can cause branches to split and break. Pirone and others (1988) note that certain deciduous trees are more liable than others to such damage. These include

Ash	Red maple
Catalpa	Siberian elm
Empress tree	Silk tree
Hickory	Tulip tree
Horse chestnut	Yellowwood

Snow can accumulate on the branches of evergreen trees and shrubs to the point where they break or split out. You can minimize damage by carefully knocking the snow off with a broom, flexible rake, or pole. If the branches are heavily loaded, touching them from below will remove the snow with less strain on the plant. The species most often damaged by snow are juniper, yew, and hemlock (Pirone and others 1988). If transplanted to a heavy snow area, moderate to large conifers from areas of little or no snow may be more subject to limb breakage than trees native to the area. Many conifers growing in snow areas can shed and withstand considerable snow loads because their branches became more pendulous from snow loads while the trees were young.

If temperatures rise above freezing, ice-covered branches may be spared the breakage that often occurs when winds follow an ice storm. Hose the most heavily coated branches with a coarse spray of water until most but not all of the ice has been melted. The amount of ice will only be increased by a fine spray applied when temperatures are at or below freezing.

FURTHER READING

See Further Reading section at the end of Chapter 23.

CHAPTER 21

Diseases _____

Fungi, bacteria, mycoplasmas, and viruses are able to invade and infect plants, causing poor growth and appearance, disruption of plant processes, distortion of certain plant parts, and even death. Some higher (green) plants can also adversely affect the function and appearance of woody plants. Nematodes are considered by some plant pathologists to be disease-causing agents, since some invade plants, particularly roots, and have adverse effects. Nematodes are animals and are often studied by entomologists or nematologists and are discussed in the next chapter.

Although some new disease agents may be introduced into a given area, most disease damage to landscape plants is associated with pathogenic agents that are endemic or have been present in the area for some time (Schoeneweiss 1978). Many examples are familiar: leaf spots and blights, rusts, mildews, vascular wilts, and trunk and branch cankers. Healthy plants are likely to become diseased if

> The disease organisms are aggressive
> The plants constitute susceptible hosts and/or are stressed
> Climate and soil conditions are favorable for disease development

Although most plant pathogens aggressively attack healthy plants, a few attack only plants that are weakened or low in vigor.

Reductions in plant vigor usually arise from exposure to stress: drought, poor soil aeration, freezing or extreme temperature fluctuation, defoliation, nutrient deficiency, chemical injury, mechanical damage, or transplant shock (Schoeneweiss 1978). One or more of these stresses may cause visible injury, even when no disease organisms are present. Stress can also have a more subtle effect; it may weaken a plant and increase its susceptibility to pathogens (Fig. 21–1) and attack by boring insects. Exposure to severe or prolonged stress may increase disease susceptibility at almost any time. In general, however, plants are most readily predisposed to infection when they encounter stress just after planting or when they age and begin to decline in vitality.

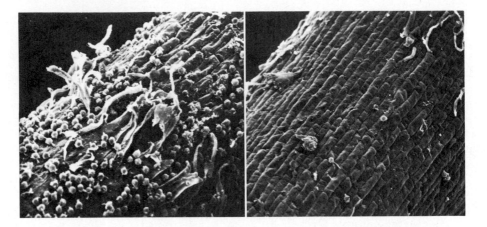

Figure 21–1 Root tip of salt-stressed chrysanthemum (left) attracts many oospores (spherical objects on root surface) of Phytophthora root rot fungus (*Phytophthora cinnamomi*) while a root tip of a nonstressed plant (right) growing in a half-strength Hoagland's solution with the same concentration of *Phytopthora* oospores attracts few oospores. Similar responses have been observed with roots of woody plants and under different types of stress. (Photo courtesy James MacDonald. 1982. *Phytopathology* 72:214–19)

Most organisms associated with stress-related disease normally grow as saprophytes on dead plant tissue. Thus, a pathogen may be present on healthy plants without causing damage until the plants become stressed. Even though unaggressive pathogens may not penetrate healthy plant surfaces directly, nearly all plants have minor wounds that can provide access. Fungus spores and bacteria can be spread by insects, wind, pruning tools, and splashing rain during wet weather.

Some of the widespread diseases are discussed to illustrate means of infection, symptoms, and measures used to control the different pathogens. Federal and state publications and commercially printed books give more detailed information on these and many other diseases. Some of these publications are listed at the end of Chapter 23. *Diseases of Trees and Shrubs* (Sinclair, Lyon, and Johnson 1987) has excellent descriptions and color photographs of diseases of landscape trees and shrubs.

Fungal parasites are the most common cause of infectious plant diseases. Fungi, no longer considered plants, have been placed in their own kingdom: *Mycota* or *Fungi*. They are filamentous nongreen organisms widely distributed on plants, in soil, and in the air. The filamentous strands (hyphae) are microscopic; mushrooms, conks, and mold are their visible reproductive stages. These structures produce millions of spores, which are dispersed by wind, rain, or insects to new hosts. Under favorable conditions of warmth and moisture, spores germinate and give rise to hyphae, which can infect susceptible plants. Thus the cycle continues.

Verticillium Wilt

Verticillium wilt is a soil-borne fungus disease known to infect several hundred woody and herbaceous species throughout the world. The causal fungus, *Verticillium dahliae* (*V. albo-atrum*), invades and destroys the water-conducting tissues. It can enter the xylem through root and trunk wounds, root grafts, and possibly intact young roots.

Symptoms. The name of the disease describes its most obvious symptom: The leaves on one or more limbs wilt. Infection often takes place in the season before symptoms appear. If the following spring is mild and soil moisture is adequate, a relatively large succulent leaf canopy can develop only to wilt during the first hot days because the large transpiring surface cannot be adequately supplied by the fungus-impaired water-conducting system. Depending on the severity of wilting, leaves may turn yellow, first at the margins and between the veins, then turn brown, and later die from the base of the plant or branch upward. If the symptoms develop slowly, the leaves will probably drop; if the leaves wilt and die quickly, they may hang on the tree for some time.

The xylem of infected plants will often turn olive green, dark brown, or black, depending on the species. If you cut diagonally through the base of a shoot or small branch, you will expose the discolored xylem (Fig. 21-2). In some infected plants, including olive, ash, and rose, xylem discoloration will be slight or nonexistent (McCain 1979b). To complicate diagnosis further, discolored xylem may indicate a number of other diseases, including Dutch elm disease. A positive diagnosis can be made only if the fungus from affected wood is cultured in a laboratory. Even when verticillium has caused xylem discoloration, it may be impossible to isolate the fungus on wilted shoots. The fungus may be lower in the branch or may no longer be viable.

One or more branches and occasionally the entire plant may wilt, depending on how far and how quickly the disease progresses. Young plants are more susceptible than old and may be more severely injured. Mildly affected plants will usually recover from a verticillium attack.

Treatment. You may select plants that are resistant or immune to verticillium wilt. A number of state Cooperative Extension publications list susceptible and resistant plants (Himelick 1969, McCain 1979b). As far as is known, conifers and

Figure 21-2 Stripping the bark from a branch infected with verticillium wilt or cutting diagonally through the branch will usually expose dark, discolored streaks in the young xylem. (Photo courtesy Robert Raabe, University of California, Berkeley)

broadleaved evergreen plants are resistant to verticillium (Pirone 1978b). In California, McCain discovered that olive trees under sod culture are less severely affected by verticillium wilt than olive trees grown in weed-free groves. Sod or grass culture might reduce the severity of this disease in other species.

No cure is known for verticillium wilt once a plant is infected. You can aid the recovery of mildly affected plants by fertilizing with nitrogen (10 to 20 g/m^2; 2 to 4 lbs/1000 ft^2) as early in the season as possible and by ensuring that the plant has sufficient water. This will stimulate the growth of additional xylem to replace some of the wood that has become plugged. Some plants may recover from initial infections and escape further difficulties; others may continue to lose branches in succeeding years but will remain essentially viable; yet others will be so seriously affected they must be removed. When trees are to be saved, delay the removal of affected branches until you can determine the amount and quality of regrowth. Removing dead or wilted branches will not remove the fungus from the plant, although it will improve appearance.

After removing seriously affected and dead plants, replant with resistant species. If you must replant susceptible species, remove as much of the root system of diseased plants as possible. Fumigating the soil with chloropicrin will be partially effective against verticillium (Sinclair and Johnson 1975); do not get too close to adjacent plants. Soil solarization may have promise, even in existing plantings (see Chapter 22). The fungus, however, forms resting black structures (microsclerotia) that are capable of surviving in soil for many years; some apparently survive fumigation. When a newly infected tree grows among a group of healthy trees of the same species, Tattar (1989) recommends cutting any possible root grafts to reduce the spread of disease. Cutting roots, however, could open healthy roots to infection.

Armillaria Root Rot

The common names of this soil-borne fungus disease are shoestring root rot, oak root fungus, mushroom root rot, honey fungus, and armillaria root rot. The organism, *Armillaria mellea,* attacks a wide range of landscape, orchard, and forest plants throughout the world. Although it infects healthy plants of susceptible species, the disease is often associated with plants of low vigor (Wargo 1980).

Symptoms. Affected trees and shrubs usually exhibit a general loss of vigor. Symptoms are similar to those caused by poorly aerated soil or an impaired root or trunk system. Plant decline is often accompanied by yellowing of foliage and leaf drop. Infected plants may die back slowly, a branch at a time, but they often wilt and die quickly if the fungus invades and girdles the trunk collar.

The most reliable signs of armillaria are fan-shaped plaques (mycelium) of white- or cream- (honey)-colored fungus tissue (Fig. 21-3) (McCain and Raabe 1972). These appear between the bark and wood in roots and trunk just at or near the soil surface. The fungus most commonly enters a susceptible plant when a rhizomorph (a dark, root-like fungus structure) contacts a root or when roots of a healthy plant come in contact with an infected one. Rhizomorphs (which resemble shoe-

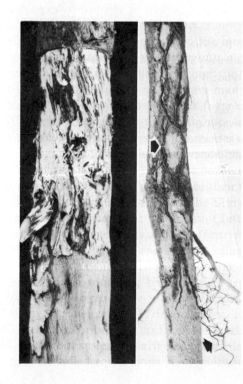

Figure 21-3 *Armillaria mellea* may weaken and kill only part of a tree, cause a general weakening, or kill a tree quickly. Diagnostic signs of the fungus disease are the white- or cream-colored fan-shaped mycelium between the bark and wood on the main roots and trunk base (left) and the string-like rhizomorphs along the roots or in the soil (right). (Photos courtesy Robert Raabe, University of California, Berkeley)

strings, hence the name "shoestring root rot") grow on the surface of affected roots and for short distances into the soil (Fig. 21–3).

Other wood-rotting fungi are sometimes mistaken for *Armillaria mellea.* These other fungi are usually secondary invaders, however, that spread throughout the bark and sometimes into the wood, causing it to crumble easily. Armillaria-affected trees retain firm wood except in very advanced stages of the disease, when wood becomes wet and soggy (McCain and Raabe 1972). In the late summer or early fall (as late as early winter in milder climates), honey-colored mushrooms may appear at the base of infected plants, although they are not common in arid regions. Many other fungi, not necessarily pathogenic, also form mushrooms at the base of plants. Armillaria mushrooms have white gills and spores; the gills are attached to and extend down the stem a short distance.

Treatment. You can select plants for resistance to armillaria root rot. Consult state or regional publications for recommendations (Raabe 1979). The disease is often associated with plants of low vigor or plants under stress. As a preventive measure, the health and vigor of uninfected plants can be improved by fertilization with nitrogen (10 to 20 g/m^2; 2 to 4 lbs/1000 ft^2), wise irrigation, and soil aeration. When you aerate the soil, take care not to spread infested soil and roots to areas that are clean. The best way to improve aeration with minimal root disturbance is to scarify the surface soil, apply a light organic mulch to reduce further compaction, minimize foot and vehicular traffic under the trees, and prevent excess water from accumulating on or in the soil, particularly near the trunk.

Progress of the disease may be slowed or stopped by exposing the root collar and large roots to the air (Sinclair, Lyon, and Johnson 1987, Pirone and others 1988, Tattar 1989). Armillaria infection has been held in check for more than 40 years in mature live oak trees in southern California by exposing the root collar and buttress roots (Wesley Wilbur, University of California, Riverside, 1990 pers. comm.).

All soil removed should be taken to a disposal site or a place where infested soil can be fumigated. Remove roots with rotted bark, remove affected firm bark up to healthy wood and bark that shows no fungal mats, and cut away any rotten wood. You need not sterilize cutting tools (Robert Raabe 1981 pers. comm.), but all wood, bark, and rhizomorphs removed should be collected and thoroughly dried or burned. The bark and wood surfaces exposed to the air will dry, decreasing the chances of further infection. Keep the roots dry and exposed to the air as long as is practical. In areas where exposed roots may freeze place clean soil around the roots in late fall. To keep the root collar surface dry, slope the bottom of the hole away from the trunk. If the excavated area must be filled, place about 250 mm (10 in.) of coarse gravel around the trunk from the bottom of the hole to 50 mm (2 in.) above the original ground level. Fill the hole up to the gravel surface and slope away to the original soil level at the edge of the hole so that water will not accumulate at the treetrunk.

Only valuable trees may be worth the efforts to save them. The roots of severely infected trees will rot, making the trees hazardous. Remove severely infected and dead plants and as much of their roots as possible, at least those that are

25 mm (1 in.) in diameter and larger. Soil disturbed when the plant and roots are removed should be kept in the infested area and fumigated or dried to kill the fungus. Alternatively, remove soil from the infested area to a depth of at least 600 mm (2 ft). Take care not to drop any soil en route to the disposal site, lest you spread the disease further. Choose the disposal site carefully so that spread of the fungus will not cause problems.

When only one tree in a group has been killed or severely infected, remove the tree and infected roots as described above, even if infected roots are found under the canopies of seemingly healthy adjacent trees. Soil can be fumigated with methyl bromide to within two-thirds of the canopy radius on one side of a living tree. The soil should be dry and the surface friable or cultivated as deeply as possible before fumigation, but, even at best, control will probably not be complete.

Dutch Elm Disease

Dutch elm disease (DED) is one of the most widely known and one of the most destructive plant diseases of the twentieth century (Sinclair and Campana 1978). The disease appeared in 1918 in Europe, and the causal fungus, *Ceratocystis ulmi,* was identified in 1922 in the Netherlands. The fungus, introduced into North America on veneer logs imported from Europe, caused disease that was discovered in Ohio in 1930 (Schreiber and Peacock 1974). By 1980, infected elms were found in almost all of the contiguous United States, although the disease appears to be less destructive in the Southwest, where no indigenous elms are found, where Asiatic species constitute an unusually high proportion of elms grown, and where summer temperatures and humidities are unfavorable to the fungus (Campana 1978).

The DED fungus is transmitted in North America by two insects: the smaller European elm bark beetle, *Scolytus multistriatus,* which preceded the fungus from Europe, and the elm bark beetle, *Hylurgopinus rufipes,* which is native to much of the midwestern and northeastern United States and southeastern Canada (Johnson and Lyon 1988). The bark beetles breed and lay eggs in the inner bark of weakened, dying, and dead elm trees. After the larval and pupal stages, adult beetles emerge and fly to healthy elms (or healthy parts of the same elm), on which they feed briefly, and to weakened elms, in which they breed. If the elm in which the beetles breed is infected with DED, their progeny will carry the fungus to healthy trees, usually a year later. Since DED weakens and kills elms, an epidemic may easily be started. The fungus can also spread from one elm tree to another through root grafts, which are common in street trees.

Strains of *C. ulmi* vary in virulence. England has had two DED epidemics: The first followed the introduction of the fungus in 1927, and a second occurred in the 1970s, caused by a more virulent import from North America. There are 20 to 30 species of elm; the American elm, English elm, and the European white elm are highly susceptible to DED, whereas a high proportion of the Asiatic species, particularly the Siberian and Chinese elms, are fairly resistant (Sinclair 1978). No elm species has been found to be immune. Plants in related genera, *Zelkova* and *Planera,* have become diseased when artificially inoculated with the fungus (Schreiber and Peacock 1974). *Zelkova carpinifola* has been severely damaged by natural infection in Iran (Sinclair 1978), but *Zelkova serrata* is considered at least as resistant as Siberian and Chinese elms (Svihra 1980).

Symptoms. The symptoms of DED, their sequence, and the rate of development depend on the species of elm, tree nutrition, soil moisture, temperature, site of inoculation (VanAlfen and MacHardy 1978), and age. Healthy elms are most commonly infected in spring and early summer, when elm bark beetles feed in the crotches of one- and two-year-old twigs (*S. multistriatus*) (see Fig. 22–3) or in branches (*H. rufipes*). These beetles make wounds in the outermost sapwood as they feed and introduce spores of the disease fungus into these wounds.

The first observable symptom is usually the sudden discoloration or drooping of leaves at the tip of one or more branches (Fig. 21–4). The leaves may then yellow, roll upward at the edges, and turn brown, or dry rapidly, turn dull green, and then brown. When the leaves yellow or curl, they usually drop quite soon; but when leaves shrivel and die rapidly, they may remain attached for several weeks. Shoot tips may also wilt and droop; this is called *flagging*. Small flags develop after elm twigs are inoculated by European elm beetles, but large flags develop after inoculation by native elm bark beetles (VanAlfen and MacHardy 1978). Trees infected in early summer may be killed the first year; others may live for several years; and some may recover.

Figure 21–4 An early symptom of Dutch elm disease is *flagging* (wilting) of leaves on one or more branches; the rest of the foliage may appear normal. (Photos courtesy USDA)

Infected elm twigs usually have brown discoloration just beneath the bark (Fig. 21–5). Cut a twig diagonally with a knife: If the tree is infected, dark spots or a continuous brown layer in the current season's growth ring will usually be exposed. Other vascular wilt diseases, however, cause many of the same symptoms, so only laboratory identification will be reliable (Schreiber and Peacock 1974).

Treatment. No single method has successfully controlled DED, so the primary emphasis is on prevention. To combat Dutch elm disease

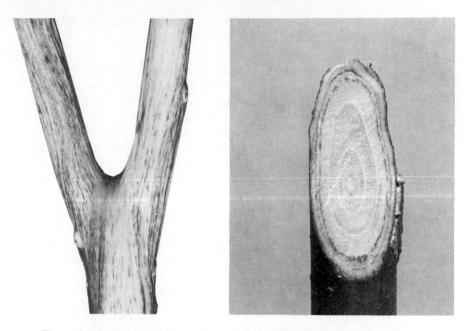

Figure 21–5 If you peel the bark from branches infected with Dutch elm disease (left) or cut diagonally through them (right), you will expose brown streaks. This symptom can be easily confused with that of verticillium wilt. (Photos courtesy USDA)

Quickly prune out infected branches
Implement beetle-trap-tree technique (Lanier 1989)
Eliminate beetle breeding sites
Protect healthy elms from feeding beetles
Prevent underground transmission of the fungus
Plant trees that are resistant to the disease (Schreiber and Peacock 1974)

Leaf Diseases

Leaves not only reflect the health of plants but can be directly injured or disfigured by a variety of causes, including fungi. Most leaf diseases of deciduous woody plants, however, are not serious. Plants may be weakened by leaf diseases, but the effects usually end with leaf fall, unless premature defoliation depletes food reserves.

In contrast, the leaf diseases of evergreens often severely affect plant health and can even kill trees and shrubs. Because evergreens normally hold individual leaves for several years, infection and loss of leaves have great physiological and visual effects. Evergreens are slow to recover from leaf drop, since several years of foliage may be lost at a single time. Evergreens with severe leaf infections are also more susceptible to insects, environmental stress, and other pathogens.

Symptoms. The leaf diseases of deciduous plants have been grouped into six categories, according to symptoms (Tattar 1989).

Leaf spot: Dead area on leaf, well distinguished from healthy tissue
Leaf blotch: Dead area on leaf, often diffusing into healthy tissue
Anthracnose: Irregular dead areas on leaf margins, sometimes killing entire leaves; shoots and small twigs often injured
Powdery mildew: White to gray-white fungus on leaf and shoot surfaces
Leaf blister: Leaf spot or blotch that is swollen or raised, producing a blister-like area on upper surface
Shot hole: Holes caused when dead areas inside leaf spots drop out

The leaf diseases of evergreens can also be grouped (Tattar 1989).

Leaf spot: Broadleaved evergreens may show spots of infection similar to those on deciduous plants.
Needle spot, needle cast, needle blight: Fungus infections of conifer needles may progress through a series of increasingly severe stages. *Needle spots* can occur at random along a needle. The spots may enlarge, coalesce, and eventually kill the entire needle, which may then drop; this last stage is called *needle cast.* When all or most of the plant is severely affected, the symptom may be called *needle blight. Needle cast* and *needle blight* are also used interchangeably.
Needle rust: Swollen white or orange blisters may form on the needles, sometimes causing needles to drop.

Treatment. Obviously, the best long-term control involves selecting species or cultivars resistant to known leaf diseases. Resistant cultivars have been selected from a number of susceptible species. Many existing landscape plants, however, are susceptible to one or more leaf diseases. Control measures can be recommended when the health of the plants is poor or the appearance of an important planting is jeopardized. Since evergreens are usually more seriously affected, they should be given priority if control measures must be limited.

As for most diseases, protection against infection is the only effective control. Infection for most leaf diseases occurs during warm wet weather in spring, when the pathogens become active and developing leaves and shoots are most susceptible. If fungicides are to be effective, they must be applied during or before bud break and about every two weeks thereafter; this will keep newly developing foliage covered until the temperatures become hot or the danger of wet weather is past, usually at the beginning of summer. After symptoms occur, control measures may reduce subsequent symptoms but will not improve the leaves already infected. As temperatures rise and if wct weather subsides, infection will decrease and normal growth will prevail. Leaves infected earlier, however, may drop throughout the summer.

You can reduce conditions favorable to leaf infection by planting and pruning to permit greater sun penetration and air movement through susceptible plants. This is especially important for evergreens, since they may have a relatively dense canopy of leaves in the early spring, when chances of infection are greatest. If foliage dries quickly and humidity within the leaf canopy is lowered, the possibility of leaf infection may be decreased.

There are from 125 to more than 300 powdery mildew species (their taxonomy is still evolving) on more than 7000 host plants (Sinclair, Lyon, and Johnson 1987). Some plants are more susceptible to powdery mildew while leaves and shoots are developing; their mature leaves and fruit become more resistant. The mature foliage of other species, however, is more susceptible. Pruning, fertilizing, and other practices that stimulate and prolong growth encourage many powdery mildews. These diseases are also favored by dry weather with warm days and cool nights, particularly in regions with a definite dry season. Although rain inhibits them, high humidity, including dew and fog, enhances infection (Sinclair, Lyon, and Johnson 1987). For these mildews, such as *Uncinula necator* on vinifera grapes, sprinkler irrigation during dry weather should not be a problem. You need to know with which powdery mildew you are dealing.

For leaf diseases in general, it is not practical to prune off infected twigs and burn them and fallen leaves because many fungus mycelia and spores will remain and can be spread by rain or blown great distances by wind. Pruning off infected small branches can improve the appearance of most affected plants, however. If only the leaves on a few twigs have been seriously infected, their removal could retard further infection, particularly if it is combined with other control measures.

Cankers

Sinclair, Lyon, and Johnson (1987) devote 103 pages of their 496-page book to canker-causing diseases; 92 of the 103 pages are about diseases caused by fungi. Cankers are lesions caused by fungal invasion of bark which usually kills phloem, cambium, and the outermost xylem. The bark in an infected area may shrink, crack, and expose the wood beneath (Pirone 1978b). Cankers, depending on the fungus, can occur on trunk, branches, twigs, or even roots. Infection occurs through wounds, including leaf and fruit scars, branch stubs, and sunburn lesions.

Symptoms. Most trunk and branch cankers can be placed in one of three groups (Tattar 1989). *Target cankers,* such as Nectria canker, are roughly circular with much callus over the canker surface and at its edges. Infection spread is relatively slow. The radial growth of the plant is about the same as that of the canker. *Diffuse cankers,* such as Cytospora canker, are elongated ovals with little callus at the margins. The pathogen spread is faster than in target cankers. The cankers enlarge faster than the radial growth of the plant and can girdle after several years. *Canker blights,* such as chestnut blight, are circular to elliptical with little or no callus and can increase rapidly. Branches and even entire plants can be girdled and killed in one season (Tattar 1989).

Treatment. Infection by many canker-causing fungi can be minimized by guarding against wounding, pruning during dry weather, and wise fertilizing and watering to hasten wound closure. Branch cankers should be pruned out. Tattar recommends cutting out cankers on the trunk and main branches by removing the bark at least 25 mm (1 in.) beyond cankered tissue. Tattar states that "Cutting tools should be surface-sterilized after each cut." Canker-rot weakening of a tree can be delayed by keeping the tree vigorous so that it grows faster than the fungus. In most cases, however, it is safest to remove trees weakened by cankers.

Canker-rots, such as Hispidus canker, not only form bark cankers but the fungi also invade the xylem. The cankers may be small, but internal decay may make a tree structurally unstable. Some *Phytophthora* species can cause collar and root rots which kill roots and may create a hazard even though the top of the tree does not appear to be in difficulty (see Chapter 17).

BACTERIA

Bacteria are microscopic organisms, devoid of chlorophyll, and usually single-celled; they obtain their food from living or dead organic matter. Like fungi, most bacteria are beneficial. They both decompose organic matter and convert complex chemical compounds to more simple forms. Bacteria also synthesize gaseous nitrogen into a usable nutrient and can be used in many industrial processes. Some species, however, cause diseases in plants and animals.

Bacteria are smaller than fungi and can penetrate healthy plants through natural openings, such as stomata, lenticels, the terminal openings of leaf veins, the glands in flowers, and wounds. Some bacteria are spread by contaminated pruning and grafting tools. Insects transmit certain bacterial pathogens.

Fire Blight

Fire blight frequently attacks plants in the rose family. Certain pears and quince are extremely susceptible; apple, crabapple, cotoneaster, and pyracantha species are also often damaged. Fire blight is caused by *Erwinia amylovora,* which survives the summer and winter in blighted twigs and cankers. During warm, moist spring weather, the bacteria ooze from the holdover cankers in small sticky drops. They are carried to blossoms and tender shoots by insects and splashing rain and from blossom to blossom by bees and flies. Temperatures between 20° and 30°C (70° and 85°F) and high humidity caused by dew, rain, fog, or irrigation favor development of the disease (McCain 1975).

Symptoms. Fire blight infection causes a sudden wilting, shriveling, and blackening of blossoms, tender shoots, and small fruit (Fig. 21–6). Affected parts look as though they have been scorched by fire—hence the name of the disease. Leaves are killed so quickly that they remain on dead shoots throughout the summer. Infection often progresses down a shoot but may be confined to the shoot if it is smaller than the branch to which it is attached (Shigo 1986a). If the shoot is almost the same size as the one to which it is attached, infection is more likely to move into the mother limb.

On large limbs, infection may produce sunken dark cankers. Only a few shoots on a plant may become infected; but if cankers enlarge, they can completely girdle branches or the main trunk, killing large portions of the plant.

Figure 21–6 When blossoms and young shoots of plants in the rose family are infected with fire blight bacteria, they wilt, shrivel, and turn black in the spring (left). Infection may be carried systemically to older branches, causing cankers of dead bark and wood (rough bark area) (right).

Treatment. Prune out affected shoots or branches as soon as possible; this should stop the spread of the infection in that area and reduce the inoculum available to infect healthy blossoms or shoots. Cut well below (300–450 mm or 12–18 in.) the infected area, whose edge will usually be easily seen. It is commonly recommended that you dip the tool into a disinfectant after each pruning cut. However, Robert Raabe, plant pathologist at the University of California, Berkeley, has found that contaminated pruning shears, whether they have cut through a diseased shoot or been dipped in a bacterial slurry, will not infect healthy woody pear shoots (1981 pers. comm.). Prunings should be burned or removed from the vicinity of susceptible plants.

If possible, remove fire blight cankers before the next spring. Prune off branches on which cankers occur. Shrubs with trunk cankers may not be worth saving. When cankers occur on the trunk or scaffold branches of large trees, Tattar (1989) recommends removing the bark from an area that includes healthy bark 75 mm (3 in.) on either side and 300 mm (12 in.) above and below the canker. The exposed wood can be left untreated. Inspect the wound later to see whether any further infection occurs; if so, it should be treated as described above.

Fire blight occurs most frequently on succulent growth. Where the disease has been a problem, cultural practices (particularly nitrogen fertilization) should be designed to avoid vigorous growth, especially early in the growing season, when conditions most favor infection. Blossoms are the most susceptible part of the plant. They can be protected from infection by sprays that contain copper or streptomycin. Make the first application before more than 10 percent of the plant has bloomed; repeat at four- to five-day intervals until bloom is over.

In areas where fire blight is known to be a problem, select species or clones that are known to be resistant or nearly so. The callery pear is quite resistant, compared to the common fruiting species.

Crown Gall

The bacterium *Agrobacterium tumefaciens* causes galls on many woody and herbaceous plants throughout the world. Most species of the rose family are susceptible, as are many other landscape plants.

Symptoms. Crown galls start out as small swellings on the bark of the lower trunk and roots, although they can occur higher on the plant, particularly on poplar and willow. These swellings continue to enlarge, engulfing more of the stem or root (Fig. 21-7). At first, the galls resemble callus tissue similar to a potato tuber in density but with a tougher texture. As the galls enlarge, their surfaces become rough and fissured. The outer portions of older galls may slough off.

Burls and similar growths more closely resemble normal wood tissue than do crown galls. Galls on black walnut, however, may be difficult to distinguish from burls, which often form on the trunks of old trees. The internal cavities of old galls are sometimes lined with dead bark and decayed tissue, whereas burls are solid (Ross and others 1970).

Crown gall bacteria commonly present in the soil enter plants through wounds. Young plants are particularly susceptible during handling in the nursery, planting, and early care in the landscape. Fumigating soil before planting can decrease the incidence of crown gall (Moore 1979). The bacteria or the plasmids that they form stimulate large numbers of undifferentiated cells, which form the galls. These galls can interfere with the movement of water and nutrients in the xylem. Crown gall is particularly serious on young plants, since one infection can engulf a main trunk, branch, or root within a year or two, seriously weakening or killing the plant. Crown gall infections on large trees seldom cause serious reduction of growth.

Treatment. Crown gall can be largely prevented if you select plants free of galls and minimize injury to the roots and trunk during planting. Most nurseries produce clean stock.

A non–disease-producing strain of crown gall bacteria, *Agrobacterium radiobacter* 'K 84,' is effective in protecting most young fruit and landscape plants from infection (New and Kerr 1972). The *A. radiobacter* 'K 84' bacterium is marketed as Galltrol A® and Norbac 84®. Crown gall bacteria resistant to 'K 84' have been reported (Moore 1979). There appear to be strains of the bacteria that are resistant, and there are plants, such as cherry, that are particularly susceptible. Milton Schroth (University of California, Berkeley, 1990 pers. comm.) states, however, that wounds must be thoroughly treated within four hours of wounding (eight to 12 if it is cold). 'K 84' can be used as a dip or a spray. If protection has not been complete, next time use 'K 84' more concentrated than recommended, applying it thoroughly soon after wounding or digging. Higher concentrations should not injure plants, but care must be taken as when handling any pesticide.

Young plants that do poorly in the landscape should be inspected for crown

Figure 21–7 Crown galls were on roots of the young almond (upper left) when dug in a nursery; the infected plant is doomed to poor growth and a possible early death; roots to the right are not infected. (Photo courtesy Milton Schroth, University of California, Berkeley) A gall on the trunk has seriously weakened this young flowering cherry tree (lower left), while the large gall on a major limb of an elm has not jeopardized more than the limb on which it occurs (lower right).

gall. You may need to expose the trunk and main roots for a complete inspection. A 25-mm (1-in.) diameter stream of water at a pressure of about 4 kg/cm² (60 psi) will remove soil from the base of a tree with little or no injury. Any galls exposed should be allowed to dry for several days. Painting the galls and 10 to 15 mm (0.5 in.) of surrounding healthy bark with Gallex® (formerly called Bacticin) has eradicated over 95 percent of treated galls on several fruit species (Ross and others 1970). Large galls may require two applications. Leave the treated areas exposed for several weeks. You can treat galls at any time, but if there is little danger of winter injury to exposed roots, treatment during the dormant season will cause the least disturbance to maintenance practices and use of the landscape.

Wetwood (Slime Flux)

The most common evidence of wetwood is fluid seeping or fluxing from wounds in the trunk or large branches. Wetwood is chronic and contributes to the decline of many trees, especially older trees growing under adverse conditions. Although almost all old elms and poplars are infected, wetwood is common in aspen, hemlock, maples (including box elder), mulberry, oak, and white pine (Sinclair, Lyon, and Johnson 1987). Wetwood is less frequent in many other species.

Symptoms. Wetwood occurs when bacteria invade and grow in parenchyma of the spring wood of trunks and large branches. Affected wood becomes dark brown and water-soaked; when the wood is cut, sap oozes out. Discolored spring wood occurs in the annual rings, primarily in the heartwood and older sapwood. Fluxing commonly occurs from spring through autumn, with the most active period in the heat of summer, when the bacteria are most active. In mild climates, some fluxing may occur in winter.

The bacteria ferment organic constituents in the sap, producing gas, primarily methane, that can develop considerable pressure. Carter (1975) reports pressures up to 4 kg/cm² (60 lbs/in.²). More common pressures, however, are 0.4 to 0.8 kg/cm² (5–10 lbs/in.²). These pressures force fermented sap and gas out of wounds in the trunk and branches. The sap is colorless or tan but darkens upon exposure to air and wets the bark below wounds. Growth of bacteria and fungi in the sap often produces foamy gray or slimy brown masses at and below wounds (Fig. 21–8). This is commonly called slime flux.

Fermented sap is more alkaline than sap from healthy wood but is toxic enough to retard or prevent callus formation at the base of wounds. Slime flux can kill leaves, shoots, and grass on which it may drip. When dry, slime flux leaves a gray or white encrustation on the bark. In addition, if current season wood is infected, leaves may yellow, wilt, and drop; branches may die; and the entire tree may decline in health and vigor. Brown streaking in the wood of young branches can easily be confused with similar symptoms of other wilt diseases.

Carter (1975) cautions that not all trees with flux are affected with wetwood. Healthy trees of a number of species, including some susceptible to wetwood, flux profusely from pruning wounds, particularly if large limbs are removed before growth begins in the spring (see Chapter 15).

Figure 21-8 Sap oozes out of an elm trunk crotch. The lighter colored bark beyond the dark wet area has been discolored from earlier heavy flows of sap.

Wounds limited to the bark may also produce slime flux, but that should not be confused with the flux that results from wetwood. Wounds associated with wetwood extend through the bark and into old sapwood.

Treatment.   No chemical or biological treatment is known that will control wetwood in infected trees. Removing dead wood will improve the appearance of a tree. Nitrogen fertilization usually stimulates growth which may help overcome adverse effects of wetwood.

An old recommended practice is to bore a hole below a fluxing wound to reduce or stop the flow of sap (Carter 1975, Pirone 1978a). Detailed directions are given by Carter. However, Shigo (1979b) and others state that any benefits of boring a hole to drain slime flux are more than offset by the possibility of spreading wetwood and decay to tissues surrounding the drain holes. In fact, Shigo (1986b) points out that the high moisture content of wetwood protects wood from decay.

VIRUSES

Viruses are submicroscopic in size and consist of a nucleic acid core with a protein covering. Living as parasites in plants and animals, they can reproduce themselves only inside living tissue. A virus stimulates the host cell to create more viruses identical to itself. The virus and its multiplication interfere with some of the functions of plant cells. Viruses are usually spread from infected to healthy plants by sucking insects (aphids and leaf hoppers) and rarely by pruning tools. Budding or grafting infected scions onto healthy rootstocks or healthy scions onto infected rootstocks will transmit viruses. Few viruses are transmitted through seed.

Symptoms. Symptoms of viral diseases commonly fall into one or more of the following categories (Tattar 1989):

Lack of chlorophyll formation in normally green tissue
Inhibition of vegetative growth
Distorted growth of all or part of the plant
Necrotic areas

Even though viruses may dwarf or seriously weaken a plant, they are seldom fatal. Plants weakened by viruses, however, may be more susceptible to other diseases and to physiological stress.

The viral diseases of fruit species have been studied more intensively than those of landscape plants. Some viral diseases have become serious in certain fruit species, primarily in cultivars that have been propagated vegetatively for a long time. Many of the flowering fruit trees used in the landscape are infected with the same viruses.

Viruses are named or described by their symptoms and the host infected. Some infect more than one species of plant and may cause either similar or different symptoms in the various species. The identity of a virus may be determined by grafting infected buds onto cultivars that produce characteristic symptoms in response to known viruses. Some of the more common viral diseases of trees have been listed by Tattar (1989) (Table 21–1).

Treatment. The best protection against viral diseases is to select virus-free or virus-indexed plants. Rather complex procedures have been established by some nurseries and governmental agencies to ensure the propagation of material that is free of the more serious viruses. Heat treatment of plants will inactivate viruses and allow disease-free foundation plants to be propagated for nursery "mother" blocks. Healthy "mother" plants are maintained in isolated blocks and are periodically indexed to ensure that they are still free from disease. Plants that are grown to be sold are propagated with care to prevent infection.

Experience with flowering cherries indexed for the more serious viruses illustrates the importance of virus-free plants: The percentage of successfully budded Yoshino cherry trees increased from 40 to 100 when virus-indexed trees were used (Leiser and Nyland 1972). Shoot growth was 2.5 times greater on the virus-indexed trees than on trees infected with necrotic ringspot (Fig. 21–9).

Mature trees infected with a virus will usually live for many years and will be satisfactory landscape features. Optimum levels of moisture and nitrogen nutrition should increase plant vigor and tolerance to other diseases and should prolong the life of trees.

MYCOPLASMAS

In 1967, some plant diseases thought to be viral were found to be caused by mycoplasma-like organisms, as yet unnamed. Mycoplasmas cause diseases in animals; somewhat similar organisms cause viral-like diseases in plants. They will be

TABLE 21-1 SOME OF THE MORE COMMON VIRAL DISEASES OF TREES
(After Tattar, T. A. 1989. *Diseases of Shade Trees,* 2nd ed. New York: Academic Press)

Host	Disease
Apple	Apple mosaic
	Chlorotic leaf spot
	Stem pitting
	Flat limb
Ash	Ash ringspot (tobacco ringspot virus)
	Ash line pattern (tobacco mosaic virus)
Birch	Birch line pattern (apple mosaic virus)
Black locust	Black locust witches'-broom (tomato spotted wilt virus)
Cherry	Stem pitting of *Prunus* (tomato ringspot virus)
	Line pattern
	Prunus necrotic ringspot
	Raspleaf
Elder	Tobacco ringspot virus on elder
Elm	Elm mosaic virus
	Elm ringspot virus
	Elm scorch virus[a]
	Elm zonate canker virus
Lilac	Lilac ringspot virus
Maple	Maple mosaic (maple mottle)
	Peach rosette virus on maple
	Tobacco ringspot virus on maple
Oak	Oak ringspot
Peach	Ringspot
	Yellowbud mosaic (stem pitting of *Prunus,* tomato ringspot virus)
	Mosaic
Pear	Stony pit
	Vein yellows and red mottle
	Ring pattern mosaic
Plum	Stem pitting of *Prunus* (tomato ringspot virus)
	Sour cherry yellows (prune dwarf virus, *Prunus* necortic ringspot virus)
Poplar	Poplar mosaic virus

[a]Stipes and Compana (1981) report a bacterium is causal agent.

called mycoplasmas here. No fungi or bacteria were associated with these diseased plants, but the disease-causing agents were transmissible by grafts and by insect vectors, characteristics of viral diseases. In contrast to viruses, which resemble organic chemicals, mycoplasmas are similar to bacteria but have only a plasma membrane with no cell wall. Although mycoplasmas have limited mobility, they can move throughout a plant in the phloem. Disease-causing mycoplasmas are parasites of plants and animals and multiply in living cells by division, like bacteria. The most common vectors of mycoplasmas are leaf hoppers. Mycoplasmas usually incubate 10 to 20 days within the insect's body before they can be transmitted to a new host (Tattar 1989). Once an insect is infected, it can transmit mycoplasmas the rest of its life.

Mycoplasmas

Figure 21–9 This Yoshino cherry tree, which is free of the recurrent necrotic ring spot virus (left), is more vigorous than the same clone infected with the virus (right). The trees have grown three years from being budded on young established rootstocks; the marker stick is 1800 mm (6 ft) tall. (Photos courtesy Andrew Leiser and George Nyland, University of California, Davis)

Symptoms. The multiplication and spread of mycoplasmas block the movement of essential organic materials in the vascular system and can cause proliferation of shoots and roots, commonly known as witches'-broom or hairy root. Roots are affected first because they are soon weakened by lack of food from the tops. Active phloem is killed in the small roots and then in progressively larger roots. Later, the upper parts of plants undergo a progressive decline that usually results in death. Plant decline may be accompanied by any or all of the following: leaf chlorosis, weak growth of many shoots (witches'-broom), branch dieback, lack of flowering and fruiting, phloem necrosis, and proliferation of roots. Some plant diseases known or believed to be caused by mycoplasmas are listed in Table 21–2.

Treatment. Prevention is the only effective way to have plants free of diseases caused by mycoplasmas. Select clean plants and take care during vegetative propagation to use only healthy grafting materials. It is not practical to try to control mycoplasmas by controlling the leafhoppers and psyllids that can transmit

TABLE 21-2 SOME COMMON MYCOPLASMA-LIKE DISEASES OF TREES
(After Tattar, T. A. 1989. *Diseases of Shade Trees,* 2nd ed. New York: Academic Press)

Host	Disease	Host	Disease
Apple	Proliferation disease	Hickory (native species)	Pecan bunch disease
Cherry	Rusty mottle		
	X-disease	Black walnut	Walnut bunch disease
Black locust	Black locust witches'-broom	Butternut	Walnut bunch disease
Peach	Little peach	English walnut	Walnut bunch disease
	Peach rosette	Japanese walnut	Walnut bunch disease
	X-disease	Willow	Willow witches'-broom
	Yellows		
Pear	Pear decline		
Pecan	Pecan bunch disease		

them. Even though tetracycline antibiotic injections result in remission of symptoms for a few years, they do not rid the plant of the mycoplasma (Tattar 1989).

MISTLETOES

Mistletoes are perennial evergreen parasites that grow on the stems and branches of trees and shrubs (Fig. 21–10). Most of the 1300 species or forms of mistletoe are tropical, but the common mistletoes of Europe (*Viscum album*) and the United States (*Phoradendron flavescens*) and a number of other species are found up to latitude 40° and at high elevations closer to the equator (Gill and Hawksworth 1961). *V. album* damages trees from the Mediterranean to Germany but is not a

Figure 21–10 These branches of green ash are infected with *Phoradendron flavescens* var. *macrophyllum;* note the swelling of the branches at the infection sites and the differences in stem diameter above and below the mistletoe (right).

significant pest farther north in Europe. *Phoradendron* does not occur in Canada, although several species infect both hardwoods and conifers in the United States. When winter temperatures sink to $-18°C$ ($0°F$) or below for several days, they have been observed to kill the aerial parts of *P. flavescens* in Illinois. The northern limit of *Phoradendron,* from east to west, is considered to be: southern New Jersey ($40°N$), southern Kansas ($37°N$), and northern Oregon ($45°N$) (Sinclair, Lyon, and Johnson 1987). *P. flavescens* is most serious as a pest in the western states but is also a problem on Florida pecans and Texas citrus (Gill and Hawksworth 1961).

Of greater economic importance are the dwarf mistletoes (primarily *Arceuthobium vaginatum* and *A. campylopodum*), which destroy many conifer species. Some trees are killed, most are stunted, and heart-rotting fungi easily become established.

Trees growing in the open, in woods with open canopies, or near berry-producing plants are infected most frequently by leafy mistletoes depending on the feeding and roosting habits of birds. Dwarf mistletoes discharge their seeds with considerable force, throwing them up to 9 m (30 ft), and strong winds have been known to carry these seeds 400 m (0.25 mile) (Gill and Hawksworth 1961). The seeds of mistletoe are covered with a mucilaginous layer that sticks to birds and small animals for further seed dispersal and to the bark of potential hosts. When the seed of leafy mistletoe attaches to bark, it germinates, and its radicle grows along the bark surface. When it reaches a suitable point, it forms a *holdfast* from which a primary *haustorium* can grow into host tissue. Haustoria function like roots, absorbing water and nutrients from plant xylem. Haustoria invade the cortical region of the bark and form a close union with the host xylem, or a strand from the original holdfast may grow along the bark to form a new plant at a second location. Berry-producing trees and shrubs near susceptible trees have been observed by the author to increase the number of infections. As might be expected, proximity to trees with mistletoe also increases the likelihood of infestation.

Signs and Symptoms. Leaves vary from the minute scales of the dwarf mistletoes to those more than 25 mm (1 in.) long, which many associate with the Christmas season. As is typical of xerophytes, the leaves are thick and leathery, with sunken stomata. Some mistletoes are rather specific and grow on only a single genus; others occur on a wide range of genera. Even though they are completely parasitic, leafy mistletoes manufacture much of their own food while dwarf mistletoes do not. All require water and nutrients from the host plant.

Mistletoes damage forest, fruit, and landscape trees and shrubs. The amount of damage varies with the species of mistletoe, its longevity, and the intensity of its parasitism. The effects of mistletoes may include any of the following: reduced vigor, poor fruit or seed crops, malfunction of woody tissues, production of galls, sparse foliage, top death, predisposition to insect and disease attack, and premature death (Gill and Hawksworth 1961).

Treatment. You can keep mistletoe establishment in new plantings to a minimum by selecting trees that are immune or highly resistant to mistletoe infestation. It is particularly important to remove mistletoe before it produces seed. Removal is easiest during the dormant season, when the green mistletoe is easily seen growing on bare branches. **You can remove mistletoe most completely by pruning out infected tree limbs.** Cut off an affected limb at least 300 mm (1 ft) below the point of mistletoe attachment, preferably at the branch attachment.

If mistletoe grows on main branches or the trunk, so that branch removal would markedly damage tree structure or appearance, a common practice is to cut the mistletoe off flush with the bark. Approximately 75 percent of the *P. macrophyllum* that were cut to 50-mm (2-in.) stubs in Modesto ash and honey locust were still growing two years later (Lichter, Reid, and Berry 1991). In contrast, regrowth from 50-mm *P. villosum* stubs in blue oak was only 20 percent after two years. Pruning is not the answer for control, but it prevents fruiting for at least two seasons.

More than 9000 of the approximately 150,000 public trees in Sacramento, California, were infected with mistletoe in 1977; 95 percent of affected trees were Modesto or Arizona ash (Torngren, Perry, and Elmore 1980). In 1980, the city began a major removal and pruning project. Seriously infected trees, with permission of the adjacent property owner, were removed; if not removed, the trees were pruned severely and infected branches were removed or cut back. Mistletoe on the trunk and main scaffolds was cut off. Less severely infected trees were pruned with similar severity. In addition, the city lent pole pruning equipment to citizens who pruned mistletoe out of private and public trees. The purpose was to improve tree health and to minimize further infection.

The program is still being carried on. Heavily infested trees that were not pruned back are severely stressed or have died (Martin Fitch, Sacramento, CA, 1990 pers. comm.). Trees that were severely stubbed back increased in vigor with low-level infection; however, the weak structure from the severe heading was not acceptable. Trees that could be properly thinned in removing the mistletoe are doing well. If a tree can only be freed of mistletoe by severe heading, the tree is replaced ("replaced" is a more positive term than "removed").

A common recommendation is to cut the mistletoe flush with the bark, wrap the area of attachment with a band of black polyethylene wide enough to exclude light, and tie it with twine or flexible tape (Johnson 1977). If light is excluded, most haustoria systems are thought to die within two years; it is difficult, however, to keep the infected area covered. Lichter, Reid, and Berry (1991) found that within a year, ants, scale, and mealy bugs accumulated under the plastic. Even if effective, wrapping mistletoe cuts is time-consuming and is particularly unsightly in winter when the wraps loosen and blow in the wind.

Dormant application of a 2,4-D (2,4-dichlorophenoxyacetic acid) amine and dicamba mixture as either a paint or a foam (Super D Weedone Foam Weed Control®) is recommended to control *P. flavescens* on limbs too large to be removed (Torngren, Perry, and Elmore 1980). Mistletoe clumps should be broken or headed back to 45- to 90-mm (1.5- to 3-in.) stubs. Cover the freshly cut stub thoroughly with the 2,4-D herbicide, taking care to confine the material to the area immediately around each stub. Treat mistletoe growths less than 75 mm (3 in.) long with herbicide, but do not cut them back or remove them. Herbicidal action is quite slow: Mistletoe growth will initially stop, and later the treated stub will turn brown. The stub should be dead in six to 10 months. Monitor treated trees annually to remove new infections and retreat persistent infections.

Although Torngren and Chan (1978) report that in California tests 90 percent or more of *P. flavescens* was controlled by dormant applications of 2,4-D and dicamba mixture to flush cuts and short stubs, mistletoe reappeared from many of the treated sites. New mistletoe strikes occurred along a branch away from the original location of the treated stub (Martin Fitch 1990 pers. comm.). The new strikes are thought to have come from haustoria that were not killed by the treatments. The mixture of 2,4-D and dicamba is not as promising as originally thought.

Ethephon (Florel®) is showing promise as a control of mistletoe. A spray of 10 percent ethephon in late winter to 50-mm (2-in.) stubs of *P. macrophyllum* and *P. villosum* gave better than 90 percent control through the second year (Lichter, Reid, and Berry 1991). Unfortunately, it was not compared to 2,4-D foam.

Equally interesting is complete repression of mistletoe stubs the growing season after spray painting them with black asphalt. Unfortunately, the trees were severely pruned before second-year results could be taken. This approach appears to be worth further investigation.

Control measures of common mistletoes must be evaluated over several years; two years does not appear to be enough. I have observed an explosion of mistletoe strikes in a few Modesto ash a few years after mistletoe had been pruned out. New mistletoe from old haustoria would seem a reasonable explanation, but no internal examination was made. Although examples are limited, no similar observations have been made for wrapped mistletoe sites.

Dwarf mistletoe of conifers is spread and intensified primarily by explosive fruits rather than by animal vectors. Once the parasite is eradicated or materially reduced in an area, there is little chance for reinfection from the outside (Gill and Hawksworth 1961). Therefore, the removal of infected limbs and trees has proven more effective with dwarf mistletoes than with *P. flavescens*.

Ethephon spraying stimulates abscission of dwarf mistletoe shoots and temporarily prevents spread of the disease by delaying fruiting for two to four years (Livingston and others 1985). This could become a reasonable practice in valuable forest stands and forested residential and recreation areas.

FURTHER READING

See Further Reading section at end of Chapter 23.

CHAPTER 22

Insects
and
Related Pests

Some insects pollinate crop plants or serve as sources of food, medicine, and dyes. Others may seriously damage crop, forest, and landscape plants and transmit serious plant and animal disease pathogens. And still others parasitize other species of insects. The life cycle of most insects and mites is short, and many produce several generations a year and exceptionally large populations when conditions are optimal. Several factors account for these large numbers and persistent survival: a high rate of reproduction, small size, hard external skeletons, and, in many insects, an ability to fly and an inactive or resting stage in the life cycle. Of the more than one million species of insects and mites in the world, about 2500 species damage ornamental plants in the United States alone (Johnson and Lyon 1988).

Insects (*Insecta*) and mites (*Arachnida*) are arthropods, which have an external skeleton, a segmented body, and jointed legs. Insects have three pairs of legs; a body composed of head, thorax, and abdomen; and usually wings (Fig. 22–3). Mites (spiders) have four pairs of legs; bodies with abdomen, a fused head, and thorax (cephalothorax); and no wings (Fig. 22–6). Many insects go through a complete metamorphosis, or four stages of development: egg, larva, pupa, and adult. Plant injury may be caused by the larva or by the adult (Table 22–1). Mites and other insects go through an incomplete metamorphosis, involving only egg, nymph, and adult stages. Control measures depend on the number of individuals present at the most vulnerable stage in a pest's life cycle.

Insects can be conveniently separated into those with chewing mouthparts and those with sucking mouthparts; they differ in their manner of feeding, the type of food they eat, and the type of plant injury they cause. They also respond to different kinds of control measures. Most larvae and some adults have chewing mouthparts

TABLE 22-1 SYMPTOMS AND SIGNS OF INSECT AND MITE ATTACK (Reprinted from Johnson, W. T., and H. H. Lyon. *Insects That Feed on Trees and Shrubs,* 2nd ed. Copyright © 1976, 1988 Cornell University. Used by permission of the publisher, Cornell University.)

Category	Pests often responsible
I. Chewed foliage or blossoms	Larvae of moths and butterflies Sawfly larvae Beetle larvae or adults Tree crickets, grasshoppers, and walkingsticks Snails and slugs
II. Bleached, bronzed, silvered, stippled (flecked), streaked, or mined leaves	Leafhoppers Lace bugs Plant bugs Thrips Aphids Psyllids Spider mites Leaf miners
III. Distortion (swelling, twisting, cupping) of plant parts	Thrips Aphids Eriophyid mites Gall makers Psyllids
IV. Dieback of twigs, shoots, or entire plant; stems, branches, and exposed roots sometimes with holes in bark; wood dust, frass, gum, or pitch may issue from holes	Wood borers Bark beetles Scale insects Gall makers Root-feeding beetle larvae
V. Presence of insect or insect-related, products on plants Honeydew and subsequent sooty mold	Aphids Soft scales Leafhoppers Mealybugs Psyllids Whiteflies
Fecal specks on leaves	Lace bugs Greenhouse thrips Some leaf beetles Some plant bugs Some sawfly adults
Tents, webs, silken mats	Tent caterpillars Leaf tiers Webworms
Bags and cases	Bagworms Case bearers
Spittle	Spittlebugs
Cottony fibrous material	Adelgids Mealybugs Some aphids Some scales Some whiteflies Flatids
Slime	Snails Slugs
Pitch tubes	Some bark beetles
Pitch or gum masses and sap flow	Larvae of certain moths Larvae of certain beetles Larvae of certain midges

and can be particularly destructive. Usually larger than insects with sucking mouthparts, they eat holes or bore in the plant. (Snails and slugs, although they are not insects, have rasping habits and can cause a loss of tissue on landscape plants.) Sucking insects insert their mouthparts into plant tissue and suck the juice. Other sucking pests include mites and nematodes. Insects commonly transmit the causal agents of plant and animal diseases. They can also seriously damage or weaken plants, making them more subject to further insect or disease attack.

A few insects are discussed in this chapter to illustrate their life cycles, the damage they do, and current control measures. Many pests can be successfully controlled by cultural practices which should always be considered before using insecticides (see Chapter 23). Control practices should be based on local or regional recommendations, conditions, experience, and regulations.

CHEWING INSECTS

Insect larvae and adults with chewing mouthparts feed on various plant parts. Caterpillars, cutworms, hornworms, slugs, and worms eat holes in leaves or along the margins. Leaf miners and skeletonizers are larvae that feed primarily between leaf veins, leaving only the epidermis. Leaf miners feed between the two leaf surfaces; skeletonizers may leave only the upper epidermis or none at all. Larvae that feed on roots are usually referred to as grubs, maggots, weevils, and wireworms. Beetles and grasshoppers are adult insects that eat above-ground parts, principally the leaves. The larvae of some species live in the soil, feeding on plant roots and later emerging as adults to eat above-ground parts. Borers are the larval stage of any insect that feeds inside roots, trunks, branches, or shoots of plants. Some adult beetles may also bore into plants.

The Japanese Beetle

Popillia japonica, the Japanese beetle, causes extensive damage to plants in eastern North America. A native of Japan, it was first found in this country in New Jersey in 1916. The beetles have spread through much of the eastern United States. Their continued spread has been limited to the north by winter cold and to the west possibly by lack of rain in late summer (Johnson and Lyon 1988). Favorable conditions exist in many parts of the Pacific coast states, particularly in irrigated areas. Several infestations have been discovered and eradicated in California since 1961.

The Japanese beetle spends about 10 months of the year in the ground as a white grub that measures about 25 mm (1 in.) long at maturity. The grubs feed on plant roots, particularly grasses, in late summer, early fall, and again in the spring. They cause the most serious root damage in late summer. Adult beetles emerge from the soil in late spring and early summer, depending on the climate (Fig. 22–1). Their period of greatest activity lasts four to six weeks, after which they gradually disappear. In the summer, the females lay eggs in small groups 25 to 50 mm (1–2 in.) deep in the soil, usually in turf sod. The eggs hatch within two weeks, and the new grubs begin feeding on roots.

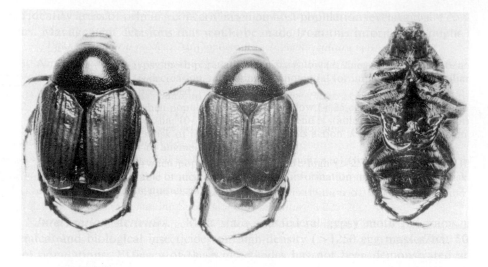

Figure 22–1 Adult Japanese beetles are about 10 mm (0.4 in.) long with shiny, metallic green bodies, coppery-brown front wings, and six small patches of white hair along the sides and back of the body, under the wing edges. (Photo courtesy Leland Brown, University of California, Riverside)

Symptoms. Larval injury to the roots manifests itself as poor plant growth, sometimes leading to death. Grubs severely injure turf and shallow-rooted annuals, but not deeper-rooted healthy plants. Adult Japanese beetles will feed on more than 275 different plant species (USDA 1973b) but seldom bother conifers (Becker 1938). They often congregate and feed on flowers, leaves, and fruit of plants exposed to the sun. They skeletonize the foliage, leaving only a lacework of veins. Beetles are most active during the warmer parts of the day. They usually begin by feeding on the upper and outer parts of plants and work downward and inward (Becker 1938). A severely attacked tree or shrub may lose most of its leaf surface in a short time.

Treatment. A number of natural phenomena govern the extent and severity of a Japanese beetle infestation. Dry summer weather may destroy many eggs and kill young grubs. Conversely, wet summer weather favors egg and grub development and usually results in greater numbers of adult beetles. Beetles and grubs are subject to several diseases, to insect parasites, and to several other natural enemies, including birds, moles, skunks, and shrews.

When adult beetles are active, commercial aircraft are routinely sprayed with insecticides before they leave the eastern United States. Many beetles still reach west coast destinations and must be intercepted either by pheromone (sex attractant) traps or physical searches. Insecticidal sprays are repeated, and many dead beetles have been found on a number of treated airplanes. Federal and state governments have initiated cooperative efforts to retard the spread of Japanese beetles into new areas; they obtain information on the distribution of the insect through the extensive use of traps along the margins of infested areas, at airports, and at other key areas (USDA 1973b).

Japanese beetle grubs can be brought under long-term control if milky disease (*Bacillus popilliae*) spore dust is applied to infested soil areas. This bacterial spore dust, available commercially, can be applied any time the ground is not frozen. Large areas must be treated, since beetles from adjacent untreated areas can fly in to attack the tops of plants. Adult beetles are not affected by milky disease. Control may take several years, since the bacteria must increase and spread in the soil so as to infect newly hatching grubs. During this period, chemical insecticide should not be applied to the soil; although it will reduce the grub population to a low level, it will also delay or prevent the build-up of milky disease bacteria (USDA 1973b).

A few insecticides, when applied according to label directions, will protect turf for one year. Adult Japanese beetles seldom become a problem when an aggressive integrated pest management strategy is practiced over a large, contiguous area (see Chapter 23). Spray thoroughly as soon as beetles appear and before damage is done. Treat only plants that need protection, but treat them repeatedly throughout the period that beetles are active. Follow all label directions on any pesticide.

Insecticides will not fully protect rose blossoms, which open quickly and are especially attractive to beetles. When beetles are most abundant, nip the rose buds and spray to protect the leaves. When the beetles become scarce, let the bushes bloom.

When only a few small plants need protection, you can physically remove beetles by shaking the plant or individual branches early in the morning, when it is cool and the beetles are quiet. Collect the beetles on a sheet placed under the plant and put them in a bucket of water containing a little kerosene. Do this every day because more beetles will fly in.

Plants in poor health are particularly susceptible to attack by beetles (USDA 1973b). Keep plants in vigorous condition with proper fertilization and other necessary cultural practices. The odor of ripe or diseased fruit attracts beetles, which then attack sound fruit. Therefore, remove ripe fruit from the plants and ground.

The Smaller European Elm Bark Beetle

The smaller European elm bark beetle (*Scolytus multistriatus*), introduced into the eastern United States in 1904 near Boston, had been found by 1974 in all but four of the contiguous United States (Arizona, Florida, Montana, and North Dakota) (Johnson and Lyon 1976). It is expected to become established throughout the elms' northern range in North America (Johnson and Lyon 1988). Although this boring insect can seriously weaken trees directly, its function as the principal vector of the fungus (*Ceratocystis ulmi*) that causes Dutch elm disease constitutes its main threat. The bark beetle attacks all species of elm and zelkova (Brown and Eads 1966).

The adult beetle is shiny, reddish brown or black, and about 3 mm (0.12 in.) long. It feeds primarily on the two- to four-year-old twig crotches of living elms (Schreiber and Peacock 1974). Beetles breed in weakened, dying, or dead elm wood whose bark is intact. Although the twigs of healthy elms are readily attacked for feeding, vigorous trees are rarely exploited for breeding purposes.

The adult beetle and larvae make a distinctive gallery of tunnels between the bark and the wood. The tunnel made by the adult female follows the grain of the wood and is usually 25 to 50 mm (1–2 in.) long (Fig. 22–2). Eggs are laid individually in cavities created along the tunnel and packed with some of the frass, the refuse left behind by boring insects, produced by the tunneling. The hatching larvae tunnel between the bark and wood away from the original tunnel and bore into the bark to pupate. As the adult emerges from the pupa, it chews its way through the bark to the surface. Overwintering larvae complete their development during the spring and emerge from the wood as adults at about the time elms begin to leaf out. In summer, the development from egg to adult is complete in about six weeks. Two to three generations occur each year.

Figure 22–2 Smaller European elm bark beetle engraving patterns. The bark has been removed to show adult egg-laying tunnels parallel to the wood grain and fanning out from these the many larval tunnels of increasing diameter. (Photo courtesy Leland Brown, University of California, Riverside)

Symptoms. The smaller European elm bark beetle confines its activities to species of elm and zelkova. It can seriously damage and even kill weakened trees. A heavy infestation can girdle a treetrunk with its system of galleries, disrupting movement of water and food in the tree. The first symptoms that may be noticed are a general weakening of the tree and a lack of foliage on some of the smaller limbs (Brown and Eads 1966). The bark of affected trees will exhibit many "shot holes" on the more severely injured branches or on the trunk. Much paprika-like frass will appear in the limb crotches, clinging to the roughened bark, or on the soil around the trunk. Peeling back bark with shot holes (Fig. 22–2) will reveal single tunnels, parallel to the wood grain, and many smaller debris-filled tunnels running perpendicular to the single tunnels. In summer, adult beetles will be found on the bark surface or in their tunnels. They may also feed in the crotches of small healthy elm twigs (Fig. 22–3).

Figure 22–3 A smaller European elm bark beetle adult and a feeding site in a twig crotch. (Photos courtesy Wayne Sinclair, Cornell University, and Illinois Natural History Survey, respectively)

The wilting, yellowing, and loss of foliage caused by Dutch elm disease may be the first prominent evidence of attack by the elm bark beetle. In such cases, beetle damage is insignificant compared with Dutch elm disease itself, which is usually fatal. In arid regions, however, the smaller European bark beetle can be directly fatal to elms suffering from drought.

Treatment. Populations of elm bark beetles can be reduced and kept at low levels by keeping trees vigorous and free of weak and dead branches. Beetles are reluctant to bore into the bark of healthy elm trees. To reduce breeding sites, burn or bury branches removed from trees, or remove the bark and burn or bury it. Destroy or debark elm stumps and wood piles. In order to minimize beetle infestation and infection by the Dutch elm disease fungus into areas where the disease is absent, elm firewood stacked for winter burning should be covered with clear plastic sheeting in the summer (Svihra 1987).

Do not prune elms during beetle emergence in the spring and summer, since fresh pruning cuts attract beetles. If wounds must be made a short time before or during beetle emergence, paint them with a thin coat of asphalt, which may reduce the attractiveness to beetles. Urea or nitrate fertilization and appropriate irrigation will help revitalize a tree weakened by elm bark beetles and will reduce the likelihood of further infestations. Methoxychlor has been applied as a dormant spray in the spring before beetle emergence (Schreiber and Peacock 1974). Different insecticides have also been registered for use in certain areas. Thorough coverage of the bark is essential, even on the tops of the trees which are not easy to reach. Sanitation practices leading to a reduction of breeding sites for beetles are most effective and also less risky to mammals and humans.

Chewing Insects

Like insects with chewing mouthparts, sucking insects attack all parts of plants, though they usually prefer leaves and developing shoots. They may stunt and deform new growth, curl leaves, form galls, and seriously weaken plants. They may cause mature leaves to yellow or glaze and, if infestation is heavy, to dry up.

Aphids

Aphids are soft-bodied small insects with sucking mouthparts. Some have wings; others do not. Species may be black, brown, green, pink, purple, red, or yellow. Most aphids have naked bodies, although many secrete a white, cottony, wax-like material. In most areas, overwintering eggs hatch into wingless females that, without fertilization, give birth to living females. The ability of females to reproduce without mating is termed *parthenogenesis*. Young are brought forth throughout the summer, accounting for rapid increases in aphid populations. Winged females are produced later in the summer or when crowding occurs. In late fall, sexual males and females are born. After mating, the female deposits one or more overwintering eggs (Essig 1958), and the cycle continues the following year.

Symptoms. Aphids live in colonies on the bark, leaves, blossoms, fruit, and roots of most plants at one time or another (Fig. 22–4). Most species excrete large amounts of honeydew, which coats the host plants and often drips onto objects below. The honeydew attracts ants, flies, and bees and serves as a medium for the growth of sooty mold fungus. In most cases, injury from aphids is inconspicuous, but the plant gradually weakens as its sap is sucked. Young growth may be stunted and deformed.

Figure 22–4 Rose aphids (*Macrosiphum rosae*) on a young stem include wingless stem mothers and young. Note the spherical black parasitized aphids. (Photo courtesy Leland Brown, University of California, Riverside)

Treatment. A number of natural enemies prey on aphids. Important predators are the larvae of green lacewings (Neuroptera), ladybird beetles (*Coleoptera*), and syrphid flies (*Diptera*). Lacewing eggs are available commercially for biological control of aphids.

Knock aphids off infested plants with a spray of water; use a hose for a few small plants or a power sprayer for more numerous, larger plants. A mild soap solution is more effective than water alone and will kill some of the aphids that may remain on plants.

Control ants in the landscape. Ants often "domesticate" aphids and use their honeydew as a source of food. Keep ants out of trees and shrubs by spraying the ground and the base of plants with diazinon. Alternatively, band the trunks of trees and shrubs with a sticky material, such as Stik-Em® or Tanglefoot®, that traps crawling insects.

Aphids can be controlled chemically with any one of a number of contact insecticides or systemics that move within plants and poison sucking insects. Systemics can be watered into the soil for root uptake; this has proved successful with small shrubs such as roses.

The San Jose Scale

The San Jose scale (*Quadraspidiotus perniciosus*) afflicts a wide range of broad-leaved woody plants in many parts of the world. In about 1870, this pest was introduced from the Orient into California; from there, it quickly spread and began to afflict deciduous fruit trees throughout the United States (Essig 1958). Until the newer synthetic organic insecticides became available in the late 1940s, the San Jose scale was one of the most common, destructive, and widely distributed pests to infest woody plants (Johnson and Lyon 1988). If not controlled, it can kill plants.

Most scale insects are very small. The San Jose scale is an armored insect, characterized by a protective shell or scale that covers its body (Fig. 22-5). The young, called *crawlers,* are born alive or from eggs and are able to crawl consider-

Figure 22-5 An old heavy infestation of San Jose scale on an apple twig. Several females (large) and many males (small, lighter gray) are present. (Photo courtesy Leland Brown, University of California, Riverside)

able distances to find a feeding site. During the first molt, they lose legs, antennae, and anal filaments and assume a stationary form, consisting primarily of mouth-parts (Essig 1958). The scale covering is black at first and gray when more developed and has a tiny, yellowish, nipple-like formation on the insect's back. Armored scales, in contrast to unarmored or soft scales, produce almost no honeydew. As many as five overlapping generations of scales develop in a single season (Johnson and Lyon 1988). The bark of shoots, limbs, and trunk is often completely encrusted with scales, which suck sap from the plant. Small scales, gray to black, can be seen on bark and on fruit surfaces. The bodies under the scale shell are bright yellow.

Symptoms. Plants seriously attacked by San Jose scale will weaken and die unless the pests are controlled. Leaves on weakened branches may drop prema-turely, and seriously affected branches may be killed back. Dark discolored areas may appear on infested young bark and the tissue below. The scale has an apparent toxic effect on plant tissue (Essig 1958).

Treatment. Even though several beneficial insects prey on the San Jose scale, they can seldom reduce scale numbers during periods of rapid population increase. Susceptible plants should be fertilized sparingly with nitrogen or not at all; scale insect population can be double on vigorous plants what it is on plants of low vigor (Johnson 1982). Dormant oil sprays may be used on deciduous plants as late in winter as possible but before the buds begin to swell (Moore, Davis, and Koehler 1979). Spray broadleaved evergreen plants, tender, dormant deciduous plants, and plants in leaf with a refined summer oil to control scale insects with little or no injury to the host plant. To increase their effectiveness, fortify either the dormant or the summer oil sprays with a systemic insecticide recommended for the region. During the spring growing season, apply the first spray when you first see the tiny crawlers, or immature scales, on the plants.

Spider Mites

Mites have eight legs when mature and have the other distinguishing characteristics of spiders (Fig. 22-6), rather than those of insects. Of the plant-feeding mites, spider mites are the most injurious to agricultural crops and landscape plants throughout the world (Essig 1958). Many species of spider mites attack woody plants.

As their name implies, mites are extremely small. Adult spider mites range in length from 0.20 to 1.0 mm (0.01–0.04 in.). The various species may be yellow, green, red, or brown; the "red spider" is familiar to most horticulturists. Many of these mites also spin extensive webs, which protect them and their eggs. As adults, the mites hibernate in the soil in cold regions, and in trash and plant crevices in warmer areas. Adults appear in late spring. Each female lays 50 to 60 eggs, primarily on leaves. The eggs usually hatch in three days, the mites reach maturity in another 10 to 12 days, and they live but one to two weeks in the summer. Needless to say, mite populations can increase rapidly under favorable conditions.

Symptoms. With a pair of needle-like teeth (*chelicerae*), the spider mite punctures and drains the epidermal cells of the host plant. Injured leaf, fruit, or

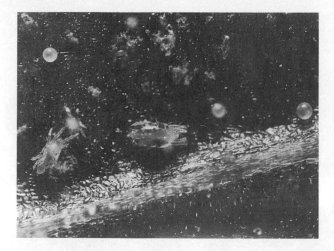

Figure 22-6 Pacific spider mites, *Tetranychus pacificus,* shown near vein of a bean leaf used in rearing. To left: slender, adult male near small, fat nymph; adult female in center; several shiny, translucent, spherical eggs are seen, without webbing; near top: whitish surface cells, lacking chlorophyll, from mite feeding. (Photo courtesy Leland Brown, University of California, Riverside)

shoot surfaces have a flecked or stippled appearance. Leaves become yellow or bronze and drop; fruit may be scarred; weakened plants can be killed. Hot, dry weather and a dusty environment favor mite build-up and injury to plants. Some mite species, however, are most destructive during the cooler weather of spring or fall.

Treatment. To minimize spider mite build-up, keep plants vigorous and well watered, and the landscape free of dust. Sprinkler irrigating or spraying susceptible plants with water will help. In arid regions, mites usually disappear with the first fall rains.

Several insects and mites prey on spider mites but effectively reduce their numbers only after leaves are severely damaged (Johnson and Lyon 1988). Unfortunately, sprays often do more harm to mite predators than to mites.

Galls

Tiny wasps and flies commonly cause a wide range of swollen abnormalities, or *galls,* on plants. In North America alone, more than 800 different kinds of galls are found on oak trees and more than 400 on beech (Schuder 1978). Galls are found on almost all species of landscape plants but are particularly numerous on rose, willow, walnut, pine, elm, and birch. Practically all plant parts may be attacked by insects that stimulate galls. Slightly more than half of the different galls are found on leaves; about 14 percent are found on twigs and branches; 7 to 9 percent each are found on buds, roots, flowers, and fruits.

Each species of insect causes a particular type of deformity on a particular plant part on a single or closely related plant species. Each type of gall has a distinct shape, size, structure, and color (Fig. 22-7). The insect involved can usually be identified entirely on the basis of the gall produced.

When a wasp lays eggs, plant tissue is stimulated into abnormal growth that usually surrounds or engulfs the eggs. The gall protects and feeds the larvae develop-

Figure 22–7 Willow apple galls caused by the sawfly (*Pontania pacifica*) on leaves of willow (left). "Oak apples" up to 50 mm (2 in.) in diameter induced on a branch of valley oak by the gallwasp *Andricus californicus* (upper right). Walnut purse galls are caused by an eriophyid mite (lower right). (Photos courtesy Carlton Koehler, University of California, Berkeley, 1980)

ing within. After wasp adults emerge from galls, they usually lay eggs in the early spring that cause gall formation and larval growth to occur when plant growth is most active. One generation per year is normal. Some wasp species produce alternating generations, so the adults that emerge from a gall are different from the adults that laid the eggs. The new adults usually lay eggs on a different plant part; the resulting larvae and adults are similar to those that existed two generations before.

Symptoms. Galls seldom cause serious injury, although twig galls produced by two species of *Callirhytis* in eastern North America can seriously injure or kill oak trees. The horned oak gall, caused by *C. cornigera,* can injure black, blackjack, pin, scrub, and water oak. *C. punctata* produces gouty oak galls on the twigs of black, pin, red, and scarlet oak (Johnson and Lyon 1988); these galls, up to 50 mm (2 in.) in diameter, often grow together and engulf more than 300 mm (12 in.) of a twig. Felt (1940) saw one red oak that bore an estimated 20,000 gouty oak galls. Beyond the galls, branches may die back.

Treatment. Control measures are seldom taken against gall-causing insects. Plants are sometimes sprayed with a dormant lime sulfur or a dormant miscible oil to reduce the number of overwintering insects. Valuable trees and those heavily infested with gall insects should also be sprayed with carbaryl (Sevin®) or methoxychlor-Kelthane just after midspring and again a month later (Pirone and others 1988). Dead and heavily infested branches should be removed and destroyed before the adults emerge in early spring.

NEMATODES

Plant-parasitic nematodes are microscopic worms 0.5 to 5 mm (0.02–0.2 in.) long. Unlike most other nematodes, they have a stylet (which resembles a hypodermic needle) in their mouth which they use to pierce cell walls. They can then feed and move into the penetrated tissue. Mature nematodes may be cylindrical or pear-like in shape, while most of the young are slender and worm-like. They may live in the soil and feed on roots or may spend most of their adult lives inside roots. Once nematodes have entered roots, they spread easily and are difficult to control. Some insects are parasitized by nematodes.

Symptoms. Nematodes attack almost any plant part but frequently attack woody plants only in the roots. The pinewood nematode (*Bursaphelenchus lignicolus*), however, invades the vascular system of twigs, limbs, and trunks in pine trees. Foliar nematodes have been reported on woody plants. Nematodes feed most commonly in or on the tips of small roots. The injury they cause reduces absorption and movement of water and nutrients. The most common symptoms are therefore typical of many other soil and root disorders: poor growth, leaf chlorosis, early leaf drop, and so forth. Diagnosis in woody plants is difficult without an examination of the roots and soil.

Root-knot nematodes are among the most damaging and attack the greatest number of plants; *Meloidogyne incognita* alone infests more than 1700 species (Streets 1984). Root-knot nematodes enter roots, causing the cells and tissue to enlarge. The smaller infested roots will have many small to medium-sized swellings or galls (Fig. 22–8). Large swellings differ from crown gall in that they extend along the root instead of adopting a spherical shape and are associated with many smaller knots. *M. incognita* is most common in areas with mild winters, while *M. hapla* infests crops and landscape plants in the northern United States and Canada. Nematodes are not particularly active below soil temperatures of 18°C (65°F) (Streets 1984). Porous sandy and loam soils favor their spread and activity. When they enter the soil, nematodes essentially live in the film of water surrounding soil particles; thus wet soils favor nematode infestation (Tattar 1989).

Meadow or root-lesion nematodes (*Pratylenchus* spp.) also attack a wide range of plants, entering the small roots and feeding in the cortical tissue of the young roots. The injured tissue dies, forming a dark-colored lesion visible on the surface of the root. Young roots can be girdled or seriously injured and the upper plant condemned to poor growth.

Stubby root nematodes (*Trichodorus* sp.) feed externally on root tips, causing the tips to thicken or become corky. As secondary roots begin growth, they too are attacked. The root systems of affected plants, including the feeder roots, are usually sparse. The curtailed root system results in a weakened top.

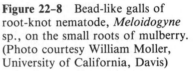

Figure 22–8 Bead-like galls of root-knot nematode, *Meloidogyne* sp., on the small roots of mulberry. (Photo courtesy William Moller, University of California, Davis)

Treatment. Check a planting site for root-knot nematodes by examining the roots of the most susceptible plants. If no susceptible plants grow on the site, test the soil in containers in which you plant seeds or seedlings of fast-growing species like beans, squash, or tomatoes. After the above-ground portion has grown for at least three or four weeks, carefully remove the roots, wash them, and examine them for root-knot galls. The number and size of galls will give a rough estimate of nematode infestation (Streets 1984).

If the soil is free of nematodes, be sure that planting stock is free of them also. Most nurseries use nematode-free soil if government regulations require pest-free plants. Even so, before accepting plants, check the roots for nematodes. If you find any root-knot galls, reject all plants from that source.

Although many landscape plants are susceptible to root-knot nematodes, a number are resistant or tolerant. If nematodes are a problem in your area, check local government publications or talk with extension personnel about lists of susceptible and resistant or tolerant plants. Such lists are provided by Streets (1984), Lear and Johnson (1975), and Flint (1990). You may use resistant rootstocks for plants that would otherwise be susceptible, such as Nemaguard or S-37 rootstocks for flowering and fruiting peaches.

Biological control, to date, has not proven successful. Marigolds are being more extensively studied as a suppressant of root-knot nematodes. Marigolds may inhibit some species but are a ''very good host to the northern root-knot nematode, *Meloidogyne hapla*'' (Flint 1990).

Isolated planting areas (containers, interior planters, roof gardens, and the like) can be freed of nematodes by heat sterilization or chemical fumigation before planting. Place pest-free soil in self-contained planters to ensure healthy plants and to minimize future maintenance. Steam or electric heat will pasteurize the soil. Temperatures above 45° C (110° F) will kill nematodes; 60° C (140° F) will kill most soilborne pathogens and insect organisms but will kill few beneficial microorganisms. Solar heat may be used when space and time allow. When moist soil is covered with clear polyethylene film and exposed to the summer sun for four to six weeks, nematodes, verticillium wilt fungus, and some annual weeds may be killed to a depth of 300 mm (12 in.). In open landscapes, nematode populations can be effectively reduced for a year or two to help new plantings get off to a vigorous start. Nematodes from adjacent untreated soil move back into the treated areas, and nematode kill will be uncertain if infested large roots (more than 25 mm or 1 in. in diameter) remain in solarized or fumigated soil.

Selective nematicides such as Nemacur® and Vydate® L may help landscape plantings get off to a better start by "providing season-long protection." Nematode control is not complete; annual application probably will be necessary. Such a program, however, reduces beneficial organisms and can lead to nematode resistance. In addition, some plant species are sensitive. Experiments indicate that application through drip irrigation systems may be most cost effective since water is needed to carry a selective nematicide into the soil and/or slow its loss from the soil; it is also applied where the most effective roots are located.

DBCP fumigants have proven extremely effective against nematodes and may be applied without plant injury to certain plantings (Lear and Johnson 1975). Their use, however, is no longer permitted in the United States because underground water supplies have become contaminated in some areas, and they have been linked to sterility of male employees at plants where the chemicals were manufactured.

Preplanting fumigation with nematicides such as methyl bromide and Telone® in sufficient concentrations are effective in killing nematodes and most other organisms, including plants and seeds. Their use should be considered for infested soil to be used in planters or in an uninfested planting, an infested plant-free area that is to be planted, or for eradicating or slowing a new infestation in an existing planting.

Almost all of these fumigants require soil preparation, application by injection—in water or under an airtight tarp, and waiting at least 48 hours before planting. Almost all of these chemicals must be applied by a government certified applicator. Regulations may also limit where these chemicals can be used. **If in doubt, check with the proper authorities as to the use of a pesticide. Read and follow label directions.**

FURTHER READING

See Further Reading Section at the end of Chapter 23.

CHAPTER 23

Integrated Plant Management (IPM+)

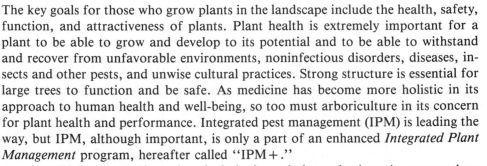

The key goals for those who grow plants in the landscape include the health, safety, function, and attractiveness of plants. Plant health is extremely important for a plant to be able to grow and develop to its potential and to be able to withstand and recover from unfavorable environments, noninfectious disorders, diseases, insects and other pests, and unwise cultural practices. Strong structure is essential for large trees to function and be safe. As medicine has become more holistic in its approach to human health and well-being, so too must arboriculture in its concern for plant health and performance. Integrated pest management (IPM) is leading the way, but IPM, although important, is only a part of an enhanced *Integrated Plant Management* program, hereafter called "IPM+."

An IPM+ approach needs to include site and plant selection; site preparation; planting and early care; managing nutrient, water, and aeration levels; pruning; monitoring plant performance; preventing or moderating plant problems; and knowing when plants should be replaced. Such an approach will improve landscape performance and function; increase the efficiency and effectiveness of maintenance; preserve more existing plants; and provide horticulturalists with opportunities to serve clients in the planning, development, and maintenance phases of their landscapes. Most horticulturists, particularly commercial arborists, work primarily with mature landscape plantings often to the detriment of the plants, the plant owners, and the horticulturists—much of their work is correcting problems that could have been easily prevented or minimized. An IPM+ approach should be a "Win, Win, Win situation."

THE ESSENCE OF INTEGRATED PLANT MANAGEMENT

An enhanced *Integrated Plant Management* (IPM+) program can be started in any landscape at any stage of development. However, the earlier it is begun, the more effective it will be. The person to carry out such a program could be an arborist, gardener, horticulturist, urban forester, or landscape contractor (hereafter called an *"IPM+ arborist"*).

The developer, the planner, the architect, and the landscape architect should consult with an IPM+ arborist so that a project site can be wisely developed. An IPM+ arborist would see that

Quality existing plants were identified

Grading plans, building and road placement, utility runs, special landscape features, and construction sites were located to preserve the greatest number of existing plants and provide suitable new planting areas

Endangered plants were moved to new locations

Construction procedures were specified to preserve existing plants and future planting sites

Construction was overseen to ensure that all workers understood the importance of the plants to be saved and that existing plants were maintained in good health during construction

Landscape architects and home owners would benefit from information supplied by an IPM+ arborist concerning

The planning, installation, and early care of new landscapes

The soil, drainage, and utility and pavement placement

Appropriate plant species, plant quality, planting procedures, and early care

An IPM+ arborist could

Ensure that nursery plants meet plant specifications as to root circling and tree health, vigor, and structure

Inspect planting and staking practices

At the beginning of an IPM+ program on a property, an IPM+ arborist would

Inventory existing plants as to species, location, size, condition, and care needed to return plants to good health and safety

Inspect tree root-collars for depth of planting, circling and kinked roots, presence of girdling rope or wire, and the condition of the main roots

Determine the extent, capacity, and condition of the irrigation system and water quality

Examine the soil as to its texture, structure, degree of compaction, uniformity with depth, and presence of soil pests

Show and explain to the property owner the condition of the landscape—what needs to be done to bring the plantings to different levels of well-being, the possible levels of continuing IPM maintenance programs, and the costs involved in the different possibilities

As an IPM+ arborist carries out a maintenance program

Plants would be monitored as to vigor, structure, safety, appearance, and the presence of diseases and insect pests, as well as beneficial organisms

Appropriate IPM procedures could be performed to keep plants at desired vigor and disease organisms and pests at acceptable levels

Timely pruning of mature trees and training of young trees could be performed to ensure strong structure and effective landscape function and appearance

Staking, mulching, and irrigation systems could be properly maintained

Written and possibly photographic records would be kept of monitoring observations, work performed, assessment of results, and recommendations for future work; all notations should be dated and the reporter identified.

Other services could be performed as requested

An IPM+ approach to landscape planning, design, installation, and maintenance holds great promise for all involved. Much of the information presented in this book can be applied to this endeavor. We all need to become more knowledgeable and versatile in our ability to install the right plants in the right environments in the right way so that wise maintenance can be more effectively and efficiently performed.

INTEGRATED PEST MANAGEMENT

An Integrated Pest Management program is a key part of an Integrated Plant Management undertaking. The main approach to pest control has been a chemical one since the 1940s, when the chlorinated hydrocarbons, the organophosphates, the carbamates, and a host of fungicides, herbicides, and other pesticides were first developed. The effectiveness of many chemicals has led to the reduction or discontinuance of some earlier control procedures—plant sanitation, crop rotation, encouragement of natural enemies, and other cultural strategies.

Unfortunately, heavy dependence on chemical control has caused problems in certain situations. Some major pests have developed resistances to previously effective insecticides. As of 1975, 75 percent of the most serious agricultural insect pests in California had developed resistance to one or more major insecticides. In some cases, the most effective chemicals caused resistant individuals to predominate all the sooner, resulting in a resistant strain. Although the most noteworthy examples involve insects, other pests develop resistance also.

Other problems have emerged from primary reliance on chemicals. After treatment with broad-spectrum pesticides, pest populations, particularly insects, will sometimes drop drastically and then surge to levels higher than before. Resurgence occurs because the pesticides kill not only the pest but its natural predators as well. If predators are not killed outright, many starve or emigrate out of the immediate area in search of food. The surviving pests can then multiply rapidly, since their food supply, unlike that of their predators, has not been interrupted. Some organisms that have not previously been problems will increase to damaging levels when their predators are decimated. For example, mites may increase dramatically on plants that have been sprayed with certain insecticides for the control of other pests.

In the face of pest resistance, pest resurgence, and secondary pest outbreaks caused by very effective pesticides, growers have often increased pesticide application in efforts to obtain more complete control. Increases in both concentration and frequency not only intensify the problems but hasten environmental contamination by the more persistent chemicals. DDT is the notorious example. It has been used effectively against many insects that transmit serious animal and plant diseases; damage many crop, forest, and landscape plants; and annoy people. One of its advantages as an insecticide—persistence—has proved to be a great disadvantage. DDT residue is everywhere, carried by wind and water to places remote from the sites of application. In addition, DDT and related chlorinated hydrocarbons accumulate in the food chain and may be concentrated in this fashion to hazardous levels.

The United States banned DDT for most uses in 1972 and has banned or severely restricted the use of aldrin, dieldrin, endrin, heptachlor, and chlordane. Restrictions affect the sale and application of many pesticides; keep up-to-date on the status of pesticides used. In the United States, **pesticides are to be used only on the plants or sites listed on the label, only against the listed pests, and only in the specified concentrations and at specified times.** Even though pesticide residues do not cause the problems on landscape plants that they cause on edible crops, the use of hazardous chemicals in the landscape is severely restricted because of the proximity of people, pets, and nontargeted plants.

For a number of reasons, primary reliance on pesticides or any other single pest control method has usually not been effective. Using a combination of techniques to control or reduce pests, while minimizing undesirable effects, is far from a new idea but is now receiving renewed interest. This approach has been greatly strengthened by a broader understanding of the interdependence among living organisms. A landscape planting, an urban forest, an agricultural area, or an entire region must be considered as a unique ecosystem. It may not be possible to reduce the population of one species without changing other components of the system. This concept suggests that even pests may be accepted within certain economic or aesthetic bounds. This approach to pest control is called Integrated Pest Management; it really is more appropriate to call it Integrated Plant Management.

Flint and van den Bosch (1981) outlined the essentials of an integrated pest management (IPM) approach. It is assumed that the pests are known to cause serious injury or effects on the host plant or its surroundings. Adapted to managed landscape plantings, IPM includes the following concepts:

> A plant or landscape planting is a component of a functioning ecosystem, even though the planting has been planned and developed. Hence, actions should be designed to develop, restore, preserve, or augment natural checks and balances, not necessarily to eliminate pest species.

The mere presence of a pest organism does not necessarily constitute a pest problem. Acceptable population and damage levels must be determined.

All possible pest control options should be considered before action is taken. Techniques employed should be as compatible with each other as possible.

An IPM program should be based on certain guidelines (Flint and van den Bosch 1981).

Understand the biology of the plants involved, especially the manner in which they are influenced by the surrounding ecosystem.

Identify the key pests, know their biology, recognize the kind of damage they inflict, and study the economic and aesthetic consequences of control measures.

Identify the key environmental factors that impinge (favorably or unfavorably) upon pest species and potential pest species in the ecosystem.

Consider concepts, methods, and materials that, individually and in concert, will facilitate permanent suppression or restraint of pest and potential pest species.

Structure a program so that it can be adjusted to meet change or varying situations.

Seek the weak links in the life cycle and population structure of important pest species, and direct control practices as narrowly as possible at those weak links. Avoid broad impact on the plant ecosystem.

Whenever possible, employ methods that preserve, complement, and augment the biological dynamics that characterize the ecosystem.

Whenever possible, diversify the ecosystem.

Monitor pests, natural enemies, and tree health regularly.

Anticipate unforeseen developments, move with caution, and be aware of the complexity of the landscape ecosystem and the changes that can occur within it.

Aesthetic Injury Threshold

One of the tenets of IPM is that certain levels of injury or pest populations are to be expected. The aesthetic injury threshold is the highest level of pest habitation or damage that would be acceptable to most of the people who use the affected area (Olkowski 1973). Such levels may be difficult to determine because of the variability of plants, climates, and human values or tolerances. Landscapes are used and viewed from both close and distant ranges. Higher levels of plant injury will be more acceptable in a "far" landscape.

Some pests can cause death or serious injury, but many only change the appearance of plants or create products that annoy people or cause damage to nearby vegetation. Plants can become quite unattractive when leaves are damaged by defoliating insects and diseases. A number of aphids, scales, and leaf hoppers excrete honeydew, which can drip on sidewalks and cars. Caterpillars may leave the plants on which they originate to injure adjacent plants. Occasionally, a large insect population may disturb viewers even though it causes little or no harm to host plants. Acceptable levels of injury or pest presence can usually be increased if the public is informed about pest management programs and their consequences to plant and human health.

Control Threshold

Control measures must usually be taken before the injury threshold is reached. This requires information about the pest's life cycle and monitoring of the plants, the pest, and its predators. Monitoring and sampling techniques have been designed for many agricultural insect pests. Some are being developed and incorporated into control programs for pests of landscape plants. For example, there are monitoring systems for fire blight (caused by *Erwinia amylovora*) and the bark beetles that transmit Dutch elm disease (*Ceratocystis ulmi*). These systems help arborists differentiate between the mere presence of a pest and densities high enough to cause unacceptable damage. Needless to say, such information is likely to be developed only for pests that cause serious damage to valuable plants that are cultivated in large enough numbers to make the effort cost-effective.

Integrated Pest Management Options

Not all measures for controlling a pest or minimizing damage need be delayed until the control threshold approaches. Some of the most economical and effective management tactics are preventive and should be employed even before pests are observed. A regular field-monitoring program must be used to determine the continuing efficacy of preventive tactics; if pests rise above control thresholds, more immediately effective measures, such as pesticide application, may be appropriate. Integrated pest management programs should, however, employ a variety of practices in concert to obtain the least disruptive, most effective, long-lasting pest control at an acceptable cost. Five major types of control measures—regulatory, genetic, biological, cultural, and chemical—contribute to an integrated pest management program. Much of the following discussion is based on *Introduction to Integrated Pest Management* (Flint and van den Bosch 1981).

Regulatory Control. An individual arborist has little or no choice in the initiation or implementation of regulatory controls, which are usually carried out by governmental agencies. Such controls can exert great influence on pest populations in a given area. Pests may be kept out of an area through quarantine and inspection; outbreaks of new pests in small areas may be eradicated; large-scale cooperative efforts (such as removal of trees infected with Dutch elm disease) may be properly supervised and carried out. Not only should such programs be supported to achieve the greatest good for the greatest number of people, but we must obey quarantines that prohibit carrying or shipping plant material into our country or across certain state borders unless it has been inspected and certified by appropriate officials.

Genetic Control. Control of pests by genetic means takes two forms: (1) host resistance—selecting or developing a plant that is resistant to particular pests; and (2) autocidal control—modifying the genetic makeup of the pest population so that it becomes self-destructive and cannot survive.

Finding or developing *host resistance* is the most successful and ecologically sound way of avoiding problems caused by insects, disease, and nematode pests (see Table 23–1). An analysis of more than 30,000 plants in the mid-Atlantic United

TABLE 23-1 TREE SPECIES WITH THE HIGHEST AND LOWEST INCIDENCE OF PROBLEMS (DISEASE, INSECTS, ENVIRONMENTAL STRESS) IN SIX MARYLAND COMMUNITIES IN 1982 (Raupp and Noland 1984)

Species	Percent most problem prone	Species	Percent least problem prone
Peach	100[a]	Black locust	0
Crabapple	78	Black gum	0
Apple (fruit)	67	White oak	3
Flowering cherry	31	Tulip poplar	4
Dogwood	28	Hickory	4

[a] Peach trees averaged 1.1 problems per tree.

States revealed that species of *Malus, Pyracantha, Cornus, Prunus,* and *Rosa* were the most problem prone while those of *Viburnum, Taxus, Forsythia* were least susceptible (Raupp and others 1985). Select plant species and cultivars with proven resistance to noxious pests associated with the growing area. Various regional publications list plant species and their relative susceptibility or resistance to common pests, particularly nematodes and vectors for disease. Grafting may protect certain susceptible plants from soil-borne diseases or nematodes. A flowering peach, for example, can be grown on a rootstock resistant to nematodes.

Normally, you would take advantage of pest resistance in perennial plants by proper selection before planting. It may be wise, however, to replace problem-prone plants with resistant species or varieties. The temporary loss in aesthetic value of mature plants may be outweighed by the long-term benefits of reduced maintenance and attractive plants in the future.

Within a susceptible species, individuals may be resistant or relatively resistant to a given pest. Such plants can be propagated vegetatively to preserve that resistance. They can also be used in breeding programs that will convey this resistance to other plants with more desirable characteristics or, vice versa, will improve certain characteristics in the resistant selection.

Breeding for pest resistance is a long-term program, particularly when perennial plants are involved. What is more disheartening, a number of pests have evolved strains able to overcome the host plant's mechanism of resistance—much as pests have developed their own resistance to pesticides. A plant naturally resistant to one serious pest may be susceptible to others, so that careful testing must precede the release of a new selection for widespread planting. Oriental pear, for example, was used as a rootstock to impart fire blight resistance to pear fruit trees but predisposes the tops to "hard end" fruit and pear decline.

Host plant resistance may be due to physiological factors (such as toxic compounds within the plant that inhibit the pest), mechanical factors (such as cuticle too thick or tough for the pest to penetrate), or the host's ability to survive despite damage from pest populations. Host resistance is usually most effective and long-lasting if it relies on more than one gene and preferably more than one character. Pests are less likely to develop strains able to overcome such "polygenic" resistance.

Host resistance and the enhancement of other characteristics may be more likely in the future as gene transfer becomes better understood and techniques improve.

Autocidal control causes the pest to contribute to its own demise. The best-known example is control of the screwworm, a serious livestock pest in the southern and southwestern United States: Releasing large numbers of laboratory-propagated sterile males into the environment has eradicated the screwworm in Florida. This technique has also eradicated two species of fruit flies on the island of Rota in the Marianas (Flint and van den Bosch 1981) and has been used against invasions of Mediterranean fruit flies in California.

Several factors are essential for successful eradication by the sterile male technique: (1) The females should mate only once; (2) the area of release must be geographically isolated so that migrants of the species will not come in from untreated areas; (3) the sterilized males must be sexually competitive with normal males; (4) the pest must be amenable to laboratory rearing, and such rearing must not have a debilitating effect on its field performance or survival; and (5) the budget for the program must be extremely large (Flint and van den Bosch 1981). A successful autocidal control program must eradicate the pest. If a few females continue to produce offspring, or if migrants come from untreated areas, the expensive program must be continued or abandoned.

Biological Control. Control by natural enemies can be effective, long-lasting, economical, and minimally disruptive of the ecosystem. Natural enemies include organisms that feed on pests (predators), parasitize pests (parasitoides), or displace pests (competitors). When managing any landscape, you should assess present and potential plant pests and their natural enemies. Assessment is complicated by the fact that most landscapes have been greatly changed from their original natural state. Exotic plants may be accompanied by their native pests but not the pests' natural enemies or may be susceptible to organisms present in the new environment that do not damage native plants.

In essentially undisturbed native plantings and in wisely managed landscapes, many potential pests are kept at safe levels by other organisms, the food supply, or other environmental factors. The importance of these natural controls is seldom appreciated until the balance is upset by a catastrophe (flood, fire, major land development); some ill-advised cultural practice; or a toxic, broad-spectrum pest control program. In such situations, pests that are seemingly new rise to injurious levels or major pests become even more damaging.

Aphids, scales, mealybugs, and other insects that suck plant juices are often controlled by predators and parasitoids, except when their natural enemies are disturbed. **Natural enemies can be conserved by using narrow-spectrum, contact pesticides applied only to plants with damaging pest levels.** Most insects that cause damage do not move much; predators are quite mobile. Even though predators and parasitoids may be killed on the plants sprayed with a short-residual spray, predators and parasitoids on nearby, unsprayed plants can move in quickly to keep the pest population under control.

Reducing dust and controlling ants also favor the natural enemies of pests. Ants commonly feed on insect honeydew and protect pests by killing their enemies.

Control ants by trimming branches to eliminate "ant bridges" and applying sticky material to trunks, or by placing ant stakes or other enclosed insecticide baits around the base of plants.

Classic biological control involves the deliberate introduction and establishment of natural enemies where they did not previously exist. Such programs are used mainly to control exotic pests that have been introduced without their key natural enemies. The exotic species may be relatively unknown as pests in their native habitats, where they are naturally kept within bounds. Because of the absence of natural enemies, exotic species comprise a high percentage of major pests in the United States. Seventeen of the 28 most serious pest arthropods (insects, mites, and so forth) in the United States are of exotic origin (Glass 1975).

The introduction of natural enemies has been used successfully against about 100 insect and weed pests throughout the world (van den Bosch and Messenger 1973). Most strains of crown gall, caused by *Agrobacterium tumefaciens,* (see Chapter 21) can be prevented if the wounds, seeds, or plants are thoroughly treated with a closely related bacterium, *A. radiobacter* 'K 84' (New and Kerr 1972). Under favorable conditions, ladybird (*Coccinellidae*) and lacewings (*Chrysopa* spp.) can be effective predators, greatly reducing aphid infestation.

Certain practices preserve and augment biological control activity. Parasitoids of the red-humped caterpillar larva (*Schizura concinna*), a serious defoliator of California highway landscapes, were strengthened when they were given access to the nectar of nearby flowering shrubs. State workers changed aphid control on those shrubs from a broadspectrum insecticide to a dilute soap spray (Pinnock and others 1978) so that the parasitoids were not killed when they sought food in the shrubs. Flowering plants can also be deliberately established as food sources for insect predators and parasitoids. Certain cultural operations can favor desirable organisms over pests.

Cultural Control. A number of maintenance practices or modifications of them can make the environment unfavorable for pest reproduction, movement, or survival. Other mechanical or physical practices may specifically combat plant pests. Such controls, including some of the oldest cultural practices known, are usually more effective for preventing pest build-up than for correcting an existing pest problem. Pest control or reduction is often the byproduct of a practice employed for another primary purpose, such as irrigation and fertilization used to maintain or improve plant growth and appearance, that may enhance the plants' ability to withstand infection by unaggressive pathogens. In other cases, pest build-up may be prevented more effectively if the timing or specific method of a cultural practice is altered; selective pruning, used in the spring to open up a tree of shrub to sunlight and air movement, can reduce the severity of some leaf diseases by reducing humidity within the plant canopy. Some of the cultural techniques devised for agriculture (crop rotation, harvest procedures, and so forth) are not readily adapted to diverse perennial landscape plantings. Others are broadly effective for minimizing pest build-up and damage. Timing is critical to the success of much cultural pest control. Certain practices will be described here to indicate the range of possibilities.

Maintaining plant vigor, as already mentioned, not only enhances plant appearance but can increase tolerance to damage and infestation by borers. The goal is moderate plant vigor through careful management of nutrition, aeration, and moisture. In certain species, extreme vigor can increase susceptibility to diseases (fire blight by *Erwinia amylovora*) or certain pests, including mites and aphids and scale insects. The optimum vigor will depend on the plants, the environment, and the potential pests.

Pest-free plants should be selected for planting so pests are not introduced at the planting site. Plants and the soil in which they are moved should be free of pathogens, insects, nematodes, and weed pests. For some species (flowering cherry, for example), virus-indexed plants are available, and their superior performance and appearance in the landscape justify the additional cost (see Fig. 21-9).

Pruning can reduce the likelihood of disease infection and pest infestation, or stop or slow them. Brown (1972) describes pruning treatments to guard against or eliminate 20 diseases and pests; Svihra (1992) describes 13. Remove weak and dead wood, the potential brooding sites of boring insects. Thin out the dense head of a plant to reduce certain leaf diseases (such as anthracnose and powdery mildew) and to increase spray penetration. Prune off infected or infested branches and shoots to remove pests and minimize continued build-up. Reduce the possibilities of wood decay by training and pruning young trees so that large pruning cuts will not be needed later to correct poor tree structure. Make pruning cuts properly to promote rapid closure and further reduce the likelihood of decay.

Sanitation in and around plants can remove or destroy the breeding, refuge, and overwintering sites of pests. Sanitation which usually involves removing infected, dead, and fallen twigs, leaves, and fruit is effective against various insects, plant pathogens, rodents, and nematodes.

Cultivation can serve as a form of sanitation in that it can mutilate and bury plant residues that would otherwise be habitats for pests. Cultivation for pest control is usually most feasible in extensive landscapes with deep-rooted plants, where root injury from cultivation can be minimized, and mechanized equipment can be used more easily. At the same time, cultivated soil can increase certain pest problems. Rain or sprinkler drops can splash mud and the spores of fungus pathogens from bare soil onto the canes and foliage of shrubs, thereby increasing the likelihood of disease.

Mulching can often reduce splashing and the drops of mud that would protect spores deposited on plant surfaces. Mulches can effectively eliminate or reduce weeds in landscape plantings (see Chapter 14). If mulch is composed of infected or infested plant material, however, it can be a source of disease inoculum or insect build-up. Fortunately, reports do not indicate that mulch is a significant contributor to pest problems on plants. Mulch can also be sprayed to inactivate or minimize pests. Each planting complex must be analyzed to determine whether cultivation or mulching would favor an increase or decrease in the pests that are present.

Burning of grain and rice stubble both eliminates the straw, which uses nitrogen if it is left to decompose, and controls serious pests that would otherwise overwinter in the stubble. Similar burning in the landscape is usually not desirable, although more extensive areas can be "control burned" to kill undesirable brush, small trees, and weeds so that range species of plants can be seeded or hazard from wildfire reduced. When permitted, prunings, leaves, and even entire plants can be burned to destroy tissue infected with diseases such as fire blight, Dutch elm disease, and Armillaria root rot (see Chapter 21).

Species diversification in the landscape can prevent the kind of devastation that occurred in many communities when Dutch elm disease (see Chapter 21) and phloem necrosis struck; elms accounted for more than 75 percent of the landscape trees in many cities. Should a serious malady strike, single-species plantings are particularly vulnerable; a high density of one susceptible species favors insect and pathogen spread. Some cities and agencies ensure diversification by limiting the trees of any one species that may be planted in a given population. A cultivar chosen for the genetic variation it will introduce may of course be susceptible to unsuspected problems in the future, even if it has superior characteristics. **For adaptation and survival of a species or planting in an extensive urban area, genetic variation is extremely important; it is less critical in single, small landscape plantings.**

Species diversification can provide alternate food sources and refuge for the natural enemies of pests. Leius (1967) found that unsprayed apple and pear orchards in Canada encouraged more parasitism on tent caterpillar (*Malacosoma americana*) eggs and codling moth (*Cydia* [*Laspeyresia*] *pomonella*) larvae when they had ground cover high in nectar-producing flowers. These kinds of interactions need more study so that landscapes can be planned with such built-in protection.

Timing of planting can be adjusted to avoid periods of insect infestation; this technique has been used successfully with some field and vegetable crops but not with woody perennial plants. Timing can be important, however, in protecting susceptible plants from spring frosts and frozen or water-logged soil.

Eradication, the death or permanent removal of all individuals of a pest species, may focus on a plant, a landscape, or a region. Eradication of a pest from a region normally requires governmental action in order to be effective. Depending on the size of the area, the number of plants involved, the pest, and the expected speed of reinfestation, eradication can involve intensive spraying or fumigation, removal of all susceptible plants, the release of sterile male insects, or a combination of these methods.

Several things can be done to eliminate pests at the plant or planting level. An infected plant part can be pruned off and burned so that the pathogen does not spread; this constitutes eradication only if all of the pathogen is eliminated. Soil can be sterilized by heat or steam as well as by chemicals. Although it may not be feasible in existing plantings or large areas, soil sterilization can effectively eliminate a number of soil-borne problems in small outdoor areas surrounded by paving or buildings, individual planters, or nursery or greenhouse soil, where young plants may be quite susceptible. It must be remembered that sterilization kills not only pests but their natural enemies and other beneficial species. A variation on conventional heat treatment is solar treatment of soil (see Chapter 22).

Monitoring is a critical component of pest management for trees and shrubs. Regular inspection of plants for insect, disease, and cultural problems allows a landscape manager to pinpoint control actions (Raupp 1985). Population levels of predators and parasitoids as well as insect pests within a landscape will indicate whether or not control measures are needed. It is essential to be able to distinguish pests from beneficial insects, particularly in their immature stages; Raupp (1990) has done this for some of the more commonly encountered larvae. If needed, sprays can be timed for maximum effectiveness using the fewest applications. Raupp (1985) reports this approach to monitoring and treating only plants with damaging pest levels reduced the number of trees sprayed by 83 and 93 percent respectively in several communities in Maryland and California.

Trapping devices, such as collection boxes or sticky surfaces, are usually used in combination with bait or some other attractant such as pheromones, which are insect sex attractants.

Steve Dreistadt, University of California, Davis, (1990 pers. comm.) suggests timing spray applications for scale insects before crawlers begin to emerge in the spring. Tightly encircle several twigs or small branches with 25-mm (1-in.) wide bands of transparent tape that is sticky on both sides. Choose representative plants convenient to monitor; place tags near the tapes to easily locate them. Inspect the tapes weekly; replace them when they become fouled. Use a hand lens to distinguish crawlers (which have appendages) from pollen and dust. Apply a horticultural oil and/or other insecticide when a sharp increase in crawler number occurs.

Growth and development of insects and plants can vary as much as two to three weeks from the "normal" time, depending on whether temperatures are above or below average. Insects often become a problem about the same time that certain plants in the area begin to grow or bloom. Orton (1989) has recorded some of these plant-insect relationships for the midwestern United States that can be used to help monitor pests and time control measures.

Similarly, working with lilac borer, *Podosesia syringae,* in Kentucky, Potter and Timmons (1983) found that a close relationship exists between the amount of heat accumulated (called *degree days,* °D) above a certain threshold temperature and the different stages in the development of a particular insect species. Dreistadt and Dahlsten (1990) found a strong positive correlation between heat accumulation (above 11°C [52°F]) and apparent damage to elm foliage from the elm leaf beetle, *Xanthogaieruca luteoia,* in California. In regions where heat accumulation data for local insect pests are available from extension advisory offices, more precise timing of needed sprays should be possible.

Traps can be used not only to monitor population levels but also as a means of control. Sticky, yellow surfaces are sometimes used in small areas to trap whiteflies, *Trialeurodes* spp. Mechanical traps can also be used to control rodents.

Chemical Control. As stated earlier, the application of chemicals has been the pest control method most commonly used in developed countries. Even though highly toxic, long-lasting pesticides may still be needed for some pests, new materials and approaches are rapidly changing control methods of many landscape and agricultural pests. Improved horticultural oils (such as Sunspray 6E®), insecticidal soaps (such as Safer® Soap), bacterial preparations (such as *Bacillus thuringiensis* [Bt]), and entomogeneous nematodes are proving effective (Nielsen 1990, Baxendale and Johnson 1990).

The specially refined horticultural soaps and oils have short periods of toxicity; the oils evaporate quickly, particularly in warm, dry weather. The soaps are effective when liquid but ineffective when dry. **Thorough coverage is essential because these sprays are effective for only a short period.** Spraying only plants that have control threshold levels of pests greatly reduces the number of plants treated. So, even if beneficial insects are killed by a spray, beneficials nearby can move into the treated areas unharmed. Remember that most sucking insects move little, while the beneficial insects are quite mobile. Also the oils and soaps can improve the effectiveness of some pesticides so that they can be used at much lower concentrations (Funk 1988, Miller 1989).

Entomogeneous nematodes and bacterial preparations have specific hosts so they endanger few beneficial insects.

These newer pesticides have low human toxicities (Table 23-2). The oils and soaps are phytotoxic to only a few of the species to which they have been applied (Davidson, Gill, and Raupp 1990, Nielsen 1990). When first using a new pesticide or treating a species for the first time, treat only a portion of the plants or planting. Observe the treated plant for several days for signs of phytotoxicity.

Choosing the appropriate formulation of pesticide and the method and timing

TABLE 23-2 ANIMAL TOXICITY OF SOME COMMON PESTICIDES AND OTHER CHEMICALS IN RELATION TO DOSE (Litewka and Stimmann [1979], Meister [1982])

Compound	Acute toxicity		Chronic toxicity	Soil persistence
	Oral LD_{50}[a]	Dermal LD_{50}	No effect at:	
	mg/kg	mg/kg	mg/kg/day	
B. thuringiensis	>8500			
Chlordane	250–590	690–840	1.0	6 month half-life
DDT	113–118	2510	0.05	Persistent
Diazinon	300–400	600–2000+	0.1[b]	4–6 week half-life
Horticultural oils[c]	>4300			Evaporates
Insecticidal soaps[d]	>16,900	>2000		When wet
Lindane	60–230	180–1000	1.25	Persistent
Malathion	1000–1375	4444	0.2	2 weeks
Nicotine[e]	50–60	140		
Parathion	3–13	7–21	0.05[b]	Not persistent
Sevin	500–850	4000+		
Simazine	1025–5000+	8160	50.0	Persistent
2,4-D	300	1500		1–4 weeks
Aspirin	300			
Caffeine	78			

[a] LD_{50} indicates the dose (mg/kg) necessary to kill 50 percent of the test animals under prescribed experimental conditions. The lower the LD_{50}, the more toxic the compound. The lowest concentration cited is listed. >500 mg considered safe.

[b] Amounts greater than this will reduce cholinesterase.

[c] Worthing and Walker (1983).

[d] Safer® Insecticide Concentrate Technical Bulletin. 1989. Newton, MA.

[e] Smoking kills an estimated 390,000 people in the United States annually (U.S. Surgeon General's Cancer Prevention Study #2. 1989. As cited in the American Cancer Society *Facts and Figures—1990.*

of application are important for effective pest control, for the safety and well-being of people and protected plants, and for the stability of the ecosystem.

Toxicity of pesticides to animals is expressed as the LD (lethal dose). The LD_{50} of a chemical indicates the dose necessary to kill 50 percent of the test animals under prescribed experimental conditions. Toxicity tests are carried out on a variety of laboratory animals, most commonly on rats. LD_{50} is expressed as a ratio of milligrams (mg) of the chemical per kilogram (kg) of animal body weight. Such figures cannot be precisely extrapolated to a lethal dose for humans, but they do provide a relative measure of chemical hazard (Table 23–2). The lower the LD_{50}, the more toxic the chemical. Chemicals with an LD_{50} of 500 and above are considered quite safe; the lethal oral dose of these formulations would range from 30 to 480 ml (1 oz–1 pt) or 450 mg (1 lb) (Stimmann 1977). A person is very unlikely to accidently ingest even 30 ml of a chemical.

LD_{50} values usually refer to ingestion of the chemical (*oral*). Similar values may be determined for *dermal* toxicity (absorption through the skin) and *inhalation* toxicity. For safe handling and application of pesticides, these latter two toxicities are more important than the ingestion toxicity. LD_{50} values are for *acute* toxicity, symptoms that occur within 24 hours of exposure. *Chronic* toxicity produces symptoms that appear more than 24 hours after exposure or, more commonly, symptoms that appear after repeated exposure to concentrations below acute toxicity. Chronic symptoms might include respiratory ailments, nervous disorders, cancer, and teratogenicity (malformation of an embryo) and might take years to develop.

Certain phytotoxic pesticides (those that injure plants) should not be used on or near susceptible plants, so read labels carefully and test on a plant if in doubt. Volatile herbicides have caused serious damage to susceptible plants several kilometers (miles) from the site of application. Ester formulations of 2,4-D applied to control weeds in California grain fields have caused serious damage to grapevines more than 20 km (12 mi) away. Amine formulations now available are much less volatile and therefore safer to use.

Labels on pesticide containers present valuable information about use, handling, and toxicity. The Federal Environmental Pesticide Control Act requires that every registered label contain specific information. For example, the label must state whether the pesticide is classified for general or restricted use. *General use* pesticides are the least hazardous to persons and the environment when used as directed; these pesticides are available for use by anyone. *Restricted use* pesticides are those that may be hazardous to the environment, the applicator, or other persons, even when label directions are followed. Restricted use pesticides are to be used only by certified applicators.

The toxicity hazard to human beings must be stated on a pesticide label. Each pesticide formulation is assigned to a toxicity category, which designates the level of hazard to human health. Categories were determined by considering the effects of the pesticides when they are ingested, inhaled, spilled on skin, or splashed in eyes (Table 23–3). The most hazardous materials are in toxicity category I; the least hazardous in category IV. The directions for use tell how to mix and apply the pesticide, where and when to apply it, how much to apply, and how often it can be applied. The label also specifies against what pests the chemical has been registered

TABLE 23-3 ACUTE TOXICITY CATEGORIES OF PESTICIDES ESTABLISHED BY THE FEDERAL ENVIRONMENTAL PESTICIDE CONTROL ACT (Stimmann 1977).

Category	Signal word required on the label	LD_{50}[a] Oral mg/kg	LD_{50}[a] Dermal mg/kg	Probable oral lethal dose less than	
I Highly toxic	DANGER— POISON Skull and crossbones	0–50	0–200	5 ml	1 tsp
II Moderately toxic	WARNING	50–500	200–2000	30 ml	1 oz
III Slightly toxic	CAUTION	500–5000	2000–20,000	480 ml (0.45 kg)	1 pt 1 lb
IV Relatively toxic	CAUTION	>5000	>20,000	>480 ml >(0.45 kg)	>1 pt >1 lb

[a] LD_{50} indicates the dose (amount of chemical/body weight) necessary to kill 50 percent of the test animals under prescribed conditions.

for use. **It is illegal to use a pesticide on a plant or a pest for which it has not been registered.** Read all information on a pesticide label before purchase or application.

Pesticide toxicity is an extremely important property and must be considered when you choose the controls to be used in a specific situation. If pesticides are part of the control program, they must be handled and applied so as to ensure minimal risks to humans and other animals.

Highly selective pesticides are of course the most desirable. 2,4-D can eliminate many broadleaved weeds without harming turf but can severely damage trees and shrubs if they have low foliage near the treated lawn. Preemergence herbicides (such as Surflan® and Treflan®) will control most emerging seedlings but do not injure established plants because little or none of the chemical leaches to their root zone. Where possible, choose insecticides that have narrow ranges of effectiveness or minimal residual properties; this will cause as little harm as possible to natural pest enemies, other beneficial insects, and nontargeted species.

Timing and methods of application can maximize pesticide effectiveness and minimize adverse reactions. Chemicals used to control disease are usually most effective when applied as protectants before the disease organism invades the plant or begins to increase. Contact and systemic herbicides are most effective and most easily applied while weeds are small. Obviously, preemergent herbicides should be applied before weed seeds germinate. Time insecticide applications to coincide with the most vulnerable part of the pest's life cycle. Many insecticides are effective only during the early active stages of the cycle. For example, *Bacillus thuringiensis* is effective only when recently sprayed foliage is eaten by young larvae of susceptible insect species. Proper timing of pesticide application requires population sampling to determine when the most vulnerable stage of the next generation will occur and whether pest numbers even warrant control measures.

Pesticides can be confined to a localized area by various methods of application. Seed treatment for disease, insect, and nematode control requires only a fraction of the pesticide that would be needed for soil application. Dormant applications usually require less spray than those performed during the growing season. Certain insects can be attracted to baited traps and poisoned there. Pests often concentrate in certain parts of a landscape, so that treatment can be limited to areas of heavy infestation. Again, periodic observations are most helpful for locating the areas of highest population. Shallow-rooted plants can be treated with soil drenches against certain pests. Trunk and root injections eliminate spraying close to homes and places of activity; such injections are effective only against certain insects and diseases, however, and repeated applications, even on an annual basis, inflict wounds that can lead to trunk decay (see Chapter 18).

You can minimize the drift of spray material by spraying when it is calm, usually early morning or evening, with equipment adjusted to discharge large spray droplets. Large droplets settle more quickly, reducing the area affected. Droplet size can be increased if the pressure on hydraulic sprayers is reduced; this of course reduces the distance the spray can be projected.

In summary, **for effective chemical pest control, you must apply the right material at the correct concentration to the right place at the right time.** Von Rumker and coworkers (1975) estimate that less than 1 percent of a foliar insecticide application may reach the insects to be controlled. On tall trees, little spray reaches the tops; most falls on lower branches or the trunk. For more thorough and uniform coverage, direct most sprays, particularly high-pressure sprays, to the treetops or to the highest portion of the target area.

Certification of commercial pesticide applicators and advisers is required by federal and state laws in the United States. Only licensed agricultural pest control operators can apply *restricted use* pesticides for hire, and any person who makes recommendations about pesticide use must be licensed as an agricultural pest control adviser. Agricultural use includes uses in parks, golf courses, roadside landscapes, cemeteries, schoolyards, private landscapes, and other similar areas. State agencies administer the registration and safe use of pesticides and the certification of pest control operators and advisers. To be certified, an individual must demonstrate knowledge about pesticide labeling, safety, use, application, handling, pests, environmental effects, equipment, and laws and regulations. Specific regulations govern requests for permission to apply a pesticide, ways of handling and using a pesticide, and records that must be kept.

AN EXAMPLE OF AN INTEGRATED PEST MANAGEMENT APPROACH—THE GYPSY MOTH

The goal of integrated pest management should be to reduce pests or damage to an acceptable level while guarding the safety of workers, other people, and desirable animals and plants, controlling effectiveness and cost, and preserving the stability of the landscape ecosystem. Such an approach may involve only one relatively simple measure, such as selecting pest-resistant plants or applying a thick mulch around shrubs to control weeds. On the other hand, acceptable control may require complex coordinated measures.

Gypsy moth caterpillars (*Lymantria dispar*) defoliate millions of trees annually in the northeastern United States, eastern Canada, and Europe. They are considered the most serious threat to oak forests in the United States (Raupp, Davidson, and Wood 1987–1988). Probably no tree pest has received more publicity or provoked more expensive control efforts than the gypsy moth (Johnson and Lyon 1988). The insects were brought into Massachusetts in the late 1860s in an attempt to establish silk moths in New England. Within 12 years after they had escaped, gypsy moths had become a serious nuisance: Trees were defoliated by midspring; streets and walks under trees were slippery with crushed caterpillars. The gypsy moth became such a problem that cities established positions for arborists who in many communities are still called moth wardens. Because its natural enemies were absent, the gypsy moth spread rapidly and has infested more than 50 million hectares (200,000 square miles) of forest land in the northeastern United States (Johnson and Lyon 1988).

Adult female gypsy moths have whitish wings with a 50 mm (2 in.) span and wavy, dark bands across the forewings. The males are dark brown. The females deposit eggs on branches, posts, buildings, automobiles, camping trailers, or almost any surface. Masses of 100 to 600 eggs are covered with buff-colored hairs (Johnson and Lyon 1988). The larvae hatch in early to midspring and within a few days climb into trees to feed on foliage. As they mature, the caterpillars feed primarily at night and descend to spend the day in protected places. Mature caterpillars attain lengths of up to 65 mm (2.5 in.). The larval stage lasts about seven weeks, after which the caterpillars find protected places and pupate. Moths emerge in the early summer; the female does not fly. Adults die without feeding after they mate and deposit eggs.

Symptoms

Defoliated trees with severely chewed leaves are the obvious mark of the gypsy moth. The decline and death of many oaks in New England have been associated with defoliation by the gypsy moth (Houston 1980). Most deciduous trees can withstand one or two years of infestation before serious decline takes place, but conifers will die after one complete defoliation (Johnson and Lyon 1988). Trees are weakened most severely if they put out one or more flushes of regrowth after defoliation early in the growing season.

Treatment Approaches

The early approach to gypsy moth control by state and federal agencies was aerial spraying of forested areas to eliminate infestations. This and other suppression measures provided only temporary (one- to three-year) control (Reardon and others 1987). Later, several demonstration and management projects in Maryland focused on integrated approaches to the gypsy moth problem (Raupp and Noland 1984).

Gypsy moth populations can be extensive, requiring the combined efforts of private property owners as well as public agencies in order to keep insects and damage at acceptable levels. Organized efforts to eradicate isolated gypsy moth infestations are conducted by some states distant from the widespread populations in the eastern United States. Tactics that might be used both by property owners, their arborists or pest control contractors, and public agencies are presented first, followed by those that usually require public agency capability. The following information is based largely on material by Raupp, Davidson, and Wood (1987–1988) and Reardon and others (1987).

General IPM Tactics for Gypsy Moth

Plant Resistant Plants. As would be expected, some species of trees and shrubs are preferred over others by gypsy moths (Table 23–4). It would be wise to select species least preferred by gypsy moths that were adapted to the site and fulfill the desired landscape function. Replace heavily attacked plants to minimize control measures, defoliated plants, and larvae invasion. Less-damaged plants can be interplanted with more resistant ones.

Keep Plants Healthy. Plants that are stressed are more vulnerable to effects of defoliation and secondary organisms that frequently follow gypsy moth infestation. Optimum soil conditions of adequate water, aeration, and nutrients are the best insurance for plants to withstand gypsy moth infestation. Minimize other stresses and secondary infections.

Destroy Egg Masses. Where infested plants are fairly isolated, egg mass destruction may reduce caterpillar numbers the following spring. Before spring, remove egg masses from infested plants, structures, garden furniture, and yard litter. Scrape them into bags for disposal or into soapy water. Handle egg masses with gloves. Be careful not to inhale hair or dust; some people are allergic to the hairs on egg masses.

By midspring, minimize hiding places for females seeking pupation or egg-laying sites. Eliminate or cover loose rocks, boards, firewood, and so forth.

TABLE 23-4 THE HOSTS OF GYPSY MOTH LARVAE, IN ORDER OF PREFERENCE (LEAVES) (Adapted from Johnson, W. T., and H. H. Lyon. *Insects That Feed on Trees and Shrubs.* 2nd ed., revised. © 1971, 1988, 1991. Ithaca, NY: Cornell University Press)

Preferred hosts	Less preferred hosts	Occasional hosts	Least preferred hosts
alder	elm	beech	ash
apple	black gum	white cedar	balsam fir
basswood	hickory	hemlock	butternut
hawthorn	maple	pine	black walnut
oak	sassafras	spruce	catalpa
some poplars			redcedar
willow			dogwood
			holly
			locust
			sycamore
			yellow poplar

Band Trees. Caterpillars can be killed by banding trunks of susceptible trees. Many young larvae have to climb trees to reach foliage; older larvae usually spend the day on the ground and the night in a tree. A *barrier band* around a trunk can prevent larvae from reaching the top of a tree. The most effective ones are pliable so they fit snugly into the crevices of rough bark and have a coating on the outer surface that remains sticky for a long time. Larvae either starve below the bands or become trapped in the coating. Barrier bands should be in place by late spring and inspected weekly to repair damage and reapply sticky material.

Sticky material can be applied directly to the bark of trees, but some oil-based materials can injure bark, particularly smooth-barked trees. Some materials stain bark which some people find unacceptable.

A *hiding band* around a treetrunk acts as a refuge for larvae where they can be removed and destroyed. A strip of burlap (or similar material) about 450 mm (18 in.) wide is folded over a stout string around the trunk. This creates a double flap 225 mm (9 in.) wide that provides shelter for migrating larvae. The larvae should be removed and destroyed every day or two; otherwise, the hiding bands may attract more caterpillars.

If a burlap band is used with a barrier band, it should be placed at least 450 mm (18 in.) above the barrier band. Wear gloves when handling gypsy moths in any stage to avoid skin irritation.

Trap Males and Disrupt Mating. Pheromone (sex attractant) traps, used to monitor insect build-up, can disorient male moths. When relatively large amounts of pheromone are placed in a small area, males fail to find females with whom to mate. This technique is known as mating disruption and has been used to manage gypsy moths, but only when populations are extremely low.

Use Insecticides. The cheapest, easiest, and most effective way to control gypsy moths on small trees and shrubs is for a property owner to apply an appropriate insecticide. A number of chemical insecticides may be registered in the region for control of gypsy moths; but check with the local extension agent or supplier for those materials registered in your area. The chemical insecticides will probably kill most of the beneficial insects. *Bacillus thuringiensis* (Bt), however, is a bacterial insecticide that kills caterpillars only; it is sold under several trade names, such as Agway®, BT®, Dipel®, and others. Bt must be sprayed on second to third instar larvae (after first and second molting or shedding of skin) when oak leaves are usually half opened. Older caterpillars are not easily controlled by these bacteria.

Larger trees should be treated by an arborist or pest control contractor. They are able to use highly effective insecticides available only to licensed pest control applicators. Diflubenzuron (Dimilin®) is considered 90 percent effective and has little effect on parasitoids and predators. Where spraying is not desirable, professionals can inject systemic insecticides such as Bidrin® (dicrotophos) into treetrunks. These chemicals move into the leaves where they kill caterpillars for about a month.

In late spring or early summer, large caterpillars may leave trees and shrubs and invade other garden areas and structures. These 50-mm (2-in.) long larvae are difficult to kill with sprays, and more will come the next day. They can be washed off buildings and pavement with a hose. Insecticides are not advised.

Part of a More Complete IPM Program

Attempting to manage gypsy moths at an acceptable level should be just part of an overall IPM program for a property or agency. The main task is regular, skilled monitoring to know the status of all the potentially harmful insects and diseases as well as the beneficials

Public IPM Intervention

The most successful way to reduce gypsy moth populations in heavily infested areas is through community-wide action. This may involve state and federal agencies as well as municipal governments. In such situations, a coordinating team representing the cooperating agencies would organize the IPM program and its execution.

The term *intervention* is commonly used instead of *control,* since the approach of integrated pest management is to maintain pests and diseases at acceptable levels using economical procedures least damaging to the ecological system. An inventory of the area to be managed is essential to be able to plan and carry out an effective program.

Decisions regarding the need for and the selection of intervention activities for gypsy moth control would be based on inventory and monitoring information that would include

Egg-mass density and quality
Larval and pupal counts
Male moth captures
Population trends (egg mass, male moth)
Parasitoid and disease incidence
Size and location of infestation(s)
Defoliation estimates and projections
Proximity to other infested areas
Stand susceptibility
Vegetative cover type
Land-use category (forest, park, town)

With this information presented by geographical area, the management team can identify areas of primary concern based on host population levels and site conditions. Management decisions that would be made from this information might be

1. *No action*—when gypsy moth population density is low (<25 egg masses/ha; 10/acre) and the population is decreasing, indicating low potential for impact. Only surveillance activities would be prescribed.
2. *Preventive action*—when population densities are low (<25 egg masses/ha) to moderate (25–250 egg masses/ha; 10–100/acre) and the trend is stable or increasing, and other information indicates risk of impact or spread. This action is to prevent expansion of pest populations, or to augment natural enemies.
3. *Suppressive action*—when population densities are high (>250 egg masses/ha; 100/acre), the trend is stable or increasing, and other information indicates a major risk of economic damage or nuisance.

Intervention Activities. Most state and federal gypsy moth programs use chemical and biological insecticides on high-density (>1250 egg masses/ha; 500/acre) populations. Efficacy of these insecticides has not been demonstrated adequately at low gypsy moth densities.

Chemical insecticides are applied by air to areas with densities greater than 2500 egg masses/ha (1000/acre), at least a 10-fold increase in population trends, and risk of serious defoliation. Dimilin® is the insecticide of choice based on its mode of action, efficacy, low application dosage, persistence, and lack of effect on parasitoids and predators of the gypsy moth. Its use is restricted to nonaquatic habitats. The ground application of chemical insecticides such as Dimilin® is restricted to urban residential areas with high-value trees or where defoliation and/or larval nuisance impacts are anticipated.

Microbial insecticides have been the major intervention tactic in the Maryland project reported (Raupp, Davidson, and Wood 1987–1988). *Bacillus thuringiensis* (Bt) and the gypsy moth nucleopolyhedrosis virus (NPV) are applied by air at a range of host densities and population trends. The gypsy moth NPV is specific to gypsy moth and is usually the major factor in causing the natural collapse of dense populations. Both microbials remain viable on foliage for three to five days, depending upon the weather following application.

Biological controls have met with mixed success. In 1980, Nichols reported that eight exotic parasitoids and one predator had become established in Pennsylvania. He stated that at that time, 92 percent of the two million ha (five million acres) of infested forests in the state were under natural or biological control. Michael Raupp, University of Maryland, (1990 pers. comm.) indicated that parasitoids have been only about 20 percent effective in Maryland. They have not been able to control population explosions, though they become effective by the time the population began to decline.

Genetic control is being evaluated against low level populations by releasing F_1 sterile eggs. The F_1 adults that develop are sterile. Normal females that mate with F_1 sterile males produce eggs that do not hatch.

Entomogenous nematodes have not proven to be effective against large-stage larvae as they use resting niches, such as burlap bands.

Summary

In regions with extensive gypsy moth distribution, the combined cooperation of public agencies and property owners offers the best opportunity for effective management of gypsy moths.

An integrated approach to pest management holds promise for safer, more effective, and less expensive pest control programs than have been used in the past. Pest management for landscape plants is much more complex than for a single agricultural crop because landscapes are usually small and contain a variety of plant species intimately associated with homes and human activity. Proximate plantings may involve several different individuals or agencies, both public and private, in any efforts to control pest populations. Under such circumstances, an integrated pest management program must be a community effort, particularly if it emphasizes biological control measures. The resolution of many difficult pest problems awaits further study and implementation. On the other hand, much can be done in private and public landscapes to improve the health and appearance of landscape plants.

FURTHER READING

AGRIOS, G. N. 1988. *Plant Pathology* 3rd ed. San Diego, CA: Academic Press.

BENNETT, G. W., and J. M. OWENS, eds. 1986. *Advances in Urban Pest Management*. New York: Van Nostrand Reinhold.

COUNCIL ON ENVIRONMENTAL QUALITY (CEQ). 1972. *Integrated Pest Management*. Washington, D.C., U.S. Government Printing Office.

FLINT, M. L., and R. VAN DEN BOSCH. 1981. *Introduction to Integrated Pest Management*. New York: Plenum Press.

GLASS, E. H., ed. 1975. *Integrated Pest Management: Rationale, Potential, Needs, and Implementation*. Entomological Society of America.

GREEN, J. L., O. MALOY, and J. CAPIZZI. 1990. *Patterns: Signs to Direct the Diagnosis of Plant Damage*. Ornamentals NorthWest Newsletter 13(6):1–24.

HUFFAKER, C. B., ed. 1973. *Biological Control*. New York: Plenum.

JOHNSON, W. T., and H. H. LYON. 1988. *Insects That Feed on Trees and Shrubs*. 2nd ed. Ithaca, NY: Comstock.

MANION, P. D. 1981. *Tree Disease Concepts*. Englewood Cliffs, NJ: Prentice Hall.

PHILLIPS, D. H., and D. A. BURDEKIN. 1982. *Diseases of Forest and Ornamental Trees*. London: Macmillan.

PIRONE, P. P. 1978. *Diseases and Pests of Ornamental Plants,* 5th ed. New York: John Wiley.

PIRONE, P. P., and others. 1988. *Tree Maintenance,* 6th ed. New York: Oxford University Press.

SHIGO, A. L. 1991. *Modern Arboriculture: A Systems Approach to Trees and Their Associates*. Durham, NH: Shigo and Trees, Associates.

SINCLAIR, W. A., H. H. LYON, and W. T. JOHNSON. 1987. *Diseases of Trees and Shrubs*. Ithaca, NY: Comstock.

SMITH, M. D. 1989. *The Ortho Problem Solver.* 3rd ed. San Francisco: Ortho Books.

SMITH, W. H. 1970. *Tree Pathology: A Short Introduction.* San Diego: Academic Press.

STREETS, R. B. 1984. *The Diagnosis of Plant Disease.* Tucson, AZ: University of Arizona Press.

TATTAR, T. A. 1989. *Diseases of Shade Trees.* 2nd ed. San Diego: Academic Press.

Appendixes

APPENDIX ONE

Common and Botanical Names of Plants

Common name	Botanical name	Common name	Botanical name
Abelia, Glossy	*Abelia × grandiflora*	Ash, Modesto	*Fraxinus velutina* var. *glabra* 'Modesto'
Acacia	*Acacia*		
Ailanthus	*Ailanthus altissima*	Ash, Shamel	*Fraxinus uhdei*
Alder	*Alnus*	Ash, White	*Fraxinus americana*
Alfalfa	*Medicago sativa*	Aspen	*Populus*
Almond	*Prunus dulcis*	Azalea	*Rhododendron*
Apple	*Malus sylvestris*	Basswood	*Tilia*
Apple, Crab	*Malus*	Beach Grass	*Ammophila*
Apricot	*Prunus armeniaca*	Bean	*Phaseolus vulgaris*
Arborvitae	*Thuja*	Beech	*Fagus*
Arborvitae, Oriental	*Platycladus orientalis*	Bentgrass, Creeping	*Agrostis stolonifera*
Ash	*Fraxinus*	Bindweed	*Convolvulus*
Ash, Arizona	*Fraxinus velutina*	Birch	*Betula*
Ash, Green	*Fraxinus pennsylvanica* var. *lanceolata*	Birch, Canoe	*Betula papyrifera*
		Birch, European White	*Betula pendula*

(cont.)

Common name	Botanical name	Common name	Botanical name
Bougainvillea	*Bougainvillea*	Elm, Red	*Ulmus rubra*
Broom	*Cytisus*	Elm, Siberian	*Ulmus pumila*
Buckeye, California	*Aesculus californica*	Elm, Smooth-Leafed	*Ulmus carpinifolia*
Buddleia	*Buddleia*	Empress Tree	*Paulownia tomentosa*
Buffalograss	*Buchloe dactyloides*	Escallonia	*Escallonia*
Butternut	*Juglans cinerea*	Eucalyptus	*Eucalyptus*
Buttonwood	*Platanus* or *Conocarpus*	Euonymus	*Euonymus*
Callistemon	*Callistemon*	Euonymus, Gold Spot	*Euonymus japonica*
Camellia	*Camellia*		'Gold Spot'
Camphor Tree	*Cinnamomum camphora*		
Carambola	*Averrhoa carambola*	Ferns	Many genera and species
Carob	*Ceratonia siliqua*	Fescue	*Festuca*
Catalpa	*Catalpa speciosa*	Fescue, Alta	*Festuca arundinacea*
Ceanothus	*Ceanothus*	Fig	*Ficus*
Ceanothus,	*Ceanothus thyrsiflorus*	Fig, Creeping	*Ficus pumila*
Blueblossom		Fig, Indian Laurel	*Ficus microcarpa*
Cedar	*Thuja* or *Cedrus*	Fir	*Abies*
Cedar, Deodar	*Cedrus deodara*	Fir, Balsam	*Abies balsamea*
Cedar, Incense	*Calocedrus decurrens*	Fir, Douglas	*Pseudotsuga menziesii*
Cedar, Red	*Juniperus virginiana*	Fir, White	*Abies concolor*
Cedar, Western Red	*Thuja plicata*	Flannel Bush	*Fremontodendron*
Cedar, White	*Chamaecyparis thyoides*		*californicum*
	or *Thuja occidentalis*	Forsythia	*Forsythia*
Cherry	*Prunus*	Forsythia, Weeping	*Forsythia suspensa*
Cherry, Barbados	*Malpighia glabra*	Ginkgo	*Ginkgo biloba*
Cherry, Flowering	*Prunus serrulata*	Golden Rain Tree	*Koelreuteria paniculata*
Cherry, Sour	*Prunus cerasus*	Golden-chain Tree	*Laburnum anagyroides*
Cherry, Surinam	*Eugenia uniflora*	Grape	*Vitis vinifera*
Cherry, Sweet	*Prunus avium*	Grapefruit	*Citrus × paradisi*
Cherry, Yoshino	*Prunus × yedoensis*	Guava	*Psidium*
Chestnut	*Castanea*	Gum, Black	*Nyssa sylvatica*
Chestnut, Horse	*Aesculus hippocastanum*	Gum, Blue	*Eucalyptus globulus*
Chestnut, Sweet	*Castanea sativa*	Gum, Red	*Eucalyptus camaldulensis*
Chrysanthemum	*Chrysanthemum*	Gum, Silver-dollar	*Eucalyptus polyanthemos*
Citrus	*Citrus*	Hackberry	*Celtis*
Clematis	*Clematis*	Hawthorn	*Crataegus*
Corn	*Zea mays*	Hazel, Witch	*Hamamelis*
Cotoneaster	*Cotoneaster*	Hemlock	*Tsuga*
Cottonwood	*Populus*	Hibiscus	*Hibiscus*
Cryptomeria	*Cryptomeria japonica*	Hibiscus, Chinese	*Hibiscus rosa-sinensis*
Cycad	*Cycas*	Hickory	*Carya*
Cypress	*Cupressus*	Hickory, Shagbark	*Carya ovata*
Cypress, Bald	*Taxodium distichum*	Holly	*Ilex*
Cypress, False	*Chamaecyparis*	Holly, Dwarf Burford	*Ilex cornuta* 'Burfordii
Dogwood	*Cornus*		Nana'
Dogwood, Eastern	*Cornus florida*	Holly, English	*Ilex aquifolium*
Elder	*Sambucus*	Holly, Japanese	*Ilex crenata*
Elm	*Ulmus*	Honeysuckle	*Lonicera*
Elm, American	*Ulmus americana*	Hornbeam	*Carpinus*
Elm, Asiatic	*Ulmus pumila*	Horsechestnut	*Aesculus hippocastanum*
Elm, Chinese	*Ulmus parvifolia*	Ice Plant	*Mesembryanthemum*
Elm, English	*Ulmus procera*	Ivy	*Hedera*
Elm, European White	*Ulmus laevis*	Ivy, English	*Hedera helix*
		Jacaranda	*Jacaranda*

Common name	Botanical name	Common name	Botanical name
Jasmine	*Jasminum*	Oak, Pin	*Quercus palustris*
Jasmine, Crape	*Ervatamia coronaria*	Oak, Red	*Quercus rubra*
Jasmine, Star	*Trachelospermum jasminoides*	Oak, Sawtooth	*Quercus acutissima*
		Oak, Scarlet	*Quercus coccinea*
Jessamine, Orange	*Murraya paniculata*	Oak, Scrub	*Quercus ilicifolia*
Juniper	*Juniperus*	Oak, Sessile	*Quercus petraea*
Juniper, Pfitzer	*Juniperus chinensis* 'Pfitzerana'	Oak, Silk	*Grevillea robusta*
		Oak, Valley	*Quercus lobata*
Larch	*Larix*	Oak, Water	*Quercus nigra*
Laurel, English	*Prunus laurocerasus*	Oak, White	*Quercus alba*
Lemon	*Citrus limon*	Oleander	*Nerium oleander*
Lilac	*Syringa*	Olive	*Olea europaea*
Lime Tree	*Tilia × vulgaris*	Olive, Swan Hill Fruitless	*Olea europaea* 'Swan Hill'
Linden	*Tilia*		
Liquidambar	*Liquidambar styraciflua*	Olive, Russian	*Elaeagnus angustifolia*
Locust	*Robinia*	Onion	*Allium cepa*
Locust, Black	*Robinia pseudoacacia*	Orange	*Citrus sinensis*
Locust Frisia	*Robinia pseudoacacia* 'Frisia'	Orange, Navel	*Citrus sinensis*
		Pagoda Tree, Japanese	*Sophora japonica*
Locust, Honey	*Gleditsia triacanthos*	Palms	numerous genera
Locust, Sunburst Honey	*Gleditsia triacanthos* f. *inermis* 'Sunburst'	Palm, Areca	*Areca catechu*
		Palm, Cabbage or Sabal	*Sabal palmetto*
Loquat	*Eriobotrya japonica*		
Magnolia	*Magnolia*	Palm, Canary Island Date	*Phoenix canariensis*
Magnolia, Southern	*Magnolia grandiflora*		
Mango	*Mangifera indica*	Palm, Coconut	*Cocos nucifera*
Manzanita	*Arctostaphylos*	Palm, Date	*Phoenix dactylifera*
Maple	*Acer*	Palm, Mexican Fan	*Washingtonia robusta*
Maple, Japanese	*Acer palmatum* or *Acer japonicum*	Palm, Paurotis	*Acoelorrhaphe wrightii*
		Palm, Pygmy Date	*Phoenix roebeleniipp*
Maple Norway	*Acer platanoids*	Palm, Queen	*Arecastrum romanzoffianum*
Maple, Red	*Acer rubrum*		
Maple, Silver	*Acer saccharinum*	Palm, Royal	*Roystonea*
Maple, Sugar	*Acer saccharum*	Palm, Sabal or Cabbage	*Sabal palmette*
Mayten, Chile	*Maytenus boaria*		
Mesquite	*Prosopis*	Palm, Senegal Date	*Phoenix reclinata*
Mesquite, Honey	*Prosopis glandulosa*	Palm, Spindle	*Hyophorbe perschasselti*
Mirror Plant	*Coprosma repens*	Palo Verde	*Cercidium*
Mistletoe, Dwarf	*Arceuthobium*	Peach	*Prunus persica*
Mistletoe, Leafy	*Phoradendron* or *Viscum*	Pear	*Pyrus communis*
Mulberry	*Morus alba*	Pear, Bradford	*Pyrus calleryana* 'Bradford'
Mulberry, Fruitless	*Morus alba* 'Stribling'		
Myrtle, Crape	*Lagerstroemia indica*	Pear, Callery	*Pyrus calleryana*
New Zealand Christmas Tree	*Metrosideros excelsus*	Pear, Evergreen	*Pyrus kawakamii*
		Pear, Oriental	*Pyrus pyrifolia*
Oak	*Quercus*	Pecan	*Carya illinoensis*
Oak, Black	*Quercus velutina*	Pepper Tree, California	*Schinus molle*
Oak, Blackjack	*Quercus marilandica*	Petunia	*Petunia × hybrida*
Oak, Blue	*Quercus douglasii*	Photinia, Chinese	*Photinia serrulata*
Oak, Cork	*Quercus suber*	Pine	*Pinus*
Oak, Holly	*Quercus ilex*	Pine, Aleppo	*Pinus halepensis*
Oak, Interior Live	*Quercus wislizenii*	Pine, Black	*Pinus nigra*

(cont.)

Common name	Botanical name	Common name	Botanical name
Pine, Coulter	*Pinus coulteri*	Sequoia	*Sequoia sempervirens*
Pine, Eastern White	*Pinus strobus*		or *Sequoiadendron*
Pine, Lodgepole	*Pinus murrayana*		*giganteum*
	Syn.: *P. contorta* var.	Sequoia, Giant	*Sequoiadendron*
	latifolia		*giganteum*
Pine, Monterey	*Pinus radiata*	Silk Tree	*Albizia julibrissin*
Pine, Ponderosa	*Pinus ponderosa*	Sorghum	*Sorghum vulgare*
Pine, Red	*Pinus resinosa*	Spruce	*Picea*
Pine, Scotch	*Pinus sylvestris*	Spruce, Black	*Picea mariana*
Pine, Slash	*Pinus caribaea*	Spruce, Norway	*Picea abies*
Pine, Sugar	*Pinus lambertiana*	Spruce, Sitka	*Picea sitchensis*
Pine, White	*Pinus strobus*	Spruce, White	*Picea glauca*
Pineapple	*Ananas comosus*	Squash	*Cucurbita maxima*
Pistache	*Pistacia*	Sweetgum	*Liquidambar styraciflua*
Pistache, Chinese	*Pistacia chinensis*	Sycamore	*Platanus*
Pittosporum	*Pittosporum*	Sycamore, American	*Platanus occidentalis*
Pittosporum, Japanese	*Pittosporum tobira*	Sycamore, California	*Platanus racemosa*
Plane, London	*Platanus × acerifolia*	Tea, Victoria	*Leptospermum*
Plane Tree	*Platanus*		*laevigatum*
Plum	*Prunus*	Tobacco	*Nicotiana tabacum*
Plum, Flowering	*Prunus*	Tomato	*Lycopersicon esculentum*
Podocarpus	*Podocarpus*	Tree-of-heaven	*Ailanthus altissima*
Poinsettia	*Euphorbia pulcherrima*	Tulip Tree	*Liriodendron tulipifera*
Poplar	*Populus*	Tung	*Aleurites montana*
Poplar, Lombardy	*Populus nigra* 'Italica'	Tupelo	*Nyssa sylvatica*
Poplar, Yellow	*Liriodendron tulipifera*	Virginia Creeper	*Parthenocissus*
Potato	*Solanum tuberosum*		*quinquefolia*
Privet, Glossy	*Ligustrum lucidum*	Walnut	*Juglans*
Privet, Wax-leaf	*Ligustrum japonicum*	Walnut, Black	*Juglans nigra*
Pyracantha	*Pyracantha*	Walnut, English	*Juglans regia*
Quack Grass	*Agropyron repens*	Walnut, Japanese	*Juglans ailanthifolia*
Quince	*Cydonia oblonga*	Walnut, Northern	*Juglans hindsii*
Raspberry	*Rubus idaeus*	California Black	
Redbud	*Cercis*	Willow	*Salix*
Redwood	*Sequoia sempervirens*	Willow, Weeping	*Salix babylonica*
	or *Sequoiadendron*	Wisteria	*Wisteria*
	giganteum	Xylosma, Shiny	*Xylosma congestum*
Redwood, Coast	*Sequoia sempervirens*	Yellowwood	*Cladrastis lotea*
Redwood, Dawn	*Metasequoia*	(American)	
	glyptostroboides	Yew	*Taxus*
Rice	*Oryza sativa*	Yew, Irish	*Taxus baccata* 'stricta'
Rose	*Rosa*	Yew, Japanese	*Taxus cuspidata*
Rose, Rambler	*Rosa*	Zelkova	*Zelkova*
Rye	*Secale cereale*	Zelkova, Japanese	*Zelkova serrata*
Sassafras	*Sassafras albidum*	Zelkova, Sawleaf	*Zelkova serrata*

Flood-Tolerant Woody Plants in the Contiguous United States

The species listed withstood 180 or more days of water covering the soil under trees. Observations were made in the 10 U.S. Corps of Engineers Divisions in the contiguous United States (Whitlow and Harris 1979).

Botanical name	Common name
Acer negundo	Box Elder
Acer rubrum	Red Maple
Acer saccharinum	Silver Maple
Carya aquatica	Water Hickory
Carya illinoensis	Pecan
Carya ovata	Shagbark Hickory
Cephalanthus occidentalis	Buttonbush
Cornus stolonifera	Red-osier Dogwood
Crataegus mollis	Red Hawthorn
Diospyros virginiana	Persimmon
Eucalyptus camaldulensis	Red Gum
Forestiera acuminata	Swamp Privet
Fraxinus pennsylvanica	Green Ash
Gleditsia aquatica	Water Locust
Gleditsia triacanthos	Honey Locust
Ilex decidua	Deciduous Holly
Liquidambar styraciflua	Liquidambar
Nyssa aquatica	Water Tupelo
Planera acquatica	Water Elm
Platanus × *acerifolia*	London Plane
Platanus occidentalis	Sycamore
Populus deltoides	Eastern Cottonwood
Quercus bicolor	Swamp White Oak
Quercus lyrata	Overcup Oak
Quercus macrocarpa	Bur Oak
Quercus nuttallii	Nuttall's Oak
Quercus palustris	Pin Oak
Salix spp.	Willow
Salix alba var. *tristis*	Golden Weeping Willow
Salix exigua	Narrow-leaf Willow
Salix hookeriana	Hooker Willow
Salix lasiandra	Pacific Willow
Salix nigra	Black Willow
Taxodium distichum	Bald Cypress
Ulmus americana	American Elm
Washingtonia robusta	Mexican Fan Palm

APPENDIX THREE

Tolerance of Woody Plants to Landfill Conditions

Species tolerant of landfill conditions	
Botanical name	Common name
Acacia latifolia	Broadleaf Acacia
Acer rubra	Red Maple
Arbutus unedo	Strawberry Tree
Eucalyptus lehmannii	Bushy Yate
Fraxinus pennsylvania	Green Ash
Ginkgo biloba	Ginkgo
Gleditsia triacanthos	Honey Locust
Grevillea robusta	Silk Oak
Liquidambar styraciflua	Liquidambar
Melaleuca quinquenervia	Cajeput Tree
Myoporum laetum	False Sandalwood
Myrica pensylvanica	Bayberry
Nyssa sylvatica	Black Gum
Picea abies	Norway Spruce
Pinus halepensis	Aleppo Pine
Pinus pinea	Italian Stone Pine
Pinus strobus	White Pine
Pinus thunbergii	Japanese Black Pine
Pittosporum undulatum	Victorian Box
Platanus occidentalis	American Sycamore
Populus hybrid	Hybrid Poplar (rooted cuttings)
Quercus palustris	Pin Oak
Schinus molle	California Pepper Tree
Taxus cuspidata	Japanese Yew
Ulmus parvifolia	Siberian Elm

Species sensitive to landfill conditions	
Botanical name	Common name
Abies concolor	White Fir
Acer saccharum	Sugar Maple
Carya ovata	Shagbark Hickory
Cornus florida	Dogwood
Malus	Apple
Picea glauca	White Spruce
Picea pungens	Colorado Blue Spruce
Pinus resinosa	Red Pine
Populus nigra 'Italica'	Lombardy Poplar
Prunus	Black Cherry
Pseudotsuga menziesii	Douglas Fir
Quercus velutina	Black Oak
Salix sp.	Willow
Sorbus aucuparia	Mountain Ash
Taxus cuspidata	Japanese Yew
Thuja	Arborvitae
Tilia americana	American Basswood
Ulmus americana	American Elm
Ulmus fulva	Slippery Elm

REFERENCES

FLOWER, F. B., E. F. GILMAN, and I. A. LEONE. 1981. Landfill Gas: What It Does to Trees and How Its Injurious Effects May Be Prevented. *J. Arboriculture* 7(2):43–52.

LEONE, I. A., and others. 1977. Damage to Woody Species by Anaerobic Landfill Gases, *J. Arboriculture* 3(12):221–225.

VAN HEUIT, R. E. 1979. *Sanitation Districts of Los Angeles County: Generation of Methane Gas in Sanitary Landfills and Its Effect on Growth of Plant Materials.* Unpublished. Delivered at 1979 meetings of Amer. Soc. Consulting Arborists. (Abstract in *Horizons.* 1980. Washington, D.C.: Amer. Assoc. Nurserymen, Sept. 1.)

APPENDIX FOUR

Sources of Other Plant Lists

Lists of plants recommended for specific purposes or situations must be used with caution. In most cases, compilations are based on field observations and individual judgments rather than actual screening tests. Frequently, a compilation is based entirely on other lists whose constraints and caveats have been excluded. Plants of one species can vary greatly in their susceptibility or resistance to environmental, disease, and insect problems. Even plants of the same clone may respond differently to a particular situation depending on prior events; a plant that has been stressed is more susceptible to certain diseases than identical plants that have not been stressed. Plant lists are helpful primarily as a starting point, after which you should obtain local information about promising plants.

The plant lists in Appendixes Two and Three are not widely available in the literature. More extensive lists of plants with certain attributes appear in other books; this appendix is a guide to some of these resources, which you may want to consult.

REFERENCES

CLOUSTON, B., ed. 1979. *Landscape Design with Plants*. London: Heinemann.

DUNMIRE, J. R. 1988. *Sunset New Western Garden Book*. Menlo Park, CA: Lane Publishing Co.

FERGUSON B. 1982. *All About Trees*. San Francisco: Ortho Books.

HILLIER, H. C. 1973. *Hillier's Manual of Trees and Shrubs*. London: David & Charles.

HOYT, R. S. 1978. *Ornamental Plants for Subtropical Regions*. Anaheim, CA.: Livingston.

LORD, E. E. 1979. *Shrubs and Trees for Australian Gardens*. 4th ed. Melbourne: Lothian.

MAINO, E., and F. HOWARD. 1955. *Ornamental Trees*. Berkeley: University of California Press.

MCMINN, H. E., and E. MAINO. 1963. *An Illustrated Manual of Pacific Coast Trees*. Berkeley: University of California Press.

MULLINS, M. G. 1979. *Reader's Digest Illustrated Guide to Gardening*. Sydney: Reader's Digest Services.

WYMAN, D. P. 1971. *Wyman's Gardening Encyclopedia*. New York: Macmillian.

	Clouston (1979)	Dunmire (1988)	Ferguson (1982)	Hillier (1973)	Hoyt (1978)	Lord (1979)	Maino and Howard (1955)	McMinn and Maino (1963)	Mullins (1979)	Wyman (1971)
Plant Characteristics										
Adaptable to acid soil				X	X				X	X
Adaptable to alkaline soil				X				X	X	
Adaptable to wet soil		X	X	X	X	X		X	X	X
Adaptable to dry soil	X	X					X	X	X	X
Adaptable to clay soil				X					X	
Adaptable to shallow soil				X	X					
Adaptable to saline soil					X					
Tolerant of smoke, dust, and abuse	X		X	X	X					
Tolerant of street conditions	X		X		X		X	X	X	X
Tolerant of seacoast conditions		X	X	X	X	X		X	X	X
Tolerant of shade		X			X		X		X	X
Tolerant of heat		X	X			X		X	X	
Tolerant of frost				X	X		X			
Pest resistant			X		X			X		X
Lawn and shade plantings		X						X		
Erosion controlling	X	X		X	X			X		
Sand holding								X		
Windbreaks	X	X			X		X	X		X
Rapid growth		X	X	X		X		X		
Slow growth					X					
Landscape and aesthetic effects	X	X	X	X	X	X	X	X	X	X

APPENDIX FIVE

Specifications for Acceptance of Nursery Trees at the Time of Delivery

Purpose: To obtain vigorous, healthy trees which can be easily trained into attractive trees with structurally strong roots and crowns.

Specifications: (The buyer should choose and/or modify the appropriate sections depending on the species, the landscape site, and the intended function of the tree.)

I. All trees shall be true to type or name as ordered or shown on the plans and shall be individually tagged or tagged in groups by species and cultivar (variety).

II. All trees shall be healthy, have a form typical for the species or cultivar, be well-rooted, and be properly trained. These characteristics are described in sections III through VII.

III. All trees shall comply with Federal and State laws requiring inspection for plant diseases and pest infestations. Inspection certificates required by law shall accompany each shipment of plants. Clearance from the County Agricultural Commissioner as required by law, shall be obtained before planting trees delivered from outside the County in which they are to be planted.

IV. The rootball of all trees shall be moist throughout and the crown shall show no signs of moisture stress.

V. Tree crown: (round headed) broadleaved, decurrent trees

   A. Crown has a single, straight trunk that has not been headed or that could be pruned to a leader

      1. Potential lateral scaffolds (height of lowest scaffold depends on landscape use)

         a. Small-growing trees (crape myrtle, flowering fruit trees): At least 50 mm (2 in.) apart vertically, which could be trained in the landscape to 3 to 7 branches 100 mm (4 in.) or more apart vertically. Large-growing trees (ash, oak, callery pear): At least 150 mm (6 in.) apart vertically, which could be trained in the landscape to 5 to 9 branches 450 mm (18 in.) or more apart vertically.

         b. Radially distributed around the trunk

         c. Not more than two-thirds ($\frac{2}{3}$) the diameter of the trunk, measured 25 mm (1 in.) above the branch

         d. Free of included bark in attachments (bark embedded between the trunk and a lateral)

      2. No laterals below the lowest potential scaffold should be larger than one-fourth ($\frac{1}{4}$) the trunk diameter at point of attachment

These specifications have in part been adapted from the *Standard Specifications,* January 1981, of the California Department of Transportation, Sacramento.

3. Each tree must be able to comply with A1 and A2 above without having or having had to remove, now or within the previous growing season (at least six months), more than 25 percent of the branches of size similar to or larger than those of the potential scaffold branches.

B. The minimum acceptable length of the most recent season's shoots should be specified, for example, shoots of small-growing trees (that is, red maple, red oak, ginkgo) might be 300 mm (12 in.); for large-growing trees the minimum acceptable length might be 450 mm (18 in.) and preferably 0.6 to 1.0 m (24–36 in.).

C. It would be desirable to have
   1. The tree stand upright without support.
   2. Small ($< \frac{1}{4}$ diameter of trunk) temporary branches along the trunk below the scaffolds

VI. Tree crown: broadleaved or coniferous, excurrent (central trunk) trees

A. Crown has a single, straight trunk with no double leaders (codominant stems) or vigorous, upright branches competing with the leader

B. Radial and vertical distribution of branches to form a symmetrical crown

VII. Roots: container, boxed, or balled & burlapped trees regardless of species or mature size

A. Check that the tree is free of roots visibly circling the trunk, and free of "knees" (roots) protruding above the soil.

B. If in a tapered container, slip the root ball out; the root-ball periphery should be free of circling roots larger than 6 mm ($\frac{1}{4}$ in.) in diameter and a bottom mat of roots 6 mm ($\frac{1}{4}$ in.) or larger (the acceptable diameter of circling peripheral roots depends on species and size of the root ball).

C. Untie the tree trunk from the stake; the trunk should not touch the top rim of the container.

D. Tip the root ball or container on its side and with a small jet of water expose the roots within 50 mm (2 in.) of the trunk to a depth of 65 mm (2.5 in.) below the topmost root attached to the trunk. The trunk and main root(s) should be free of circling roots and kinks. Replace soil washed from around the trunk with a similar soil mix (less than ten [10] percent of the total root-ball volume should need to be added).

E. If the trees pass the above inspections, further inspect the roots by removing the soil from the roots of not less than two (2) trees nor more than two (2) percent of the total number of trees of each species or variety from each source. The trunk and main roots shall be free of circling and kinked roots. Circling roots at the periphery of the root ball shall not be reason for rejecting a tree unless the circling roots are large for the species and shoot growth is not acceptable for the species (*see* VII-B).

VIII. In case the sample trees inspected are found to be defective, the buyer reserves the right to reject the entire lot or lots of trees represented by the defective samples. Any plants rendered unsuitable for planting because of this inspection will be considered as samples and will not be paid for.

IX. The buyer shall be notified when plants are to be shipped at least ten (10) days prior to the actual shipment date.

Pruning Guidelines
for Landscape Trees

Purpose: To develop and preserve tree structure and health. These guidelines are presented as working guidelines, recognizing that trees are individually unique in structure, form, and growth response—not only between, but also within species and cultivars. The appropriate sections should be chosen and/or modified depending on the species, the landscape site, the intended function of the tree, the present age and condition of the tree, and the desired severity of pruning.

I. PRUNING TECHNIQUES

A. Types of Pruning Cuts

1. *Thinning.* A thinning cut removes a branch at its point of attachment or shortens it or the leader to a lateral large enough (at least one-half the diameter of the cut being made) to assume the terminal role (this latter cut is known as *drop crotching*) (Figs. 15–6 and 15–12). Thinning cuts can effectively direct growth and retain the natural form of the tree.

2. *Heading.* A heading cut removes a branch to a stub, a bud, or a lateral branch not large enough to assume the terminal role (Figs. 15–3 and 15–4).

B. Location of Pruning Cuts

1. When removing a live branch, make the pruning cut either just outside the branch bark ridge and collar, or just outside the branch bark ridge and through the outside half of the collar (Fig. 15–7). If no collar is visible, as in birch and alder, the angle of the cut should approximate the angle formed by the branch bark ridge and the axis of the trunk or the branch cut to.

2. When removing a dead branch, the final cut should be made outside the collar of live callus tissue even if the collar has grown out along the dead branch (Fig. 15–16).

3. To prevent tearing or stripping the bark when removing a large branch, remove most of the branch by making two cuts, the first one from the bottom and

Sections I and III are adapted from the Pruning Standards of the Western Chapter of the International Society of Arboriculture (Perry 1988).

the second from above, close to the first cut. The final cut is made as described in B1 (Fig. 15–14).

4. When reducing the length of a branch or the height of a leader, the final cut should be made just beyond (without violating) the branch bark ridge of the branch being cut to. To minimize the possibility of the branch splitting out, the cut should approximately bisect the angle formed by the branch bark ridge and an imaginary line perpendicular to the trunk or the branch cut (Fig. 15–7).

C. Structural Considerations

1. A goal of structural pruning is to maintain the size of lateral branches to less than three-fourths ($\frac{3}{4}$) the diameter of the trunk or parent branch. If the branch is codominant or close to the size of the trunk or parent branch, thin the laterals on the competing branch 15 to 25 percent, particularly near the terminal. Thin the parent branch less, if at all, in order to keep it dominant.

2. On large-growing trees, branches that are more than one-third the diameter of the trunk should be spaced at least 600 mm (24 in.) apart on center. If this is not possible because of the present size of the tree, such branches should have 15 to 25 percent of their foliage thinned, particularly near the terminals.

3. Before beginning work in a tree, inspect the root-collar of the tree for signs of weak or broken roots, for the presence of decay or cavities, for adventitious roots, or for other signs of a hazardous condition. If the tree appears to be hazardous, it may be wise to remove the tree or at least obtain a second opinion.

D. Pruning Severity

1. The removal of many small branches rather than a few large branches requires more time but will produce a more natural appearance, force fewer watersprouts, lengthen the time until the next pruning, and help maintain the vigor and structure of mature trees. On mature trees the maximum size (base diameter) of any occasional undesirable branch that may be left within a tree crown—such as 15 mm ($\frac{1}{2}$ in.), 30 mm (1 in.), or 50 mm (2 in.) branch diameter—establishes the detail or degree of pruning desired and should be specified before work begins.

2. No more than one-third ($\frac{1}{3}$) of the live foliage of a tree should be removed at one time without good reason.

E. General Considerations

1. Pruning cuts should be clean and smooth with the bark at the edge of the cut firmly attached to the wood.

2. Large or heavy branches that cannot be thrown clear should be lowered on ropes to prevent injury to the tree and other property.

3. Wound dressings and tree paints have not been shown to be effective in preventing or reducing decay. They are not recommended for preventing decay, but asphalt tree paint can minimize borer attacks on fresh pruning wounds of several species (elm, pecan, and oak), and fungicidal paints are effective for certain canker-

causing disease organisms. If asphalt wound dressing is used for cosmetic purposes or to discourage borer attack, it should be applied in a thin coat at the time of pruning.

4. Climbing techniques

    a. Climbing and pruning practices should not injure the tree except for the pruning cuts.

    b. The use of climbing spurs or gaffs should be avoided. Their use may be considered only when branches are more than throw-line distance apart. In such cases, the spurs should be removed as soon as the climber is tied in.

    c. Spurs may be used to reach an injured climber or when removing a tree.

    d. Rope injury to thin barked trees from loading out heavy limbs should be avoided by installing a block in the tree to carry the load. This technique may also be used to reduce injury to a branch crotch from a climber's line.

II. TRAINING YOUNG TREES

Young trees can be trained to grow into structurally strong trees well suited to the site. These trees will fulfill their intended function sooner, should require little corrective pruning as they mature, and will substantially reduce pruning costs over the life of the trees. Young trees of large, mature size should have a sturdy, tapered trunk with well-spaced branches smaller in diameter than the trunk.

These standards apply primarily to large-growing decurrent (round-headed) trees. Trees that will become decurrent, such as most oaks, elms, and callery pears, seldom have lateral shoots on current-season's growth. Early training of central-leader trees (excurrent), such as liquidambar and most conifers, may be less important except to remove low branches and those competing with the leader.

A. Trunk Development of Young Trees

1. For most trees, maintain a single, fairly straight trunk. Only head the leader if (a) in summer, it is very vigorous and it is desirable to correctly position the lowest main branch or to space other main branches at least 150 to 450 mm (6–18 in.) apart vertically (depending on mature tree size), or (b) when dormant, to remove a tuft of terminal twigs so that more vigorous shoots can be chosen the next growing season.

2. More potential main branches than will eventually be needed should be left on a newly transplanted tree the first year or two. This will ensure a better selection of main branches and rapid early growth of a tree.

Potential main branches can be spaced 150 to 300 mm (6–12 in.) apart vertically the first year or two. By the fourth year, however, large-growing trees should be thinned to 5 to 9 branches at least 450 mm (18 in.) or farther apart, earlier if 4 below is fulfilled. Remove first the largest of the potential branches that you know will not be kept.

A potential main branch and main branches should be one-half ($\frac{1}{2}$) the diameter of or smaller than the trunk diameter immediately above the branch. If a branch is too large, thin laterals on it or, if too few laterals, head the branch.

3a. Select small *temporary branches* along the trunk below the desired height of the lowest main branch and between main branches. It is more important to have temporary branches below the lowest main branch than above to protect the trunk and increase its caliper and taper. Temporary branches should be kept for 2 to 4 years unless they become more than 15 mm ($\frac{1}{2}$ in.) in diameter.

Preferred vertical spacing is 100 to 150 mm (4–6 in.) but none within 150 mm (6 in.) of potential main branches. Select the least vigorous shoots for temporary branches. If larger-than-desired branches need to be kept as temporaries, head them back to 2 or 3 buds. It is important to have some on the trunk side facing the afternoon sun (equator-west side). Included bark in and angle of the attachment of temporary branches is not important because they will be removed.

3b. During the growing season, temporary branches should be kept shorter than 300 mm (12 in.). If still growing, cut back by at least half. Temporary branches may need two to four prunings during a growing season, depending on tree vigor. This can be quickly and inexpensively done and provides opportunity to inspect for other problems.

At the first dormant pruning, thin out one-fourth ($\frac{1}{4}$) to one-third ($\frac{1}{3}$) of the temporary branches. Leave them uniformly spaced, removing the largest or cutting them back to 2 to 3 buds. At the next dormant prunings, remove about the same number of temporary branches as removed the first year. In most situations, by the fourth dormant pruning all of the temporary branches should be removed.

4. One half ($\frac{1}{2}$) of the foliage should be on branches (temporary and main) arising in the lower two-thirds ($\frac{2}{3}$) of a tree. This will increase trunk taper and more uniformly distribute branch weight and wind stress along the trunk.

B. Main Branch Selection of Young Trees

1. The height of the lowest main branch will depend on the function of the tree: is the tree to screen an unsightly view, be a wind break, shade a patio, be a street tree, or whatever?

2. More potential main branches can be left than will be needed to ensure a selection of well-placed and vigorous main branches. Potential main branches should be spaced 150 to 300 mm (6–12 in.) apart vertically. By the fourth or fifth year, the main branches (at least in the lower part of a tree) should be selected and be at least 150 to 450 mm (6–18 in.) or more apart, depending on mature tree size.

3. Select 5 to 9 main branches (3 to 5 on small-growing trees) to give radial distribution for light penetration and tree balance. None should be directly over another unless it is 3 m (10 ft) or more above the lower branch.

C. Developing Strong Branch Structure of Young Trees

1. Main branches should be one-half ($\frac{1}{2}$) the diameter or smaller than that of the trunk immediately above the branch (See A2). For strength, the relative size of a branch to the trunk is more important than its angle of attachment.

2. No main branch attachments should have included bark.

3. Encourage laterals along main branches, but none larger than one-half ($\frac{1}{2}$)

the diameter of the main branch. The lateral closest to the trunk should be at least 600 mm (2 ft) out on the main branch.

III. TYPES OF MATURE TREE PRUNING

A. Crown Cleaning

Crown cleaning or cleaning out is the removal of dead, dying, diseased, crowded, weakly attached, and low-vigor branches and watersprouts from a tree crown.

B. Crown Thinning

Crown thinning includes crown cleaning and the selective removal of branches to increase light penetration and air movement through the crown (Fig. 15–34). Increased light and air maintain and stimulate interior foliage, which in turn improves branch taper and strength. Thinning reduces the wind-sail effect of the crown and the weight of limbs. Thinning the crown can emphasize the structural beauty of trunk and branches as well as improve the growth of plants beneath the tree by increasing light penetration. When the crowns of mature trees are thinned, seldom should more than one-third ($\frac{1}{3}$) of the live foliage be removed.

After pruning, at least one-half ($\frac{1}{2}$) of the foliage should be on branches that arise in the lower two-thirds ($\frac{2}{3}$) of the tree. Likewise, when laterals are thinned from a limb, an effort should be made to retain inner lateral branches, that is, one-half ($\frac{1}{2}$) of the foliage on laterals along the inner two-thirds ($\frac{2}{3}$) of the branch. Trees and branches so pruned will have stress evenly distributed throughout the tree and along the branches.

An effect known as *lion's tailing* results from pruning out (stripping) the inside lateral branches. Lion's tailing, by removing most of the inner foliage, displaces the weight to the ends of the branches and may result in weakened branch structure, limb breakage, sunburned branches, and watersprouts.

C. Crown Reduction

Crown reduction is used to reduce the height and/or spread of a tree (Fig. 15–35). Thinning cuts are most effective in maintaining the structural integrity and natural form of a tree and in delaying the time when it will need to be pruned again, thereby reducing maintenance costs. The lateral to which a branch or trunk is cut should be at least one-half ($\frac{1}{2}$) the diameter of the cut being made.

D. Crown Restoration (Renewal)

Crown restoration can improve the structure and appearance of a tree that has been topped or severely pruned using heading cuts. One to three sprouts on main branch stubs should be selected to reform a more natural appearing crown; all except one will be headed to one or two buds to speed wound closure. Selected vigorous sprouts may need to be thinned to a lateral—or even headed—to control length growth in order to ensure adequate attachment for the size of the sprout. Restoration will require several prunings over a number of years.

E. Crown Raising

Crown raising removes the lower branches of a tree in order to provide clearance for buildings, vehicles, pedestrians, and vistas (Fig. 15–33). It is important that a tree have at least one-half ($\frac{1}{2}$) of its foliage on branches that originate in the lower two-thirds ($\frac{2}{3}$) of its crown to ensure a well formed, tapered structure and to uniformly distribute stress within a tree (Fig. 3–8).

F. Utility-Line Clearing

Utility-line clearing can combine all five types of pruning just described to direct tree growth away from utility lines (Fig. 15–38). Detailed studies of pruning techniques, tree responses, and costs have shown that the previously recommended pruning techniques can be used to advantage in utility-line pruning. To achieve clearance, make thinning cuts primarily back to medium-size branches within the crown. Remove the entire branch if more than 50 percent of live foliage needs to be removed from it.

Severe height reduction (including heading) occasionally may be the only alternative when a tree is directly under lines; it may be best to replace the tree.

Plant growth regulators can usually be used to advantage on many of the fast-growing species. Vegetation control along transmission rights-of-way is usually best done with chemicals.

G. View Restoration

Crown thinning, reduction, and raising can be used to enhance or restore a scenic view or light penetration to the property. Views can be attractive when seen through sparsely-leaved branches. Care must be taken to not overprune.

Bibliography

AGRIOS, G. N. 1969. *Plant Pathology*. San Diego, CA: Academic Press.

ALBERTY, C. A., H. M. PELLETT, and D. H. TAYLOR. 1984. Characterization of Soil Compaction on Vertical and Horizontal Root Distribution. *J. Environ. Hort.* 5(1):33–36.

ALJIBURY, F. K., J. L. MEYER, and W. E. WILDMAN. 1979. *Managing Compacted and Layered Soils*. Univ. Calif. Agr. Sci. Leaflet 2635.

AMERICAN HORTICULTURAL SOCIETY. 1980. Are Wood Ashes Good for the Soil? *Area Nursery and Garden Store Newsletter* (Ohio State University) 22(4):8.

———. 1988. New Research on Mulches. *Amer. Horticulturist,* Nov., p. 11.

AMERICAN NATIONAL STANDARDS INSTITUTE (ANSI Z60.1). 1986. *American Standard for Nursery Stock*. Washington, DC: American National Standards Institute.

———. (ANSI Z133). 1988. *American National Standard for Tree Care Operations*. Washington, DC: American National Standards Institute.

AMERICAN NURSERYMAN. 1970. Steel-Clad Balled Trees. *American Nurseryman* 131(1):14.

ANDERSON, R. F. 1960. *Forest and Shade Tree Entomology*. New York: John Wiley.

ANDRESEN, J. W. 1977. Survey of Growth and Survival of Trees Indicated No Detrimental Effects Caused by High-Pressure Sodium Street Lighting. *Utility Arborist Assn. Newsletter* 7(4):1–2.

ARBORICULTURAL ASSOCIATION. 1976. *Trees: A Guide to Pruning*. Romsey, Hants, England: Arboricultural Association.

ARBORICULTURAL ASSOCIATION. 1990. *Amenity Valuation of Trees and Woodlands*. Romsey, Hants, England: Arboricultural Assoc.

ARMSTRONG, N. 1925. Open Cavities. *Proc. Natl. Shade Tree Conf.* 1:104–8.

ASHBAUGH, F. A. 1968. Edison Electrical Institute Tree Growth Control Project, 1968. *Proc. Inl. Shade Tree Conf.* 44:169–72.

634

AYERS, R. S. 1977. Quality of Water for Irrigation. *J. Irrigation and Drainage Div., Amer. Soc. Civil Engr.* 103(IR2):135–54.

BACKHAUS, R. A., R. M. SACHS, and J. L. PAUL. 1980. Morphactin Transport and Metabolism in Tree Bark Tissues: Comparative Studies in Vitro. *Physiologia Plantarum* 50:131–36.

BAILEY, L. H. 1916. *The Pruning Manual.* London: Macmillan.

———. 1949. *Manual of Cultivated Plants.* New York: Macmillan.

BAILEY, L. H. E. Z. BAILEY, and STAFF OF THE L. H. BAILEY HORTORIUM. 1976. *Hortus Third.* New York: Macmillan.

BAINBRIDGE, D. 1990. Tubex Tree Shelters for Tree Establishment on Extreme Arid Sites. *Tree Shelters* (Tubex, St. Paul, MN), Winter–Spring, 1990–91, pp. 12–13.

BAKER, K. F. 1957. *The U. C. System for Producing Healthy Container-Grown Plants.* Calif. Agr. Exp. Sta. Manual 23.

BAKKEN, S. 1986. *Tree Hazard Control Program: Guidelines and Standards for the California Department of Parks and Recreation.* Sacramento: Calif. Dept. of Parks and Recreation.

BANCROFT-WHITNEY. 1967. *Deering's California Codes: Agricultural Code Annotated.* Div. 18, Ch. 5, Art 6. San Francisco: Bancroft-Whitney.

BARFIELD, B. V., and J. F. GERBER, eds. 1979. *Modification of the Aerial Environment of Plants.* St. Joseph, MI: Amer. Soc. of Agr. Engineers.

BARROWS, W. 1970. Some Practical Steps to Quality Control of New Street Plantings. *Int'l Shade Tree Conf. West Chapter Newsletter,* Sept., pp. 12–15.

BARTLETT, F. A. 1958. Pruning Trees. In *Plants & Gardens Pruning Handbook* (Brooklyn Botanic Garden) 14(3):199–203.

BAXENDALE, R. W., and W. T. JOHNSON. 1990. Efficacy of Summer Oil Spray on Thirteen Commonly Occurring Insect Pests. *J. Arboriculture* 16:89–94.

BEAN, W. J. 1950. *Trees and Shrubs Hardy in the British Isles.* 3 vols. London: Butler and Tanner.

BEAUCHAMP, K. H. 1955. Tile Drainage: Its Installation and Upkeep. In *Water: Yearbook of Agriculture,* ed. A. Stefferud. Washington, DC: U.S. Government Printing Office, pp. 508–20.

BECKENBACH, J., and J. H. GOURLEY. 1932. Some Effects of Different Cultural Methods upon Root Distribution of Apple Trees. *Proc. Amer. Soc. Hort. Sci.* 29:202–4.

BECKER, W. B. 1938. *Leaf-Feeding Insects of Shade Trees.* Mass. Agr. Exp. Sta. Bul. 353.

BEN-JAACOV, J., G. NATANSON, and A. HAGILADI. 1980. The Use of a Wind-Controlled Overhead Irrigation System to Prevent Damage by Wind-Borne Salts. *J. Amer. Soc. Hort. Sci.* 105(6):833–35.

BENNETT, J. P. 1931. *The Treatment of Lime-Induced Chlorosis with Iron Salts.* Univ. Calif. Agr. Exp. Sta. Circ. 321.

BENSON, L., and R. A. DARROW. 1954. *Trees and Shrubs of the Southwestern Deserts.* Tucson: University of Arizona Press.

BERDAHL, P., and others. 1978. *California Solar Data Manual.* Berkeley: Lawrence Berkeley Lab., University of California.

BERNATZKY, A. 1978. *Tree Ecology and Preservation.* New York: Elsevier Publishing.

BERNSTEIN, L. 1964. *Salt Tolerance of Plants.* USDA Inf. Bull. 283.

BERNSTEIN, L., and L. E. FRANCOIS. 1973. Sensitivity of Alfalfa to Salinity of Irrigation and Drainage Water. *Soil Sci. Soc. Amer. Proc.* 37:931–43.

BEY, C. F. 1980. Growth Gains from Moving Black Walnut Provenances Northward. *J. Forestry* 78(10):640–41.

BICKELHAUPT, R. E. 1983. Typar Solves Weed and Root Problems. Clinton, IA: *Bickelhaupt Arboretum Letter,* Dec.

BIDDLE, P. G. 1979. Tree Root Damage to Buildings: An Arboriculturist's Experience. *Arboricultural J.* 3(6):397–412.

BINGHAM, C. 1968. *Trees in the City.* Chicago: Amer. Soc. Planning Off. Report 236.

BINGHAM, F. T. 1966. Phosphorus. In *Diagnostic Criteria for Plants and Soils,* ed. H. D. Chapman. Riverside, CA: H. D. Chapman, pp. 324–61.

BINNS, W. O., H. INSLEY, and J. B. H. GARDINER. 1983. Nutrition of Broadleaved Amenity Trees: I. Foliar Sampling and Analysis for Determining Nutrient Status. *Arb. Res. Note* 50/83/555:1–4.

BIRAN, I., and A. M. KOFRANEK. 1976. Evaluation of Fluorescent Lamps as an Energy Source for Plant Growth. *J. Amer. Soc. Hort. Sci.* 101(6):9–17.

BLACK, M. 1978. Reducing Vandalism. *Western Chapter News* (Intl. Soc. Arboriculture), July–August, p. 42.

BLAIR, G. D. 1940. *Tree Clearance for Overhead Lines.* Chicago: Electrical Publications, Inc.

BLAZE, K. L. 1990. *PIRM Handbook of Background Information and Operating Instructions.* Ringwood, Victoria: Biosensors.

BONNER, I., and A. GALSTON. 1952. *Principles of Plant Physiology.* San Francisco: W. H. Freeman.

BOODY, L., and A. D. M. RAYNER. 1983. Origins of Decay in Living Deciduous Trees: The Role of Moisture Content and a Re-Appraisal of the Expanded Concept of Tree Decay. *New Phytol.* 94:623–41.

BORST, H. L., and R. WOODBURN. 1942. The Effect of Mulching and Methods of Cultivation on Runoff and Erosion from Muskingham Silt Loam. *Agr. Engineering* 23:19–22.

BOWEN, G. D. 1987. The Biology and Physiology of Infection and Its Development. In *Ecophysiology of VA Mycorrhizal Plants,* ed. G. R. Safir. Boca Raton, FL: CRC Press.

BOWMAN, D. C., R. Y. EVANS, and J. L. PAUL. 1990. Fertilizer Salts Reduce Hydration of Polyacrylamide Gels and Affect Properties of Gel-Amended Container Media. *J. Amer. Soc. Hort. Sci.* 115(3):382–86.

BOYCE, S. G. 1954. The Salt Spray Community. *Ecology Monographs* 24:29–67.

BOYNTON, D., and G. H. OBERLY. 1966. Apple Nutrition. In *Nutrition of Fruit Crops: Temperature, Subtropical, Tropical,* ed. N. F. Childers. New Brunswick, NJ: Horticultural Pub., Rutgers University, pp. 1–50.

BRADFORD, G. R. 1966. Boron. In *Diagnostic Criteria for Plants and Soil,* ed. H. D. Chapman. Riverside, CA: Chapman, pp. 33–61.

BRADY, N. C. 1990. *The Nature and Property of Soils,* 10th ed. New York: Macmillan.

BRAMBLE, W. C., W. R. BYRNES, and R. J. HUTNIK. 1990. Resistance of Plant Cover Types to Tree Seedling Invasion on an Electric Transmission Right-of-Way. *J. Arboriculture* 16:130–35.

BRANSON, R. L. 1977. Make Softened Water Suitable for Watering Plants. *Univ. Calif. Soil and Water Newsletter* 32:1.

——. 1980. Wood Ashes as a Soil Additive. *Soil and Water* (Univ. Calif. Coop. Ext.) 45:2–3.

BRIDGEMAN, P. H. 1976. *Tree Surgery.* London: David & Charles.

BRIDEL, R., B. L. APPLETON, and C. E. WHITCOMB. 1983. Planting Techniques for Tree Spade Dug Trees. *J. Arboriculture* 9:282–84.

BRITISH BROADCASTING CORPORATION (BBC). 1979. The Effect of Trees on Television Reception. *Arboriculture Res. Note* 14:1–2.

BRITISH STANDARDS INSTITUTE. 1966. *Recommendations for Tree Work.* London: British Standards Inst. 3998.

——. 1989a. *British Standard Guide for Trees in Relation to Construction.* London: British Standards Inst. 5837.

——. 1989b. *British Standard Recommendations for Tree Work.* London: British Standards Inst. 3998.

BRITTON, J. C. 1990. Root Crown Examinations for Disease and Decay. *J. Arboriculture* 18(7):v.

——. 1991. *Pruning Guidelines.* Urbana, IL: Intl. Soc. Arboriculture.

BROECKER, W. S. 1970. Enough Air. *Environment* 12(7):28–31.

——. 1981. Handbook on Pruning. *Plants & Gardens* 37(2).

BROSCHAT, T. K. 1991. Effects of Leaf Removal on Survival of Transplanted Sabal Palms. *J. Arboriculture* 17:32–33.

BROSCHAT, T. K., and A. W. MEEROW 1990. Palm Nutrition Guide. *Nursery Digest* (Cooper City, FL) 24(4):1–4.

BROWN, C. L. 1971. Primary Growth. In *Trees: Structure and Function,* ed. M. Zimmermann and C. L. Brown. New York: Springer-Verlag, pp. 1–66.

BROWN, C. L., R. G. McALPINE, and P. P. KORMANIK. 1967. Apical Dominance and Form in Woody Plants: A Reappraisal. *Amer. J. Bot.* 54(2):153–62.

BROWN, G. E. 1972. *The Pruning of Trees, Shrubs, and Conifers.* London: Faber and Faber.

BROWN, I. R. 1987. Suffering at the Stake. In *Advances in Practical Arboriculture,* ed. D. Patch. For. Com. Bull. 65. London: Her Majesty's Stationery Office, pp. 85–90.

BROWN, L. R., and C. O. EADS. 1966. *A Technical Study of Insects Affecting the Elm Tree in Southern California.* Calif. Agr. Exp. Sta. Bull. 821.

BURGER, D. W., P. SVIHRA, and R. W. HARRIS. 1991. Treeshelter Use in Producing Container-Grown Trees in the Nursery. *HortSci.* 25:(in press).

BUTIN, H., and A. SHIGO. 1981. *Radial Shakes and "Frost Cracks" in Living Oak Trees.* USDA For. Serv. Res. Paper NE–478.

CABORN, J. M. 1965. *Shelterbelts and Windbreaks.* London: Faber and Faber.

CALIFORNIA FERTILIZER ASSOCIATION. 1985. *Western Fertilizer Handbook,* 7th ed. Sacramento, CA: Calif. Fertilizer Assoc.

CAMPANA, R. J. 1978. Comparative Aspects of Dutch Elm Disease in Eastern North America and California. *California Plant Pathology* (University of California) 41:1–4.

CAPEL, J. A. 1987. *A Guide to Tree Pruning,* ed. D. R. Helliwell. Advisory Booklet. Ramsey, England: Arboricultural Assoc.

CARPENTER, E. D. 1970. Salt Tolerance of Ornamental Plants. *American Nurseryman* 131(2):12, 54–71.

CARTER, J. C. 1975. *Diseases of Midwest Trees*. Urbana: University of Illinois.

CATHEY, H. M. 1975. Comparative Plant Growth-Retarding Activities of Ancymidol with APC, Phosfon, Chlormequat and SADH on Ornamental Plant Species. *HortSci.* 10:204–16.

CATHEY, H. M., and L. E. CAMPBELL. 1975a. Effectiveness of Five Vision-Lighting Sources on Photo-Regulation of 22 Species of Ornamental Plants. *Amer. Soc. Hort. Sci.* 100(1):65–71.

———. 1975b. Security Lighting and Its Impact on the Landscape. *J. Arboriculture* 1:181–87.

CHADWICK, L. C. 1939. The Distribution of Roots of Moline Elm in Relationship to Fertilizer Application. *Proc. Natl. Shade Tree Conf.* 15:38–51.

———. 1941. *Fertilization of Ornamental Trees, Shrubs, and Evergreens*. Ohio Agr. Exp. Sta. Bull. 620.

———. 1971. 3000 Years of Arboriculture: Past, Present and Future. *Arborist's News* 36(6):73a–78a.

CHAN, F. J., R. W. HARRIS, and A. T. LEISER. 1977. *Direct Seeding Woody Plants in the Landscape*. Univ. Calif. Agr. Sci. Leaflet 2577.

CHANCELLOR, W. J. 1976. *Compaction of Soil by Agricultural Equipment*. Univ. Calif. Agr. Sci. Bull. 1881.

CHANDLER, W. H., and D. S. BROWN. 1951. *Deciduous Orchards in California Winters*. Calif. Agr. Ext. Serv. Circ. 179.

CHANDLER, W. H., and R. D. CORNELL. 1952. *Pruning Ornamental Trees, Shrubs and Vines*. Calif. Agr. Ext. Circ. 183.

CHANEY, W. R. 1986. Anatomy and Physiology Related to Chemical Movement in Trees. *J. Arboriculture* 12:85–91.

CHANEY, W. R., and T. T. KOZLOWSKI. 1977. Patterns of Water Movement in Intact and Excised Stems of *Fraxinus americana* and *Acer saccharum* Seedlings. *Ann. Bot.* 41:1093–1100.

CHANG, J. H. 1968. *Climate and Agriculture*. Chicago: Aldine.

CHAPMAN, H. D. 1960. *Leaf and Soil Analyses in Citrus Orchards*. Calif. Agr. Ext. Serv. Manual 25.

———. 1966a. Calcium. In *Diagnostic Criteria for Plants and Soils,* ed. H. D. Chapman. Riverside, CA: H. D. Chapman, pp. 65–92.

———. 1966b. Zinc. In *Diagnostic Criteria for Plants and Soils,* ed. H. D. Chapman. Riverside, CA: H. D. Chapman, pp. 484–99.

CHITTENDON, F. J., ed. 1951. *Dictionary of Gardening*. Oxford: Clarendon Press.

CLARK, F. E. 1957. Living Organisms in the Soil. In *Soil: The 1957 Yearbook of Agriculture*. Washington, DC: U.S. Government Printing Office.

CLOUSTON, B., ed. 1979. *Landscape Design with Plants*. London: Heinemann.

COLE, L. C. 1968. Can the World Be Saved? *Biol. Sci.* 18:679–84.

COLLINS, J. F. 1920. *Tree Surgery*. USDA Farmers' Bull. 1178.

COLORADO SPRINGS FIRE DEPT. 1989. *Protecting Your Home from Wildfire*. Colorado Springs Fire and Park and Recreation Depts.

CONKLIN, E. L. 1978. Interior Landscaping. *J. Arboriculture* 4(4):73–79.

CONNELL, J. H., and others. 1979. Gaseous Ammonia Losses Following Nitrogen Fertilization. *Calif. Agr.* 33(1):11-12.

CONOVER, C. A. 1978. Plant Conditioning: It's Here. *New Horizons.* Washington, DC: Hort. Res. Inst., pp. 6-9.

CONOVER, C. A., and R. T. POOLE. n.d. *Fertilization of Indoor Foliage Plants.* Apopka, FL: Agr. Res. Ctr., pp. 130-33.

CONOVER, C. A., R. T. POOLE, and R. W. HENLEY. 1975. Growing Acclimatized Foliage Plants. *Florida Foliage Grower* 12(9):1-7.

COOK, D. I., and D. F. VAN HAVERBEKE. 1971. *Trees and Shrubs for Noise Abatement.* U.S. Forest Serv. Res. Bull. 246.

COOKE, R. C., and A. D. M. RAYNER. 1984. *The Ecology of Saprotrophic Fungi.* London and New York: Longman.

COOL, R. A. 1976. Tree Spade vs. Bare Root Planting. *J. Arboriculture* 2(5):92-95.

COPELAND, E. B. 1906. On the Water Relations of the Coconut Palm. *Philippine J. Sci.* 1:6-57.

CORLEY, W. L. 1984. Soil Amendments at Planting. *J. Environ. Hort.* 2(1):27-30.

COSGROVE, T. J. 1989. The Greenhouse Effect: Perceptions and Misperceptions. *J. Arboriculture* 15:285-89.

COSTELLO, L., and J. L. PAUL. 1975. Moisture Relations in Transplanted Container Plants. *HortSci.* 10:371-72.

COSTELLO, L. R., and A. M. BERRY. 1991. California Tree Failure Report Program: An Overview. *J. Arboriculture* 17(9): 250-55.

COSTELLO, L. R., N. P. MATHENY, and J. R. CLARK. 1991. *Estimating Water Requirements of Landscape Plantings.* Univ. Calif. Agr. and Natl. Res. Leaflet 21493.

COSTELLO, L. R., and J. D. PETERSON. 1989. Decay Detection in Eucalyptus: An Evaluation of Two Methods. *J. Arboriculture* 15:185-88.

COUNCIL ON ENVIRONMENTAL QUALITY. 1972. *Integrated Pest Management.* Washington, DC: U.S. Government Printing Office.

COUTTS, M. P., and J. J. PHILIPSON. 1980. Mineral Nutrition and Tree Growth. In *Mineral Nutrition of Fruit Trees,* ed. D. Atkinson, J. E. Jackson, R. O. Sharples, and W. M. Waller. London: Butterworths, pp. 123-26.

CREECH, J. L., and W. HAWLEY. 1960. Effects of Mulching on Growth and Winter Injury of Evergreen Azaleas. *Proc. Amer. Soc. Hort. Sci.* 75:650-57.

CRIPE, R. E. 1978. Lighting Protection for Trees. Reported by J. A. Weidhaas, Jr. *American Nurseryman* 148(9):58-60.

CTLA. 1992. *Guide for Plant Appraisal.* Urbana, IL: Intl. Soc. Arboriculture.

CURRY, E. A., and M. W. WILLIAMS. 1984. A Comparison of Methods of Application of Growth Retardants to Apple and Pear Trees. *Proc. Plant Growth Regul. Soc. Amer.* 20:151.

CUTLER, D. F., and I. B. K. RICHARDSON. 1989. *Tree Roots and Buildings,* 2nd ed. Harlow, England: Longman Scientific and Technical.

DADDOW, R. L., and G. E. WARRINGTON. 1984. The Influence of Soil Texture on Growth-Limiting Soil Bulk Densities. Proc. Conv. Soc. Amer. For., pp. 252-56.

D'AMBROSIO, D. P. 1990. Crown Density and Its Correlation to Girdling Root Syndrome. *J. Arboriculture* 16:153–57.

DAVENPORT, D. C., R. M. HAGAN, and P. E. MARTIN. 1969. Antitranspirants: Uses and Effects on Plant Life. *Calif. Turfgrass Culture* 19(4):25–27.

DAVEY, K. L. 1967. Pruning Mature Trees. *Western Landscaping News* 7(12):14.

DAVIDSON, J. A., S. A. GILL, and M. J. RAUPP. 1990. Foliar and Growth Effects of Repetitive Summer Horticultural Oil Sprays on Trees and Shrubs Under Drought Stress. *J. Arboriculture* 16:77–81.

DAVIES, R. J. 1987. Do Soil Ameliorants Help Tree Establishment. *Arbor. Research Note* 69.

DAVIS, S. H., JR. 1971. *What's Wrong with My Tree?* NJ Coop. Ext. Serv. Leaflet 156-A.

DAVIS, W. B., and others. 1974. *Landscaping in Containers without Natural Drainage.* Univ. Calif. Agr. Sci. Leaflet 2577.

DEERING, R. B. 1955. Effective Use of Living Shade. *Calif. Agr.* 9(9):10–11.

DEERING, R. B., and F. A. BROOKS. 1953. Landscaping for Summer Shade. *Calif. Agr.* 7(5):11, 16.

DEL MORAL, R., and C. H. MULLER. 1970. Alleopathic Effects of *Eucalyptus camaldulensis. Amer. Midland Naturist* 83:254–82.

DEPARTMENT OF THE ENVIRONMENT. 1988. *The Effects of the Great Storm.* U.K. Dept. Environ.; Min. Agric., Fisheries, and Food; and For. Com. London: Her Majesty's Stationery Office.

DICKEY, R. D. 1977. *Nutritional Deficiencies of Woody Ornamental Plants Used in Florida Landscapes.* Fla. Agr. Exp. Bull. 791.

DOMIR, S. C. 1978. Chemical Control of Tree Height. *J. Arboriculture* 4(7):145–53.

DONSELMAN, H., and T. BROSCHAT. 1984. Transplanting Palms Successfully. *Florida Nurseryman,* Jan., pp. 11–12.

DOORENBOS, J., and W. O. PRUITT. 1977. *Crop Water Requirements, 2nd ed. Irrigation and Drainage.* Paper No. 24. Rome: Food and Agr. Org. of the United Nations.

DREISTADT, S. H., and D. L. DAHLSTEN. 1990. Relationships of Temperature to Elm Leaf Beetle (Coleoptera: Chrysomelidae) Development and Damage in the Field. *J. Econ. Entomol.* 83:837–41.

DULING, J. Z. 1969. Recommendations for Treatment of Soil Fills around Trees. *Arborist's News* 34(6):1–4.

DUNMIRE, J. R. 1988. *Sunset New Western Garden Book.* Menlo Park, CA: Lane Publishing Co.

DUNN, N. P. 1975. *The Mechanics of Shade Tree Evaluation.* Memphis, TN: Private publication.

EATON, F. M. 1966. Chlorine. In *Diagnostic Criteria for Plants and Soils,* ed. H. D. Chapman. Riverside, CA: H. D. Chapman, pp. 98–135.

EBENRECK, S. 1989. The Values of Trees. In *Shading Our Cities,* ed. G. Moll and S. Ebenreck. Washington, DC: Island Press, pp. 49–57.

ENVIRONMENTAL CONSULTANTS. 1990. *Tree Growth Regulator Study—Final Report.* Hagerstown, MD: Potomac Edison Company, pp. 1–24.

ENVIRONMENTAL PROTECTION AGENCY (EPA). 1977. *Economic Analysis, Root Control, and Backwater Flow Control as Related to Infiltration/Inflow Control.* Cincinnati: Environmental Protection Technology Series EPA 600/2–77–017a.

EPSTEIN, E. 1972. *Mineral Nutrition of Plants: Principles and Perspectives.* New York: John Wiley.

ERICSSON, T. 1981. Effects of Varied Nitrogen Stress on Growth and Nutrition in Three *Salix* Clones. *Physiologia Plantarum* 51:423–29.

ERICSSON, T., and T. INGESTAD. 1988. Nutrition and Growth of Birch Seedlings at Varied Relative Phosphorus Addition Rates. *Physiologia Plantarum* 72:227–35.

ESSIG, E. D. 1958. *Insects and Mites of Western North America.* New York: Macmillan.

EVANS, J., and C. W. SHANKS. 1907. Treecholters. *4rh Res. Note* 63/87/SILS:1–4.

EVANS, P. E., and J. E. KLETT. 1984. The Effects of Dormant Pruning Treatments on Leaf, Shoot, and Root Production from Bare-Root *Malus sargentii. J. Arboriculture* 10:298–302.

———. 1985. The Effects of Dormant Branch Thinning on Total Leaf, Shoot, and Root Production From Bare-Root *Prunus cerasifera* 'Newportii.' *J. Arboriculture* 11:149–51.

FARQUHAR, G. D. 1979. Carbon Assimilation in Relation to Transpiration and Fluxes in Ammonia. In *Photosynthesis and Plant Development,* ed. R. Marcelle, H. Clijsters, and M. Van Pouche. The Hague: Dr. W. Junk, bv.:321–28.

FAYLE, D. C. F. 1968. *Radial Growth in Tree Roots: Distribution, Timing, Anatomy.* University of Toronto, Forestry Tech. Rep. 9.

———. 1976. Stem Sway Affects Ring Width and Compression Wood Formation in Exposed Root Bases. *Forest Sci.* 22(2):193–94.

———. 1981. What Grade Changes Do to Tree Roots. *The Buckeye Arborist,* May–June, pp. 4–6.

FEDERER, C. A. 1971. *Effects of Trees in Modifying Urban Microclimate.* Proc. Symp. on the Role of Trees in the South's Urban Environment (USDA Forest Service).

FELIX, R., and A. L. SHIGO. 1977. Rots and Rods. *J. Arboriculture* 3(10):187–90.

FELT, E. P. 1940. *Plant Galls and Gall-Makers.* Ithaca, NY: Comstock.

FERERES, E., ed. 1981. *Drip Irrigation Management,* ed. E. Fereres. Univ. Calif. Agr. Sci. Leaflet 21199.

FERGUSON, B. 1982. *All About Trees.* San Francisco: Ortho Books.

FERGUSON, B. K. 1987. Water Conservation Methods in Urban Landscape Irrigation: Exploratory Overview. *Water Resources Bull.* (23):147–52.

FEUCHT, J. R. 1988. Early Results of Colorado Study on Wire Baskets Recommend Cautious, Thoughtful Use. *Pac. Coast Nur.* 47(4):57–58.

FEUCHT, J. R., and J. D. BUTLER. 1988. *Landscape Management.* New York: Van Nostrand Reinhold.

FISHER, M. E., E. SATCHELL, and J. M. WATKINS. 1975. *Gardening with New Zealand Plants, Shrubs, and Trees.* Auckland and London: William Collins.

FLEMER, W., III. 1986. Planting Trees and Shrubs in Island Beds. *Grounds Maintenance* 21(9):96–97.

FLINT, M. L. 1990. *Pests of the Garden and Small Farm: A Grower's Guide to Using Less Pesticide.* Univ. Calif. Agr. and Natl. Resources Pub. 3332.

FLINT, M. L., and R. VAN DEN BOSCH. 1981. *Introduction to Integrated Pest Management.* New York: Plenum Press.

FOSTER, R. S. 1978. *Landscaping That Saves Energy Dollars.* New York: McKay.

FRAEDRICH, B. R., D. C. BOOTH, and E. T. SMILEY. 1989. *Pest Management Recommendations 1990.* Charlotte, NC: Bartlett Tree Research Lab., p. 116.

FRANCIS, H. L., J. R. BREECE, and R. L. BALDWIN. 1974. *Christmas Tree Farming with Monterey Pines in Southern California.* Univ. Calif. Agr. Ext. AXT-n 200.

FRANÇOIS, L. E. 1980. *Salt Injury to Ornamental Shrubs and Ground Covers.* USDA Home and Garden Bull. 213.

FRANÇOIS, L. E., and R. A. CLARK, 1978. Salt Tolerance of Ornamental Shrubs, Trees, and Iceplant. *J. Amer. Soc. Hort. Sci.* 103(2):280–83.

———. 1979. Boron Tolerance of Twenty-Five Ornamental Shrub Species. *J. Amer. Soc. Hort. Sci.* 104(3):319–22.

FREE, M. 1961. *Plant Pruning in Pictures.* Garden City, NY: Doubleday.

FRYDENLUND, M. M. 1977. *Lightning Protection for Home, Farm and Family.* Harvard, IL: Lightning Protection Inst.

FUCHS, A. D. 1988. A Comparison of Three Different Trunk Injection Systems for Use with Plant Growth Regulators. *J. Arboriculture* 14:94–98.

FULLER, R. G., D. E. BELL, and H. E. KAZMAIER. 1965. Tree Wound Dressing for Sprout Control in Pole Line Clearing. *Edison Electric Inst. Bull.* 33:290–94.

FUNK, R. 1988. Davey's Plant Health Care. *J. Arboriculture* 14:285–87.

GALE, J., and R. M. HAGAN. 1966. Plant Antitranspirants. *Ann. Rev. Plant Physiol.* 17:269–82.

GALLAGHER, P. W., and T. D. SYDNOR. 1983. Electrical Resistance Related to Volume of Discolored and Decayed Wood in Silver Maple. *HortSci.* 18:762–64.

GANS, H. J. 1967. *The Levittowners.* New York: Pantheon Books.

GARDNER, V. R., F. C. BRADFORD, and H. D. HOOKER, JR. 1939. *The Fundamentals of Fruit Production.* New York: McGraw-Hill.

GARDNER, W. R., and M. FIREMAN. 1958. Laboratory Studies of Evaporation from Soil Columns in the Presence of a Water Table. *Soil Sci.* 85:244–49.

GARNER, J. H. B. 1974. The Death of Woody Ornamental Plants Associated with Leaking Natural Gas. *Arborist's News* 39(12):13–17.

GARROT, D. J., JR., and others. 1990. Quantification of Pecan Water Stress for Irrigation Scheduling. *Proc. 24th West. Pecan Conf.* Las Cruces: New Mexico State University.

GEIGER, R. 1961. *The Climate Near the Ground.* Cambridge: Harvard University Press.

GILBERTSON, P. 1987. *The Survival of Trees in Urban Areas—A Biological, Social and Economic Analysis.* Ph.D. Thesis. University of Liverpool.

GILBERTSON, P., A. D. KENDLE, and A. D. BRADSHAW. 1987. Root Growth and the Problems of Trees in Urban and Industrial Areas. In *Advances in Practical Arboriculture,* ed. D. Patch. For. Com. Bull. 65. London: Her Majesty's Stationery Office, pp. 59–66.

GILL, L. S., and F. G. HAWKSWORTH. 1961. *The Mistletoes: A Literature Review.* USDA Forest Serv. Pest Leaflet 147.

GILLIAM, C. H., D. C. FARE, and J. T. EASON. 1988. Control of *Acer rubrum* Growth with Flurprimidol. *J. Arboriculture* 14:99–105.

GILMAN, E. F., I. A. LEONE, and F. B. FLOWER. 1987. Effect of Soil Compaction and Oxygen Content on Vertical and Horizontal Root Distribution. *J. Environ. Hort.* 5(1):33–36.

GLASS, E. H., ed. 1975. *Integrated Pest Management: Rationale, Potential, Needs, and Implementation.* Washington, DC: Entomological Soc. of Amer.

GLOCK, W. S., and S. H. AGERTER. 1962. Rainfall and Tree Growth. In *Tree Growth,* ed. T. T. Kozlowski. New York: The Ronald Press Company.

GOLDHAMER, D. A. 1989. *Drought Irrigation Strategies for Deciduous Orchards.* Univ. Calif. Agr. Sci. Leaflet 21453.

GOLDHAMER, D. A., and R. L. SNYDER. 1989. *Irrigation Scheduling.* Univ. Calif. Agr. Sci. Leaflet 21454.

GOODIN, J. R., and V. T. STOUTEMYER. 1962. Carob Tree Growth Stimulated with Gibberellin. *Calif. Agr.* 16(9):4–5.

GOUIN, F. R. 1979. Anti-Desiccants for Protecting Evergreens. *News and Views* (Amer. Hort. Soc.) 21(1):6.

———. 1983. Girdling by Roots and Ropes. *J. Environ. Hort.* 1:50–52.

GOWANS, K. D. 1970. Water Pollution Considerations in Landscape Management. *Proc. Calif. Park and Recreation Administrators Inst.* 11(20):1–5.

GOWANS, K. D., and H. L. HALL. 1971. *Providing Drainage for Ornamental Trees in Layered Soils.* Univ. Calif. Agr. Sci. Leaflet 2575.

GRANT, J., D. A. GOLDHAMER, and M. GAGNON. 1986. *The Water Budget Method: Irrigation Scheduling.* Univ. Calif. Agr. Sci. Leaflet 21419.

GRANT, J. A., and C. L. GRANT. 1955. *Trees and Shrubs for Pacific Northwest Gardens.* San Carlos, CA: Brown & Nourse.

GRANT, R. 1990. Largest Landfill Gas Recovery Unit Opens in Meadowlands District News Release. Lynhurst, NJ: Hackensack Meadowlands Dev. Com., April.

GRAY, D. H., and A. T. LEISER. 1982. *Biotechnical Slope Protection and Erosion Control.* New York: Van Nostrand Reinhold.

GREEN, J. L. 1978. Plant Injury Caused by Sawdusts and Barks. *Ornamentals Northwest Newsletter* (Oregon State University), Dec.-Jan. 1977–78, pp. 8–12.

GREEN, J. L., O. MALOY, and J. CAPIZZI. 1990. Patterns: Signs to Direct the Diagnosis of Plant Damage. *Ornamentals Northwest* 13(6):1–24.

GROUNDS, R. 1973. *The Complete Handbook of Pruning.* New York: Macmillan.

HACKETT, W. P. 1976. Control of Phase Change in Woody Plants. *Acta Horticultural* 56:143–54.

HALFACRE, R. G., and J. A. BARDEN. 1979. *Horticulture.* New York: McGraw-Hill.

HALLIWELL, B., J. TURPIN, and J. WRIGHT. 1979. *The Complete Handbook of Pruning,* 2nd ed. London: Ward Lock.

HAMILTON, W. D. 1976. Street Tree Root Problem Survey. *Landscape Supervisors Forum Newsletter, May 11,* Hayward: Calif. Coop. Ext., pp. 1–3.

———. 1977. *Preventing Formation of Ginkgo Fruit.* Hayward: Calif. Coop. Ext., p. 1.

———. 1978. Drip Irrigation I, II, & III. *Growing Points.* Hayward: Calif. Coop. Ext., April, pp. 3–4; May, pp. 2–4; June, pp. 1–4.

———. 1983. Roots Pot Bound—Trees Grew. *Growing Points.* Univ. Calif. Coop. Ext. 21(4):2.

———. 1988. Significance of Root Severance on Performance of Established Trees. *J. Arboriculture* 14:288–92.

HAMILTON, W. D., and K. D. GOWANS. 1970. Fire Protection in the Bay Area. *Growing Points.* Univ. Calif. Coop. Ext., Oct., pp. 1–8.

HAMILTON, W. D., and D. MARLING. 1981. Large Tree Cavity Work and Cabling. *J. Arboriculture* 7(7):180–82.

HARRIS, J. R., and E. F. GILMAN. 1991. Production Method Affects Growth and Root Regeneration of Leyland Cypress, Laurel Oak and Slash Pine. *J. Arboriculture* 17(3):64–69.

HARRIS, R. W. 1962. Water: Hazardous Necessity. *Proc. Intl. Shade Tree Conf.* 38:182–88.

———. 1966. Influence of Turfgrass on Young Landscape Trees. *Proc. XVII Intl. Hort. Cong.* 1:81.

———. 1967. Factors Influencing Root Development of Container-Grown Trees. *Proc. Intl. Shade Tree Conf.* 43:304–14.

———. 1972. High-Temperature Limb Breakage. *Proc. Intl. Shade Tree Conf.* 48:133–34.

———. 1983. Summer Branch Drop. *J. Arboriculture* 9:111–13.

———. 1989. Arboriculture: World Glimpses and Ideas. *J. Arboriculture* 15:62–66.

HARRIS, R. W., and L. BALICS. 1963. Strong Branch Structure for Modesto Ash. *Calif. Agr.* 17(2):10–11.

HARRIS, R. W., and R. H. COPPOCK. 1977. *Saving Water in Landscape Irrigation.* Univ. Calif. Agr. Sci. Leaflet 2976.

HARRIS, R. W., and W. B. DAVIS. 1976. *Planting Landscape Trees.* Univ. Calif. Agr. Sci. Leaflet 2583.

HARRIS, R. W., and W. D. HAMILTON. 1969. Staking and Pruning Young *Myoporum laetum* Trees. *J. Amer. Soc. Hort. Sci.* 94:359–61.

HARRIS, R. W., A. T. LEISER, and W. B. DAVIS. 1976. *Staking Landscape Trees.* Univ. Calif. Agr. Ext. Leaflet 2576.

HARRIS, R. W., D. LONG, and W. B. DAVIS. 1967. *Root Problems in Nursery Liner Production.* Univ. Calif. Agr. Sci. Leaflet 2563.

HARRIS, R. W., J. L. PAUL, and A. T. LEISER. 1977. *Fertilizing Woody Plants.* Univ. Calif. Agr. Sci. Leaflet 2958.

HARRIS, R. W., R. M. SACHS, and R. E. FISSELL. 1971. Control of Trunk Sprouts with Growth Regulators. *Calif. Agr.* 25:1–3.

HARRIS, R. W., and others. 1969. *Pruning Landscape Trees.* Univ. Calif. Agr. Sci. Leaflet 2574.

HARRIS, R. W., and others. 1971. Root Pruning Improves Nursery Tree Quality. *J. Amer. Soc. Hort. Sci.* 96:105–8.

HARRIS, R. W., and others. 1972. Spacing of Container-Grown Trees in the Nursery. *J. Amer. Soc. Hort. Sci.* 97(4)503–6.

HARTIG, R. 1878. *Die Zersetzungserscheinungen des Holzes der Nadelbäume und der Eiche [Decay Phenomena in the Wood of Conifers and Oaks].* Berlin: Springer.

HARTMAN, F. O. 1980. Removing Flowers and Fruits from Trees. *Landscape Facts.* Ohio State Univ. Coop. Ext. Serv. CP14-80:1–2.

HARTMANN, H. T. 1967. 'Swan Hill': A New Ornamental Fruitless Olive for California. *Calif. Agr.* 21(1):4–5.

HARTMANN, H. T., D. E. KESTER, and F. J. DAVIES, JR. 1990. *Plant Propagation: Principles and Practices,* 5th ed. Englewood Cliffs, NJ: Prentice Hall.

HARTMANN, H. T., A. M. KOFRANEK, V. E. RUBATZKY, and W. J. FLOCKER. 1988. *Plant Science: Growth, Development, and Utilization of Cultivated Plants,* 2nd ed. Englewood Cliffs, NJ: Prentice Hall.

HAUPT, E. H. 1980. The Private Tree Worker and Energized Lines. *J. Arboriculture* 6(4):93–95.

HAUSENBUILLER, R. L. 1985. *Soil Science: Principles and Practices,* 3rd ed. Dubuque, IA: Wm. C. Brown.

HAYES, G. 1979. *Solar Access Law.* Cambridge, MA: Balinger.

HEAD, G. C. 1968. Seasonal Changes in the Diameter of Secondarily Thickened Roots of Fruit Trees in Relation to Growth of Other Parts of the Tree. *J. Hort. Sci.* 43:275–82.

HEERMANN, D. F., and R. A. KOHL. 1980. Fluid Dynamics of Sprinkler Systems. In *Design and Operation of Farm Irrigation Systems,* ed. M. E. Jensen. St. Joseph, MI: Amer. Soc. Agr. Engr., pp. 583–618.

HEINICKE, D. R. 1966. The Effect of Natural Shade on Photosynthesis and Light Intensity in Red Delicious Apple Trees. *Proc. Am. Soc. Hort. Sci.* 88:1–8.

HEISLER, G. M. 1982. Reductions of Solar Radiation by Tree Crowns. In *The Renewable Challenge. Proc. Inter. Solar Energy Soc.,* Houston, TX, pp. 133–38.

HELALIA, A. M., and J. LETEY. 1988. Cationic Polymer Effects on Infiltration Rates with Rainfall Simulator. *Soil Sci.* 52(1):247–50.

HENDRICKSON, A. H. 1918. What Size Nursery Trees? *Calif. State Dept. of Agr. Monthly Bull.* 7:171–74.

HEPTING, G. H. 1935. *Decay Following Fire in Young Mississippi Delta Hardwoods.* USDA Tech. Bull. 494.

———. 1968. Diseases of Forest and Tree Crops Caused by Air Pollutants. *Phytopathology* 58:1098–101.

HERRINGTON, L. P. 1974. Trees and Acoustics in Urban Areas. *J. Forestry* 72(8):462–65.

HERRINGTON, L. P., G. E. BERTOLIN, and R. E. LEONARD. 1972. Microclimate of a Suburban Park. In *Proc. Conf. on Urban Environment and 2nd Conf. on Biometeorology* (Amer. Meteorological Soc., Worcester, MA), pp. 43–44.

HERSHEY, D. R., and J. J. PAUL. 1983. Ion Absorption by a Woody Plant with Episodic Growth. *HortSci.* 18(3):357–59.

HICKMAN, G. W., J. CAPRILE, and E. PERRY. 1989. Oak Tree Hazard Evaluation. *J. Arboriculture* 15:177–94.

HILLEL, D. 1980. *Applications of Soil Physics.* San Diego, CA: Academic Press.

HILLIER, H. C. 1973. *Hillier's Manual of Trees and Shrubs.* London: David & Charles.

HIMELICK, E. B. 1969. *Tree and Shrub Hosts of Verticillium albo-atrum.* Ill. Natural History Survey Biol. Notes 66.

———. 1970. Frost Cracks on London Plane Tree an Important Outdoor Winter Thermometer. *Arborist's News* 35(1):5–6.

———. 1991. *Transplanting Manual for Trees and Shrubs.* Urbana, IL: Intl. Soc. Arboriculture.

HIMELICK, E. B., and D. W. CEPLECHA. 1976. Dutch Elm Disease Eradication by Pruning. *J. Arboriculture* 2(5):81–84.

HITCHCOCK, A. E., W. CROCKER, and P. W. ZIMMERMAN. 1932. Toxicity of Illuminating Gas in Soils. *Proc. Natl. Shade Tree Conf.* 9:34–36.

HOCKS, J. 1972. Changes in Composition of Soil Air Near Leaks in Natural Gas Mains. *Soil Sci.* 13:46–54.

HODEL, D. R. 1986. Drought Tolerance of Selected Non-Irrigated Trees. *Landscape and Irrigation,* Aug., pp. 66–69.

HODGE, S. 1990. Organic Soil Amendments for Tree Establishment. *Arbor. Res. Note* 86–90–ARB:1–4.

HOITINK, H. A. J., and H. A. POOLE. 1977. Composted Bark Media for Control of Soil Borne Plant Pathogens. *Ohio Florists' Assn. Bull.* 567, pp. 10–11.

HOLEWINSKI, D. E., J. W. ORR, and J. P. GILLON. 1983. Development of Improved Tree-Trimming Equipment and Techniques. *J. Arboriculture* 9:137–40.

HOLMES, F. W. 1986. *Drought After-Effects.* Notes from the Shade Tree Labs. Univ. Mass. Coop. Ext., Dec.

HORTICULTURAL RESEARCH INSTITUTE. 1971. *A Technical Glossary of Horticultural and Landscape Terminology.* Washington, D.C.: Horticultural Research Institute.

HOUSER, J. S. 1937. Borer Control Experiments. *Proc. Natl. Shade Tree Conf.* 13:159–68.

HOUSTON, D. R. 1980. Effects of Defoliation on Trees and Shrubs. In *Gypsy Moth Compendium,* ed. P. M. Wargo. USDA Tech. Bull. 1584:217–18.

HOYT, R. S. 1978. *Ornamental Plants for Subtropical Regions.* Anaheim, CA: Livingston.

HUDLER, G. 1981. Salt Injury to Roadside Plants. *Grounds Maintenance* 16(2):80–84.

HUDLER, G. W., and M. A. BEALE. 1981. Anatomical Features of Girdling Root Injury. *J. Arboriculture* 7(2):29–32.

HUDSON, B. O. 1981. Maintaining Safety While Maintaining Trees. *Arbor Age* 1(1):14–15.

HULL, J., JR., and L. N. LEWIS. 1959. Response of One-Year-Old Cherry and Mature Bearing Cherry, Peach and Apple Trees to Gibberellin. *Proc. Amer. Soc. Hort. Sci.* 74:93–100.

HUTCHINSON, F. E. 1968. The Relationship of Road Salt Applications to Sodium and Chloride Ion Levels in the Soil Bordering Major Highways. In *Proc. Symp. Pollutants in the Roadside Environment* (Univ. Conn.), ed. E. D. Carpenter, pp. 24–35.

IDSO, S. B., and others. 1981. Normalizing the Stress Degree Day Concept for Environmental Variability. *Agr. Meteor.* 24:45–55.

INGESTAD, T., and A.-B. LUND. 1979. Nitrogen Stress in Birch Seedlings I. Growth Technique and Growth. *Physiologia Plantarum* 45:137–48.

INTERAGENCY AGRICULTURAL INFORMATION TASK FORCE. n.d. (after 1974). *Drought Tips: Common Irrigation Problems: Some Solutions.* Davis: Univ. Calif. Land, Air, and Water Resources Ext.

IRVINE, City of. 1975. *Comparative Costs of Park Maintenance.* Unpublished report. City of Irvine, CA.

IYER, J. G., R. B. CAVEY, and S. A. WILDE. 1980. Mycorrhizae: Facts and Fallacies. *J. Arboriculture* 6(8):213–20.

JACKS, G. V., W. D. BRIND, and R. SMITH. 1955. *Mulching.* Commonwealth Bureau of Soil Sci. Tech. Comm. 49.

JACOBS, M. R. 1939. *A Study of the Effect of Sway on Trees.* Australian Commonwealth Forest Bull. 26.

JANICK, J. 1986. *Horticultural Sciences,* 4th ed. New York: Freeman.

JEFFERS, W. A., and R. ABBOTT. 1979. New System of Guying Trees. *J. Arboriculture* 5(6):121–23.

JENSEN, I. B., and R. L. HODDER. 1979. *Tubelings, Condensation Traps, Mature Tree Transplanting and Root Sprigging Techniques for Tree and Shrub Establishment in Semiarid Areas.* Mont. Agr. Exp. Sta. Res. Report 141, Vol. 2.

JOHNSON, C. 1977. *Mistletoe Control in Shade Trees.* Univ. Calif. Agr. Sci. Leaflet 2571.

JOHNSON, C. M. 1966. Molybdenum. In *Diagnostic Criteria for Plants and Soils,* ed. H. D. Chapman. Riverside, CA: H. D. Chapman, pp. 286–301.

JOHNSON, D. W. 1981. *Tree Hazards: Recognition and Reduction in Recreation Sites.* U.S. Forest Serv., Forest Pest Mgmt. Tech. Rpt. R 2–1.

JOHNSON, M. S. 1984. The Effects of Gel-Forming Polyacrylamides on Moisture Storage in Sandy Soils. *J. Sci. Food Agr.* 35:1196–200.

JOHNSON, W. T. 1982. The Scale Insect, a Paragon of Confusion. *J. Arboriculture* 8(5):113–23.

JOHNSON, W. T., and H. H. LYON. 1988. *Insects That Feed on Trees and Shrubs,* 2nd ed. Ithaca, NY: Comstock.

JOHNSTONE, R. A. 1988. Economics of Utility Lateral Trimming. *J. Arboriculture* 14:74–77.

KELLOGG, A. 1882. *Forest Trees of California.* Sacramento, CA: State Printing Office.

KELLY, S. 1969. *Eucalypts.* Melbourne: Thomas Nelson.

KENWORTHY, A. L. 1953. Depletion of Soil Moisture in a Mature Apple Orchard with a Sod-Mulch System of Soil Management. *Mich. Agr. Sta. Quarterly Bull.* 36:39–45.

KEPNER, R. A. 1950. *The Principles of Orchard Heating.* Calif. Agr. Exp. Sta. Cir. 400.

KIMBALL, M. H., and F. A. BROOKS. 1959. Plant Climates of California. *Calif. Agr.* 13(5):2–7.

KIMBALL, S. L. 1990. The Physiology of Tree Growth Regulators. *J. Arboriculture* 16:39–41.

KING, G. C., C. BEATLY, and M. MCKENZIE. 1970. *Polyurethane for Filling Tree Cavities.* Univ. Mass. Pub. 58.

KOPINGA, J. 1985. Site Preparation Practices in the Netherlands. *Metria* 5:72–84.

KOZLOWSKI, T. T., ed. 1971. *Growth and Development of Trees,* Vol. 1. San Diego, CA: Academic Press.

KOZLOWSKI, T. T., and W. J. DAVIES. 1975. Control of Water Balance in Transplanted Trees. *J. Arboriculture* 1:1–10.

KOZLOWSKI, T. T., P. J. KRAMER, and S. G. PALLARDY. 1991. *The Physiological Ecology of Woody Plants.* San Diego, CA: Academic Press.

KOZLOWSKI, T. T., and C. H. WINGET. 1964. Diurnal and Seasonal Variation in Radii of Tree Stems. *Ecology* 45:149–55.

KRAEBEL, C. J. 1936. *Erosion Control of Mountain Roads.* USDA Cir. 380.

KRAMER, P. J. 1969. *Plant and Soil Water Relationships: A Modern Synthesis.* New York: McGraw Hill.

KRAMER, P. J., and T. T. KOZLOWSKI. 1979. *Physiology of Woody Plants.* San Diego, CA: Academic Press.

KUBLER, H. 1988. Frost Cracks in Stems of Trees. *Arboricultural J.* 12:163–75.

LABANAUSKAS, C. K. 1966. Manganese. In *Diagnostic Criteria for Plants and Soils,* ed. H. D. Chapman. Riverside, CA: H. D. Chapman, pp. 264–85.

LANDSBERG, H. E. 1970. Climates and Urban Planning. In *Urban Climates.* Geneva, Switzerland: World Meteorological Org. Tech. Note 108, pp. 364–74.

LANGE, A. H., C. L. ELMORE, and A. B. SAGHIR. 1973. *Diagnosis of Phytotoxicity from Herbicides in Soils.* Univ. Calif. Agr. Ext. Leaflet TA–69.

LANIER, G. N. 1989. Trap Trees for Control of Dutch Elm Disease. *J. Arboriculture* 15:105–11.

LANPHEAR, F. O. 1971. Urban Vegetation: Values and Stresses. *HortSci.* 6:332–34.

LARSON, P. R. 1965. Stem Form of Young Larix as Influenced by Wind and Pruning. *Forest Sci.* 11(4):412–24.

LATHROP, J. K., and R. A. MECKLENBURG. 1971. Root Regeneration and Root Dormancy in *Taxus. J. Amer. Soc. Hort. Sci.* 96(1):111–14.

LEAF, A. L. 1968. K, Mg, and S Deficiencies in Forest Trees. In *Forest Fertilization: Theory and Practice.* Knoxville: Tennessee Valley Authority, pp. 88–122.

LEAR, B., and D. E. JOHNSON. 1975. *Controlling Nematodes in the Home Garden.* Univ. Calif. Agr. Sci. Leaflet 2112.

LEDIN, R. B. 1961. Pruning Palms. *Amer. Hort. Magazine* 40:142–43.

LEIGHTON, G. M., R. D. HARTER, and G. R. CROMBIE. 1978. *Sewage Sludge Composting in Small Towns.* Univ. N.H. Sta. Bull. 508.

LEISER, A. T., and J. D. KEMPER. 1968. A Theoretical Analysis of a Critical Height of Staking Landscape Trees. *Proc. Amer. Soc. Hort. Sci.* 92:713–20.

———. 1973. Analysis of Stress Distribution in the Sapling Tree Trunk. *J. Amer. Soc. Hort. Sci.* 98(2):164–70.

LEISER, A. T., and G. NYLAND. 1972. Unpublished report.

LEISER, A. T., and others. 1972. Staking and Pruning Influence Trunk Development of Young Trees. *J. Amer. Soc. Hort. Sci.* 97(4):498–503.

LEIUS, K. 1967. Influence of Wild Flowers on Parasitism of Tent Caterpillar and Codling Moth. *Canadian Entomology* 99:444–46.

LEONARD, O. A., D. E. BAYER, and R. K. GLENN. 1974. Control of Tree Roots. *Weed Sci.* 22:516–20.

LEONARD, O. A., and W. A. HARVEY. 1965. *Chemical Control of Woody Plants.* Calif. Agr. Exp. Sta. Bull. 812.

LEONARD, O. A., W. B. MCHENRY, and L. A. LIDER. 1974. Herbicide Residues in Soil of the Vine Row 21 Months Following 9 Successive Annual Applications. *Proc. California Weed Conf.,* Sacramento, CA: 26:115–22.

LEONARD, O. A., and N. R. TOWNLEY. 1971. Control of Tree Roots in Sewers and Drains. *Calif. Agr.* 25(11):13–15.

LEONE, I. A., and others. 1977. Damage to Woody Species by Anaerobic Landfill Gases. *J. Arboriculture* 3(12):221–25.

LERMAN, S. L., and E. F. DARLEY. 1975. Particulates. In *Responses of Plants to Air Pollution,* ed. J. B. Mudd and T. T. Kozlowski. San Diego, CA: Academic Press, pp. 141–58.

LICHTER, J. M., M. S. REID, and A. M. BERRY. 1991. New Methods for Control of Leafy Mistletoe (*Phoradendron* spp.) on Landscape Trees. *J. Arboriculture* 17:127–30.

LINDOW, S. E. 1980. New Method of Frost Control Through Control of Epiphytic Ice Nucleation Active Bacteria. *Calif. Plant Pathology* 48:1–5.

LITEWKA, J., and M. W. STIMMANN. 1979. *Pesticide Toxicities*. Univ. Calif. Agr. Sci. Leaflet 21062.

LIVINGSTON, W. H., and others. 1985. Effective Use of Ethylene-Releasing Agents to Prevent Spread of Eastern Dwarf Mistletoe in Black Spruce. *Can. J. For. Res.* 15(5):872–76.

LOCKE, L. F., and H. V. ECK. 1965. *Iron Deficiency in Plants*. USDA Home and Garden Bull. 102.

LONG, D. 1961. Developing and Maintaining Street Trees. *Proc. Intl. Shade Tree Conf.* 37:172.

LORD, E. E. 1979. *Shrubs and Trees for Australian Gardens,* 4th ed. Melbourne: Lothian.

LOVELADY, S. M. 1965. Handsome Factories Yield Unexpected Joys. *Wall Street Journal,* Dec. 1, p. 1.

LULL, H. W. 1959. *Soil Compaction on Forest and Range Lands.* USDA Forest Serv. Misc. Pub. 768.

LUMIS, G. P. 1990. Wire Baskets: A Further Look. *Amer. Nurseryman* 172(4):128–31.

LUMIS, G. P., G. HOFSTRA, and R. HALL. 1973. Sensitivity of Roadside Trees and Shrubs to Aerial Drift of Deicing Salt. *HortSci.* 8(6):475–77.

LUMIS, G. P., and S. A. STRUGER. 1988. Root Tissue Development Around Wire-Basket Transplant Containers. *HortSci.* 23(2):401.

MCALISTER, E. J. 1985. A Method of Assigning a Monetary Value to "Amenity Trees." *J. Australian Inst. Horticulture,* Aug., pp. 83–85.

MCCAIN, A. H. 1975. *Fire Blight of Fruits and Ornamentals.* Univ. Calif. Agr. Sci. Leaflet 2715.

———. 1979a. *Sycamore Anthracnose.* Univ. Calif. Div. Agr. Sci. 2618.

———. 1979b. *Verticillium Wilt.* Univ. Calif. Agr. Sci. Leaflet 2592.

MCCAIN, A. H. and R. D. RAABE. 1972. *Armillaria Root Rot.* Univ. Calif. Agr. Ext. OSA 80.

MCCARTHY, L. 1991. IFA Nurseries Flexibility Exercises Cut Injury Time Loss and Lower Insurance Costs. *Pacific Coast Nurseryman* 50(1):8.

MCCLEMENTS, J. K. 1988. *Sunset Pruning Handbook.* Menlo Park, CA: Lane Publishing Co.

MCCLINTOCK, E., and A. T. LEISER. 1979. *An Annotated Checklist of Woody Ornamental Plants of California, Oregon, and Washington.* Univ. Calif. Agr. Sci. Pub. 4091.

MACDANIELS, L. H. 1932. Factors Affecting the Breaking Strength of Apple Tree Crotches. *Proc. Amer. Soc. Hort. Sci.* 29:44.

MCDONALD, A. J. S., A. ERICSSON, and T. LOHAMMAR. 1986. Dependence of Starch Storage on Nutrient Availability and Photon Flux Density in Small Birch (*Betula pendula* Roth). *Plant, Cell, & Environ.* 9:433–38.

MACDONALD, J. D. 1982. Effect of Salinity Stress on Development of Phytophthora Root Rot of Chrysanthemum. *Phytopathology* 72:214–19.

MACGREGOR, J. 1971. Why the Wind Howls Around Those Plazas Close to Skyscrapers. *Wall Street Journal,* Feb. 18, p. 1.

MCMINN, H. E. 1939. *An Illustrated Manual of California Shrubs.* San Francisco: J. W. Stacy.

MCMINN, H. E., and E. MAINO. 1963. *An Illustrated Manual of Pacific Coast Trees.* Berkeley: University of California Press.

McPHERSON, E. G., ed. 1984. A Methodology for Locating and Selecting Trees for Solar Control in Utah. *Proc. 1981 Ann. Meet. Amer. Sect. Intl. Solar Energy Soc.* Newark, DE, pp. 369–73.

McQUILKIN, W. E. 1950. Effects of Some Growth Regulators and Dressings on the Healing of Tree Wounds. *J. Forestry* 48:423–28.

MADER, D. L., and R. N. COOK. 1982. Soil Fertility for Urban Trees. In *Urban Forest Soils, A Reference Workbook,* ed. P. J. Craul. Syracuse, NY: SUNY-CESY, pp. 4–1 to 4–28.

MAFTOUN, M., and W. L. PRITCHETT. 1970. Effects of Added Nitrogen on the Availability of Phosphorus to Slash Pine on Two Lower Coastal Plain Soils. *Soil Sci. Soc. Amer. Proc.* 34(4):685–90.

MAGGS, D. H. 1960. The Effect of Number of Shoots on the Quantity and Distribution of Increment in Young Apple Trees. *An. Bot. N.S.* 24:345–55.

MAGILL, A. W. 1970. *Five California Campgrounds: Conditions Improve after 5 Years of Recreational Use.* USDA Forest Serv. Res. Paper PSW-62.

MAINO, E., and F. HOWARD. 1955. *Ornamental Trees.* Berkeley: University of California Press.

MAIRE, R. G. 1976. *Landscape for Fire Protection.* Univ. Calif. Agr. Sci. Leaflet 2401.

MAIRE, R. G., and R. L. BRANSON. 1972. Salinity Tolerance of Landscape Plants. Los Angeles County Agr. Ext. Serv. Progress Report, May 11.

MANION, P. D. 1981. *Tree Disease Concepts.* Englewood Cliffs, NJ: Prentice Hall.

MAROTZ, G. A., and J. C. COINER. 1973. Acquisition and Characterization of Surface Material Data for Urban Climatological Studies. *J. Applied Meteorology* 12:919–23.

MARSH, R. E., and M. W. CUMMINGS. 1976. *Pocket Gopher Control with Mechanical Bait Applicator.* Univ. Calif. Div. Agr. Sci. Leaflet 2699.

MARSHALL, R. P. 1931. *The Relation of Season of Wounding and Shellacking to Callus Formation in Tree Wounds.* USDA Tech. Bull. 246.

MARTIN, W. E., J. VLAMIS, and N. W. STICE. 1953. Field Correction of Calcium Deficiency on a Serpentine Soil. *Agronomy J.* 45:204–8.

MARX, D. H. 1973. Mycorrhizae and Feeder Root Disease. In *Ectomycorrhizae,* ed. G. C. Marks and T. T. Kozlowski. San Diego, CA: Academic Press, pp. 351–82.

MATHENY, N. P. 1989. Preserving Trees Affected by Development. In *A Technical Guide to Community and Urban Forestry in Washington, Oregon, and California,* ed. R. Morgan. Portland, OR: World Forestry Center, pp. 34–41.

MATHENY, N. P., and J. R. CLARK. 1992. *Photographic Guide to the Evaluation of Hazard Trees in Urban Areas.* Urbana, IL: Intl. Soc. Arboriculture.

MATTHEWS, W. E. 1988. After the Hurricane. *Arboricultural Assoc. News,* Winter, 1987/88, p. 1.

MAUGH, T. H., II. 1979. SO$_2$ Pollution May Be Good for Plants. *Science* 205:383.

MAYNE, L. S. 1975. Cabling and Bracing. *J. Arboriculture* 1(6):101–6.

MAYNE, L. S. 1982. Specifications for Construction Around Trees. *J. Arboriculture* 8:289–91.

MEE, T. R., and J. F. BARTHOLIC. 1979. Man-Made Fog. In *Modification of the Aerial Environment of Crops,* ed. B. J. Barfield and J. F. Gerber. St. Joseph, MI: Amer. Soc. Agr. Engineers, pp. 334–52.

MEISTER, E. L., JR. 1982. *Farm Chemicals Handbook.* Willoughby, OH: Meister.

MENNINGER, E. A. 1964. *Seaside Plants of the World*. New York: Hearthside Press.

MERCER, P. C. 1979. Attitudes to Pruning Wounds. *Arboricultural J.* 3:457–65.

MERRILL, J., and M. SOLOMONSON. 1977. Mycorrhizae and Their Horticultural Importance. *Univ. Wash. Arboretum Bull.* 40(2):11–15.

MERRILL, W., D. H. LAMBERT, and W. LIESE. 1975. In *Phytopathological Classics* 12. [Translation of *Wichtige Krankheiten der Waldbäume* or *Important Diseases of Forest Trees*. R. Hartig. 1874. Berlin: Springer.] St. Paul, MN: Amer. Phytopath. Soc., unnumbered introduction.

MESKIMEN, G. 1970. Combating Grass Competition for Eucalyptus Planted in Turf. *Tree Planter Notes* 21(4):3–5.

MESSENGER, S. 1984. Treatment of Chloritic Oaks and Red Maples by Soil Acidification. *J. Arboriculture* 10:122–28.

METCALF, L. J. 1975. *The Cultivation of New Zealand Trees and Shrubs*. Wellington, N.Z.: A. H. and A. W. Reed.

MEYER, M. M., and H. B. TUKEY JR. 1965. Nitrogen, Phosphorus and Potassium Plant Reserves and the Spring Growth of *Taxus* and *Forsythia*. *Proc. Amer. Soc. Hort. Sci.* 87:537–44.

MIDDLETON, J. T., E. F. DARLEY, and R. F. BREWER. 1958. Damage to Vegetation from Polluted Atmosphere. *J. Air Pollution Control Assn.* 8:7–15.

MIKOLA, P. 1973. Application of Mycorrhizal Symbiosis in Forestry Practice. In *Ectomycorrhizae: Their Ecology and Physiology,* ed. G. C. Marks and T. T. Kozlowski. San Diego, CA: Academic Press, pp. 383–411.

MILLER, F. D. 1989. The Use of Horticultural Oils and Insecticidal Soaps for Control of Insect Pests of Amenity Plants. *J. Arboriculture* 15:257–62.

MILLER, R. W. 1988. *Urban Forestry: Planning and Managing Urban Greenspaces*. Englewood Cliffs, NJ: Prentice Hall.

MILLER, V. J. 1959. Crotch Influence on Strength and Breaking Point of Apple Tree Branches. *Proc. Amer. Soc. Hort. Sci.* 73:27–32.

MILLS, L. J., and K. RUSSELL. 1981. *Detection and Correction of Hazard Trees in Washington's Recreation Areas*. Dept. Natl. Res. Rpt. No. 42.

MONCK, J. W. 1980. *Root Growth in Sewers*. Speech at American Public Works Assoc. Convention in Kansas City.

MOORE, E. O. 1981. A Prison Environment's Effect on Health Care Service Demands. *J. Environ. Systems* 11:17–34.

MOORE, L. S. 1979. Research Update on Crown Gall and Hairy Root Diseases in the Northwest. *Ornamentals Northwest Newsletter* 3(5):17–19.

MOORE, W. S., C. S. DAVIS, and C. S. KOEHLER. 1979. *Scale Insects and Their Control*. Univ. Calif. Agr. Sci. Leaflet 2237.

MORELL, J. D. 1984. Parkway Tree Augering Specifications. *J. Arboriculture* 10:129–32.

MORGAN, W. C. 1965. New Irrigation and Aerification Methods. *Calif. Turfgrass Culture* 15(2):11–15.

MORLING, R. J. 1963. *Trees: Including Preservation, Planting, Law, Highways*. London: Estates Gazette.

MORRIS, L. A., and R. F. LOWERY. 1988. Influence of Site Preparation on Soil Conditions Affecting Stand Establishment and Tree Growth. *South J. Applied For.* 12(3):170–78.

MUIRHEAD, D. 1961. *Palms.* Globe, AZ: D. S. King.

MULLINS, M. G. 1979. *Reader's Digest Illustrated Guide to Gardening.* Sydney: Reader's Digest Services.

MURPHY, R. C., and W. E. MEYER. 1969. *The Care and Feeding of Trees.* New York: Crown.

MYERS, M. K., and H. C. HARRISON. 1988. Evaluation of Container Plantings in an Urban Environment. *J. Arboriculture* 14:293–97.

NATIONAL ARBORIST ASSOCIATION (N.A.A.). 1987. *National Arborist Association Standards.* Amherst, NH: National Arborist Association.

NATIONAL FIRE PROTECTION ASSOCIATION (NFPA). 1980. *Lightning Protection Code 1980.* Boston: National Fire Protection Association.

NEEL, P. L. 1967. Factors Influencing Trunk Development of Landscape Trees. *Proc. Intl. Shade Tree Conf.* 43:293–303.

———. 1971. Experimental Manipulation of Trunk Growth in Young Trees. *Arborist's News* 36(3):25a–31a.

NEEL, P. L., and R. W. HARRIS. 1971. Motion-Induced Inhibition of Elongation and Induction of Dormancy in Liquidambar. *Science* 173:58–59.

NEELY, D. 1970. Healing of Wounds on Trees. *J. Amer. Soc. Hort. Sci.* 95(5):536–40.

———. 1972. Hints on Diagnosis of Tree Problems. *Proc. Intl. Shade Tree Conf.* 48:33–37.

———. 1976. Iron Deficiency Chlorosis of Shade Trees. *J. Arboriculture* 2(7):128–30.

———. 1979. Tree Wounds and Wound Closure. *J. Arboriculture* 5(6):135–40.

———. 1988a. Tree Wound Closure. *J. Arboriculture* 14:148–52.

———. 1988b. Wound Closure Rates on Trees. *J. Arboriculture* 14:250–54.

———. 1988c. Closure of Branch Pruning Wounds with Conventional and 'Shigo' Cuts. *J. Arboriculture* 14:261–64.

———. 1991. Water Transport at Stem-Branch Juncture in Woody Angiosperms. *J. Arboriculture* 17:285–90.

NEELY, D., and E. B. HIMELICK. 1987. Fertilization and Watering Trees. *Ill. Natural History Survey Bull.* 52:1–20.

NEELY, D., E. B. HIMELICK, and W. R. CROWLEY, JR. 1970. Fertilization of Established Trees: A Report of Field Studies. *Ill. Natural History Survey Bull.* 30(4):235–66.

NEW, P. B., and A. KERR. 1972. Biological Control of Crown Gall: Field Measurements and Glasshouse Experiments. *J. Applied Bacteriology* 35:279–87.

NEWMAN, C. J. 1963. Transplanting Semi-Mature Trees. In *Trees,* ed. R. J. Morling. London: Estates Gazette, pp. 29–39.

NG, H. 1990. *Advanced Tree Trimming Equipment.* Electric Power Res. Inst. Report EL-6901.

NICHOLS, J. O. 1980. *The Gypsy Moth.* Bur. Forestry Dept., Environmental Resources. Harrisburg, PA, pp. 1–34.

NIELSEN, D. G. 1990. Evaluation of Biorational Pesticides for Use in Arboriculture. *J. Arboriculture* 16:82–88.

NIELSON, D. G., and others. 1985. Common Street Trees and Their Pest Problems in the North Central United States. *J. Arboriculture* 11:225–32.

NOLAND, T. L., and T. T. KOZLOWSKI. 1979. Influence of Potassium Nutrient on Susceptibility of Silver Maple to Ozone. *Can. J. For. Res.* 9:501–3.

652   Bibliography

OHR, H. D. 1989. Diseases Can Diminish Palm Tree Value. *Arbor Age* 9(9):26, 32.

OKE, T. R. 1972. *Evapotranspiration in Urban Areas and Its Implications for Urban Climate Planning.* WNO-CIB Intl. Colloq. on Building Climatology Proc.

OLKOWSKI, W. C. 1973. *A Model Ecosystem Management Program.* Proc. Tall Timbers Conf. on Animal Control by Habitat Management 5:103–17.

OPITZ, K. W. 1970. *Spray to Prevent Fruit Set on Ornamental Olive Trees.* Univ. Calif. Agr. Sci. Leaflet 2479, Rev.

ORTON, D. A. 1989. *Coincide: The Orton System of Pest Management.* Flossmoor, IL: Plantsmen's Publications.

OSBORNE, R. 1975. *Garden Trees.* Menlo Park, CA: Lane.

OSTER, J. D., and M. J. SINGER. 1984. *Water Penetration Problem in California Soils,* Dept. Land, Air, and Water Resources Ppr. No. 10011. Univ. Calif. Davis.

OTTOSON, R. 1976. The Effect of Soil Residual Herbicides on Landscape Planting and Maintenance. *Proc. Calif. Weed Conf.* 28:153–56.

OUTCALT, S. I. 1972. A Reconnaissance Experiment in Mapping and Modeling the Effect of Land Use on Urban Thermal Regimes. *J. Applied Meteorology* 11:1369–73.

PAINE, L. A. 1971. *Accident Hazard Evaluation and Control Decisions on Forested Recreation Sites.* U.S. Forest Serv. Res. Ppr. PSW-68.

PARKER, E. R., and R. W. SOUTHWICK. 1941. Manganese Deficiency in Citrus. *Proc. Amer. Soc. Hort. Sci.* 39:51–58.

PATCH, D., W. O. BINNS, and D. F. FOURT. 1984. *Nutrition of Broadleaved Amenity Trees. II—Fertilizers.* Arb. Res. Note 52–84 SSS.

PATCH, D., M. P. COUTTS, and J. EVANS. 1984. Control of Epicormic Shoots on Amenity Trees. *Arb. Res. Note* 54–84.

PATTERSON, J. C. 1977. Soil Compaction—Effects on Urban Vegetation. *J. Arboriculture* 3:161–67.

PAYNE, B. R. 1973. The Twenty-Nine Tree Home Improvement Plan. *Natl. History* 82(9):74–75.

PERRY, E. J., ed. 1988. *Pruning Standards.* St. Helena, CA: West. Ch. Intl. Soc. Arbor.

PERRY, T. O. 1982. The Ecology of Tree Roots and the Practical Significance Thereof. *J. Arboriculture* 8:197–211.

PETERSEN, A., and D. ECKSTEIN. 1988. Road Trees in Hamburg—Their Present Situation of Environmental Stress and Their Future Chance of Recovery. *Arboricultural J.* 12:109–18.

PETERSON, J. T. 1969. *The Climate of Cities: A Survey of Recent Literature.* Raleigh, NC: Natl. Air Pollution Control Adm., U.S. Public Health Serv.

PHILIP, J. R., and D. A. deVRIES. 1957. Moisture Movement in Porous Material Under Thermal Gradients. *Trans. Amer. Geophys. Union* 38:222–32.

PHILIPSON, J. J., and M. P. COUTTS. 1977. The Influence of Mineral Nutrition on the Root Development of Trees. II. The Effect of Specific Nutrient Elements on the Growth of Individual Roots of Sitka Spruce. *J. Exper. Botany* 28(105):864–71.

PINNOCK, D. E., and others. 1978. Integrated Pest Management in Highway Landscapes. *Calif. Agr.* 32(2):33–34.

PIRONE, P. P. 1978a. *Tree Maintenance,* 5th ed. New York: Oxford University Press.

——. 1978b. *Diseases and Pests of Ornamental Plants,* 5th ed. New York: John Wiley.

PIRONE, P. P., and others. 1988. *Tree Maintenance,* 6th ed. New York: Oxford University Press.

POTTER, D. A., and G. M. TIMMONS. 1983. Forcasting Emergence and Flight of the Lilac Borer (Lepidoptera: Sessiidae) Based on Pheromone Trapping and Degree-Day Accumulations. *Environ. Entomol.* 12:400–403.

POTTER, M. J. 1987. Shelter Questions and Answers. *For. & British Timber,* Oct., pp. 28–29.

——. 1989. Treeshelters: Their Effects on Microclimate and Tree Establishment. *International Conference on Fast Growing and Nitrogen-Fixing Trees.* Phillips University, Marburg, Germany, Oct.

POWERS, R. F. 1979. Nutrient Deficiency Symptoms in Conifers. In *Principles of Silviculture,* 2nd ed., ed. T. W. Daniel, J. A. Helms, and F. S. Baker. New York: McGraw-Hill.

——. 1981a. *Nutritional Characteristics of Ponderosa Pine and Associated Species.* Ph.D. Dissertation, University of California, Berkeley.

——. 1981b. Response of California True Fir to Fertilization. In *Forest Fertilization Conference,* ed. S. P. Gessel, R. M. Kennedy, and W. A. Atkinson. University of Washington, Institute of Forest Resources Contrib. 50:95–101.

PRATT, D. 1982. Plants That Fight Back. *West Land. News* 22(7):32, 34, 39.

PREAUS, K. B., and C. E. WHITCOMB. 1980. Transplanting Landscape Trees. *J. Arboriculture* 6:221–23.

PRIDHAM, A. M. S. 1938. Growth of Pin Oak (*Quercus palustris*): Report of Seven Years' Observations. *Proc. Amer. Soc. Hort. Sci.* 35:739–41.

PRITCHETT, W. L. 1979. *Properties and Management of Forest Soils.* New York: John Wiley.

PROEBSTING, E. L. 1935. Field and Laboratory Studies on the Behavior of NH_4 Fertilizer with Special Reference to the Almond. *Proc. Amer. Soc. Hort. Sci.* 33:46–50.

——. 1958. *Fertilizers and Cover Crops for California Orchards.* Calif. Agr. Ext. Serv. Cir. 466.

PRUITT, W. O. 1971. *Factors Affecting Potential Evapotranspiration.* West Lafayette, IN (Purdue University): Proc. Third Intl. Sem. Hydrology Professors, pp. 82–102.

PRYKE, J. F. S. 1979. Trees and Buildings. *Arboricultural J.* 3(6):388–96.

QUICK, J., and J. M. RIBLE. 1967. *Soil Analysis.* Calif. Agr. Ext. OSA 98.

RAABE, R. D. 1974. A Look at Rapid Composting. *Calif. Hort. J.* 35(1):17–18.

——. 1979. *Resistance or Susceptibility of Certain Plants to Armillaria Root Rot.* Univ. Calif. Agr. Sci. Leaflet 2591.

RAE, W. A. 1969. Large Tree Moving by Frozen Root Balls. *Trees,* Jan.–Feb., pp. 8–9.

RAUPP, M. J. 1985. Monitoring: An Essential Factor to Managing Pests of Landscape Trees and Shrubs. *J. Arboriculture* 11:349–55.

——. 1990. Recognizing the Larvae of Key Pests and Beneficials Found on Woody Landscape Plants. *J. Arboriculture* 16:49–54.

RAUPP, M. J., J. A. DAVIDSON, and F. E. WOOD. 1987–1988. *The Gypsy Moth and the Homeowner.* Univ. Maryland Coop. Ext. Serv. Fact Sheet 242, Rev.

RAUPP, M. J., and R. M. NOLAND. 1984. Implementing Landscape Plant Management Programs in Institutional and Residential Settings. *J. Arboriculture* 10:161–69.

RAUPP, M. J., and others. 1985. The Concept of Key Plants in Integrated Pest Management for Landscapes. *J. Arboriculture* 11:317–22.

REARDON, R., and others. 1987. Development and Implementation of a Gypsy Moth Integrated Pest Management Program. *J. Arboriculture* 13:209–16.

REECE, R. A. 1979. Trees and Insurance. *Arboricultural J.* 3(7):492–99.

REHDER, A. 1940. *Manual of Cultivated Trees and Shrubs Hardy in North America,* 2nd ed. New York: Macmillan.

REIL, W. O. 1979. Pressure-Injecting Chemicals into Trees. *Calif. Agr.* 6:16–19.

REUTHER, W., and C. K. LABANAUSKAS. 1966. Cooper. In *Diagnostic Criteria for Plant and Soils,* ed. H. D. Chapman. Riverside, CA: H. D. Chapman, pp. 157–79.

REUTHER, W., and others. 1981. *Irrigating Deciduous Orchards.* Univ. Calif. Agr. Sci. Leaflet 21212.

RICH, A. E. 1971. Salt Injury to Roadside Trees. *Proc. Intl. Shade Tree Conf.* 47:77a–79a.

RICH, S. 1975. Air Pollution and Agricultural Practices. In *Responses of Plants to Air Pollution,* ed. J. B. Mudd and T. T. Kozlowski. San Diego, CA: Academic Press, pp. 335–60.

RICHARDS, S. J., and A. W. MARSH. 1961. Irrigation Based on Soil Suction Measurements. *Soil Sci. Soc. Amer. Proc.* 25:65–69.

RICHARDSON, S. D. 1958. Bud Dormancy and Root Development in *Acer saccharinum.* In *The Physiology of Forest Trees,* ed. K. V. Thimann. New York: Ronald Press, pp. 409–25.

RICKMAN, R. W. 1979. *Bad with the Good.* USDA Agr. Res. 28(3):15.

RILEY, C. C., and W. R. OKIE. 1983. Peach Tree Pruning Time in Relation to Susceptibility and Spread of *Botryosphaeria dothidea. Phytopathology* 73:798.

RISHBETH, J. 1970. The Role of Basidiospores in Stump Infection by *Armillaria mellea.* In *Root Diseases and Soil-Borne Pathogens,* ed. T. A. Toussoun and others. Berkeley: University of California Press, pp. 141–46.

ROBBINS, K. 1986. *How to Recognize and Reduce Tree Hazards in Recreation Sites.* U.S. Forest Serv. NA-FR-31.

ROBBINS, W. W., T. E. WEIER, and C. R. STOCKING. 1950. *Botany: An Introduction to Plant Science,* 2nd ed. New York: John Wiley.

ROBINETTE, G. O. 1972. *Plants/People/and Environmental Quality.* Washington, DC: U.S. Natl. Park Serv.

ROLF, K. 1990. Soil Improvement and Increased Growth Response from Subsoil Cultivation. *J. Arboriculture* 16:307.

ROSS, N., and others. 1970. *Reducing Loss from Crown Gall Disease.* Calif. Agr. Exp. Sta. Bull. 845.

RUBENS, J. M. 1978. Soil Desalination to Counteract Maple Decline. *J. Arboriculture* 4:33–42.

RUSHFORTH, K. D. 1979. Summer Branch Drop. *Arboriculture Res. Note* (British Dept. of the Environment), Dec., pp. 1–2.

RUSSEL, J. C. 1939. The Effect of Surface Cover on Soil Moisture Losses by Evaporation. *Proc. Soil Sci. Soc. Amer.* 4:65–70.

RUTH, W. A., and V. W. KELLEY. 1932. *A Study of the Framework of the Apple Tree and Its Relation to Longevity.* Ill. Agr. Exp. Sta. Bull. 376:509–637.

SACHS, R. M., and W. P. HACKETT. 1972. Chemical Inhibition of Plant Height. *HortSci.* 7:440–47.

SACHS, R. M., T. KRETCHUN, and T. MOCK. 1975. Minimum Irrigation Requirements for Landscape Plants. *J. Amer. Soc. Hort. Sci.* 100(5):499–502.

SACHS, R. M., and others. 1986. Chemical Control of Tree Growth by Bark Painting. *J. Arboriculture* 12:284–91.

———. 1990. Slow Release Formulations of Growth Retardants. *Proc. 2nd Annual Conf. West. Plant Growth.* Regulator Soc. San Jose, CA: pp. 80–84.

SAFIR, G. R., and C. E. NELSEN. 1980. Water and Nutrient Uptake by Vesicular-Arbuscular Mycorrhizal Plants. In *Role of Mycorrhizal Associations in Crop Production.* Proc. of a Colloq. at Rutgers University, New Brunswick, NJ.

SANDELL, P., and P. KUBE. n.d. *Dryland Tree Establishment—Central Australia.* Forestry Unit, Conservation Commission, Alice Springs, N.T.

SANDFORT, S. S., and R. C. RUNCK, III. 1986. Trees Need Respect, Too! *J. Arboriculture* 12:141–45.

SANTAMOUR, F. S., JR. 1972. Shade-Tree Improvement Research at the U.S. National Arboretum. *Proc. Intl. Shade Tree Conf.* 48:132–33.

———. 1979a. Inheritance of Wound Compartmentalization of Soft Maples. *J. Arboriculture* 5(10):220–25.

———. 1979b. Root Hardiness of Green Ash Seedlings from Different Proveniences. *J. Arboriculture* 5(12):276–79.

SARGENT, C. S. 1926. *Manual of the Trees of North America.* Boston: Houghton.

SCARLETT, A. L., and C. L. WAGENER. 1973. *Managing and Marketing California Forest-Grown Christmas Trees.* Calif. Agr. Ext. Serv. AXT-182.

SCHIECHTL, H. M. 1978. Umweltverträgliche Abhangsbefestigung [Environmentally Compatible Slope Stabilization]. Organ der Deutschen Gesellschaft für Erd und Grundbau [German Nat'l. Soc. Soil Mechanics and Foundation Engineering]. *Geotechnik* 1:10–21.

SCHMID, J. A. 1975. *Urban Vegetation.* Univ. Chicago Dept. of Geography Res. Pap. 161.

SCHOENEWEISS, D. F. 1973. Diagnosis of Physiological Disorders of Woody Ornamentals. *Proc. Intl. Shade Tree Conf.* 49:33a–38a.

———. 1978. The Influence of Stress on Diseases of Nursery and Landscape Plants. *J. Arboriculture* 4(10):217–25.

———. 1981. Infectious Diseases of Trees Associated with Water and Freezing Stress. *J. Arboriculture* 7:13–18.

———. 1982. Prevention and Treatment of Construction Damage. *J. Arboriculture* 8:169–75.

SCHREIBER, L. R., and J. W. PEACOCK. 1974. *Dutch Elm Disease and Its Control.* USDA For. Serv. Inf. Bull. 193.

SCHUDER, D. L. 1978. Identifying and Controlling Insect Galls. *American Nurseryman* 148(6):14, 115–19.

SCHULTZ, H. B., and J. V. LIDER. 1968. *Frost Protection with Overhead Sprinklers.* Calif. Agr. Ext. Serv. Leaflet 201.

SCHÜTTE, K. H. 1966. Trace Element Deficiencies in Cape Vegetation. *J. S. African Bot.* 26:145–49.

SCOTT, D. H. 1973. *Air Pollution Injury to Plant Life.* Washington, DC: National Landscape Assn.

SCROGGINS, T. 1971. Using Leached Cedar Tow for Packing? Be Careful! *Oregon Ornamental and Nursery Digest* 15(2):4.

SEMONIN, R. G. 1978. Severe Weather Climatology in the Midwest and Arboriculture. *J. Arboriculture* 4:128–36.

SHANKS, C. W. 1987. Treeshelters: A Guide to Their Use and Information on Suppliers. *Arb. Adv. & Info. Serv.* (England), pp. 1–7.

SHARON, E. M. 1973. Some Histological Features of *Acer saccharum* Wood Formed after Wounding. *Canadian Jour. Forest Res.* 3:83–89.

SHARON, M. F. 1987. Tree Health Management: Evaluating Trees for Hazard. *J. Arboriculture* 13:285–93.

SHEARMAN, R. C., and others. 1979. A Comparison of Turfgrass Clippings, Oat Straw, and Alfalfa as Mulching Material. *J. Amer. Soc. Hort. Sci.* 104(4):461–63.

SHIGO, A. L. 1979a. Tree Care. *J. Arboriculture* 5(9):vi.

———. 1979b. *Tree Decay: An Expanded Concept.* USDA Forest Serv. Agr. Inf. Bull. 419.

———. 1981. To Paint or Not to Paint. In *Handbook of Pruning.* Brooklyn: Brooklyn Botanic Garden, *Plants & Gardens* 37(2):20–23.

SHIGO, A. L. 1983. Targets for Proper Tree Care. *J. Arboriculture* 9(11):285–94.

———. 1985. How Tree Branches Are Attached to Trunks. *Can. J. Bot.* 63:1391–1401.

———. 1986a. *A New Tree Biology.* Durham, NH: Shigo and Trees, Associates.

———. 1986b. *A New Tree Biology Dictionary.* Durham, NH: Shigo and Trees, Associates.

———. 1989. *Tree Pruning: A Worldwide Photo Guide.* Durham, NH: Shigo and Trees, Associates.

———. 1990. *Pruning Trees Near Electric Utility Lines.* Durham, NH: Shigo and Trees, Associates.

———. 1991. *Modern Arboriculture: A Systems Approach to Trees and Their Associates.* Durham, NH: Shigo and Trees, Associates.

SHIGO, A. L., and R. CAMPANA. 1977. Discolored and Decayed Wood Associated with Injection Wounds in American Elm. *J. Arboriculture* 3(12):230–35.

SHIGO, A. L., and R. FELIX. 1980. Cabling and Bracing. *J. Arboriculture* 6(1):5–9.

SHIGO, A. L., and E. VH. LARSON. 1969. *Photo Guide to the Patterns of Discoloration and Decay in Living Northern Hardwood Trees.* USDA Forest Serv. Res. Pap. NE-127.

SHIGO, A. L., and H. G. MARX. 1977. *Compartmentalization of Decay in Trees.* USDA Forest Serv. Agr. Inf. Bull. 405.

SHIGO, A. L., W. E. MONEY, and D. I. DODDS. 1977. Some Internal Effects of Mauget Tree Injections. *J. Arboriculture* 3(11):213–20.

SHIGO, A. L., W. C. SHORTLE, and P. W. GARRETT. 1977. Genetic Control Suggested in Compartmentalization of Discolored Wood Associated with Tree Wounds. *Forest Sci.* 23:179–82.

SHIGO, A. L., and C. L. WILSON. 1977. Wound Dressings on Red Maple and American Elm: Effectiveness after Five Years. *J. Arboriculture* 3(5):81–87.

SHIGO, A. L., and others. 1979. *Internal Defects Associated with Pruned and Nonpruned Branch Stubs in Black Walnut.* USDA Forest Serv. Res. Pap. 440.

SHORTLE, W. C. 1979. New Look at Tree Care. *J. Arboriculture* 5(12):281–84.

SHURTLEFF, M. C. 1980. The Search for Disease-Resistant Trees. *J. Arboriculture* 6(9):238–44.

SHURTLEFF, M. C., and B. J. JACOBSEN. 1983. Iron Chlorosis of Wood Plants: Cause and Control. Univ. of Illinois. *Report on Plant Diseases* No. 603.

SINCLAIR, W. A. 1978. Range, Suscepts, Losses. In *Dutch Elm Disease: Perspectives after 60 Years,* ed. W. A. Sinclair and R. J. Campana. Cornell Univ. Agr. Exp. Sta. *Search Agr.* 8(5):6–8.

SINCLAIR, W. A., and R. J. CAMPANA. 1978. Development and Status of Dutch Elm Disease. In *Dutch Elm Disease: Perspectives after 60 Years,* ed. W. A. Sinclair and R. J. Campana. Cornell Univ. Agr. Exp. Sta. *Search Agr.* 8(5):5–6.

SINCLAIR, W. A., H. L. LYON, and W. T. JOHNSON. 1987. *Diseases of Trees and Shrubs.* Ithaca, NY: Cornell University Press.

SINCLAIR, W. A., and W. T. JOHNSON. 1975. *Verticillium Wilt.* Cornell Tree Pest Leaflet A-3.

SINGER, M. J., and D. N. MUNNS. 1987. *Soils: An Introduction.* New York: Macmillan.

SMILEY, E. T., and others. 1990. Evaluation of Soil Aeration Equipment. *J. Arboriculture* 16:118–23.

SMITH, D. M. 1962. *The Practice of Silviculture.* New York: John Wiley.

SMITH, E. M. 1976. Pin Oak Chlorosis: A Serious Landscape Problem. *American Nurseryman* 143(3):15, 44.

SMITH, E. M., and T. A. FRETZ. 1979. *Chemical Weed Control in Commercial Nursery and Landscape Plantings.* Ohio Coop. Ext. Serv. MM-297.

SMITH, E. M., and K. W. REISCH. 1975. Fertilizing Trees in the Landscape: Progress Report. Abstract. *J. Arboriculture* 1(4):77.

SMITH, R. C. 1977. Planting Trees with a Power Auger. *Grounds Maintenance,* May, p. 84.

SMITH, W. H. 1970. *Tree Pathology: A Short Introduction.* San Diego, CA: Academic Press.

SNYDER, R. L., and others. 1987. *Using Reference* ET$_0$ *and Crop Coefficients to Estimate Crop* ET$_c$ *for Agronomic Crops, Grasses, and Vegetable Crops.* Univ. Calif. Agr. Sci. Leaflet 21427.

SPOMER, L. A. 1983. Physical Amendment of Landscape Soils. *J. Environ. Hort.* 1(3):77–80.

STASIUK, W. N., and P. E. COFFEY. 1974. Rural and Urban Ozone Relationships. *J. Air Pollution* 24(9):819.

SPOMER, L. A. 1983. Physical Amendment of Landscape Soils. *J. Environ. Hort.* 1(3):77–80.

STERRETT, J. P. 1986. Two Groups of Plant Growth Regulators. *Utility Arborist Assoc. Newsletter.* 7(3):17–20.

STERRETT, J. P., T. J. TWORKOSKI, and P. T. KUJAWSKI. 1989. Physiological Responses of Deciduous Tree Root Collar Drenched with Flurprimidol. *J. Arboriculture* 15:120–24.

STIMMANN, M. W. 1977. *Pesticide Application and Safety Training.* Univ. Calif. Agr. Sci. Pub. 4070.

STIPES, R. J., and R. J. CAMPANA. 1981. *Compendium of Elm Diseases.* St. Paul, MN: American Phytopathological Society.

STONE, E. L. 1968. Microelement Nutrition of Forest Trees: A Review. In *Forest Fertilization: Theory and Practice.* Knoxville: Tennessee Valley Authority, pp. 132–75.

STORIE, R. E. 1932. *An Index for Rating the Agricultural Value of Soils.* Calif. Agr. Exp. Sta. Bull. 556.

STREETS, R. B. 1984. *The Diagnosis of Plant Diseases*. Tucson: University of Arizona Press.

STRIBLEY, D. P. 1987. Mineral Nutrition. In *Ecophysiology of VA Mycorrhizal Plants*, ed. G. R. Safir. Boca Raton, FL: CRC Press.

STRIBLING'S NURSERIES. 1966. New System in Tree Plantings. *Stribling's Fruit and Grape Growers Newsletter*, Jan.

STROMBERG, L. K. 1975. *Water Quality for Irrigation*. Fresno County: Calif. Coop. Ext. Serv. Pub., Nov. 18.

SVIHRA, P. 1980. *Dutch Elm Disease in California*. Univ. Calif. Agr. Sci. Leaflet 21189.

———. 1987. Suitability of Elm Firewood to Bark Beetle Attack Stored Under Polyethylene Sheeting. *J. Arboriculture* 13:164–66.

———. 1990. Purple Plum Fruit Control by Growth Inhibitors. A Look at Maintain CF-125. *Proc. West Plant Growth Regulator Soc.* 2:103–6.

———. 1992. Eradicative Pruning in Landscape Trees. *J. Arboriculture* 17:(in press).

SWANSON, B. T., and C. ROSEN. 1989. *Tree Fertilization*. Minnesota Ext. Ser. Ag-FO-2421.

SZETO, E. 1989. *Plant Growth Regulator Literature Review and Status Report*. San Ramon, CA: Pacific Gas and Electric Company. Env., Health, and Safety Rpt. 009. 4-89.4.

TALBOTT, J. A., and others. 1976. Flowering Plants as Therapeutic/Environmental Agent in a Psychiatric Hospital. *HortSci.* 11(4):365–66.

TATE, R. L. 1981. Characteristics of Girdling Roots on Urban Norway Maples. *J. Arboriculture* 7:268–70.

TATTAR, T. A. 1989. *Diseases of Shade Trees*, 2nd ed. San Diego: Academic Press.

TENNESSEE VALLEY AUTHORITY. 1968. *Forest Fertilization: Theory and Practice*. Symposium on Forest Fertilization. Muscle Shoals, AL: Tennessee Valley Authority.

TERLOUW, A. L. 1981. Lava Slag, the Elixir of Life for Street Trees? *Groen* 37(7):301–304.

THAYER, R. L., JR. 1981. *Solar Access: It's the Law*. Univ. Calif., Davis, Inst. Govt. Affairs and Ecology, Environmental Quality Series 34.

———. 1986a. Solar Access Control Strategies for Vegetation in Existing and New Developments. *J. Arch. Plan. Res.* 3:199–217.

———. 1986b. Trees and Solar Access Law: Two Case Studies. Council of Education of Landscape Architecture Conf., pp. 1–13.

THAYER, R. L., JR., and B. T. MAEDA. 1985. Measuring Street Tree Impact on Solar Performance: A Five-Climate Computer Modeling Study. *J. Arboriculture* 11:1–12.

THOMPSON, A. R. 1940. *Transplanting Trees and Other Woody Plants*. U.S. Natl. Park Serv. Tree Preservation Bull. 9.

———. 1959. *Tree Bracing*. U.S. Natl. Park Serv. Tree Preservation Bull. 3.

———. 1961. *Shade Tree Pruning*. Rpt. U.S. Natl. Park Serv. Tree Preservation Bull. 4.

THORTON, P. L. 1971. Managing Urban and Suburban Trees and Woodland for Timber Products. In *Trees and Forests in an Urbanizing Environment*. Univ. Mass., Amherst, Coop. Ext. Serv. Planning and Resource Devel. Monograph 17:129–32.

TICKNOR, R. L. 1981. Deciduous Trees Modify Temperature of Buildings. *Weeds, Trees, and Turf* 20(2):22–23, 25–26.

TIPPETT, J. T., and J. L. BARCLAY. 1987. Detection of Bark Lesions Caused by *Phytophthora cinnamomi* in *Eucalyptus marginata* with the Plant Impedance Ratio Meter and the Shigometer. *Can. J. For. Res.* 17:1228–33.

TISDALE, S. L., and W. L. NELSON. 1975. *Soil Fertility and Fertilizers,* 3rd ed. New York: Macmillan.

TODHUNTER, M. N., and W. F. BEINEKE. 1979. Effect of Fescue on Black Walnut Growth. *Tree Planter's Notes,* Summer, 20–23.

TOKLE, G. O., and J. MARKER. 1987. *Wildfire Strikes Home.* U.S. Forest Serv. Natl. Fire Prot. Assoc. No. SPP-86.

TORNGREN, T. S., and F. CHAN. 1978. *Mistletoe Control Progress Report: Sacramento County.* Sacramento County: Univ. Calif. Coop. Ext. Serv., Jan. 16.

TORNGREN, T. S., E. J. PERRY, and C. L. ELMORE. 1980. *Mistletoe Control in Shade Trees.* Univ. Calif. Agr. Sci. Leaflet 2571.

TOWNSEND, A. M. 1977. Improving the Adaptation of Maples and Elms to the Urban Environment. In *Proc. 16th Meeting Canadian Improvement Assoc.: Part 2.* Winnipeg: University of Manitoba Press, pp. 27–31.

TROUSE, A. C., JR., and R. P. HUMBERT. 1959. Deep Tillage in Hawaii: Subsoiling. *Soil Sci.* 88:150–58.

TURNER, N. C., and H. C. DEROO. 1974. Hydration of Eastern Hemlock as Influenced by Waxing and Weather. *Forest Sci.* 20:19–24.

ULRICH, R. S. 1984. View Through a Window May Influence Recovery from Surgery. *Science* 224:420–21.

———. 1986. Human Responses to Vegetation and Landscapes. *Landscape and Urban Planning* 13:29–44.

UNITED STATES DEPARTMENT OF AGRICULTURE (USDA). 1957. *Soil: The 1957 Yearbook of Agriculture.* Washington, DC: U.S. Superintendent of Documents.

———. 1969. *Pruning Ornamental Shrubs and Vines.* USDA Home and Garden Bull. 165.

———. 1973a. *Air Pollution Damages Trees.* Upper Darby, PA: USDA Forest Serv., Northeast Area.

———. 1973b. *Controlling the Japanese Beetle.* USDA Home and Garden Bull. 159.

———. 1975. *Protecting Shade Trees During Home Construction.* USDA Home and Garden Bull. 104.

———. 1979. *Seed and Planting Stock Dealers.* Washington, DC: USDA Forest Serv., FS-331.

———. 1988. *Plant Hardiness Zone Map.* USDA Misc. Pub. 1475.

UNTERMANN, R. K. 1978. *Principles and Practices of Grading, Drainage, and Road Alignment: An Ecological Approach.* Reston, VA: Reston Publishing Co.

URBAN, J. 1989. New Techniques in Urban Tree Planting. *J. Arboriculture* 15:281–84.

VAN ALFEN, N. K., and W. E. MACHARDY. 1978. Symptoms and Host-Pathogen Interactions. In *Dutch Elm Disease: Perspective after 60 Years,* ed. W. A. Sinclair and R. J. Campana. Cornell Univ. Agr. Sta. *Search Agr.* 8(5):20–25.

VAN DAM, J., J. W. MAMER, and W. W. WOOD. 1981. *Labor Requirement Analysis for Landscape Maintenance.* Univ. Calif. Agr. Sci. Leaflet 21232.

VAN DEN BOSCH, R., and P. S. MESSENGER. 1973. *Biological Control.* New York: Harper & Row.

VAN DER MEIDEN, H. A. 1957. (no title) *Ned. Bos. Tjd.* 10:229–42.

VAN DE WERKEN, H. 1981. Fertilization and Other Factors Enhancing the Growth Rate of Young Shade Trees. *J. Arboriculture* 7(2):33–37.

———. 1984a. Fertilization Practices as They Influence the Growth Rate of Young Shade Trees. *J. Environ. Hort.* 2(2):64–69.

———. 1984b. Why Use Obsolete Fertilizer Practices? *American Nurseryman* 159(7):65–71.

VAN HEUIT, R. E. 1979. *Sanitation Districts of Los Angeles County: Generation of Methane Gas in Sanitary Landfills and Its Effect on Growth of Plant Materials.* Unpublished paper delivered at 1979 meeting of Amer. Soc. of Consulting Arborists. (Abstract in *Horizons* 1980. Washington, DC: Amer. Assoc. Nurserymen, Sept.)

VANSTONE, D. E., and W. G. RONALD. 1981. Comparison of Bare-Root Versus Tree Spade Transplanting of Boulevard Trees. *J. Arboriculture* 7:271–74.

VEIHMEYER, F. J., and A. H. HENDRICKSON. 1955. Does Transpiration Decrease as the Soil Moisture Decreases? *Transactions, Amer. Geophysical Union* 36:425–48.

VERNER, L. 1955. *Hormone Relations in the Growth and Training of Apple Trees.* Univ. Idaho, College of Agr. Res. Bull. 28.

VIETS, F. G., JR. 1966a. *Zinc Deficiency in the Silver Oak.* Ann. Admin. Rept. Sci. Dept. (Tea Sect.) United Planters Assoc. S. India 1962/63, pp. 70–71. (For Abstr. 26:2320).

VON RUMKER, R. V., and others. 1975. *A Study of the Efficiency of the Use of Pesticides in Agriculture.* Washington, DC: Office of Pesticide Programs, Office of Water and Hazardous Materials, Environmental Protection Agency Report 540/9–75–025.

WAGAR, J. A. 1985a. SOLPLOT: An Aid to Controlling Sunlight and Shade. USDA For. Serv. *Forestry Res. West,* April, pp. 5–7.

———. 1985b. Reducing Surface Rooting of Trees with Control Planters and Wells. *J. Arboriculture* 11:165–71.

WAGAR, J. A., and P. A. BARKER. 1983. Tree Root Damage to Sidewalks and Curbs. *J. Arboriculture* 9:177–81.

WAGENER, W. W. 1963. *Judging Hazard from Native Trees in California Recreational Areas: A Guide for Professional Foresters.* U.S. Forest Serv. Res. Ppr. PSW-P1.

WALKER, R. E., and G. F. KAH. 1989. *Landscape Water Management Handbook.* Calif. Dept. Water Resources. Sacramento: Version 4.1.

WALLACE, A., and S. D. NELSON. 1986. Foreword. *Soil Science* 141(5):311.

WALLER, J. 1965. Cooperative Enterprises—Activities of Civic Groups, Organizations and Individuals in Beautification Programs. *Proc. Intl. Shade Tree Conf.* 41:195–201.

WALLIHAN, E. F. 1966. Iron. In *Diagnostic Criteria for Plants and Soils,* ed. H. D. Chapman. Riverside, CA: H. D. Chapman, pp. 203–12.

WALLIS, G. W., D. J. MORRISON, and D. W. ROSS. 1980. *Tree Hazards in Recreation Sites in British Columbia.* B.C. Min. Lands, Parks and Housing—Can. Forest Serv. Jt. Rpt. No. 13.

WALMSLEY, T. J. 1989. *Factors Influencing the Establishment of Amenity Trees.* Ph.D. Thesis. University of Liverpool.

WALTERS, D. T., and A. R. GILMORE. 1976. Allelopathic Effects of Fescue on the Growth of Sweetgum. *J. Chem. Ecol.* 2:469–79.

WARE, G. W. 1978. *The Pesticide Book.* San Francisco: W. H. Freeman.

WARGO, P. M. 1979. Starch Storage and Radial Growth in Woody Roots of Sugar Maple. *Can. J. For. Res.* 9:48–56.

———. 1980. *Armillaria mellea:* An Opportunist. *J. Arboriculture* 6(10):276–78.

WATER POLLUTION CONTROL FEDERATION. 1980. Operation and Maintenance of Wastewater Collection Systems. *Manual of Practice #7.*

WATKINS, J. V. 1961. *Your Guide to Florida Landscape Plants.* Gainesville: University of Florida Press.

WATSON, G. W. 1985. Tree Size Affects Root Regeneration and Top Growth After Transplanting. *J. Arboriculture* 11:37–40.

WATSON, G. W., S. CLARK, and K. JOHNSON. 1990. Formation of Girdling Roots. *J. Arboriculture* 16:197–202.

WATSON, G. W., and E. B. HIMELICK. 1982. Root Distribution of Nursery Trees and Its Relationship to Transplanting Success. *J. Arboriculture* 8:225–29.

WATSON, M. R. 1987. Use of Tree Growth Regulators at Potomac Edison. *J. Arboriculture* 13:65–69.

WEAVER, R. J. 1972. *Plant Growth Substances in Agriculture.* San Francisco: W. H. Freeman.

WEBSTER, A. D., and J. D. QUINLAN. 1984. Chemical Control of Tree Growth of Plum (*Prunis domestica* L.): Preliminary Studies with the Growth Retardant Paclobutrazol (PP333). *J. Hort. Sci.* 59(3):367–75.

Webster's Seventh New Collegiate Dictionary. 1976. Springfield, MA: Merriam.

WEIDENSAUL, T. C. 1973. Are Trees Efficient Air Purifiers? *Arborist's News* 38:85–89.

WEIER, T. E., C. R. STOCKING, and M. G. BARBOUR. 1974. *Botany: An Introduction to Plant Biology,* 5th ed. New York: John Wiley.

WEIER, T. E., C. R. STOCKING, M. G. BARBOUR, and T. L. ROST. 1982. *Botany: An Introduction to Plant Biology;* 6th ed. New York: John Wiley.

WEINBAUM, S. A., M. L. MERWIN, and T. T. MURAOKA. 1978. Seasonal Variation in Nitrate Uptake Efficiency and Distribution of Absorbed Nitrogen in Non-Bearing Prune Trees. *J. Amer. Soc. Hort. Sci.* 103(4):516–19.

WEINBAUM, S. A., and others. 1984. Effects of Time of Nitrogen Application and Soil Texture on the Availability of Isotopically Labelled Fertilizer Nitrogen to Reproductive and Vegetative Tissue of Mature Almond Trees. *J. Amer. Soc. Hort. Sci.* 109(3):339–43.

WEISER, C. J. 1970a. Cold Resistance and Acclimation in Woody Plants. *HortSci.* 5(5):403–10.

———. 1970b. Cold Resistance and Injury in Woody Plants. *Science* 169:1269–78.

WELCH, D. S. 1949. The Cause and Treatment of Cavities in Trees. *Proc. Natl. Shade Tree Conf.* 25:126–31.

WHITCOMB, C. E. 1979a. Amendments and Tree Establishment. *J. Arboriculture* 5(7):167.

———. 1979b. Factors Affecting the Establishment of Urban Trees. *J. Arboriculture* 5(10):217–19.

———. 1980. Effects of Black Plastic and Mulches on Growth and Survival of Landscape Plants. *J. Arboriculture* 6(10):10–12.

———. 1986. Solving the Iron Chlorosis Problem. *J. Arboriculture* 12:44–48.

———. 1987. *Establishment and Maintenance of Landscape Plants.* Stillwater, OK: Lacebark Publications.

WHITE, D. P. 1956. Aerial Application of Potash Fertilizer to Coniferous Plantations. *J. Forestry* 54:762–68.

WHITE, P. M. 1987. Transplanting Trees and Shrubs in the Fall. *Country Jour.*, Nov., pp. 33–35.

WHITE, R. F. 1945. *Effects of Landscape Development on the Natural Ventilation of Buildings and Their Adjacent Areas.* College Station: Texas Engr. Exp. Sta. Res. Report 45.

WHITEHEAD, F. H. 1963. The Effects of Exposure on Growth and Development. In *Water Relations of Plants,* ed. A. J. Rutter and F. H. Whitehead. New York: John Wiley.

WHITLOW, T. H., and R. W. HARRIS. 1979. *Flood Tolerance in Plants: A State-of-the-Art Review.* Vicksburg, MS: U.S. Army Engineer Waterways Exp. Sta. Tech. Report E-79-2.

WHITTAKER, R. H. 1954. The Ecology of Serpentine Soils· A Symposium. *Ecology* 35(2):258–59.

WIENER, L. 1982. Brush Away Borers. *Fruit Grower,* Nov.

WILDMAN, W. E. 1969. Water Penetration: Where Is the Restricting Layer. *Soil and Water* (Univ. Calif. Coop. Ext. Serv.) 7:1–4.

———. 1976. *Diagnosing Soil Physical Problems.* Univ. Calif. Ext. Leaflet 2664.

WILDMAN, W. E., and K. D. GOWANS. 1975. *Soil: Physical Environment and How It Affects Plant Growth.* Univ. Calif. Ext. Leaflet 2280.

WILDMAN, W. E., J. L. MEYER, and R. A. NEJA. 1975. *Managing and Modifying Problem Soils.* Univ. Calif. Ext. Leaflet 2791.

WILKES, J. 1983. Use of the Shigometer Technique to Detect Decay in Eucalypts. *Aust. For.* 46(1):35–38.

WILSON, B. F. 1984. *The Growing Tree,* rev. ed. Amherst: University of Massachusetts Press.

WILSON, C. L., and A. L. SHIGO. 1973. Dispelling Myths in Arboriculture Today. *American Nurseryman* 127:24–28.

WILSON, P. J. 1983. The Shigometer Technique in Practice. *Arboricultural J.* 7:81–85.

WITTROCK, G. I. 1971. *The Pruning Book,* Emmaus, PA: Rodale Press.

WOLVERTON, B. C., A. JOHNSON, and K. BOUNDS. 1989. *Interior Landscape Plants for Indoor Air Pollution Abatement.* Natl. Aero. and Space Adm., Stennis Space Center, MS.

WONG, T. L., R. W. HARRIS, and R. E. FISSELL. 1971. Influence of High Soil Temperatures of Five Woody Species. *J. Amer. Soc. Hort. Sci.* 96:80–83.

WONG, T. W., J. E. G. GOOD, and M. P. DENNE. 1988. Tree Root Damage to Pavements and Kerbs in the City of Manchester. *Arboricultural J.* 12:17–34.

WOODHOUSE, J. M. 1989. *Water Storing Polymers as Aids to Vegetation Establishment on Arid Soils.* Ph.D. Thesis. University of Liverpool.

WOODTLI, K. 1989. Radosan + Sanator. *J. Arboriculture* 15:253.

WORLD METEOROLOGICAL ORGANIZATION. 1971. *Meteorological Services of the World.* Geneva: World Meteo. Org. 2.

WORTHING, C. R., and S. B. WALKER. 1983. *The Pesticide Manual,* 7th ed., ed. C. R. Worthing. Croydon, England: British Crop Protection Council.

WYMAN, D. P. 1936. *Growth Experiments with Pin Oaks Which Are Growing Under Lawn Conditions.* Cornell Univ. Agr. Exp. Sta. Bull. 646.

———. 1957. Mulching Practices at the Arnold Arboretum. *Plants & Gardens* (Brooklyn Botanic Garden) 13(1):27–30.

———. 1971. *Wyman's Gardening Encyclopedia.* New York: Macmillan.

Zanetto, J. 1978. Trees for Solar Neighborhoods. *Landscape Architecture,* Nov., pp. 514–19.

Zeleny, R. J. 1988. *Wildfire Safety: Model Regulations.* Col. State Forest Serv.

Zimmermann, M. H., and C. L. Brown. 1971. *Trees: Structure and Function.* New York: Springer-Verlag.

Zisa, R. P., H. G. Halverson, and B. J. Stout. 1980. *Establishment and Early Growth on Compost Soils in Urban Areas.* USDA Forest Serv. Res. Paper NE 451.

Index

Page numbers for illustrations and tables are indicated in **boldface type.** Index topics printed in **boldface type** are covered in greater detail in the first edition of this book, *Arboriculture: Care of Landscape Trees, Shrubs, and Vines* (1983).

To assist the reader, chemical and trade names of growth regulating chemicals and herbicides and the trade names of products and equipment are listed in the index. No endorsement of named products is intended, nor is criticism implied of similar products that are not mentioned. The names of few pesticides are listed in the index.

Water movement in soil (*cont.*)
  infiltration, **157**
  soil strata effect, 158, **159**
    gravel in planting hole, 158,
      203
  vapor, 158
Watersprout, 19, **20**
Water table, 158
Wattling, **262**
Weather-induced problems, 88–
  117, 543
Weed control, 349
Wetting agent, 345
Wetwood (slime flux), 570, **571**
Wind
  control, 126
  damage, **113, 127, 461**
  effects, 92, 94, **115,** 116, 130

landscape protection, **123,** 127
load on a tree, 78
modified by plants, 123
tall building effects, **127**
Windbreak, 5, 126, **129**
  height and porosity effect, 126
Wind chill, 115
Winter burn, 89, 92
Winter chilling and rest, 100
Wire basket, 70, **208**
Wiring of small trees, 484, **485**
Wood. *See also* Xylem
  ashes, 148, 288
  chips, **359,** 360, 362
Wound(s), 24, **25, 492**
  closure, factors affecting, **492**
  compartmentalization, 24, **25,**
    492

injection, 494
  toxic chemicals, 495
protection of pruning, 388
treatment, **170,** 495–97
  graft of bark, **498**
  replacement of loose bark, 497

Xeriscape, 334
Xylem, **22**
  diffuse-porous, 23, **24**
  earlywood, 23
  heartwood, **22**
  latewood, 23
  nonporous, 23, **24**
  ring-porous, 23, **24**
  sapwood, 23

Zinc, 145, 305, 311

MEASUREMENT CONVERSIONS

To convert a measurement in one column to that in the other column, either divide or multiply as indicated by the arrows by the number in the center column.

	Divide $--------\rightarrow$ Multiply $\leftarrow --------$	
Length[a]		
millimeter, mm	25.4	*inch, in.*
millimeter, mm	303	*feet, ft*
centimeter, cm	30	*feet, ft*
meter, m	0.303	*feet, ft*
meter, m	0.91	*yard, yd*
kilometer, km	1.6	*mile, mi*
feet, ft	5260	*mile, mi*
micron, μ	1000	millimeter, mm
millimicron, mμ	1,000,000	millimeter, mm
Angstrom, Å	10	millimicron, mμ
Area		
millimeter^2, mm^2	645	*inch^2, in.^{2}*
centimeter^2, cm^2	6.45	*inch^2, in.^{2}*
centimeter^2, cm^2	929	*feet^2, ft^2*
meter^2, m^2	0.09	*feet^2, ft^2*
meter^2, m^2	0.8	*yard^2, yd^2*
kilometer^2, km^2	2.6	*mile^2, mi^2*
hectare, ha	0.4	*acre, A*
meter^2, m^2	100,000	hectare, ha
feet^2, ft^2	43,560	*acre, A*
acre, A	640	*mile^2, mi^2*
Volume		
milliliter, ml	5	*teaspoon, tsp*
milliliter, ml	15	*tablespoon, tbs*
milliliter, ml	30	*ounce, oz* (fluid)
liter, l	0.24	*cup*
liter, l	0.47	*pint, pt*
liter, l	0.95	*quart, qt*
liter, l	3.8	*gallon, gal* (U.S.)
meter^3, m^3	0.03	*feet^3, ft^3*
meter^3, m^3	0.76	*yard^3, yd^3*
liter, l	10,000	meter^3, m^3
gallon, gal (U.S.)	7.48	*feet^3, ft^3*
gallon, gal (U.S.)	1.2	*gallon, gal* (imperial)

[a] English units are in *italics*. Metric units are in standard type.

Examples: 500 mm divided by 25.4 = 19.7 in.

5 gal multiplied by 3.8 = 19 liters (l)